Universitext

J. Frédéric Bonnans · J. Charles Gilbert
Claude Lemaréchal · Claudia A. Sagastizábal

Numerical Optimization

Theoretical and Practical Aspects

Second Edition

With 52 Figures

 Springer

J. Frédéric Bonnans
Centre de Mathématiques Appliquées
Ecole Polytechnique
91128 Palaiseau
France
e-mail: Frederic.Bonnans@inria.fr

J. Charles Gilbert
INRIA Rocquencourt
BP 105
78153 Le Chesnay
France
e-mail: Jean-Charles.Gilbert@inria.fr

Claude Lemaréchal
INRIA Rhône-Alpes
655, avenue de l'Europe
Montbonnot
38334 Saint Ismier
France
e-mail: Claude.Lemarechal@inria.fr

Claudia A. Sagastizábal
On leave from INRIA Rocquencourt
Correspondence to:
IMPA
110, Estrada dona Castorina
22460-320 Jardim Botânico
Rio de Janeiro–RJ
Brazil
e-mail: sagastiz@impa.br

Original French edition "Optimisation Numérique" was published by Springer-Verlag Berlin Heidelberg, 1997.

Mathematics Subject Classification (2000): 65K10, 90-08, 90-01, 90CXX

Library of Congress Control Number: 2006930998

ISBN: 3-540-35445-X Springer-Verlag Berlin Heidelberg New York

Springer-Verlag Berlin Heidelberg New York
a member of Bertelsmann Springer Science+Bussiness Media GmbH

springer.com
© Springer-Verlag Berlin Heidelberg 2006

Cover design: Erich Kirchner, Heidelberg
Typesetting by the authors using a LaTeX macro package

Printed on acid-free paper: SPIN: 11777410   41/2141/SPi - 5 4 3 2 1 0

Preface

This book is entirely devoted to numerical algorithms for optimization, their theoretical foundations and convergence properties, as well as their implementation, their use, and other practical aspects. The aim is to familiarize the reader with these numerical algorithms: understanding their behaviour in practice, properly using existing software libraries, adequately designing and implementing "home-made" methods, correctly diagnosing the causes of possible difficulties. Expected readers are engineers, Master or Ph.D. students, confirmed researchers, in applied mathematics or from various other disciplines where optimization is a need.

Our aim is therefore not to give most accurate results in optimization, nor to detail the latest refinements of such and such method. First of all, little is said concerning optimization theory itself (optimality conditions, constraint qualification, stability theory). As for algorithms, we limit ourselves most of the time to stable and well-established material. Throughout we keep as a leading thread the actual *practical value* of optimization methods, in terms of their efficiency to solve real-world problems. Nevertheless, serious attention is paid to the theoretical properties of optimization methods: this book is mainly based upon theorems. Besides, some new and promising results or approaches could not be completely discarded; they are also presented, generally in the form of special sections, mainly aimed at orienting the reader to the relevant bibliography.

An introductory chapter gives some generalities on optimization and iterative algorithms. It contains in particular motivating examples, ranking from meteorological forecast to power production management; they illustrate the large field of branches where optimization finds its applications. Then come four parts, rather independent of each other. The first one is devoted to algorithms for unconstrained optimization which, in addition to their direct usefulness, are a basis for more complex problems. The second part concerns rather special methods, applicable when the usual differentiability assumptions are not satisfied. Such methods appear in the decomposition of large-scale problems and the relaxation of combinatorial problems. Nonlinearly constrained optimization forms the third part, substantially more technical, as the subject is still in evolution. Finally, the fourth part gives a deep account of the more recent interior point methods, originally designed for the simpler problems of linear and quadratic programming, and whose application to more general situations is the subject of active research.

This book is a translated and improved version of the monograph [43], written in French. The French monograph was used as the textbook of an intensive two week course given several times by the authors, both in France and abroad. Each topic was presented from a theoretical point of view in morning lectures. The afternoons were devoted to implementation issues and related computational work. The conception of such a course is due to J.-B. Hiriart-Urruty, to whom the authors are deeply indebted.

Finally, three of the authors express their warm gratitude to Claude Lemaréchal for having given the impetus to this new work by providing a first English version.

Notes on this revised edition. Besides minor corrections, the present version contains substantial changes with respect to the first edition. First of all, (simplified but) nontrivial application problems have been inserted. They involve the typical operations to be performed when one is faced with a real-life application: modelling, choice of methodology and some theoretical work to motivate it, computer implementation. Such computational exercises help getting a better understanding of optimization methods beyond their theoretical description, by addressing important features to be taken into account when passing to implementation of any numerical algorithm.

In addition, the theoretical background in Part I now includes a discussion on global convergence, and a section on the classical pivotal approach to quadratic programming. Part II has been completely reorganized and expanded. The introductory chapter, on basic subdifferential calculus and duality theory, has two examples of nonsmooth functions that appear often in practice and serve as motivation (pointwise maximum and dual functions). A new section on convergence results for bundle methods has been added. The chapter on applications of nonsmooth optimization, previously focusing on decomposition of complex problems via Lagrangian duality, describes also extensions of bundle methods for handling varying dimensions, for solving constrained problems, and for solving generalized equations. Also, a brief commented review of existing software for nonlinear optimization has been added in Part III.

Finally, the reader will find additional information at http://www-rocq. inria.fr/~gilbert/bgls. The page gathers the data for running the test problems, various optimization codes, including an SQP solver (in Matlab), and pieces of software that solve the computational exercises.

Paris, Grenoble, Rio de Janeiro,   *J. Frédéric Bonnans*
May 2006   *J. Charles Gilbert*
                                                             Claude Lemaréchal
                                                         Claudia A. Sagastizábal

Table of Contents

Part II Nonsmooth Optimization

Part III Newton's Methods in Constrained Optimization

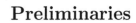

Preliminaries

1 General Introduction

We use the following notation: the working space is $\mathbb{R}^n$, where the scalar product will be denoted indifferently by (x, y) or $\langle x, y \rangle$ or $x^\top y$ (actually, it will be the usual dot-product: $(x, y) = \sum_{i=1}^n x^i y^i$); $|\cdot|$ or $\|\cdot\|$ will denote the associated norm. The *gradient* (vector of partial derivatives) of a function $f : \mathbb{R}^n \to \mathbb{R}$ will be denoted by ∇f or f'; the *Hessian* (matrix of second derivatives) by $\nabla^2 f$ or f''. We will also use continually the notation $g(x) = f'(x)$.

1.1 Generalities on Optimization

1.1.1 The Problem

Given a set X and a function $f : X \to \mathbb{R}$ (the *objective function*), we want to find $x^* \in X$ such that, for all $x \in X$, there holds $f(x) \geqslant f(x^*)$. The variable x is usually called *decision* or *control* variable.

We will consider only the case where X is a subset of $\mathbb{R}^n$, defined by *constraints*, i.e., given a number $m_I + m_E$ of functions $c_j : \mathbb{R}^n \to \mathbb{R}$ for $j = 1, \ldots, m_I + m_E$, the problem is

$$\begin{cases} \min f(x) & x \in \mathbb{R}^n \\ c_j(x) \leqslant 0 & j \in I \\ c_j(x) = 0 & j \in E. \end{cases} \qquad (P)$$

Here, I and E are two disjoint sets of integers, of cardinalities m_I and m_E respectively. We thus have m_I *inequality* constraints, indexed in I, and m_E *equality* constraints, indexed in E.

Remark 1.1. We do not consider problems of *combinatorial optimization*, where the set X is discrete, or even finite. They could be covered by our formalism via constraints of the type $x^i(1 - x^i) = 0$ (to express $x^i \in \{0, 1\}$) but this is very artificial – and not at all efficient in general. Actually, combinatorial optimization problems call for methods totally different from those presented in this book. Their intersection is not totally empty, though: §8.2 will mention the use of continuous optimization to bound the optimal value in combinatorial problems. Section 1.2.4 will give an illustrative example.

In another class of problems, the vector-variable $x \in \mathbb{R}^n$ becomes a function of time $x(t), t \in [0, T]$: these are *optimal control* problems. They are close to our formalism, possibly after discretizing $[0, T]$; in fact, examples are given in §1.2.2 and 1.2.3.

Perhaps rather paradoxically, the methods in this book extend easily to optimal control problems, while they fit very badly to combinatorial optimization. □

1.1.2 Classification

Among the various possible classifications, the following is made according to the difficulty of the problem to solve.

1. Unconstrained problems $(m_I = m_E = 0, I = E = \emptyset)$
    1.1 Quadratic problems: $f(x) = \frac{1}{2}(x, Mx) - (b, x)$ (M symmetric $n \times n$)
    1.2 Nonlinear problems: f neither linear nor quadratic.
2. Linearly constrained problems (the functions c_j are affine)
    2.1. Problems with equality constraints only $(m_I = 0, I = \emptyset)$
        2.1.1 Linear-quadratic problems: f quadratic
        2.1.2 Nonlinear problems: f neither linear nor quadratic
    2.2 Problems with inequality constraints
        2.2.1 Linear programming: f linear (needs $m_I \geqslant n - m_E$)
        2.2.2 Linear-quadratic problems: f quadratic
        2.2.3 Linearly constrained nonlinear problems.
3. Nonlinear programming
    3.1 With equality constraints only
    3.2 General nonlinear programming.

Observe that

– in optimization, the word "linear" is frequently (mis)used, instead of affine (see 2; recall that an affine function is the sum of a linear function and a constant term);
– 2.1 is the minimization in a hyperplane, isomorphic to a subspace of dimension $n - m_E$, so that 2.1 is equivalent to 1, at least theoretically;
– 1.1 reduces to solving a linear system ($Ax = b$ – at least if A is positive definite); 2.1.1 as well, in view of the preceding remark;
– 2.2 minimizes f in a convex polyhedron, the simplest being a parallelotope, defined by simple bounds: $a^i \leqslant x^i \leqslant b^i$, for $i = 1, \ldots, n$;
– 2.2 is considerably more complicated than 2.1, simply because one does not know in advance which inequalities will play a role at the optimal point. Said otherwise, there are 2^{m_I} ways of putting a problem 2.2 into the form 2.1; the question is: which is the correct one? An inequality constraint is said to be *active* at x (not necessarily optimal) when $c_j(x) = 0$. To put 2.2 into the form 2.1, one needs to know which constraints will be active at the (unknown!) optimum point.

1.2 Motivation and Examples

In this section, we show with some examples the variety of domains where one finds optimization problems considered in the present book. Since problems of the linear type (categories 2.2.1 and 2.2.2 in §1.1.2, described in the fourth part) have existed for a long time, and are well known, it is not necessary to motivate this branch. This is why the four examples below are of the "general" nonlinear type.

1.2.1 Molecular Biology

An important problem in biochemistry, for example in pharmacology, is to determine the geometry of a molecule. Various techniques are possible (X-ray crystallography, nuclear magnetic resonance,...) one of these is convenient when

– the chemical formula of the molecule is known,
– the molecule is not available, making it impossible to conduct any experiment,
– one has some knowledge of its shape and one wants to *refine* it.

  The idea is then to compute the positions of the atoms in the space that minimize the associated potential energy. Let N be the number of atoms and call $x_i \in \mathbb{R}^3$ the spatial position of the ith atom. To the vector $X = (x_1, \ldots, x_N) \in \mathbb{R}^{3N}$ is associated a potential energy $f(X)$ (the "conformational energy"), which is the sum of several terms. For example:

– Bond length: between two atoms i and j at distance $|x_i - x_j|$, there is first an energy of the type

$$L_{ij}(x_i, x_j) = \lambda_{ij}(|x_i - x_j| - d_{ij})^2 .$$

– There is also a Van der Waals energy, say

$$V_{ij}(x_i, x_j) = v_{ij}\left(\frac{\delta_{ij}}{|x_i - x_j|}\right)^6 - w_{ij}\left(\frac{\delta_{ij}}{|x_i - x_j|}\right)^{12} .$$

Here, the $\lambda_{ij}, v_{ij}, w_{ij}, d_{ij}, \delta_{ij}$'s are known constants, depending on the pair of atoms involved (carbon-carbon, carbon-nitrogen, etc.)

– Valence angle: between three atoms i, j, k forming an angle θ_{ijk} (writing down the value of θ_{ijk}, as a function of x_i, x_j, x_k, is left as an exercise!), there is an energy

$$A_{ijk}(x_i, x_j, x_k) = \alpha_{ijk}(\theta_{ijk} - \bar{\theta}_{ijk})^2 ,$$

where, here again, α_{ijk} and $\bar{\theta}_{ijk}$ are known constants.

Other types of energies may also be considered: electrostatic, torsion angles, etc. The total energy is then the sum of all these terms, over all pairs/-triples/quadruples of atoms. The important thing to understand here, is that this energy can be computed (as well as its derivatives) for any numerical values taken by the variables x_i. And this is true even if these values do not correspond to any reasonable configuration; simply, the resulting energy will then be unreasonably large (if the model is reasonable!); the optimization process, precisely, will aim at eliminating these values.

This is obviously a problem from category 1.2 in §1.1.2. Note that the objective function is disagreeable:

- With its many terms, it is long to compute.
- With its strong nonlinearities, it does not enjoy the properties useful for optimization: it is definitely not quadratic, and not even convex. Actually, in most examples there are many equilibrium points X^* (local minima); this is why the only hope is to *refine* a specific one: by assumption, some estimate X_0 is available, close to the sought "optimal" X^*. Otherwise the optimization algorithm could only find some uncontrolled equilibrium, "by chance".

Such a problem will call for methods from the first part of this book, more precisely §4.4. Actually, since nowaday's "interesting" molecules have 10^3 atoms and more, this problem is also large-scale; as a result, it will rather be necessary to use methods from Sections 5.6, 6.3, or also 6.4.

1.2.2 Meteorology

To forecast the weather is to know the state of the atmosphere in the future. This is quite possible, at least theoretically (and within limits due to the chaotic character of phenomena involved). Let $p(z,t)$ be the state of the atmosphere at point $z \in \mathbb{R}^3$ and time $t \in [0,7]$ (assuming a forecast over one week, say); p is actually a vector made up of pressure, wind speed, humidity... The evolution of p along time can be modeled: avoiding technicalities, fluid mechanics tells us that

$$\frac{\partial p}{\partial t}(z,t) = \Phi(p(z,t)), \tag{1.1}$$

where Φ is a certain differential operator. For example, (1.1) could be the Navier-Stokes equation, but approximations are generally introduced.

To forecast the weather once our model Φ is chosen, it "suffices" to integrate (1.1). For this, initial conditions are needed (the question of boundary conditions is neglected here; for example, we shall say that they are periodicity conditions, (1.1) being integrated on the whole earth). Here comes optimization, in charge of estimating $p(\cdot,0)$ via an *identification* process, which we roughly explain.

In fact, the available information also contains all the meteorological observations collected in the past, say during the preceding day. Let us denote by $\Omega = \{\omega_i\}_{i \in I}$ these observations. To fix ideas, we could say that each ω_i represents the value of p at a certain point (z_i, t_i) (but actually, only some coordinates of the vector $p(z_i, t_i)$ are observed). To take these – noisy – data into account, a natural and well-known idea is to consider the problem

$$\min_p \|p - \Omega\|, \tag{1.2}$$

(1.1) being considered as a constraint (called in this context the *state equation*).

– Observe here that our optimization problem is not posed with respect to some $x \in \mathbb{R}^n$ but to p, varying in a functional, infinite-dimensional, space. See Remark 1.1; we are dealing with an optimal control problem. Notwithstanding, any numerical implementation implies first a discretization, which reduces the problem to the framework of this book.

– Note also that (1.1) is a priori valid on the whole interval $[-1, +7]$, but (1.2) concerns $[-1, 0]$ only. Actually, optimization just deals with this latter interval; it is only for the forecast itself, after optimization is finished, that the interval $[0, 7]$ will come into play.

– Since p and Ω do not live in the same space (the number $|I|$ of observations, possibly very large, is certainly finite), Ω must first be embedded in the same function space as p. Besides, the norm $\|\cdot\|$ in (1.2) must be carefully chosen. These aspects, which concern modeling only, have a big influence on the behaviour of solution algorithms.

At this point, it is a good idea not to view (1.1), (1.2) as a nonlinearly constrained optimization problem (category 3.2 in §1.1.2), but rather as an unconstrained one (category 1.2). In fact, call $u(z) = p(z, -1)$ the state of the atmosphere at z, at initial time $t = -1$. A fundamental remark is then: assuming u to be known, (1.1) gives unambiguously $p(z, t) = p_u(z, t)$ for all z and all $t \geqslant -1$: the unknown p_u depends on the variable u *only*. Hence, the objective value in (1.2) also depends on u *only*. Our problem can therefore be formulated as $\min_u \|p_u - \Omega\|$, which means:

– to minimize with respect to u (unconstrained variable)
– the function defined by (1.2),
– where $p = p_u$ is obtained from (1.1)
– via the initial condition $p(\cdot, -1) = u$.

The actual decision variable in this formulation is u indeed: p plays only the role of a parameter, called *state variable*, while the terminology *control variable* is here reserved to u. The objective function will be denoted by $J(u)$, rather than $f(x)$. Thus, the number of variables is reduced (drastically: passing from about 10^9 for p, to about 10^7 for u alone) and, more importantly, any form of constraint is eliminated.

Remark 1.2. The "normal", *direct*, problem is to compute $p(z, t)$ from $p(z, 0)$ via (1.1). Here we solve the *inverse* problem: to compute $p(z, 0)$ from (a partial knowledge of) $p(z, t)$.

The above description is of course very sketchy and does not reveal all the difficulty of the problem. For instance: the number of observations is about 10^5, which is by far insufficient to identify the 10^7 unknowns. To orient the search toward reasonable p_u's, any a priori information on the stationary solutions to (1.1) is an important element, which is taken into account in actual implementations.   □

Here again, the methods from the first part of this book will be used. The problem is more than ever large-scale: after discretization, $u \in \mathbb{R}^{10^7}$; calling for §6.3 therefore becomes a must.

1.2.3 Trajectory of a Deepwater Vehicle

Most optimal control problems consist in optimizing a trajectory; an example is towing a submarine vehicle. Consider a deepwater observation device (the "fish"), moving close to the sea bottom, and pulled from the surface by a tug. The problem is to control the tug so that the fish makes a given maneuver, while avoiding obstacles. For example, one may ask to make a U-turn in minimal time.

Let L be the length of the pulling cable. One may assume that L is a known constant, or that the cable is inextensible; anyway L is for this problem several kilometers long, and one cannot assume that the cable behaves like a rigid rod. As a result, the fish's trajectory is a rather complicated function of the tug's. A possible model is as follows.

– Let $y(s,t) \in \mathbb{R}^3$ be the position in the sea of a point at time t and (curvilinear) coordinate $s \in [0, L]$ along the cable.
– Then $y(0, t)$ is the tug's position, it is the control variable; $y(L, t)$ is the fish's, it is the variable to be controlled.
– These two variables are not independent: from inextensibility, we have

$$\left\| \frac{\partial y}{\partial s} \right\| = 1 \qquad (1.3)$$

and y obeys the state equation

$$\frac{\partial^2 y}{\partial t^2} - \frac{\partial}{\partial t}\left(T(s,t) \frac{\partial y}{\partial s}\right) + \tau\left(\frac{\partial y}{\partial t}\right) = w. \qquad (1.4)$$

Here T is the cable's tension (unknown), w its linear weight rate and τ models the drag.
– In addition to this system of equations, there are appropriate initial and boundary conditions, among which $y(0, t) = u(t)$, which simply expresses that $y(0, \cdot)$ plays a special role (the control!).

Just as in §1.2.2, we are again faced with an optimal control problem: the objective function (for example the time needed to make a U-turn) depends on the control u implicitly, via a state (y_u, T_u), solution to a state equation. However, the situation is no longer as "simple" (!) as in §1.2.2: we still have to express that the fish must evolve above the sea bottom, which yields constraints on the state: if $\varphi(z^1, z^2)$ is the height of free water at $z \in \mathbb{R}^2$, one must impose

$$y^3(L, t) \geqslant \varphi(y^1(L, t), y^2(L, t)), \quad \text{for all } t. \tag{1.5}$$

These constraints in turn depend implicitly on u, and they are actually infinitely many (i.e. many, after discretization). As a result, it is hardly possible to "reduce" the problem with respect to u only. We now have to call for the third part of this book (constrained nonlinear optimization): the distinction between control and state variables is no longer relevant. In the sense of §1.1.1, the decision variables are now the couple (y, T), with respect to which one must

– minimize a certain function $f(y)$ (for example the time of the U-turn)
– under equality constraints $c_j(y, T) = 0$, $j \in E$, which symbolize the state equations (1.3), (1.4) (here E is big)
– and inequality constraints $c_j(y) \leqslant 0$, $j \in I$, which symbolize constraints on the state (1.5) (and I is just as big).

This example illustrates, among other things, the ambiguity which can exist concerning the decision variables: in the sense of optimal control, the control variable is u; however, the optimization algorithm "sees" as decision variable the whole of (y, T). Of course, the algorithm designer is allowed – and even strongly advised – to remember the origin of the problem, and to let $y(0, \cdot)$ play a particular role in the complete set of variables $\{(y, T)(s, t)\}_{s,t}$.

1.2.4 Optimization of Power Management

We complete this list of examples with a problem having nothing to do with the preceding : to optimize the production of electrical power plants. The following constitutes a simplest instance among realistic models. Consider a set I of power plants (hydro-electrical, thermal, nuclear or not). One wishes to optimize their production over a horizon $\{1, \ldots, T\}$, for example $T = 48$ half-hours; the demand is supposed to be known, call it $d_1, \ldots, d_T$. If p_t^i denotes the energy produced by the production unit $i \in I$ during the period t, one must first satisfy the demand constraints

$$\sum_{i \in I} p_t^i \geqslant d_t, \quad \text{for } t = 1, \ldots, T. \tag{1.6}$$

Use the notation $p^i = \{p_1^i, \ldots, p_T^i\}$ for the production-vector of unit i. To each unit is associated a production cost $c^i : \mathbb{R}^T \to \mathbb{R}$: one wishes to solve

$$\min \sum_{i \in I} c^i(p^i). \tag{1.7}$$

Besides, each unit has its own *technological constraints* describing the set D^i of possible production vectors:

$$p^i \in D^i, \quad \text{for } i \in I.$$ (1.8)

Describing the c^i's and D^i's may not be a simple task, which goes beyond our framework. We just note here their *disparity*: nuclear and hydro plants have nothing to do with each other, neither in their operation costs, nor in their constraints. For one thing, a hydro plant has basically linear characteristics (category 2.2.1 in §1.1.2), although it becomes nonlinear (category 3.2) in accurate models. By contrast, thermal plants have an important combinatorial aspect, owing to a $0-1$ behaviour: it is not possible to change their production level continuously, neither at any time.

The crude problem is to minimize (1.7) under constraints (1.6), (1.8). This problem is large-scale: as an example, the French power mix has about 200 plants working every day, which gives birth to $200 \times 48 = 10^4$ variables p^i_t (and even many more, due to combinatorics; actually, each unit i is an optimal control system, with its own additional state variables). Yet, the real difficulty of the problem is not its size but its heterogeneity: nonlinear methods of this book will fail, just as combinatorial methods.

This is why it is suitable to transform this problem. The key is to observe that, if constraints (1.6) were not present, each plant could be treated separately: one would have to solve, for each $i \in I$

$$\min c^i(q), \quad q \in D^i.$$ (1.9)

Here, the dummy variable q represents the production-vector p^i. Each of the latter problems becomes solvable, by a method tailored to each case, depending on i. Starting from this remark, a particular heuristic technique is rather well-suited for (1.6)–(1.8). More precisely, Lagrangian relaxation (§8.2) approximates a solution by minimizing a convex nonsmooth function, to be seen in Chap. 10.

1.3 General Principles of Resolution

The problems of interest here – such as those of §1.2 – are solved via an algorithm which constructs iteratively $x_1, x_2, \ldots, x_k, \ldots$ To obtain the next iterate, the algorithm needs to know some information concerning the original problem (P) of §1.1.1: essentially, the numerical value of f and c for each value of x; often, their derivatives as well.

– If there are only linear or quadratic functions, this information is globally and explicitly available in the data: a linear [resp. quadratic] function (b, x) [resp. (x, Ax)] is completely characterized by the vector b [resp. the matrix A]. As a result, categories 1.1, 2.1.1, 2.2.1, 2.2.2 of §1.1.2 make up a very particular class, and call for very particular methods, studied in the fourth part of this volume.

– By contrast, as soon as really general functions are involved, this information is computed in a *black box* (subprogram) characterizing (P), and
independent of the selected algorithm. This subprogram can be called *simulator*, since it simulates the behaviour of the problem under the action of
the decision variables (optimal or not).

Hence (and it is important to convince oneself with this truth), a computer
program solving an optimization problem is made up of *two distinct parts*:

– One is in charge of managing x and is the algorithm proper; call it (A),
as Algorithm; it is generally written by a mathematician, specialized in
optimization.

– The other, the simulator, depending on (P), performs the required calculations for each x decided by (A); it is generally written by a practitioner
(engineer, physicist, economist, etc.), the one who wishes to solve the specific optimization problem.

The distinction between (A) and (P) is not always straightforward, actually it depends on the modeling. Consider the examples of the preceding
section:

§1.2.1. There is no ambiguity in the biochemistry problem: (A) places the
atoms in the space, (P) computes the resulting energy, and perhaps
its derivatives as well: they are very useful for (A).

§1.2.2. The case of meteorology is also relatively clear: (A) decides the initial conditions (denoted by u or $p(\cdot, -1)$ rather than x); (P) inte-
grates the state equation over $[-1, 0]$, which allows the computation
of the objective function (1.2); call $J(u)$ this objective. Note that
differentiating J is now far from trivial; yet, it is certainly possible
(at least after discretization, in case of theoretical difficulties for the
continuous version). More is given on this topic in §1.6 below.

§1.2.3. In the cable problem the situation is no longer so clear-cut. In a
control-like formulation as in §1.2.2, (A) would decide the tug's trajectory, and (P) would integrate (1.3), (1.4) to obtain the fish's
trajectory; the objective value and the constraint value (1.5) would
ensue.

In the suggested "general-constrained" formulation, (A) fixes the
trajectory and tension of every point on the cable. The job of (P)
is now much more elementary: it knows the values of $(y, T)(s, t)$
for each (s, t) – they have been fixed by (A) – and it just have to
compute the values (and derivatives) of the objective, of the equality
constraints (1.3), (1.4), and of the inequality constraints (1.5).

§1.2.4. A complication appears in production optimization because the
problem is not really (1.6)–(1.8), but rather an auxiliary abstract
problem, which will be seen in §8.3.2. The objective is actually a
perturbation of (1.7), namely a *Lagrange function* incorporating the
term $\sum_t \lambda_t \left(\sum_i p_t^i - d_t \right)$; the decision variables are no longer the p_t^i's

but the λ_t's, i.e. the multipliers associated with (1.6). Thus, (A) fixes the λ_t's, while (P) solves for each i a perturbation of (1.9), namely

$$\min_{q \in D^i} c^i(q) + \sum_t \lambda_t q_t .$$

Remark 1.3. In addition to the (A)–(P) distinction, another fundamental thing to understand here is the following: for any problem considered, the only information available for (P) is the result of a numerical calculation, generally complicated; for example, the resolution of a partial differential equation, or the optimization of a number of nuclear plants, etc. Hence, (A) has to proceed by "trial and error": it assigns trial values to the decision variables x, and it corrects these values upon observation of the answer from (P); and this will repeatedly make up the iterations of the optimization process.   □

Now the current iteration of an optimization algorithm is made up of two phases: to compute a direction, and to perform a line-search.

– Computing a direction: (P) is replaced by a model (P_k), which is simpler; then (P_k) is solved to yield a new approximation $x_k + d$.
– Line-search: a stepsize $t > 0$ is computed so that $x_k + td$ is "better" than x_k in terms of (P).
– The new iterate is then $x_{k+1} = x_k + td$.

Remark 1.4. The direction is computed by solving (usually accurately) an *approximation* (P_k) of (P). By contrast, the stepsize is computed by observing the *true* (P) on the restriction of $x \in \mathbb{R}^n$ to the half-line $\{x_k + td\}_{t \in \mathbb{R}_+}$ (x_k and d fixed).

Replacing the given problem (P) by a simpler (P_k) is a common technique in numerical analysis. By contrast, the second phase which corrects $x_k + d$, is a technique specific to optimization. Its motivation is *stabilization*. All this will be seen in detail in the next chapters.   □

The next two subsections are devoted to some convergence theory tailored to optimization algorithms.

1.4 Convergence: Global Aspects

Let an optimization algorithm generate some sequence $\{x_k\}$. This algorithm is said to converge *globally* when

$\{x_k\}$ converges to "what is wished" for any initial iterate x_1.

Caution: this terminology is ambiguous because "what is wished" does not mean a solution to the initial problem (P), often called *global optimum*. Here, one rather stresses the fact that the initial iterate can be arbitrarily far from "what is wished", without impairing convergence; actually, "what is wished" generally means an x satisfying what is called the necessary optimality conditions (see below and the sections involved: §§2.2 and 13.3).

In connection with Remark 1.4, one generally has a *merit function* Θ : $\mathbb{R}^n \to \mathbb{R}$, which is minimal at "what is whished": (P) is thus equivalent to minimizing Θ over the *whole* of $\mathbb{R}^n$. The simplest example is unconstrained optimization: one must minimize f over $\mathbb{R}^n$, so one naturally takes $\Theta = f$. The word "better" introduced in §1.3 can then be given the meaning

$$\Theta(x_{k+1}) < \Theta(x_k) . \tag{1.10}$$

Then let us review the various convergence properties that an optimization algorithm may enjoy. First, a direct consequence of (1.10) is that

$$\{\Theta(x_k)\} \text{ has a limit, possibly } -\infty$$

– of course, $\Theta(x_k) \to -\infty$ reveals an ill-posed problem (P).

Minimal requirement To make things simple, let us assume that Θ is a continuously differentiable function and consider its first-order development around a given x:

$$\Theta(x + h) \simeq \Theta(x) + (\nabla\Theta(x), h) .$$

Assuming $\nabla\Theta(x) \neq 0$ and taking $h = -t\nabla\Theta(x)$ with a small $t > 0$, we obtain $\Theta(x + h) - \Theta(x) \simeq -t|\nabla\Theta(x)|^2 < 0$; as a result, x cannot minimize Θ. We say that $\nabla\Theta(x) = 0$ is an *optimality condition* for x to minimize Θ. The least property that should be satisfied by a sequence $\{x_k\}$ constructed as in §1.3 is then[1]

$$\liminf |\nabla\Theta(x_k)| = 0 ; \tag{1.11}$$

this means that the gradient $\nabla\Theta(x_k)$ will certainly have a norm smaller than ε for some finite k, no matter how $\varepsilon > 0$ is chosen. Thus, in this context, a globally convergent algorithm has to satisfy (1.11) for any starting point x_1.

It should be noted that (1.11), or even the property $\lim |\nabla\Theta(x_k)| = 0$, is fairly weak indeed: it does not tell much unless $\{x_k\}$ itself has some limit point. For example, it does not imply that $\{x_k\}$ is a *minimizing sequence*, i.e. that $\Theta(x_k) \to \inf \Theta$.

Boundedness If the original minimization problem (P) is reasonably well-posed, a reasonable merit function satisfies

$$\Theta(x) \to +\infty \quad \text{when} \quad |x| \to +\infty$$

(for example, minimizing e^x over $x \in \mathbb{R}$ is an ill-posed optimization problem: it has no solution). Together with (1.10), this property automatically guarantees that $\{x_k\}$ is a bounded sequence. As a result, $\{x_k\}$ has a cluster point; and every subsequence $\{x_k\}_{k \in K}$ is also bounded.

---

[1] The lim inf [resp. lim sup] of a numerical sequence is its smallest [resp. largest] cluster point.

Convergent sequences Assume boundedness of $\{x_k\}$. Then (1.11) guarantees the existence of a subsequence $\{x_k\}_{k \in K}$ satisfying

$$x_k \xrightarrow{k \in K} x^* \quad \text{and} \quad \nabla\Theta(x_k) \xrightarrow{k \in K} 0\,,$$

from wich continuity of $\nabla\Theta$ implies $\nabla\Theta(x^*) = 0$.

On the other hand, the monotonicity property (1.10) implies that the whole sequence $\{\Theta(x_k)\}$ tends to $\Theta(x^*)$: all cluster points of $\{x_k\}$ have the same Θ-value. Whether this value is the minimum value of Θ is more delicate.

When Θ is a convex function, the optimality condition $\nabla\Theta(x^*) = 0$ is (necessary and) sufficient for x^* to minimize Θ (use for example the well-known property $\Theta(y) \geqslant \Theta(x^*) + (\nabla\Theta(x^*), y - x^*)$ for all y). In this situation, we conclude that *all* the cluster points of $\{x_k\}$ minimize Θ; and finally, the whole of $\{x_k\}$ converges to the same limit x^* if Θ has a single minimum point x^* (for example if Θ is strictly convex).

Let us summarize our considerations: admitting that (P) can be formulated as minimizing a differentiable function Θ, the key property to be satisfied by an algorithm is (1.11). If Θ enjoys appropriate additional properties, then the limit points of $\{x_k\}$ will minimize Θ, and hence solve (P).

1.5 Convergence: Local Aspects

Now $\{x_k\}$ is assumed to have a limit x^* – which may or may not be "what is wished" – and one wants to know at what speed $x_k - x^*$ tends to 0; in particular, one tries to compare this error to an exponential function. This study is limited to large values of k (hence x_k is already close to x^*): it is only a *local* study. First recall some notation: $s = o(t)$ means that s is "infinitely smaller" than t; more precisely $\frac{s}{t} \to 0$. Here t and s are two variables (depending on a parameter x, on an iteration number k, etc.); t is scalar-valued and positive; strictly speaking, s as well; when s is vector-valued, the correct and complete notation should be $|s| = o(t)$. In practice, it is implicitly understood that $t \downarrow 0$ (say when $x \to x^*$, or $k \to +\infty$) and $s = o(t)$ means that s tends to 0 infinitely faster than t. The notation $s = O(t)$ means that s is not infinitely bigger than t: there exists a constant C such that $s \leqslant Ct$.

Consider now a sequence $\{x_k\}$ converging to x^*; two types of convergence are relevant:

Q-convergence : this is a study of the quotient $q_k := |x_{k+1} - x^*|/|x_k - x^*|$.
– *Q*-linear convergence is said to hold when $\limsup q_k < 1$.
– *Q*-superlinear convergence when $\lim q_k = 0$.
– Particular case: *Q*-quadratic convergence when $q_k = O(|x_k - x^*|)$; or equivalently: $|x_{k+1} - x^*| = O(|x_k - x^*|^2)$; roughly, the number of exact digits doubles at each iteration.

Often, "Q" is omitted: superlinear convergence implicitly means Q-superlinear convergence.

R-convergence : even though Theorems 1.7 and 1.8 below give a more natural definition, R-convergence is originally a study of the rate $r_k := |x_k - x^*|^{1/k}$.
– $\limsup r_k < 1$: R-linear convergence,
– $\lim r_k = 0$: R-superlinear convergence.

Remark 1.5. A sequence converging sublinearly to its limit (q_k or r_k tends to 1) is in practice considered as not converging at all, because convergence is so slow; an algorithm with sublinear convergence must simply be forgotten.
□

  R-linear convergence means geometric or exponential convergence: setting $r := \limsup r_k$, we have $r_k \leqslant r + \varepsilon$ for all $\varepsilon > 0$ and k large enough; this is equivalent to $|x_k - x^*| \leqslant (r + \varepsilon)^k$ (and note: $r + \varepsilon$ can be made < 1).
  Q-convergence is more powerful, in that the error at iteration $k + 1$ can be bounded in terms of the error at iteration k: if $q = \limsup q_k$,

$$|x_{k+1} - x^*| \leqslant (q + \varepsilon)|x_k - x^*|, \quad \text{for all } \varepsilon > 0 \text{ and } k \text{ large enough.}$$

In a way, Q-convergence is a Markovian concept: it only involves what happens at the present iteration. In the above writing, "iteration k [resp. $k + 1$]" can be replaced by "current iterate x [resp. next iterate x_+]" and "k large enough" by "x close enough to x^*". In plain words, Q-superlinear convergence is expressed by: if the current iterate is close to the limit, then the next iterate is infinitely closer. This is not true for R-convergence, since k plays its role in the definition of r_k, which has to be a kth root. The next result confirms that Q-linear convergence implies geometric convergence:

Theorem 1.6. *If x_k tends Q-linearly to x^*, then: for all $q > \limsup q_k$, there exists k_0 and $C > 0$ such that*

$$|x_k - x^*| \leqslant Cq^k \text{ for all } k \geqslant k_0.$$

Proof. Fix q as announced, k_0 such that

$$|x_{i+1} - x^*| \leqslant q|x_i - x^*| \text{ for } i \geqslant k_0,$$

which gives (multiplying out for $i = k_0, \ldots, k - 1$)

$$|x_k - x^*| \leqslant |x_{k_0} - x^*|q^{k-k_0} = \frac{|x_{k_0} - x^*|}{q^{k_0}}q^k$$

and the result is obtained with $C := |x_{k_0} - x^*|/q^{k_0}$.
□

Once again, this theorem does not contain all the power of Q-convergence, since it does not say that the error decreases at the rate $q < 1$ at *each* iteration.

Quite often, convergence speed is established via a study of an upper bound of the error. Q-convergence of an upper bound of $|x_k - x^*|$ becomes R-convergence for $\{x_k\}$. For example:

Theorem 1.7. *If $|x_k - x^*| \leqslant s_k$ where s_k converges Q-superlinearly to 0, then $\{x_k\}$ converges R-superlinearly to x^*.*

Proof. Fix $\varepsilon > 0$. From Theorem 1.6, there is C such that $s_k \leqslant C\varepsilon^k$ for k large enough. Hence, by assumption,

$$|x_k - x^*|^{1/k} \leqslant s_k^{1/k} \leqslant C^{1/k}\varepsilon .$$

Pass to the limit on k: $C^{1/k} \to 1$ and $\limsup |x_k - x^*|^{1/k} \leqslant \varepsilon$.   □

Actually, the converse is also true. To show it, we give a last result, stated in terms of linear convergence, to make a change:

Theorem 1.8. *Let x_k tend to x^* R-linearly. Then $|x_k - x^*|$ is bounded from above by a sequence s_k tending to 0 Q-linearly.*

Proof. Call $r < 1$ the limsup of $|x_k - x^*|^{1/k}$ and take $\varepsilon \in \,]0, 1 - r[$. For k large enough, $|x_k - x^*| \leqslant (r + \varepsilon)^k$. The sequence $s_k := \max\{|x_k - x^*|, (r + \varepsilon)^k\}$ is indeed an upper bound of $\{|x_k - x^*|\}$ and, for k large enough, $s_k = (r + \varepsilon)^k$; hence s_k answers the question.   □

These two theorems establish the equivalence between R-convergence of a nonnegative sequence tending to 0, and Q-convergence of an upper bound. This gives another definition of R-convergence, perhaps more natural than the original one; namely: $x_k \to x^*$ R-superlinearly when $|x_k - x^*| \leqslant s_k$, for some $\{s_k\}$ tending to 0 Q-superlinearly.

1.6 Computing the Gradient

As seen in §1.3, the main duty of the user of an optimization algorithm is to write a simulator computing information needed by the algorithm. It has also been said (and it will be confirmed all along this book) that the simulator should compute not only function- but also derivatives-values. This is not always a trivial task, especially in optimal control problems. Take for example the case of meteorology in §1.2.2: it is easy to understand how the objective function of (1.2) (call it f) can be computed via (1.1), for given values of the control variable $u(\cdot) = p(\cdot, -1)$; but how about the *total derivative* of f with respect to u? Since f is given *implicitly* by (1.1), one must somehow invoke the implicit function theorem, which may be tricky. Indeed, computing the

Jacobian of the operator "control variable $\mapsto$ state variable" is often out of question, and useless anyway. Here we demonstrate a technique commonly used, which involves the *adjoint equation*. For reasons to be explained in Remark 1.9 below, we do this computation in a finite-dimensional setting, even though optimal control problems are usually set in some function space.

So we consider the following situation. The control variables are $\{u_t\}_{t=1}^T$ where $u_t \in \mathbb{R}^n$ for each t. The state variables are likewise $\{y_t\}_t$ with $y_t \in \mathbb{R}^m$, given by the state equation

$$\begin{cases} y_t = F_t(y_{t-1}, u_t), & \text{for } t = 1, \ldots, T, \\ y_0 \text{ given}. \end{cases} \tag{1.12}$$

Here, for each t, F_t is a function (possibly nonlinear) from $\mathbb{R}^m \times \mathbb{R}^n$ to $\mathbb{R}^m$. Besides, a function is given, say

$$f = \sum_{t=1}^T f_t(y_t, u_t),$$

where, for each t, f_t sends $\mathbb{R}^m \times \mathbb{R}^n$ to $\mathbb{R}$. It is purposely that we do not specify formally which variables f depends on. Incidentally, note that f can be the objective function of our optimal control problem; but it can equally be a constraint, involving the state variables; for example a *final-time constraint* $c(y_T)$ (imposed to be 0, or nonnegative, etc.)

Call $v = du \in \mathbb{R}^{nT}$ a differential of u; it induces from (1.12) a differential $z = dy \in \mathbb{R}^{mT}$, and finally a differential df. To be specific, we assume the usual dot product in each of the spaces involved and we use the notation $(\cdot, \cdot)_n$ [resp. $(\cdot, \cdot)_m$] for the dot-product in $\mathbb{R}^n$ [resp. $\mathbb{R}^m$]. In the control space, the scalar product is therefore

$$(g, v) = \sum_{j=1}^T (g_t, v_t)_n.$$

Our problem is then as follows: find $\{g_t\}_{t=1}^T$ such that the differential of f is given by $df = (g, v)$. This will yield $\{g_t\}_t \in \mathbb{R}^{nT}$ as the gradient of f, considered as a function of the control variable u alone.

To solve this problem, we have from (1.12) (assuming appropriate smoothness of the data)

$$\begin{cases} z_t = (F_t)'_y(y_{t-1}, u_t) z_{t-1} + (F_t)'_u(y_{t-1}, u_t) v_t & \text{for } t = 1, \ldots, T, \\ z_0 = 0 \end{cases} \tag{1.13}$$

($z_0 = 0$ because y_0 is fixed!). In this writing, the Jacobian $(F_t)'_y(y_{t-1}, u_t)$ is an $m \times m$ matrix and $(F_t)'_u(y_{t-1}, u_t)$ is $m \times n$. We have also

$$df = \sum_{t=1}^T (\nabla_y f_t(y_t, u_t), z_t)_m + \sum_{t=1}^T (\nabla_u f_t(y_t, u_t), v_t)_n;$$

here $\nabla_y f_t(y_t, u_t) \in \mathbb{R}^m$ and $\nabla_u f_t(y_t, u_t) \in \mathbb{R}^n$. We need to eliminate z between these various relations; this is done by a series of tricks:

Trick 1. Multiply the tth linearized state equation in (1.13) by a vector $p_t \in \mathbb{R}^m$ (unspecified for the moment) and sum up. Setting $G_t := (F_t)'_y(y_{t-1}, u_t)$ and $H_t := (F_t)'_u(y_{t-1}, u_t)$, we obtain

$$\sum_{t=1}^{T}(p_t, z_t)_m = \sum_{t=1}^{T}(p_t, G_t z_{t-1})_m + \sum_{t=1}^{T}(p_t, H_t v_t)_m \,.$$

Single out $(p_T, z_T)_m$ in the lefthand side, transpose G_t and H_t, and re-index the sum in z; remembering that $z_0 = 0$, this gives

$$0 = -(p_T, z_T)_m - \sum_{t=1}^{T-1}(p_t, z_t)_m + \sum_{t=1}^{T-1}(G_{t+1}^{\top}p_{t+1}, z_t)_m + \sum_{t=1}^{T}(H_t^{\top}p_t, v_t)_n \,.$$

Trick 2. Add to the expression of $\mathrm{d}f$ and identify with respect to the z_t's. Setting $\gamma_t := \nabla_y f_t(y_t, u_t)$ and $h_t := \nabla_u f_t(y_t, u_t)$:

$$\mathrm{d}f = (-p_T + \gamma_T, z_T)_m + \sum_{t=1}^{T-1}(-p_t + G_{t+1}^{\top}p_{t+1} + \gamma_t, z_t)_m + \sum_{t=1}^{T}(H_t^{\top}p_t + h_t, v_t)_n \,.$$

Trick 3. Now it suffices to choose p so as to cancel out the coefficient of each z_t: requiring

$$\begin{cases} p_T = \gamma_T \,, \\ p_t = G_{t+1}^{\top}p_{t+1} + \gamma_t & \text{for } t = T-1, \ldots, 1 \,, \end{cases} \tag{1.14}$$

we obtain the gradient in the desired form:

$$g_t = H_t^{\top}p_t + h_t \quad \text{for } t = 1, \ldots, T.$$

The (backward) recurrence relations (1.14) form the so-called *adjoint equation*, whose solution p is the *adjoint state*.

Remark 1.9. In optimal control problems, the state variable is often given by a differential equation, say

$$\begin{cases} \dot{y}(t) = F(y(t), u(t), t) \,, & \text{for } t \in \,]0, T[\,, \\ y(0) \text{ given}, \end{cases}$$

instead of the recurrence relations (1.12). Then the "adjoint trick" can nevertheless be reproduced: multiply the above equation by a function $p(t)$ (the continuous adjoint state), integrate from 0 to T, and integrate the lefthand side by parts. The resulting adjoint equation is another differential equation, instead of (1.14).

However, the actual minimization algorithm, implemented on the computer, certainly does not solve this original problem; it can but solve some *discretized* form of it (a computer can hardly work in infinite dimension). Using a subscript δ to connote such a discretization, we are eventually faced with minimizing a certain function $f_\delta(u_\delta)$, with respect to some finite-dimensional variable u_δ. For numerical efficiency of the minimization algorithm, *it is important that the simulator computes the exact gradient of f_δ*, and not some discretized form of the continuous gradient ∇f. One way of achieving this is to carefully select the discretization scheme of the adjoint equation. But the safest approach is to *discretize first* the problem (and in particular the state equation), and then only to construct the adjoint equation *of the discretized problem*.

This is why we bothered to demonstrate the mechanism for the tedious discrete case; after this, reproducing the calculations in the continuous case is an easy exercise (only formal, though: differentiability properties of the infinite-dimensional problem must still be carefully analyzed; otherwise, difficulties may occur for $\delta \to 0$). □

Remark 1.10. The adjoint technique opens the way to the so-called *automatic* or *computational differentiation*. Indeed, consider a computer code which, taking an input u, computes an output f. Such a code can be viewed as a "control process" of the type (1.12):

- The tth line of this code is the tth equation in (1.12).
- The intermediate results of this code (the lefthand sides of the assignment statements) form altogether a "state" y, which is a function of the "control" u.
- Forming the righthand side of the adjoint equations then amounts to differentiating one by one each line of the code.
- Afterwards, solving the adjoint equations – to obtain finally the gradient ∇f – amounts to writing these "linearized lines" bottom up.

These operations are all purely mechanical and lend themselves to automatization. Thus, one can conceive the existence of a software which

- takes as input a computer code able to calculate $f(u)$ (for given u),
- and produces as output another computer code able to calculate $\nabla f(u)$ (again for given u).

It is worth mentioning that such software do not need to know anything about the problem. They do not even need mathematical formulae representing the computation of f. What they need is just the first half of a simulator; and then they write down its second half. □

Bibliographical Comments

Among other monographs devoted to optimization algorithms, [107, 27, 277, 86] can be suggested. See also [128, 160] for a style very close to users' concerns, while [239] insists more on theorems.

A function Θ for which a stationary sequence $(\nabla\Theta(x_k) \to 0)$ is not necessarily minimizing $(\Theta(x_k) \not\to \inf \Theta)$ is given in [350]. The various types of local convergence are defined and studied in [278].

As for available optimization software, the situation is rapidly evolving. First, there is the monograph [267], which reviews most individual codes and organized libraries existing in the beginning of the 90's. Generally speaking, the Harwell library has well-considered optimization codes. In fact, this library goes far beyond optimization, as it covers the whole of numerical analysis, from linear algebra to differential equations:

> http://www.cse.clrc.ac.uk/Activity/HSL.

On the other hand, the Galahad software is exclusively devoted to optimization and can normally be used for free:

> http://galahad.rl.ac.uk/galahad-www.

The Scilab environment and the Modulopt library include implementations of some of the algorithms presented in this book:

> http://www-rocq.inria.fr/scilab/scilab.html
> http://www-rocq.inria.fr/estime/modulopt.

The internet address

> http://www-neos.mcs.anl.gov/neos

collects and updates, under the name NEOS, the vast majority of software existing throughout the world, even allowing a "push-button" use of some of them.

For computational differentiation, see for example [181], [88], [151] (but the idea is much older, going back to [339, 208] and others). We mention Adolc, Adifor, Tapenade as available software; the addresses are as follows:

> http://www.math.tu-dresden.de/wir/project/adolc
> http://www-unix.mcs.anl.gov/autodiff/ADIFOR
> http://www-sop.inria.fr/tropics/tapenade/tutorial
> http://www-unix.mcs.anl.gov/autodiff/AD_Tools

Part I

Unconstrained Problems

Claude LEMARÉCHAL

In this first part, we consider the problem of minimizing a function f, defined on all of the space $\mathbb{R}^n$. We will always assume f sufficiently smooth, say twice continuously differentiable; in fact, a rather minimal assumption is that f has a Lipschitz continuous gradient.

We start with a short introductory chapter, containing in particular the gradient method, often deemed important. However we pass rapidly over it, because actually it is (or should be) never used. In contrast, the whole Chap. 3 is devoted to line-searches, a subject often neglected although it is of crucial importance in practice.

In fact, the gradient method is limited to first-order approximations, whereas efficient optimization *must* take second order into account, explicitly or implicitly; it is even fair to say that this is a necessary and sufficient condition for efficiency. Using second order amounts to applying Newton's principle. Chapter 4 starts from these premises to study the utmostly important and universally used quasi-Newton method. Conjugate gradient (Chap. 5) is given mainly for historical reasons: this method has been much used but it is now out of date. Chapter 6 is quite different: it mainly concerns methods less used these days, but which cannot be overlooked; either due to the importance of problems they treat (Gauss-Newton, Levenberg-Marquardt), or because they will become classical in the future (trust-region, various uses of Newton's principle). Besides, it outlines the traditional resolution of quadratic programs (item 2.2.2 in the classification of §1.1.2), namely by pivoting.

A short additional chapter presents an application problem: seismic reflexion tomography. It can be used to illustrate the behaviour of unconstrained optimization algorithms, and also to get familiarized with the actual writing of a nontrivial simulator.

2 Basic Methods

We start with some generalities on the unconstrained optimization problem

$$\min f(x) \quad \text{subject to } x \in \mathbb{R}^n . \tag{2.1}$$

2.1 Existence Questions

A very first condition for (2.1) to be meaningful is that f be bounded from below. Then there exists a lower bound but not necessarily an optimal solution (e.g. $f(x) = e^x$); an additional assumption is required.

The following property is usually satisfied (at least, it is reasonable): f is (continuous and) "$+\infty$ at infinity"; more precisely: $f(x) \to +\infty$ if $|x| \to +\infty$. Such a function is called *inf-compact* (cf. §1.4). Then the problem can be restricted to a bounded set, say $\{x : f(x) \leqslant f(x_1)\}$ (often called *slice* of f at level $f(x_1)$) and existence of a global minimum x^* is guaranteed: a continuous function has a minimum on a compact set.

Remark 2.1. There is a delicate point in infinite dimensions. An existence proof goes as follows:

- f bounded from below $\Rightarrow$ existence of a (finite) lower bound f^* and of a minimizing sequence $\{x_k\}$, i.e. $f(x_k) \to f^*$.
- Slice bounded $\Rightarrow \{x_k\}$ bounded $\Rightarrow$ existence of a weak cluster point x^* (in a reflexive Banach space).
- To conclude $f(x_k) \to f(x^*)$ (i.e. $f(x^*) = f^*$) one needs also the lower semi-continuity of f for the weak topology, which holds when f is convex; an assumption which thus appears naturally in infinite dimension. □

The above remark introduces two concepts which, although not fundamental, have their importance in optimization.

Definition 2.2. *A function f is* lower semi-continuous *at a given x^* when* $\liminf_{x \to x^*} f(x) \geqslant f(x^*)$.
A function f is convex *when*

$$f(\alpha x + (1 - \alpha)y) \leqslant \alpha f(x) + (1 - \alpha)f(y) \quad \text{for all } x, y, \text{ and } \alpha \in \,]0, 1[.$$

A set $C \subset \mathbb{R}^n$ is convex *when*

$$\alpha x + (1 - \alpha)y \in C \quad \text{for all } x, y \text{ in } C, \text{ and } \alpha \in]0,1[.$$

Accordingly, we will say that our general problem (P) *of §1.1.1 is* convex *if f and each c_j, $j \in I$ are convex, while $\{c_j\}_{j \in E}$ is affine.*
We also recall that a mapping $c : \mathbb{R}^n \to \mathbb{R}^p$ is affine *if there exists a linear mapping $L : \mathbb{R}^n \to \mathbb{R}^p$ such that*

$$c(x) - c(y) = L(x - y) \quad \text{for all } x, y \in \mathbb{R}^n . \qquad \qquad \square$$

2.2 Optimality Conditions

The question is now: how to recognize an optimum point? There are necessary conditions, and sufficient conditions, which are well-known:

– Necessary conditions: if x^* is optimal, then
  · 1st-order necessary condition (NC1): the gradient $f'(x^*)$ is zero;
  · 2nd-order necessary condition (NC2): the Hessian $f''(x^*)$ is positive semi-definite[1].
– Sufficient condition (SC2): if x^* is such that $f'(x^*) = 0$ and $f''(x^*)$ is positive definite, then x^* is a local minimum (i.e. $f(x) \geqslant f(x^*)$ for x close to x^*).

Example 2.3. easiest: f quadratic, i.e. $f(x) = \frac{1}{2}(x, Ax) + (b, x) + c$. Then (NC1) is the linear system $Ax + b = 0$. If A is positive definite, this system has a unique solution, which is the minimum point. If A is positive semi-definite, and if $b \in \operatorname{Im} A$, there is a hyperplane of solutions, which make up the minima. We conclude that minimizing an unconstrained quadratic function is nothing other than solving a linear system, whose matrix is symmetric, and normally positive definite. $\qquad \square$

The difference between (NC2) and (SC2) is weak, in practice negligible. An x satisfying (NC1) is called *critical* or *stationary*. If f is convex, (NC1) = sufficient condition for global minimum. The ambition of an optimization algorithm is limited to identifying stationary points; this implies $f'(x_k) \to 0$. With relation to §1.4, we will be even more modest and say that an algorithm *converges globally* when $\liminf |f'(x_k)| = 0$. Recall how this is misleading; x_k need not converge to a global minimum,... or even may not converge at all (i.e. may diverge: cf. e^x, once again).

In view of second-order conditions, the following class of functions appears naturally, for which every stationary point satisfies (SC2):

---

[1] Recall that an operator A is positive [resp. semi-]definite when $(d, Ad) > 0$ [resp. $\geqslant 0$] for all $d \neq 0$.

Definition 2.4. *The function* f *is said to be* locally elliptic *if, on every bounded set* B*, it is* C^2 *and its Hessian is positive definite; hence, there exist (in finite dimension) two positive constants* $0 < \ell(B) \leqslant L(B)$ *such that*

$$\ell(B)|d|^2 \leqslant (f''(x)d, d) \qquad and \qquad |f''(x)d| \leqslant L(B)|d| \, . \qquad \square$$

Observe that ℓ and L bound the eigenvalues of f'' on B. A locally elliptic function is convex (assuming B convex), and even locally *strongly convex* i.e.

$$f(y) \geqslant f(x) + (f'(x), y - x) + \frac{1}{2}\ell(B)|y - x|^2 \qquad (2.2)$$

for all x et y in B (obtained by integration along $[x, y] \subset B$). This relation, written at a minimum point $x = x^*$, expresses that f enjoys a *quadratic growth* near x^*. From (2.2), we also have

$$(f'(x) - f'(y), x - y) \geqslant \ell(B)|x - y|^2$$

(write the symmetric relation and add up), which expresses that f' is *strongly monotone* (locally, on B).

After these preliminaries, we turn to numerical algorithms solving our problem (2.1). Knowing that the ambition of an algorithm is to find a stationary point, i.e. to solve $f'(x) = 0$, the first natural idea is to use methods solving (nonlinear) systems of equations.

2.3 First-Order Methods

To solve a nonlinear system $g(x) = 0$, we mention two methods: Gauss-Seidel and successive approximations; but in our context, recall that $g : \mathbb{R}^n \to \mathbb{R}^n$ is not an arbitrary mapping: it is a *gradient*, of a function which must be *minimized*; thus, among the possible stationary points, those having g' positive (semi)definite are preferred.

2.3.1 Gauss-Seidel

This method can also be called "one coordinate at a time". Basically, it works as follows.

– All the coordinates are fixed, say to 0.
– The first coordinate is modified, by solving the first equation with respect to this first coordinate.
– And so on until n.
– The process is repeated.

In other words, each iteration of this algorithm consists in solving one equation with one unknown. The iterate x_{k+1} differs from x_k by one coordinate only, namely $i(k)$, the rest of the integer division of k by n.

This method is little interesting, its use is not recommended. Incidentally, observe the crook: how can we solve each of the equations in its second step? (remember Remark 1.3).

2.3.2 Method of Successive Approximations, or Gradient Method

In its crudest form, the method of successive approximations is the following. One wants to solve $g(x) = 0$ via the iterative scheme: $x_{k+1} = x_k + tg(x_k)$, where $t \neq 0$ is a fixed coefficient; the motivation is that the fixed points of $\{x_k\}$ satisfy $x = x + tg(x)$, and therefore are solutions. In general, choices of t ensuring convergence are unknown. In case where g is actually a gradient, of a function to be minimized, something can be said:

Theorem 2.5. *Suppose that, locally, g is Lipschitz continuous and strongly monotone (i.e. f is locally elliptic) and that a solution exists. Then the algorithm converges if $t < 0$ is close enough to 0.*

Proof. Setting $F(x) := x + tg(x)$, we write the algorithm in the form $x_{k+1} = F(x_k)$ and we show that F is a contraction. Let x_1 be the first iterate, x^* a solution ($g(x^*) = 0$), B the ball of center x^* and radius $|x_1 - x^*|$. Then

$$|x_2 - x^*|^2 = |x_1 - x^*|^2 + 2t(x_1 - x^*, g(x_1) - g(x^*)) + t^2|g(x_1) - g(x^*)|^2.$$

Take $t < 0$; then the assumptions give

$$|x_2 - x^*|^2 \leqslant (1 + 2\ell t + L^2 t^2)|x_1 - x^*|^2.$$

It suffices to take $t > -2\ell/L^2$ to obtain (recursively) $x_k \in B$ and $x_k \to x^*$ Q-linearly. We have shown at the same time the uniqueness of x^*.   □

Remark 2.6. The existence hypothesis is essential to have compactness of $\{x_k\}$ (without it, $g(x) = e^x$ is a counter-example). This hypothesis can be replaced by the global (instead of local) ellipticity of f; the proof still applies, and shows the existence of a (unique) solution.   □

2.4 Link with the General Descent Scheme

Now, knowing that the problem to be solved is not arbitrary, but that there is a potential function to be minimized, can we modify, improve, interpret the methods of §2.3, according to what was seen in §1.3? For this, we need to distinguish in the above two methods the calculation of a direction and the line-search. This is possible:

– To compute the direction d_k, make the change of variable $x = x_k + d$, replace $f(x_k + d)$ by $f(x_k) + (g(x_k), d)$ (valid for small $|d|$) and let d_k solve the following model-problem:

$$\min\,(g(x_k), d)\quad \|d\| \leqslant \delta \qquad\qquad (P_k)$$

(at this point, $\|\cdot\|$ represents an arbitrary norm, not necessarily the Euclidean norm $|\cdot|$).

– Then, x_{k+1} is sought along d_k, in accordance with general line-search principles.

By construction, $(g(x_k), d_k) < 0$: a *descent direction* at x_k is obtained, i.e. a d satisfying $f(x_k + td) < f(x_k)$ for some $t > 0$ (actually, for all $t > 0$ small enough).

Remark 2.7. Here the norm $\| \cdot \|$ is arbitrary. The coefficient $\delta > 0$ is essential to guarantee that (P_k) has a solution (the linear function $(g(x_k), \cdot)$ is unbounded on $\mathbb{R}^n$), but the exact value of δ does not matter: d_k depends multiplicatively on δ, and the length of the direction is irrelevant: it will be absorbed by the line-search anyway.   □

Several possibilities are obtained, depending on the choice of $\| \cdot \|$ in (P_k).

2.4.1 Choosing the ℓ_1-Norm

Suppose first $\|d\| = \sum_{i=1}^{n} |d^i|$. A graphic resolution of (P_k) gives d_k parallel to a certain basis axis (one corresponding to the largest component of the gradient). One therefore sees that this direction modifies only one coordinate of the current iterate, just as in the Gauss-Seidel method. However, the coordinates are not modified in a cyclic order, here; at each iteration, it is rather the most "rewarding" coordinate that is modified.

Let us now focus on the computation of the stepsize $t > 0$. To compute t_k along the d_k thus obtained, an immediate idea consists in minimizing the univariate merit function $q(t) := f(x_k + td_k)$. For this, one must solve $q'(t) = 0$. We have

$$q'(t) = \sum_{i=1}^{n} g^i(x_k + td_k) \frac{\mathrm{d}}{\mathrm{d}t} (x_k + td_k)^i = (g(x_k + td_k), d_k).$$

But in the present case, d_k is one of the vectors in the canonical basis (up to its sign). Hence, to solve $q'(t) = 0$ is to cancel the corresponding component of the gradient, i.e. to do precisely as in the Gauss-Seidel method.

In summary, consider the following variant of Gauss-Seidel: at each iteration, choose one index $i(k)$, corresponding to a largest component (in absolute value) of $g(x_k)$, and solve for the component $x^{i(k)}$ the equation $g^{i(k)}(x) = 0$. Seen through optimization glasses, this variant can be viewed as

– choose as direction a solution to (P_k), where $\| \cdot \|$ is the ℓ_1-norm,
– compute the stepsize by minimizing f along this direction.

Remark 2.8. In the Gauss-Seidel method, the univariate equations giving $x_{k+1}^{i(k)}$ may have several solutions. A merit of the above interpretation is to allow a choice among these solutions: at each iteration, one has not only to solve an equation, but also to decrease f. Here appears a first advantage of optimization, over general equation solving.   □

2.4.2 Choosing the ℓ_2-Norm

When $\|\cdot\|$ is the norm associated with the scalar product, look again at a graphical resolution of (P_k): as a direction, the optimal d is $d_k = -g(x_k)$; the gradient method comes up. Here again, to decrease $q(t) = f(x_k + td_k)$ provides a constructive method to compute the stepsize, while Theorem 2.5 does not give any explicit bound on t (this would require the knowledge of $|x_1 - x^*|$ and of the corresponding constants ℓ, L,...). Here lies another advantage yielded by optimization, just as in Remark 2.8.

Remark 2.9. In numerical analysis, when g does not enjoy particular properties, *stability* is always a problem: the sequence $\{x_k\}$ should at least be bounded! Here, the requirement $q(t) < q(0)$, i.e. $f(x_{k+1}) < f(x_k)$, results in a safe stabilization of $\{x_k\}$: if the problem is well-posed, f should increase at infinity. One more confirmation that forcing to zero the gradient of a function *to be minimized* is easier than solving a general system of equations; remember Remark 1.4.   □

2.5 Steepest-Descent Method

In §2.4, a family of minimization methods has been given: the direction solves a certain linearized problem at x_k, and the stepsize is computed according to the general principle $f(x_{k+1}) < f(x_k)$. The most natural idea to compute this stepsize is to minimize $q(t) = f(x_k + td_k)$ at each iteration; it is the essence of Gauss-Seidel's method anyway. This same idea can be applied with the ℓ_2-norm, which gives the following method:

(i) Compute $d_k = -g(x_k) =: -g_k$;
(ii) Compute t_k solving $\min_{t>0} f(x_k + td_k)$.

Remark 2.10. The constraint $t > 0$ plays no real role, it could be replaced by $t \geqslant 0$. Anyway, $t_k > 0$ would be obtained, because $q'(0) = (g_k, d_k) = -|g_k|^2 < 0$ (q decreases locally near $t = 0$, hence 0 cannot be a minimum of q).

Note that optimality of t_k is expressed by $q'(t_k) = 0$, which writes $(g_{k+1}, d_k) = -(d_{k+1}, d_k) = 0$ at each iteration: each direction is orthogonal to the preceding one.   □

This procedure will be called *method of steepest descent*. It therefore consists in computing the steepest-descent direction associated with the $|\cdot|$-norm (this is the gradient), and then the optimal stepsize along this direction. This method is *very bad* because it is very slow; in fact, the gradient direction is itself very bad to decrease f. It is known that $f(x - tg)$ decreases for t close to 0; but, except when x is far from a minimum point, $f(x - tg)$ starts increasing for rather small values of t already; as a result, the method is forced to take small t's, i.e. the iterates x_k cluster together; the sequence of iterates oscillates and is subject to zigzags. Simple ways of computing better directions will be seen in Chap. 4. The present method should actually be **forbidden**. Its only usefulness is to serve as a basis for all the methods actually used.

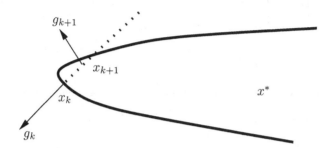

Fig. 2.1. Steepest descent is very bad

It is interesting to prove convergence of a variant of the steepest-descent method, in which the stepsize is computed in a more general way. Consider, instead of (ii):

(iii) $q(t_k) \leqslant q(t_k^*)$, where t_k^* is the smallest positive solution to $q'(t) = 0$.

We will assume that such a solution exists (a counter-example could be $q(t) = e^{-t}$).

Theorem 2.11. *Suppose $g = \nabla f$ is Lipschitz-continuous on the so-called slice $F_1 := \{x : f(x) \leqslant f(x_1)\}$. Then the method defined by (i), (iii) satisfies*
- *either $f(x_k) \to -\infty$,*
- *or $g(x_k) \to 0$.*

Proof. Preliminary remarks: at each iteration, since $q'(0) < 0$, the continuity of g, hence of q', implies that $q(t)$ decreases between 0 and t_k^* (in fact, t_k^* is the first local minimum or inflexion point of q, met in the direction of increasing t's; it is certainly not a local maximum). The segment $[x_k, x_k + t_k^* d_k]$ is therefore included in F_1; and this is true for all k. A Lipschitz constant L for g on F_1 is therefore valid for every segment $[x_k, x_k + t_k^* d_k]$.

Argument 1 First we prove that, in a neighborhood of $t = 0$, $f(x_k + t d_k)$ decreases at a non-negligible rate. More precisely:

$$\text{if } \quad 0 \leqslant t \leqslant \min\{t_k^*, \tfrac{1}{2L}\} \quad \text{then} \quad f(x_k + t d_k) \leqslant f(x_k) - \tfrac{1}{2} t |g_k|^2 .$$

For this, take z of the form $x_k + t d_k$ with $t \in \,]0, t_k^*[$. Mean-value theorem: for some z' between x_k and z (hence $z' \in F_1$),

$$f(z) = f(x_k) + (g(z'), z - x_k) = f(x_k) + (g_k, z - x_k) + (g(z') - g_k, z - x_k) .$$

Since the vectors g_k and $z - x_k$ are opposite, $(g_k, z - x_k) = -|g_k||z - x_k|$. Besides, the last scalar product above is bounded by $L|z - x_k|^2$ (apply successively Cauchy-Schwarz, and g Lipschitz on $[x_k, z']$, and $z' \in]x_k, z[$). We therefore have

$$f(z) \leqslant f(x_k) - |g_k||z - x_k| + L|z - x_k|^2 = f(x_k) - t|g_k|^2 + Lt^2|g_k|^2 \,.$$

For $t \leqslant \frac{1}{2L}$, write $Lt^2 = Lt.t \leqslant t/2$. It follows

$$f(x_k - tg_k) \leqslant f(x_k) + (-t + Lt^2)|g_k|^2 \leqslant f(x_k) - \frac{t}{2}|g_k|^2 \,.$$

Remembering that t must also be lower than t_k^*, Argument 1 is proved.

Argument 2 Now we show that the x_k's move in a non-negligible way. More precisely:

$$t_k \geqslant t_k^* \geqslant \frac{1}{L} \,.$$

Set $x_k^* = x_k + t_k^* d_k$. We have $(g(x_k^*), d_k) = 0$, which we write $(g_k^*$ is $g(x_k^*))$

$$0 = (g_k^*, x_k^* - x_k) = (g_k, x_k^* - x_k) + (g_k^* - g_k, x_k^* - x_k) \,.$$

Use the same techniques as in Argument 1:

$$0 = -|g_k||x_k^* - x_k| + (g_k^* - g_k, x_k^* - x_k) \leqslant |x_k^* - x_k|(-|g_k| + L|x_k^* - x_k|)$$

hence $|x_k^* - x_k| = t_k^*|g_k| \geqslant |g_k|/L$.

Remembering that $q(t)$ decreases on $[0, t_k^*]$, we certainly have $t_k \geqslant t_k^*$ and Argument 2 is proved.

Synthesis Take $z = x_k - \frac{1}{2L}g_k$; from Argument 2, z lies between x_k and x_k^* hence

$$f(x_{k+1}) \leqslant f(x_k^*) \leqslant f(z)$$

and, from Argument 1, $f(z) \leqslant f(x_k) - \frac{1}{4L}|g_k|^2$. We conclude

$$|g_k|^2 \leqslant 4L[f(x_k) - f(x_{k+1})] \,. \tag{2.3}$$

The theorem is then proved by summation.   □

Remark 2.12. It is important to grasp the mechanism of the above proof, rather than the inequalities themselves.

Argument 1 bounds from below the rate of decrease of f, using exclusively the definition of the direction; this argument would still hold if, instead of being collinear to $-g_k$, the direction d_k made with $-g_k$ an angle far from 90°.

Argument 2 uses essentially the definition of the stepsize, which must be large enough to give a sufficiently large decrease of f in Argument 1.

The proof also shows that $|g_k|^2$ tends to 0 at least as fast as a convergent series. This does not imply a very exciting speed, though: one could have for example $|g_k| = 1/k$, which tends to 0 sublinearly.   □

To establish a convergence speed result, additional hypotheses are necessary. The key-property in this domain is that, when leaving the optimal set, f must increase with a comparable speed in all directions. Lipschitz continuity of the gradient implies that this speed is at most quadratic: denoting by X^* the optimal set, and by f^* the corresponding optimal value, there holds

$$f(x) \leqslant f^* + \frac{1}{2} L \operatorname{dist}^2(x, X^*)$$

(obtained by integration). So we need this speed to be at least quadratic, and this is just local ellipticity of f near X^*. In a word, the adequate assumption is the following growth condition:

$$f \text{ is convex; } X^* \text{ is nonempty and, in a neighborhood of } X^*, \tag{G}$$
$$\exists \ell > 0 \text{ such that } f(x) \geqslant f^* + \tfrac{1}{2}\ell \operatorname{dist}^2(x, X^*)$$

Lemma 2.13. *Under Assumption (G), there holds in a neighborhood of X^*:*

$$\ell[f(x) - f^*] \leqslant 2|g(x)|^2 .$$

Proof. Let x^* the projection of x onto X^*. From convexity and Cauchy-Schwarz,

$$f^* \geqslant f(x) + (g(x), x^* - x) \geqslant f(x) - |g(x)| \operatorname{dist}(x, X^*)$$

i.e., using Assumption (G),

$$f(x) - f^* \leqslant |g(x)| \operatorname{dist}(x, X^*) \leqslant |g(x)| \sqrt{2 \frac{f(x) - f^*}{\ell}} . \qquad \square$$

Remark 2.14. Assumption (G), added to the Lipschitz continuity of the gradient, shows that the speeds of convergence of $f(x_k)$ to f^*, of x_k to X^*, and of $g(x_k)$ to 0 are the same. Indeed, calling x^* the projection of an arbitrary x onto X^*, one has for $|x - x^*|$ small enough:

$$|g(x) - 0|^2 \leqslant L^2 |x - x^*|^2 \leqslant 2 \frac{L^2}{\ell} [f(x) - f^*] \leqslant 4 \frac{L^2}{\ell^2} |g(x)|^2 \leqslant 4 \frac{L^4}{\ell^2} |x - x^*|^2 . \quad \square$$

Under these circumstances, the steepest-descent algorithm converges linearly:

Theorem 2.15. *Make the assumptions of Theorem 2.11 and of Lemma 2.13. Then the sequence $\varepsilon_k := f(x_k) - f^*$ tends to 0 Q-linearly.*

Proof. Copy the proof of Theorem 2.11 to obtain (2.3) and, using Lemma 2.13,

$$f(x_k) - f(x_{k+1}) \geqslant \frac{1}{4L} |g_k|^2 \geqslant \frac{\ell}{8L} [f(x_k) - f^*]$$

i.e. $\varepsilon_k - \varepsilon_{k+1} \geqslant \frac{\ell}{8L} \varepsilon_k$, or $\varepsilon_{k+1} \leqslant (1 - \frac{\ell}{8L}) \varepsilon_k$. $\qquad \square$

2.6 Implementation

When the simulator is available, to answer upon request $f(x)$ and $g(x)$, the general form of a minimization algorithm will be as follows.

Algorithm 2.16 (Schematic descent algorithm).
Step 0 (Initialization). The initial iterate x_1 and a stopping tolerance $\varepsilon > 0$ are given; set $k = 1$.
Step 1 (Stopping test). Compute $g(x_k)$; if $|g(x_k)| \leqslant \varepsilon$ stop.
Step 2 (Computing the direction). Compute the direction; for example set $d_k = -g(x_k)$ (although it is forbidden).
Step 3 (Line-search). Find an appropriate stepsize $t_k > 0$, satisfying in particular $f(x_k + t_k d_k) < f(x_k)$.
Step 4 (Loop). Set $x_{k+1} = x_k + t_k d_k$; increase k by 1 and go to 1.   □

Remark 2.17. Expliciting k is useless; one can set $d = d_k$, $t = t_k$, overwrite x_{k+1} on x_k, etc. This spares computer memory.

In this scheme, everything is clear except Step 3, which specifies neither the conditions to be satisfied by the stepsize, nor how to meet them. Computing the stepsize is actually a subalgorithm iterating on $t > 0$, during which is established a dialogue with the simulator (use of the real (P), once again). The line-search problem is important enough to motivate the full Chap. 3 by itself. Here and now, however, we can say that it is advised to test in Step 4 whether the line-search has been successful, otherwise the algorithm must be authoritatively stopped (see below the end of §3.2).   □

EXTREMELY IMPORTANT REMARK FOR ALL THE SEQUEL:
CONDITIONING

Suppose we perform a linear change of variables, say $x(y) = Ay$, or $y(x) = A^{-1}x$. We can then consider the function h defined by

$$h(y) = f[x(y)] = f(Ay).$$

Minimizing h (with respect to y) is obviously equivalent to minimizing f (with respect to x): the optima are "the same" – via the application of A. However, this equivalence is misleading because it is *grossly wrong numerically*.

This is clear for the Gauss-Seidel method, since changing one coordinate of y does not correspond to changing one coordinate of x (except if A is diagonal). As for gradient methods, use the well-known result

Proposition 2.18. *The gradient of h is given by the formula*

$$h'(y) = A^\top g(Ay).$$

Proof. By definition, we have for all $z \in \mathbb{R}^n$

$$(h'(y), z) = h(y + z) - h(y) + o(|z|) = f(Ay + Az) - f(Ay) + o(|z|)$$
$$= (g(Ay), Az) + o(|Az|) + o(|z|) = (A^\top g(Ay), z) + o(|z|)$$

where we have used the property $o(|Az|) = o(|z|)$. The result follows, since z was arbitrary. $\qquad\square$

Then observe the following fact: starting from a given x, the gradient method applied to f generates a next iterate of the form

$$x_+ = x - tg(x) \,.$$

To the initial x, there corresponds $y = A^{-1}x$; starting from this same initialization, the same gradient method applied to h generates a next iterate of the form $y_+ = y - tA^\top g(x)$, to which there corresponds

$$x'_+ = Ay_+ = A[y - tA^\top g(x)] = x - tAA^\top g(x) \,.$$

In other words, the effect of the change of variables amounts to multiplying the directions by $AA^\top$, *which changes everything!* (unless $A^\top$ is an orthogonal matrix, in which case one would have $AA^\top = I$). Using this remark, one can try and find a clever change of variables, such that gradient methods behave best numerically. This is a so-called *preconditioning*.

The simplest preconditioner is diagonal, which amounts to adjust *scale factors*: one sets

$$y_i = \frac{x_i}{\bar{x}_i} \quad i = 1, \dots, n$$

and one minimizes with respect to the variable y; it will be advantageous to take for $\bar{x}_i$ a "nominal" variation range for x_i, which makes y_i dimensionless. To choose the scale factors $\bar{x}_i$, the following rule serves as a guide. Let an increment δ be given to the variable i; it yields an increment Δ_i for the function h. One should strive to obtain all the Δ_i's of the same order of magnitude (i.e. roughly independent of i).

Bibliographical Comments

Recall again that the methods of the present chapter have a theoretical value only; they can be found in [71]; see [79] for a study of Gauss-Seidel. The steepest-descent method was proposed by Cauchy [70] to solve systems of equations via least squares.

3 Line-Searches

In this chapter, considering the problem of computing the direction as solved (but do not forget that we have seen *bad* directions only, better ones will be studied in the next chapters), we focus on the computation of the stepsize. Here appear the most serious practical difficulties, while directions are generally easy to compute, once the theory is well-mastered. A firm experience is required to write a good computer code for line-searches, which are unique to optimization, and fairly important as they guarantee stability: remember Remark 1.4.

So, we are given:
– the starting point x of the line-search;
– the direction of search d;
– a merit-function $t \mapsto q(t)$, defined for $t \geqslant 0$, representing $f(x + td)$.
  Besides, d is assumed to be a *descent direction*:

> In the following, we will always suppose $q'(0) < 0$.

Remark 3.1 (fundamental). The function q is not known explicitly, via an analytical formula, but only pointwise: the only available information is the numerical value of $q(t)$ for each numerical value of t; most often, we will also assume that the numerical value of $q'(t) = (f'(x + td), d)$ is computed at the same time. This computation is performed in the simulator, a subprogram characterizing the problem to be solved, which gives for each numerical value of $z \in \mathbb{R}^n$ the numerical values $f(z)$ and $g(z) = f'(z) \in \mathbb{R}^n$. Simply, we assume that this subprogram exists. In these circumstances, the search for a convenient t (satisfying $q(t) < q(0)$, among other things) can only be done by trials and errors: see §1.3 again, in particular Remark 1.3. □

3.1 General Scheme

To construct a line-search algorithm, one first defines a *test* with three possible exits; given $t > 0$, it answers whether

a) t is satisfactory

b) t is too large

c) t is too small.

This test is performed upon observation of $q(t)$, and possibly of $q'(t)$.

Example 3.2. To fix ideas, let us give an extremely simple example; but we will see below that it is too simple:

a) $q'(t) = 0$ (then it is normal to stop the search: t seems to minimize q);
b) $q'(t) > 0$ (then q seems to have a minimum point smaller than t);
c) $q'(t) < 0$ (then q seems to have a minimum point larger than t).

This test is illustrated by Fig. 3.1: the three circles represent those t satisfying a); the intervals I_1 and I_3 [resp. I_2 and I_4] are those where t satisfies c) [resp. b)].

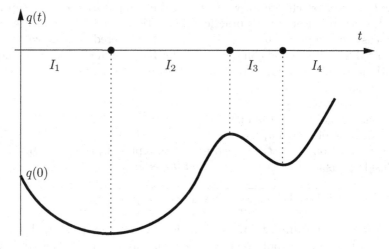

Fig. 3.1. A simplistic example

At this point, note something wrong in this example: the property $q'(t) = 0$ does not imply $q(t) < q(0)$; hence, the property of being "satisfactory" does not even imply the descent property, here; and there is something even worse, which we will see at the end of this §3.1.   □

Now we will call t_L a too small t (on the left of a desired t), t_R a too large t (on the right of a desired t). To initialize the search, 0 is obviously a t_L; if no upper bound is a priori available for the stepsize, t_R can be initialized to 0, with the convention that $t_R = 0$ means "no too large t has been found so far" (logically, t_R should be initialized at $+\infty$ but this has little meaning on a computer). Schematically, the algorithm is then the following:

Algorithm 3.3 (Schematic line-search).
Step 0. Start from an initial $t > 0$. Initialize $t_L = 0$ and t_R ($= 0$ for example).

Step 1. Test t;
    if a) terminate;
    if b) declare $t_R = t$ and go to 2;
    if c) declare $t_L = t$ and go to 2.
Step 2. If no real t_R has been found yet $(t_R = 0)$, compute a new $t > t_L$.
    Else $(t_R > 0)$ compute a new $t \in]t_L, t_R[$.
    Loop to 1.   □

Remark 3.4. In principle, choosing the initial t in Step 0 is not of line-search's concern. Logically t should be initialized by the minimization algorithm itself, which has on hand more information, once the direction is computed. Such is for example the case of Newtonian methods, to be studied in Chap. 4. If no such information is available, a possibility is to assume that q is quadratic, and that its decrease from $t = 0$ to the optimal t is $\delta = f(x_{k-1}) - f(x_k)$; in this case, q is minimized at $t = -2\delta/q'(0)$, which can serve as an initialization. This technique is often called *Fletcher's initialization*.   □

Thus, the line-search algorithm is a sequence of interpolations, reducing the *bracket* $[t_L, t_R]$, and possibly preceded by a sequence of extrapolations (as long as $t_R = 0$).

The observations below are straightforward, but fundamental for a good understanding of the mechanism.

– Extrapolations are performed until a real t_R is found (which may happen at the first try).
– Once t_R has become nonzero, it remains such; then the interpolation phase starts.
– In any case, t_L increases each time it is modified.
– As soon as t_R is nonzero, it decreases each time it is modified,
– but there always holds $t_L < t_R$.

Now, in order for the line-search to make sense and to stop after a finite number of tries, the test in Step 1 of Algorithm 3.3 must satisfy the following properties:

PROPERTY 1 The three possible exits in the test a), b), c) form a partition of $\mathbb{R}_+$ (so that every $t \geq 0$ is classified without ambiguity).

PROPERTY 2 No large t must satisfy c), that is to say: there exists $\bar{t}$ such that a) or b) holds for every $t \geq \bar{t}$ (to avoid an increase of t_L to $+\infty$).

PROPERTY 3 Every $[t_L, t_R]$ contains a nonzero interval satisfying a) (to avoid a decrease of $|t_R - t_L|$ all the way to 0).

This shows in particular that the simplistic Example 3.2, which uses only the sign of $q'(t)$, is really too simple. For one thing, a) is in general satisfied only by one point, the minimum of q: to terminate the line-search, one must try *exactly* this point (one of the three circles in Fig. 3.1). Besides, q-values are completely ignored; a local minimum can therefore be produced, possibly with a value greater than $q(0)$. On Fig. 3.1, for example, if a t is tried somewhere in the interval I_3, the search is henceforth trapped on the right of the picture and it becomes impossible to obtain $q(t) < q(0)$.

In addition to the stopping test in Step 1 of Algorithm 3.3, the line-search is characterized by the computation of the new t in Step 2. We start with a few words on this question, even though it is Step 1 that is the most important, since it conditions the convergence of the sequence x_k.

3.2 Computing the New t

Call t the stepsize-value just tested, and t_+ the value to be computed. There are two cases.

Extrapolation Consider first the case when no upper bound is available, i.e. $t_R = 0$. Then t_+ is wished to be "significantly" larger than t. In fact, one wishes

PROPERTY E Infinitely many extrapolations would imply $t_L \to +\infty$.

In this way, a real t_R will eventually be found, due to PROPERTY 2 of the test a), b), c) (§3.1).

For this, the simplest is to set $t_+ = at$, with $a > 1$ fixed, for example $a = 10$. Some more sophisticated factor can also be used. Actually, the best is to fit a cubic function, as in the case of interpolation, given below.

Interpolation If some $t_R > 0$ is already available, t_+ must be "significantly" between t_L and t_R. In fact, one wishes

PROPERTY I Infinitely many interpolations would imply $|t_R - t_L| \to 0$.

In this way, case a) will eventually occur, due to PROPERTY 3 of the test a), b), c) (§3.1).

Here again, the simplest technique is to set $t_+ = (t_L + t_R)/2$, an interpolation by *bisection*. Then the length of the bracket is halved at each try. Actually, the best is again to fit a cubic function by the technique below.

Cubic fitting In order to accelerate the line-search (which is important since it is performed at each minimization iteration), it is advised to guide the search for t_+ by the observed behaviour of q; this can be done both in case of extrapolation and interpolation. For this, the general idea is to select t-values already tested, to fit a simple function (a polynomial) coinciding with the corresponding values of q (and possibly of q'), and finally to compute t_+ minimizing this function.

When the simulator computes simultaneously function- and derivative-values (the usual case), the universal technique is the following.

– Select two values of the stepsize: the current value t and the preceding one, say t_- (with the initialization $t_- = 0$).
– With two values q and q' for each of these two points t and t_-, we therefore have 4 informations $q := q(t)$, $q_- := q(t_-)$, $q' := q'(t)$, $q'_- := q'(t_-)$ which allow the computation of the 4 coefficients of a 3rd-degree polynomial; in good cases, this polynomial has a local minimum which can be computed.
– We leave it to the reader to check that the following calculations give the result:

$$p := q' + q'_- - 3\frac{q - q_-}{t - t_-}, \quad D := p^2 - q'q'_-, \quad d := \sqrt{D}\,\mathrm{sign}\,(t - t_-).$$

If $D < 0$, no local minimum exists (then set for example $D = 0$); otherwise

$$t_+ = t + r(t_- - t) \quad \text{with} \quad r := \frac{d + p - q'}{2d + q'_- - q'}.$$

Precautions: safeguard Once again, the line-search is a subalgorithm, executed at each minimization iteration. Contenting oneself with asymptotic properties of this subalgorithm is therefore out of question. It is crucial to show that case a) occurs for sure after finitely many trials. For this, PROPERTY E ($t_L \rightarrow +\infty$) and PROPERTY I ($|t_R - t_L| \rightarrow 0$) are essential, in combination with respectively Properties 2 and 3 of the test a) b) c) (§3.1).

These two properties E and I are automatically implied respectively by the simple techniques $t_+ = at$ and $t_+ = (t_R + t_L)/2$; but they must be artificially forced if the polynomial fitting is used.

Hence, once t_+ is computed, no matter by which process, it must be forced "significantly" on the right of t_L (in extrapolation) or inside $[t_L, t_R]$ (in interpolation). For this, a *safeguard* is necessary. In practice, one does as follows: fix $a > 1$ and $\theta \in \,]0, 1/2[$, and then

– in extrapolation, replace t_+ by $\max\{t_+, at\}$;
– in interpolation, replace t_+ successively by
  $\min\{t_+, t_R - \theta(t_R - t_L)\}$, then by $\max\{t_+, t_L + \theta(t_R - t_L)\}$.

Besides, any mathematical proof needs assumptions on q and it may happen that these assumptions are not satisfied; or it may happen that the proof does not apply, due to roundoff errors; it may also happen that mistakes have been made when programming the computation of q'. To avoid an infinite loop, always to be feared, it is safe to impose emergency tests at two places:

– in extrapolation, stop authoritatively if t_L becomes very large;
– in interpolation, stop authoritatively if $|t_R - t_L|$ becomes very small.

From now on in this chapter, we will consider as solved the question of finding the new t, and we will focus on the construction of the test a) b) c) in Step 1 of the general line-search scheme (Algorithm 3.3).

3.3 Optimal Stepsize (for the record only)

The present Section 3.3 develops a technique which must indeed be avoided. We somewhat detail it for its historical and pedagogical value only: it is intuitively natural and easy to grasp. But it is actually just an introduction to the modern techniques of §3.4 (this is somehow like first-order methods of Chap. 2, which are just an introduction to modern methods of Chap. 4).

We start with a simple remark, considerably helpful for the construction of the test a) b) c): the descent property requires that a) contains in particular the property "$q(t) < q(0)$": a t such that $q(t) \geqslant q(0)$ has to be classified either in b) or in c). Now – and here appears a key-idea – since $q'(0) < 0$, we certainly have $q(t) < q(0)$ for t small enough. A t such that $q(t) \geqslant q(0)$ can therefore be *safely classified* in b) (too large).

Historically, one has first tried to search a t^* such that

$$q(t^*) < q(0) \quad \text{and} \quad q'(t^*) = 0 \,.$$

Starting from this idea the test in §3.1 was then: choose $\varepsilon > 0$ and define a) b) c) by:

a) $q(t) < q(0)$ and $|q'(t)| \leqslant \varepsilon$ (then terminate);
b) $q(t) \geqslant q(0)$ or $q'(t) > \varepsilon$ (then $t_R = t$);
c) $q(t) < q(0)$ and $q'(t) < -\varepsilon$ (then $t_L = t$).

Note that this test is substantially more sophisticated than the simple Example 3.2 (the latter did not even imply a decrease for q!). It is nevertheless out of date, now, and is replaced by modern tests of §§3.4, 3.5 below. For pedagogy, we show that it is consistent, at least theoretically.

Theorem 3.5. *Suppose that $q \in C^1$ is inf-compact (and that $q'(0) < 0$, as always). Then this line-search is finite (providing that t_+ is consistently computed, of course; see the remarks at the end of §3.2).*

Proof. We have to prove that a) occurs after finitely many tries. We will proceed by contradiction, assuming that, at each try, either b) or c) occurs.

Let us first show that the sequence of extrapolations is finite. In fact, inf-compactness implies that, for t large enough, c) cannot occur: otherwise t would tend to infinity (PROPERTY E) and continually satisfy $q(t) \leqslant q(0)$. After finitely many extrapolations, we therefore get some $t_R > 0$, which provokes a switch to the interpolation phase.

Let us now show that the sequence of interpolations is finite. Suppose it is infinite. In view of PROPERTY I, the two sequences $\{t_L\}$ and $\{t_R\}$ are adjacent[1]. They both tend to some t^*.

Each t_L satisfies c); pass to the limit in $t_L \to t^*$:

$$q(t^*) \leqslant q(0) \quad \text{and} \quad q'(t^*) \leqslant -\varepsilon.$$

Pass likewise to the limit in b):

$$q(t^*) \geqslant q(0) \quad \text{or} \quad q'(t^*) \geqslant \varepsilon.$$

The only possibility is therefore

$$q(t^*) = q(0) \quad \text{and} \quad q'(t^*) \leqslant -\varepsilon.$$

From $q(t^*) = q(0)$, the definition of c) shows that no t_L can equal t^*: indeed t_L tends to t^* but stays strictly smaller than t^*. Then, using the property $q(t_L) < q(0) = q(t^*)$, the mean-value theorem gives

$$0 < \frac{q(t_L) - q(t^*)}{t_L - t^*}.$$

Passing to the limit provides the contradiction $0 \leqslant q'(t^*) \leqslant -\varepsilon$. □

Remark 3.6. From its motivation, ε should be small; but this means nothing (is 10^{15} small? yes, compared to the number of atoms in the Universe). To give ε a meaning, it is suitable to compare $q'(t)$ with $q'(0)$, taking $\varepsilon = -mq'(0)$; here m is a dimensionless coefficient, for example $m = 10^{-3}$. Note that m cannot be 0, otherwise the proof does not work – and here precisely lies the difficulty with Example 3.2. □

3.4 Modern Line-Search: Wolfe's Rule

The essence of §3.3 is to identify a local minimum of q, which makes it tempting to choose ε very small. Yet, a) will be hard to obtain: the line-search subalgorithm will be time-consuming (in spite of every possible quality of the t_+ from §3.2). This is why more tolerant stopping tests have been devised, motivated by a triviality which must never be overlooked: it is $f(x)$, and not $q(t)$, that we wish to minimize. *Striving to minimize accurately f along the current direction, at each iteration, is therefore completely useless.*

We now describe the method seeming the most intelligent in the current state of the art, commonly called the line-search of *Wolfe*. Two coefficients $0 < m_1 < m_2 < 1$ are chosen, and cases a) b) c) are the following:

a) $q(t) \leqslant q(0) + m_1 t q'(0)$ and $q'(t) \geqslant m_2 q'(0)$	(then terminate);
b) $q(t) > q(0) + m_1 t q'(0)$	(then $t_R = t$);
c) $q(t) \leqslant q(0) + m_1 t q'(0)$ and $q'(t) < m_2 q'(0)$	(then $t_L = t$).

[1] Two real sequences $\{u_\ell\}$ and $\{v_\ell\}$ are *adjacent* when u_ℓ increases, v_ℓ decreases, $v_\ell \geqslant u_\ell$, and $v_\ell - u_\ell \to 0$; then they have a common limit.

This has a good interpretation:

- First, f is asked to decrease enough (hence x_{k+1} will not be too far from x_k).
- Second, the derivative is required to increase enough (hence x_{k+1} will not be too close to x_k).

Figure 3.2 represents a function q similar to that of Fig. 3.1, and displays again the set a) of satisfactory t's, and the set $I_1 \cup I_3$ [resp. $I_2 \cup I_4$] of t's declared too small [resp. too large]. Note: to familiarize oneself with this mechanism, one can mentally suppress the local minimum on the right of the picture, making a convex function with q; then I_2 will extend to $+\infty$, the t's declared "too large" will really be too large, i.e. on the right of *every* satisfactory t.

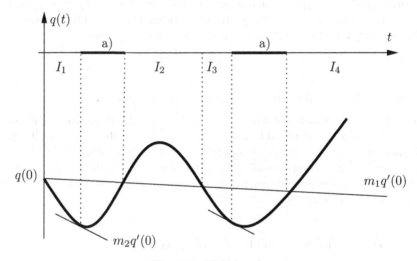

Fig. 3.2. Wolfe's rule

Theorem 3.7. *Suppose that $q \in C^1$ is bounded from below. Then Wolfe's line-search terminates.*

Proof. The proof is copied from that of Theorem 3.5. Start with extrapolations. If they were infinitely many, a sequence of stepsizes t would be constructed, tending to infinity, and such that

$$q(t) \leqslant q(0) + m_1 t q'(0)$$

hence $q(t)$ would tend to $-\infty$.

Suppose now that the sequence of interpolations is infinite. Just as in Theorem 3.5, the two sequences $\{t_L\}$ and $\{t_R\}$ are adjacent and have a common limit t^*. Then, passing to the limit in b) and c): $q(t^*) = q(0) + m_1 t^* q'(0)$.

From b), this implies in particular that no t_R can equal t^*: indeed t_R tends to t^* but stays strictly bigger than t^*.

Then write b) in the form

$$q(t_R) > q(0) + m_1 q'(0)(t^* + t_R - t^*) = q(t^*) + m_1 q'(0)(t_R - t^*)$$

and divide by $t_R - t^* > 0$ to obtain

$$\frac{q(t_R) - q(t^*)}{t_R - t^*} > m_1 q(0).$$

Passing to the limit: $q'(t^*) \geqslant m_1 q'(0) > m_2 q'(0)$. On the other hand, passing to the limit in c): $q'(t^*) \leqslant m_2 q'(0)$, contradiction.   □

Incidence on the sequence $\{x_k\}$ A whole family of minimization algorithms is now obtained, by combining the above Wolfe line-search with any kind of (descent) direction. We turn to the question whether the resulting sequence $\{x_k\}$ converges to a minimum point. Just as in Theorem 2.11, we will decompose the proof into three arguments. This proof scheme is so important that we will in turn decompose it into two formal results. We start with the third argument, called Synthesis in Theorem 2.11; it becomes the purely technical Lemma 3.8 below.

Of course, convergence cannot hold independently of the choice of d_k (the line-search will be helpless if d_k is "too orthogonal" to g_k). The angle between the direction and the gradient thus appears as essential, and we set

$$\cos \theta_k = \frac{-(g_k, d_k)}{|g_k| \cdot |d_k|}.$$

To say that d_k is a "definite" descent direction is to say that $\cos \theta_k$ is "sufficiently" positive. This concept appears to be conveniently quantified by (3.2) below.

Lemma 3.8. *Consider an algorithm to minimize f, in which there holds at each iteration:*

$$r \cos^2 \theta_k |g_k|^2 \leqslant f(x_k) - f(x_{k+1}), \tag{3.1}$$

where the coefficient $r > 0$ does not depend on k. If

$$the\ series\ \sum_{k=1}^{+\infty} \cos^2 \theta_k\ \ diverges , \tag{3.2}$$

then

- *either the objective-function tends to $-\infty$,*
- *or $\liminf |g_k| = 0$.*

Proof. Immediate: if the decreasing sequence $\{f(x_k)\}$ is bounded from below, say by f^*, then

$$\sum_{k=1}^{+\infty} \cos^2 \theta_k |g_k|^2 \leqslant \frac{1}{r} \sum_{k=1}^{+\infty} [f(x_k) - f(x_{k+1})] \leqslant \frac{1}{r}[f(x_1) - f^*] < +\infty.$$

As a result, if $|g_k|$ were bounded from below, say by $\delta > 0$, then we would have $\delta^2 \sum \cos^2 \theta_k < +\infty$, contradiction.   □

Of course, property (3.2) in this result depends on the way the direction is computed. As for property (3.1), it can be established independently of the direction: it comes just as in Arguments 1 and 2 of Theorem 2.11.

Theorem 3.9. *Consider a minimization algorithm using Wolfe's rule. If g is Lipschitz-continuous on the slice $\{x : f(x) \leqslant f(x_1)\}$, then (3.1) holds.*

Proof. Barring trivial cases where, at some iteration, $t \to +\infty$ and $q(t) \to -\infty$, Wolfe's line-search terminates on a) at each iteration (see Theorem 3.7). Then x_k stays in the slice and the g_k's satisfy the corresponding Lipschitz condition.

Argument 1 Bounding $f(x_k) - f(x_{k+1})$ from below (the stepsize is not too large). Express $q'(0)$ at iteration k:

$$t_k q'(0) = t_k (d_k, g_k) = -t_k \cos \theta_k |d_k||g_k| = -\cos \theta_k |x_{k+1} - x_k||g_k|.$$

This form directly shows that the descent property in a) implies

$$m_1 \cos \theta_k |g_k||x_{k+1} - x_k| \leqslant f(x_k) - f(x_{k+1}).$$

Argument 2 Bounding $|x_{k+1} - x_k|$ from below (the stepsize is not too small). The second half of a) gives (subtracting (g_k, d_k) from both sides)

$$(g_{k+1} - g_k, d_k) \geqslant (m_2 - 1)(g_k, d_k) = (1 - m_2) \cos \theta_k |g_k||d_k|.$$

Apply Cauchy-Schwarz and then Lipschitz properties to the left-hand side; there exists a constant L (depending on x_1) such that

$$(g_{k+1} - g_k, d_k) \leqslant L|x_{k+1} - x_k||d_k|.$$

We therefore deduce, after division by $|d_k|$:

$$(1 - m_2) \cos \theta_k |g_k| \leqslant L|x_{k+1} - x_k|.$$

To finish the proof, multiply this last inequality by $|g_k|$ and use Argument 1: (3.1) holds with $r := m_1(1 - m_2)/L$.   □

Remark 3.10. While following the scheme of Theorem 2.11, the arguments come here much more easily, thanks to the rules satisfied by the stepsize. Indeed, Theorem 2.15 can be reproduced to establish the following result, whose proof is left to the reader: let an algorithm use Wolfe's line-search, and compute the direction d_k so that $\cos \theta_k \geqslant c > 0$. This algorithm converges Q-linearly if f has quadratic growth in a neighborhood of a minimum point (cf. Theorem 2.15).   □

3.5 Other Line-Searches: Goldstein and Price, Armijo

Wolfe's rule needs the value $q'(t)$ – hence $\nabla f(x_k + td_k)$ – at each cycle in the line-search. It therefore may lose some efficiency when computing the gradient takes much more time than computing the function alone. Although rare, this situation exists, making it desirable to compute as few gradients as possible. Then the rules below become suitable: their tests a) b) c) do not require the computation of q'.

3.5.1 Goldstein and Price

Wolfe's rule used the slope $q'(t)$ of q at t. The present rule rather uses the average slope $[q(t) - q(0)]/t$ between 0 and t: it is required to lie between two given slopes m_1 and m_2, with $0 < m_1 < m_2 < 1$. In other words, the test is

$$
\begin{array}{lll}
\text{a)} & m_2 q'(0) \leqslant \dfrac{q(t) - q(0)}{t} \leqslant m_1 q'(0) & \text{(then terminate);} \\[2ex]
\text{b)} & m_1 q'(0) < \dfrac{q(t) - q(0)}{t} & \text{(then } t_R = t); \\[2ex]
\text{c)} & \dfrac{q(t) - q(0)}{t} < m_2 q'(0) & \text{(then } t_L = t).
\end{array}
$$

The analogy with Wolfe's rule becomes more visible if the test is rather written in the equivalent form

a) $q(t) \leqslant q(0) + m_1 t q'(0)$ and $\dfrac{q(t) - q(0)}{t} \geqslant m_2 q'(0)$;

b) $q(t) > q(0) + m_1 t q'(0)$;

c) $\dfrac{q(t) - q(0)}{t} < m_2 q'(0)$ [and hence $q(t) < q(0) + m_1 t q'(0)$!].

Proceeding as in Theorem 3.7, one shows rather easily that this rule yields a finite line-search.

3.5.2 Armijo

This rule is somewhat special, in that t is never declared too small, hence no extrapolation ever occurs. More precisely, one chooses $m_1 \in \,]0, 1[$ and

$$
\begin{array}{lll}
\text{a)} & q(t) \leqslant q(0) + m_1 t q'(0) & \text{(then terminate);} \\
\text{b)} & q(t) > q(0) + m_1 t q'(0) & \text{(then } t_R = t); \\
\text{c)} & \text{never.}
\end{array}
$$

Here again, it is easy to show that the resulting line-search terminates.

Note the importance of property a) above. Often called *Armijo's condition*, it appears in all modern rules – including Wolfe and Goldstein-Price. It guarantees that t is not too large and proves Argument 1 in Theorem 3.9.

Remark 3.11. Armijo's rule is dangerous: since it never increases t, it heavily relies on the initial stepsize to produce a move $x_{k+1} - x_k$ large enough.

- First, the reader can check that reproducing Theorem 3.9 amounts to assuming that the initial t is bounded away from 0. This is why Armijo's line-search is usually initialized on a constant, say $t = 1$ (independently of k).
- Yet, this constant may be hard to guess. With $n = 1$, consider the example (not elliptic, though) $f(x) = x^4/4$ and take the gradient method. Then $x_{k+1} = x_k - t_k x_k^3$. Knowing that the minimum point is $x^* = 0$, the Q-convergence quotient is $x_{k+1}/x_k = 1 - t_k x_k^2$. Unless t_k (and hence the initial stepsize) increases fairly fast, sublinear convergence $x_{k+1}/x_k \to 1$ will occur. The reader can check that Fletcher's initialization of Remark 3.4 does not help.

Therefore, this rule is usually limited to situations where, for some reason, extrapolations should be avoided (examples will appear in Sections 6.1.2 and 17.1. Its merit is mainly theoretical: it is easy to understand (and to explain) and it allows a cheap construction of implementable algorithms.   □

3.5.3 Remark on the Choice of Constants

No matter what rule is used (Wolfe, Goldstein-Price, Armijo), it is advised to take $m_1 < 1/2$, and it is advised in Wolfe or Goldstein-Price to take $m_2 > 1/2$. This is because of the following (well-known?) result:

Proposition 3.12. *Suppose q is quadratic with a minimum point t^*. Then*

$$q(t^*) = q(0) + \frac{1}{2}q'(0)t^*.$$

Proof. The quadratic function q can be written, for some $c > 0$,

$$q(t) = \frac{1}{2}ct^2 + q'(0)t + q(0)$$

in which case

$$t^* = \frac{-q'(0)}{c}, \quad q(t^*) = -\frac{1}{2}\frac{q'(0)^2}{c} + q(0) \quad \text{and} \quad \frac{q(t^*) - q(0)}{t^*} = \frac{1}{2}q'(0). \quad □$$

Thus, if q happens to be quadratic, Armijo's condition will reject the optimal stepsize if $m_1 > \frac{1}{2}$; this is clumsy.

Remark 3.13. There is a more serious argument (here we anticipate on Chap. 4, more precisely §4.7): suppose d is given by a Newtonian method, in which case $t = 1$ will be tried first. For x close to a minimum point, f is almost quadratic; $t = 1$ is almost optimal and must be accepted by the line-search; otherwise, superlinear convergence is killed.   □

Along the same lines, the descent test

$$q(t) \leqslant q(0) + m_1 t q'(0)$$

has an interesting interpretation. In fact, the term $tq'(0)$ is the linear estimate of the variation of q between 0 and t. Said otherwise: if q were an affine function, we would have $q(t) - q(0) = tq'(0)$. Armijo's test therefore consists in requiring the objective-function to decrease by at least a fraction $m_1 < 1$ of the linear decrease – a sort of *nominal* decrease.

Let us go further; if q were quadratic and if t minimized q, then the decrease of q would be $\frac{1}{2}tq'(0)$ (Proposition 3.12). Writing Armijo's test in the form

$$q(t) - q(0) \leqslant (2m_1)(tq'(0)/2) \,,$$

we see that this test consists also in requiring from the objective-function to decrease by at least a fraction $2m_1$ (< 1) of the "quadratic predicted" decrease. This observation is useful when generalizing line-searches; see in particular §6.1.

3.6 Implementation Considerations

All the necessary ingredients to set up an optimization module are now given, and numerical experiments can be conducted: the reader can for example program the gradient method ($d = -g$) with Wolfe's line-search, etc. It is strongly advised to divide the program in three blocks quite distinct, remembering §1.3. These will be respectively blocks 1.1, 2 and 1.2 in Table 3.6 (the numbering 1.1 and 1.2 suggests that these two blocks are in the responsibility of the same person: the user; using again notation from §1.3, (A) corresponds to block 2, and (P) to block 1).

Let us say it again: we use here $d = -g$, only for didactic reasons, but we remember that it is actually a numerical absurdity. The aim of Table 3.6 is to

– make concrete the modular structure of an optimization program,
– let the reader program a line-search; an excellent means to become familiar with the subject.

A last remark: the resulting program will very likely fail (at least at the first try). Experience indicates that, in 90% of cases, the mistake is not in the optimization algorithm proper (Block 2) but in the gradient computation (Block 1.2). Such a mistake can be detected rather reliably, upon observation of Wolfe's line-search. If the gradient is bugged, the following paradox is eventually observed: a sequence $\{t = t_R\}$ is produced, tending to 0 with $q(t)$ staying stubbornly larger than $q(0)$, while $q'(t)$ stays stubbornly negative.

BLOCK 1.1 – Define the problem
        (let alone the number of variables).
    – Initialize x among other things
        (tolerances for the stopping test for example).

        CALL BLOCK 2

    – Exploit the results.

BLOCK 2.1 At each iteration:
    – Perform the stopping test.
    – Compute the direction
        (for example the gradient, *faute de mieux*);
    – Initialize the stepsize $t > 0$

        CALL BLOCK 2.2

    – Pass to the next iterate.

BLOCK 2.2 – Perform a line-search
        (knowing x, d, t coming from Block 2.1).
    For this, one needs to repetitively

        CALL BLOCK 1.2

BLOCK 1.2 – Compute $f = f(x)$ and $g = g(x)$,
        (for x coming from Block 2, and
          having various data from Block 1.1).

Bibliographical Comments

Even though "modern" line-searches are now more than 30 years old – [12, 168, 361] – they are rarely pointed out in monographs; there are exceptions, such as [107], or the first edition of [128]. The viewpoints exposed here are those of [228], similar to [265].

4 Newtonian Methods

The present chapter details the *most important* approach (by far) to compute a descent direction at each iteration of a minimization algorithm. This is the quasi-Newton method, defined in §4.4. To use another direction cannot be considered without a serious motivation; this has been true for decades and will probably remain so for several more years.

4.1 Preliminaries

To solve the optimality condition $g(x) = 0$, we have seen in Chap. 2 essentially two possibilities for the direction: the gradient, or a vector of the canonical basis. Both are bad; to do better, let us recall the Newton principle. Starting from the current iterate x_k, replace g by its linear approximation:

$$g(x_k + d) = g(x_k) + g'(x_k)d + o(|d|)$$

where $g'(x_k)$ is the Jacobian of g at x_k. Using the notation from §1.3, our problem (P) is to find d such that $g(x_k+d) = 0$; to obtain the model-problem (P_k), we then neglect the term $o(|d|)$; this gives the linearized problem $g(x_k)+ g'(x_k)d = 0$. Its solution is $d^N = -[g'(x_k)]^{-1}g(x_k)$ (when $g'(x_k)$ is invertible), and the next iterate is $x^N = x_k + d^N$.

In the case of an optimization problem, g is the gradient of f, $g' = f''$ is its Hessian. Just as g was approximated to first order, f can be approximated to second order:

$$f(x_k + d) = f(x_k) + (f'(x_k), d) + \frac{1}{2}(d, f''(x_k)d) + o(|d|^2).$$

The quadratic approximation thus obtained is minimized (in the elliptic case) when its gradient vanishes: $f'(x_k) + f''(x_k)d = 0$. We realize an evidence: Newton's method on $\min f(x)$ is Newton's method on $f'(x) = 0$.

The big advantage of Newton's method is well known: it converges very fast.

Theorem 4.1. *If f'' is continuous and invertible near a solution x^*, then the convergence of Newton's method is Q-superlinear. If, in addition, $f \in C^3$, this convergence is Q-quadratic.*

Proof. Call r the remainder term in the second-order Taylor expansion, i.e. the function from $\mathbb{R}^n$ to $\mathbb{R}^n$ such that

$$[0 =] \ f'(x^*) = f'(x) + f''(x)(x^* - x) + r(x^* - x).$$

If x is the current iterate, denote by $x^N = x + d^N$ the next iterate, given by Newton's method: $0 = f'(x) + f''(x)(x^N - x)$; we obtain by subtraction

$$0 = f''(x)(x^* - x^N) + r(x^* - x).$$

In view of continuity and invertibility, this implies

$$|x^* - x^N| \leqslant M |r(x^* - x)|,$$

M being a bound for $|(f'')^{-1}|$ in a neighborhood of x^*.

By definition and continuity, $r(x^* - x) = o(|x^* - x|)$; and if f is C^3, then $r(x^* - x) = O(|x^* - x|^2)$. The conclusions follow.   □

This theorem implies by no means global convergence of Newton's method. It simply says that, if x is close to the solution, then x^N is infinitely closer. In fact, drawbacks of Newton's method are also well-known:

- in general, it diverges violently;
- in addition, it requires to compute the Hessian, and then to solve a linear system; this is heavy;
- in our situation where g is the gradient of a function to be *minimized*, another drawback is that $\{x_k\}$ will probably rush to the closest stationary point, possibly a local maximum; the descent property $f(x_{k+1}) < f(x_k)$ is not guaranteed.

4.2 Forcing Global Convergence

Just as for first-order methods, the system $g(x) = 0$ is not arbitrary: here again we have on hand a merit function, which can be forced to decrease at each iteration. We therefore call for the model technique, introduced in §1.3; see also §2.4.

- Computing the Newton estimate:

$$\min f(x_k) + (g(x_k), d) + \frac{1}{2}(d, f''(x_k)d), \quad d \in \mathbb{R}^n$$

(admitting that it makes sense), is only considered as a first phase, which is to solve a model-problem (P_k) of the original problem (P).
- The solution d^N is then considered as a direction, along which a line-search is performed to decrease the function $q(t) = f(x_k + td^N)$.
- The next iterate will then be $x_{k+1} = x_k + t_k d^N$.

In the inf-compact case, we will thus have stabilized the sequence $\{x_k\}$.

As was seen in Chap. 3, the line-search will be possible only if $q'(0) = (g_k, d^N) < 0$. Using the definition of d^N, this means that $(g_k, f_k''^{-1} g_k)$ must be positive. If f is elliptic, so will be the case and everything will go fine; otherwise, something more must be done. Actually, the present "damped Newton's method" works correctly only when f_k'' is positive definite at each iteration. This is why the modern tendency is to prefer the variant by *trust region*, mentioned later in §6.1.

Remark 4.2. It may be useful to recall that, in numerical analysis, linear systems $(Ad = -g)$ are not solved by matrix inversion $(d = -A^{-1}g)$. When A is positive definite (which is in principle the case in the present situation where $A = f_k''$), the Cholesky decomposition is used: a lower-triangular matrix L is computed, such that $A = LL^\top$. The resolution of $Ad = -g$ is then simple: solve two triangular systems $Ly = -g$, and then $L^\top d = y$. Even better: A can be decomposed as $A = LDL^\top$, where D is diagonal and L, still triangular, has now 1 as diagonal terms (the so-called *Bunch and Parlett* decomposition). □

Note also that, in contrast to first-order methods, the present $d_k = -f_k''^{-1} g_k$ yields not only a direction, but also a stepsize along this direction. Indeed, the point $x^N := x_k + d_k$, which is the Newton estimate, is in the good cases an excellent approximation of the minimum point of f. In other words, the stepsize $t = 1$ is supposedly best; it would be ideal to have $t_k = 1$, at least for large k (when Newton's estimates start to be really good). This is useful to initialize the stepsize; remember Remark 3.4.

Finally, it is easy to obtain in the elliptic case a global convergence theorem, since $\cos(d_k, g_k)$ stays away from 0 (see again Theorem 3.9).

4.3 Alleviating the Method

Consider now the second drawback of Newton's method. Rather than computing explicitly f'', and solving the corresponding linear system, another idea is to approximate directly $(f'')^{-1}$ by a matrix W, to be computed at each iteration. Then, alongside with the descent process on f, an identification process of the Hessian (or rather its inverse) is performed. The general form of the algorithm is then the following (the subscript "+" denotes the next iterate $k + 1$, absence of index denotes the current iterate k):

Algorithm 4.3 (Schematic quasi-Newton algorithm).

Step 0. An initial iterate x and stopping tolerance ε are given; an initial matrix W, positive definite, is also chosen. Compute the initial gradient $g = g(x)$.

Step 1. If $|g| \leqslant \varepsilon$ stop.

Step 2. Compute $d = -Wg$.

Step 3. Make a line-search initialized on $t = 1$, to obtain the next iterate $x_+ = x + td$ and its gradient $g_+ = g(x_+)$.

Step 4. Compute the new matrix W_+ for the next iteration and loop to 1.

$\square$

The ingredients characterizing this method are therefore:

– the line-search, which will be for example that of Wolfe (§3.4),
– the initial matrix W (which can be the identity, for want of a better idea),
– the computation of W_+ in Step 4.

Let us explain how this last matrix is computed. In the sequel, we will use the notation

$$s = s_k = x_{k+1} - x_k \quad \text{and} \quad y = y_k = g_{k+1} - g_k$$

(observe that s and y are known when W_+ must be computed).

Knowing that we want to approximate a symmetric matrix (an inverse Hessian) and to obtain descent directions, W_+ is of course required to be *symmetric positive definite*.

Besides, to give W_+ a chance to approximate an inverse Hessian, W_+ is required to satisfy $W_+ y = s$, called the *quasi-Newton*, or *secant* equation. Its explanation is as follows: the mean-value G of f'' between x and x_+ satisfies $y = Gs$. The quasi-Newton equation has therefore the effect of forcing W_+ to have the same action as G^{-1} on y, a subspace of dimension 1.

Of course, the two above requirements leave infinitely many possible *quasi-Newton matrices*.

4.4 Quasi-Newton Methods

From now on in the remaining of this chapter, $(g, x) = g^\top x$ is the standard dot-product.

A quasi-Newton method is the realization of Algorithm 4.3, where the matrices W_k are computed recursively: $W_{k+1} = W_k + B_k$, the corrections B_k being chosen so that

(i) W_k is symmetric positive definite for all k,
(ii) the quasi-Newton equation $W_{k+1} y_k = s_k$ is satisfied for all k.

Among all the possible corrections, stability reasons lead us to the additional requirement

(iii) B_k is minimal in some sense.

This still leaves a large range of possibilities, depending on the sense chosen in (iii), and a considerable number of methods have been studied.

Historically, the first was the so-called method of Davidon-Fletcher-Powell, which takes

$$W_+ = W + \frac{ss^\top}{(y,s)} - \frac{Wyy^\top W}{(y,Wy)} \,. \tag{DFP}$$

Watch the matrix writing; for example, $y^\top W$ is a row-matrix which, pre-multiplied by the column Wy, produces an $n \times n$ matrix, of rank one because its kernel is the subspace orthogonal to Wy. Thus, the correction B_k is a "small" matrix: its rank is at most 2 (its kernel being the subspace spanned by s and Wy). Then observe that W_+ is a symmetric matrix, which satisfies the quasi-Newton equation:

$$W_+y = Wy + s\frac{(y,s)}{(y,s)} - Wy\frac{(y,Wy)}{(y,Wy)} = s \,.$$

At present, specialists rather unanimously agree on a method found independently, and using different arguments, by C. Broyden, R. Fletcher, D. Goldfarb, D. Shanno:

$$W_+ = W - \frac{sy^\top W + Wys^\top}{(y,s)} + \left[1 + \frac{(y,Wy)}{(y,s)}\right]\frac{ss^\top}{(y,s)} \,. \tag{BFGS}$$

Remark 4.4. Suppose that, instead of W, one wants to compute M, aimed at approximating f''. Then M will have to satisfy the so-called "dual" quasi-Newton equation

(ii') $M_+s = y$.

Let now a quasi-Newton formula be written $W_+ = W + B(W,y,s)$; here $B(W,y,s)$ denotes the correction of W, for example (DFP) or (BFGS). Obviously, setting $M = W^{-1}$, the matrix $M_+ = M + B(M,s,y)$ (obtained by an inversion of s and y) satisfies the dual quasi-Newton equation (ii') and approximates f'', just as W_+ approximates $(f'')^{-1}$. Thus, whenever a quasi-Newton formula is invented, two such are obtained, mutually "dual": one for M and the other for W. For example, it can be shown that (DFP) and (BFGS) formulae are mutually dual; said otherwise, if $M = W^{-1}$, then

$$M_+ = M + \frac{yy^\top}{(y,s)} - \frac{Mss^\top M}{(Ms,s)} \tag{BFGS'}$$

gives the inverse of W_+ obtained by (BFGS).   □

Since a matrix is symmetric positive definite if and only if its inverse is so, positive definiteness of (DFP) or (BFGS) are equivalent properties. Another writing of (DFP) is

$$M_+ = \left[I - \frac{ys^\top}{(y,s)}\right]M\left[I - \frac{sy^\top}{(y,s)}\right] + \frac{yy^\top}{(y,s)} \,. \tag{4.1}$$

Indeed, developing the product gives

$$M_+ = M - \frac{ys^\top M + Msy^\top}{(y,s)} + \frac{(Ms,s)}{(y,s)^2} yy^\top + \frac{yy^\top}{(y,s)}$$

which is the dual of (BFGS).

With these preliminaries, the following result shows that (BFGS) and (DFP) preserve positive definiteness, providing that (y,s) be positive:

Theorem 4.5. *Suppose M is positive definite. Then $(y,s) > 0$ is a necessary and sufficient condition for (4.1) to give a positive definite matrix M_+.*

Proof. First, M_+ is obviously symmetric if M is so (compute $M_+^\top$). Also, observe from the quasi-Newton equation $M_+s = y$ that the condition is necessary.

Now take $u \neq 0$ and set $v = u - (y,u)s/(y,s)$. Then

$$(M_+u, u) = (Mv, v) + \frac{(y,u)^2}{(y,s)}$$

which is a sum of two nonnegative terms. If the first is nonzero, we are done. Otherwise $v = 0$ (M is positive definite); hence $u = (y,u)s/(y,s)$ is collinear to s; in addition, the coefficient (y,u) is nonzero (otherwise u would be 0). Altogether, the second term in the expression of (M_+u, u) is nonzero; it is actually positive by assumption.   □

Note that the property $(y,s) > 0$ is automatically guaranteed by Wolfe's line-search (§3.4); a curious "a posteriori" motivation.

Remark 4.6. The one-dimensional case is interesting: if $n = 1$, the quasi-Newton equation defines a unique W_+ or M_+ (either one is a positive number) by $W_+ = s/y$. Starting from two initial iterates x_1 and x_2, the algorithm reduces to

$$x_{k+1} = x_k - \frac{x_k - x_{k-1}}{g_k - g_{k-1}} g_k$$

known as the secant method, or "regula falsi": the tangent to the graph of g (which is used in Newton's method) is replaced by the secant between x and x_-. Locally, this method converges Q-superlinearly and its order of convergence (the speed at which q_k converges to 0) can even be explicitly given: it can be shown that $|x_+ - x^*|/|x - x^*|^r$ is bounded, where $r = (\sqrt{5} + 1)/2$ is the golden section.

The opposite case (big n) also deserves comment. A drawback of Newtonian methods (including the present quasi-Newton variant) is the necessity of storing an $n \times n$ matrix (what if $n = 10^5$!?). Yet, as long as k is small, computing the direction d_k of a quasi-Newton method needs only $2k$ vectors s_i and $y_i, i = 1, \ldots, k$. In the case of a really large-scale problem, at least a few iterations can be performed without computing explicitly the whole matrix W_k: just develop the product $W_k g_k$ in terms of these k vector pairs. This point will be seen in more detail in §5.6 and 6.3.   □

4.5 Global Convergence

Basically, there exists one single theorem of global convergence, which concerns BFGS.

When Wolfe's line-search is used, Lemma 3.8 and Theorem 3.9 suggest that the whole issue is to bound the cosine between $-d_k$ and g_k, i.e. the condition number of W_k, or of $M_k = W_k^{-1}$. This is the essence of the proof of Theorem 4.9 below. Note that it is not simple, since the condition number has to involve the term (y_k, s_k) (see Theorem 4.5), which in turn depends on the line-search; the properties of BFGS formula and of Wolfe rule must therefore interfere in the proof; they cannot be used separately as in §3.4.

First, we admit without proof the following preliminary result, to be read with (BFGS') in mind:

Lemma 4.7. *For the BFGS formula approximating the Hessian:*

$$M_+ = M + \frac{yy^\top}{(y, s)} - \frac{Mss^\top M}{(Ms, s)} \,,$$

the trace and determinant of M_+ are given by:

$$\operatorname{tr} M_+ = \operatorname{tr} M + \frac{|y|^2}{(y, s)} - \frac{|Ms|^2}{(Ms, s)} \qquad and \qquad \det M_+ = \det M \frac{(y, s)}{(Ms, s)} \,.$$

If f is convex and has a gradient locally Lipschitzian with constant L (i.e. $|y| \leqslant L|s|$), then $|y|^2 \leqslant L(y, s)$. □

For the proofs below, we recall that the arithmetic mean is greater than the geometric mean: if $a_1, \ldots, a_k$ are k positive numbers, then

$$\left(\frac{1}{k} \sum_{i=1}^{k} a_i \right)^k \geqslant \prod_{i=1}^{k} a_i \,.$$

The cosine between d_k and $-g_k$ is given by the extreme eigenvalues of M_k; roughly speaking, the largest is the trace and the smallest is the determinant. We start with bounding these two quantities.

Lemma 4.8. *Suppose f is convex and has a locally Lipschitzian gradient. If M_1 is positive definite and if $(y_k, s_k) > 0$ at each iteration, the BFGS formulae satisfy*

$$\operatorname{tr} M_{k+1} \leqslant Ak \,, \tag{4.2}$$

$$\det M_{k+1} \geqslant c \prod_{i=1}^{k} \frac{|g_i|^2 (y_i, s_i)}{C(g_i, s_i)^2} \,, \tag{4.3}$$

where A, c and C are positive constants.

Proof. Lemma 4.7 provides an immediate bound on the trace of M_k:

$$\operatorname{tr} M_+ \leqslant \operatorname{tr} M + \frac{|y|^2}{(y,s)} \leqslant \operatorname{tr} M + L$$

and (4.2) is easily deduced by summation (knowing that $\operatorname{tr} M_1$ is a constant).
Calling c the determinant of M_1, we first have (Lemma 4.7):

$$\det M_+ = c \prod_{i=1}^{k} \frac{(y_i, s_i)}{(M_i s_i, s_i)}. \tag{4.4}$$

Besides, take again the trace-relation:

$$\operatorname{tr} M - \operatorname{tr} M_+ + L \geqslant \frac{|Ms|^2}{(Ms,s)}$$

and, by summation,

$$\operatorname{tr} M_1 + kL \geqslant \sum_{i=1}^{k} \frac{|M_i s_i|^2}{(M_i s_i, s_i)}$$

(we have neglected the positive term $\operatorname{tr} M_+$). Knowing that $\operatorname{tr} M_1$ is a constant, we can therefore write $C \geqslant \frac{1}{k} \sum \frac{|M_i s_i|^2}{(M_i s_i, s_i)}$ and we have the same bound on the geometric mean:

$$C^k \geqslant \prod_{i=1}^{k} \frac{|M_i s_i|^2}{(M_i s_i, s_i)}.$$

Multiply this last inequality by (4.4):

$$C^k \det M_+ \geqslant c \prod_{i=1}^{k} \frac{|M_i s_i|^2 (y_i, s_i)}{(M_i s_i, s_i)^2};$$

the result comes, dividing by C^k and using $M_i s_i = -t_i g_i$.   □

We are now in a position to give the global convergence result:

Theorem 4.9. *Suppose f is convex with a Lipschitzian gradient on the slice $\{x : f(x) \leqslant f(x_1)\}$. Then the BFGS algorithm with Wolfe's line-search and W_1 positive definite satisfies:*
   – either the objective function tends to $-\infty$,
   – or $\liminf |g(x_k)| = 0$.

Proof. Take again (4.3) and bound the right-hand side. For the numerator, the second half of Wolfe's rule gives

$$(y_i, s_i) = (g_{i+1}, s_i) - (g_i, s_i) \geqslant (m_2 - 1)(g_i, s_i) = (1 - m_2)(-g_i, s_i),$$

which eliminates one of the two terms (g_i, s_i) in the denominator. As for the second term, the descent-test gives

$$f(x_{i+1}) - f(x_i) \leqslant m_1(g_i, x_{i+1} - x_i)$$

i.e.

$$(-g_i, s_i) \leqslant \frac{1}{m_1}[f(x_i) - f(x_{i+1})].$$

In summary, we therefore obtain

$$\det M_+ \geq c\alpha^k \frac{\prod_{i=1}^{k} |g_i|^2}{\prod_{i=1}^{k}[f(x_i) - f(x_{i+1})]},$$

where α is the constant $m_1(1 - m_2)/C$. Use again the inequality of arithmetic and geometric means:

$$\det M_+ \geqslant c\alpha^k \frac{\prod_{i=1}^{k} |g_i|^2}{\left(\frac{1}{k}\sum_{i=1}^{k}[f(x_i) - f(x_{i+1})]\right)^k} = c\left[\frac{\alpha k}{f(x_1) - f(x_+)}\right]^k \prod_{i=1}^{k} |g_i|^2.$$

If f is bounded from below, $\alpha/[f(x_1) - f(x_+)] \geqslant \beta > 0$; compare this inequality with (4.2) using again the means-inequality:

$$c(\beta k)^k \prod_{i=1}^{k} |g_i|^2 \leqslant \det M_+ \leq \left(\frac{1}{n}\operatorname{tr} M_+\right)^n \leqslant \left(\frac{1}{n}Ak\right)^n \leq Pk^n;$$

here, P is the constant $(A/n)^n$. If $|g_i|^2 \geqslant \varepsilon > 0$ for all i, then $c(\beta\varepsilon k)^k \leqslant Pk^n$, which is impossible when $k \to +\infty$ (an exponential grows faster than a polynomial). $\qquad\square$

4.6 Local Convergence: Generalities

Remember that Newtonian methods are designed to converge fast, hence the need for this section. Yet, studying the local convergence of quasi-Newton methods is intricate, and we will pass rapidly over some details particularly tedious.

A first result gives a general criterion for superlinear convergence, when solving $g(x) = 0$: $|g|$ must decrease infinitely faster than the move from one iterate to the next.

Lemma 4.10. *Consider a mapping g from $\mathbb{R}^n$ to $\mathbb{R}^n$ and let a sequence $\{x_k\}$ tend to x^* such that $g(x^*) = 0$. Assume that g' is continuous in a neighborhood of x^*, and that $g'(x^*)$ is invertible. Then*

$$q_k := \frac{|x_{k+1} - x^*|}{|x_k - x^*|} \to 0 \quad \Longleftrightarrow \quad \frac{|g(x_{k+1})|}{|x_{k+1} - x_k|} \to 0.$$

Proof. We use again the abridged notation x and x_+ (which is possible thanks to the Markovian character of Q-convergence). Call G the mean-value of g' between x_+ and x^*, so that

$$g(x_+) = 0 + G(x_+ - x^*).$$

By continuity, G is bounded and "uniformly invertible": there exist two positive numbers ℓ and L such that, for k large enough (hence x_+ close enough to x^*),

$$\ell |g_+| \leqslant |x_+ - x^*| \leqslant L |g_+|$$

(L is the norm of G^{-1}, ℓ is the inverse of the norm of G).

Dividing the second inequality by $|x_+ - x|$ and using the triangle inequality:

$$\frac{L|g_+|}{|x_+ - x|} \geqslant \frac{|x_+ - x^*|}{|x_+ - x|} \geqslant \frac{|x_+ - x^*|}{|x_+ - x^*| + |x - x^*|} = \frac{q}{1 + q} .$$

Hence, if $|g_+|/|x_+ - x|$ tends to 0, q tends to 0: the condition is sufficient.

Conversely, write

$$\frac{|g_+|}{|x_+ - x|} = \frac{|g_+|}{|x_+ - x^*|} \frac{|x_+ - x^*|}{|x - x^*|} \frac{|x - x^*|}{|x_+ - x|} .$$

The first quotient is smaller than $1/\ell$. The second is q, which tends to 0 by assumption. It suffices to show that the third is bounded. But its inverse is

$$\frac{|x_+ - x|}{|x - x^*|} \geqslant \frac{|x - x^*| - |x^* - x_+|}{|x - x^*|} = 1 - q ,$$

which is bounded from below, for example by $\frac{1}{2}$, when $q \to 0$. $\square$

Let us come back to our quasi-Newton method $x_{k+1} = x_k - t_k M_k^{-1} g_k$ and let us admit for the moment that $t_k = 1$ for large k. In view of Theorem 4.1, superlinear convergence can be expected if the matrix $M_k - f''(x_k)$ tends to 0. However this is far too demanding, since the behaviour of M_k^{-1} has an importance only on g_k, a subspace of dimension 1: M_k^{-1} can behave arbitrarily on the rest of $\mathbb{R}^n$, the algorithm will not even notice it. In fact, there is a remarkable criterion, said of Dennis and Moré, which refines Theorem 4.1.

Theorem 4.11 (criterion of Dennis and Moré). *Let the nonlinear function g from $\mathbb{R}^n$ to $\mathbb{R}^n$ and the sequence $\{x_k\}$ satisfy the assumptions of Lemma 4.10. Suppose in addition that $\{x_k\}$ is generated with the help of matrices M_k by the formula*

$$x_{k+1} = x_k - M_k^{-1} g_k .$$

Then x_k converges Q-superlinearly (to x^*) if and only if

$$v_k := [M_k - g'(x^*)] \frac{x_{k+1} - x_k}{|x_{k+1} - x_k|} \to 0 .$$

Proof. By definition,

$$[M - g'(x^*)](x_+ - x) = -g - g'(x^*)(x_+ - x) = g_+ - g - g'(x^*)(x_+ - x) - g_+.$$

Call G the mean-value of g' between x and x_+, i.e. $g_+ - g = G(x_+ - x)$, which gives

$$[M - g'(x^*)](x_+ - x) = [G - g'(x^*)](x_+ - x) - g_+ .$$

We deduce the two inequalities

$$\frac{|g_+|}{|x_+ - x|} \leqslant |G - g'(x^*)| + |v| \quad \text{and} \quad |v| \leqslant |G - g'(x^*)| + \frac{|g_+|}{|x_+ - x|}$$

which proves the result, because $|G - g'(x^*)| \to 0$ by continuity. □

4.7 Local Convergence: BFGS

Theorem 4.11 gives a Q-superlinear convergence criterion for quasi-Newton matrices in methods without line-search (we have assumed $t_k = 1$). Here, superlinear convergence of a quasi-Newton algorithm will be established in two steps:
– to show that the matrices M_k satisfy the criterion of Dennis-Moré,
– to show that the stepsize $t_k = 1$ eventually satisfies Wolfe's rule.

Recall that, when studying local convergence, x_k is a priori supposed to tend to a minimum point x^* of f, at which additional assumptions can be accepted.

The proofs below make an extensive use of the angle between the gradient and the direction:

$$\cos \theta_k := \frac{-(g_k, d_k)}{|g_k|\,|d_k|} = \frac{-(g_k, s_k)}{|g_k|\,|s_k|} = \frac{(M_k s_k, s_k)}{|M_k s_k|\,|s_k|}.$$

To start, we establish a lemma concerning the line-search, which completes Theorem 3.9.

Lemma 4.12. *Let an algorithm using Wolfe's rule generate a sequence $\{x_k\}$ converging to a point x^*. Suppose that, in a neighborhood of x^*, f is strongly convex and g is Lipschitzian, with respective constants ℓ and L. Then, for x_k close enough to x^*,*

$$\frac{1 - m_2}{L}|g(x_k)|\cos\theta_k \leqslant |s_k| \leq 2\frac{1 - m_1}{\ell}|g(x_k)|\cos\theta_k .$$

Proof. The left inequality is nothing other than Argument 2 in Theorem 3.9. For the right inequality, strong convexity gives

$$f(x) + (g, s) + \frac{1}{2}\ell|s|^2 \leqslant f(x_+)$$

which, added to the descent test $f(x_+) \leqslant f(x) + m_1(g, s)$, gives:

$$(g, s) + \frac{1}{2}\ell|s|^2 \leqslant m_1(g, s),$$

or

$$\frac{1}{2}\ell|s|^2 \leq (m_1 - 1)(g, s) = (1 - m_1)|g|\,|s|\cos\theta. \qquad \square$$

To continue, we need a key-result, expressing that $\cos\theta_k$ stays away from 0 for a non-negligible number of iterations.

Lemma 4.13. *Under the hypotheses of Lemma 4.12, suppose that the algorithm uses the BFGS direction. Then there exists $\gamma > 0$ such that, for all k, $\cos\theta_i \geqslant \gamma$ for at least half of the indices $i = 1, \ldots, k$.*

Proof. Start from Lemma 4.8. Using in (4.3) the second half of Wolfe's rule $(y_i, s_i) \geqslant (1 - m_2)(-g_i, s_i)$, we have for a certain constant D

$$\prod_{i=1}^{k} \frac{|g_i|^2}{(-g_i, s_i)} \leqslant D^k \det M_+ \leqslant D^k \left(\frac{1}{n}\operatorname{tr} M_+\right)^n \leqslant D^k \left(\frac{1}{n}Ak\right)^n.$$

Here, we have used successively the means-inequality, and the bound (4.2) of Lemma 4.8. By definition of θ_i and from Lemma 4.12, this gives

$$\prod_{i=1}^{k} \frac{1}{\cos^2\theta_i} \leqslant \left(2\frac{1 - m_1}{\ell}D\right)^k \left(\frac{1}{n}Ak\right)^n \leqslant E^k$$

for a certain constant E. We see that, on the average, the $\cos\theta$'s are far from 0. The rest is purely technical and is left to the reader. $\qquad \square$

Lemma 4.14. *Under the hypotheses of Lemma 4.13, the series $\sum |x_k - x^*|$ converges.*

Proof. Starting from the descent-test, use successively Lemmas 4.12 and 2.13:

$$f(x) - f(x^*) \geqslant f(x_+) - f(x^*) + m_1|g|\,|s|\cos\theta$$
$$\geqslant f(x_+) - f(x^*) + \frac{m_1(1 - m_2)}{L}|g|^2\cos^2\theta$$
$$\geqslant f(x_+) - f(x^*) + m_1(1 - m_2)\frac{\ell}{2L}\cos^2\theta[f(x) - f(x^*)].$$

We deduce

$$f(x_+) - f(x^*) \leqslant \left(1 - m_1(1 - m_2)\frac{\ell}{2L}\cos^2\theta\right)[f(x) - f(x^*)].$$

This inequality holds at each iteration. In view of Lemma 4.13, we therefore can write for at least half of the iterations between 1 and k:

$$f(x_{i+1}) - f(x^*) \leqslant \kappa[f(x_i) - f(x^*)] \quad \text{for a certain } \kappa < 1.$$

Since, on the other hand, $\{f(x_i)\}$ is decreasing, we see that $f(x_k)$ converges R-linearly to $f(x^*)$. From strong convexity (Remark 2.14), x_k also converges R-linearly to x^*: $\{|x_k - x^*|\}$ is a geometric sequence, whose sum is finite.   □

Having this last result, we can at last show that the BFGS matrix allows superlinear convergence. Of course, we need for this a second-order assumption on f; indeed, we need "slightly more" than the mere existence of $f''(x^*)$ (positive definite).

Theorem 4.15. *Assume that the BFGS algorithm with Wolfe's line-search converges to x^*, in a neighborhood of which f is strongly convex and has a Lipschitzian Hessian. Then the criterion of Dennis-Moré of Theorem 4.11 is satisfied.*

Proof. First, it is easily checked that the hypotheses of Lemma 4.12 (and hence of Lemmas 4.13, 4.14) are satisfied: the series $\sum |x_k - x^*|$ converges.

On the other hand, using again the notation x for x_k, call G the mean-value of f'' between x and x_+: we have

$$y = Gs = f''(x^*)s + [G - f''(x^*)]s .$$

The Lipschitz property gives rather easily

$$|G - f''(x^*)| \leqslant C \max\{|x - x^*|, |x_+ - x^*|\} ,$$

from which we can write that

$$\frac{|y_k - f''(x^*)s_k|}{|s_k|} \leqslant C \max\{|x - x^*|, |x_+ - x^*|\} ,$$

which tends to 0 as fast as a convergent series (Lemma 4.14).

Then comes a general result: this last property, in conjunction with BFGS formula, guarantees the criterion of Dennis-Moré. The proof is just as tedious as the previous ones, but much longer; it is not given. Its essence is to make a change of variables, so that $f''(x^*)$ becomes the identity matrix. Using bounding techniques of the same type as in the preceding results, one then shows that $\cos \theta_k \to 0$, which is the desired property indeed (after the change of variables, the negative gradient *is* Newton's direction!)   □

To conclude, it remains to make sure that the stepsize $t_k = 1$ is accepted by (Wolfe's) line-search; otherwise Theorem 4.15 is killed. Thinking of the matter, one realizes that Wolfe's rule is just a filter which, given an initial point x, accepts a candidate $x_+ = x + s$ if and only if

$$f(x_+) \leqslant f(x) + m_1(g, x_+ - x) \quad \text{and} \quad (g_+, x_+ - x) \geqslant m_2(g, x_+ - x). \quad (4.5)$$

Whether s is set to td, possibly with $d = -Wg$, is just a matter of notation.

Theorem 4.16. *Suppose f has a minimum point x^* with a Hessian $f''(x^*)$ positive definite. Then there exist $\delta > 0$ and $\varepsilon > 0$ such that: if $|x - x^*| \leqslant \delta$, then every x_+ satisfying $|x_+ - x^*| \leqslant \varepsilon|x - x^*|$ is accepted by Wolfe's rule, providing that $0 < m_1 < \frac{1}{2}$ and $m_2 > 0$.*

Proof. We use the notation $M = f''(x^*)$. Multiplying by $s = x_+ - x$ the development $g(x) = M(x - x^*) + o(|x - x^*|)$, we obtain

$$(g, s) = (M(x - x^*), s) + (o(|x - x^*|), s)$$
$$= (M(x - x^*), x^* - x) + (M(x - x^*), x_+ - x^*) + (o(|x - x^*|), s).$$

It is easy to see that $|s| = O(|x - x^*|)$ when $|x_+ - x^*| = o(|x - x^*|)$; as a result, we therefore have

$$(g, s) = -(M(x - x^*), x - x^*) + o(|x - x^*|^2). \tag{4.6}$$

Now call $\ell > 0$ and L the extreme eigenvalues of M. Add m_1 times (4.6) to the development

$$f(x) = f(x^*) + \frac{1}{2}(M(x - x^*), x - x^*) + o(|x - x^*|^2)$$

to obtain

$$f(x) + m_1(g, s) = f(x^*) + (\tfrac{1}{2} - m_1)(M(x - x^*), x - x^*) + o(|x - x^*|^2)$$
$$\geqslant f(x^*) + (\tfrac{1}{2} - m_1)\ell|x - x^*|^2 + o(|x - x^*|^2).$$

On the other hand,

$$f(x_+) = f(x^*) + \tfrac{1}{2}(M(x_+ - x^*), x_+ - x^*) + o(|x_+ - x^*|^2)$$
$$\leqslant f(x^*) + \tfrac{1}{2}L|x_+ - x^*|^2 + o(|x_+ - x^*|^2)$$
$$\leqslant f(x^*) + \tfrac{1}{2}L\varepsilon^2|x - x^*|^2 + o(|x_+ - x^*|^2).$$

We therefore see by subtraction that the first half of (4.5) is satisfied if $|x-x^*|$ is small enough and if $\varepsilon^2 < 2(1/2 - m_1)\ell/L$.

For the second half of (4.5), take again (4.6):

$$m_2(g, s) \leqslant -m_2\ell|x - x^*|^2 + o(|x - x^*|^2).$$

On the other hand, write again (4.6) replacing x by x_+ (check that there is no problem with the $o(\cdot)$):

$$(g_+, s) = -(M(x_+ - x^*), x_+ - x^*) + o(|x - x^*|^2)$$
$$\geqslant -L|x_+ - x^*|^2 + o(|x - x^*|^2)$$
$$\geqslant -L\varepsilon^2|x - x^*|^2 + o(|x - x^*|^2).$$

We again see by subtraction that the second half of (4.6) is satisfied for $\varepsilon^2 < m_2\ell/L$. □

Piecing together all these results, we now realize that BFGS + Wolfe line-search does converge superlinearly. For the reader's convenience, we state the summarizing theorem:

Theorem 4.17. *Consider the BFGS algorithm with Wolfe's line-search, used with $0 < m_1 < \frac{1}{2}$ and $m_1 < m_2 < 1$, the line-search being initialized with $t = 1$ (at least after finitely many iterations). Assume that the generated sequence $\{x_k\}$ converges to x^* such that $f'(x^*) = 0$, and in a neighborhood of which f is strongly convex and has a Lipschitzian Hessian.*

Then the convergence is superlinear: $\dfrac{|x_{k+1} - x^*|}{|x_k - x^*|} \to 0$ *(remember also Remark 2.14).*

Proof. Call M_k the matrix used at each iteration: $\{M_k\}$ is given by (BFGS'), say. From Theorem 4.15,

$$[M_k - f''(x^*)] \frac{x_{k+1} - x_k}{|x_{k+1} - x_k|} \to 0$$

so that, from Theorem 4.11,

$$\frac{x_k - M_k^{-1} g_k - x^*}{x_k - x^*} \to 0 \,.$$

Thus, we can apply Theorem 4.16: for k large enough, $x_+ := x_k - M_k^{-1} g_k$ is accepted by the line-search and we do have $x_{k+1} = x_+$, which is infinitely better than x_k.   □

Bibliographical Comments

The pioneering paper concerning quasi-Newton is [96] which, hard to read, has remained little known for 30 years, until [97] let it "see the light of the sun". Nevertheless, [96] was deemed important enough by the experts of the time to motivate [132], which served as a "publicity agent" during these 30 years. Papers on the subject are too many to be enumerated, the synthetic study [105] is a landmark. We mention [106] for a nice and convincing derivation of BFGS.

Both global and local convergence theorems are due to M.J.D. Powell: [290]. We mention also the proof of [66], more elegant but requiring strong convexity, even for global convergence; there the complete proof of Theorem 4.15 can be found. The second part of Lemma 4.7 is proved and explained in [196; §E.4]. It is amusing to mention that Theorem 4.9 is called Lemma by Powell in his original preprint; he was reluctant to publish [290], wanting to get rid of the convexity assumption, which he found artificial. Powell has also shown in [289] the global convergence of DFP; there, the line-search is assumed exact, an assumption which seems necessary: [294].

The criterion of Dennis-Moré (Theorem 4.11) is given in [104]. To prove first R-linear convergence (Lemma 4.14), and then Q-superlinear convergence (Theorem 4.15) may seem artificial – not mentioning other proofs in §4.7. Yet, this is the only known way, since 1976. All this demonstrates enough how hard is the convergence study of quasi-Newton (global as well as local).

5 Conjugate Gradient

The conjugate gradient method is also aimed at accelerating the methods of Chap. 2. Its first motivation is to solve in n iterations a linear system with symmetric positive definite matrix (or, equivalently, to minimize in n iterations a quadratic strongly convex function on $\mathbb{R}^n$), without storing an additional matrix, without even storing the matrix of the system. In fact, to solve $Ax + b = 0$ (A symmetric positive definite), the conjugate gradient method just needs a "black box" (a subroutine) which, given the vector u, computes the vector $v = Au$. Naturally, this becomes particularly interesting when, while n is large, A is sparse and/or enjoys some structure allowing automatic calculations. Typical examples come from the discretization of partial differential equations.

5.1 Outline of Conjugate Gradient

Take again our general problem: we want to minimize a function f, having on hand the "black box" (simulator) computing $f(x)$ and $g(x) = f'(x)$. Starting from the initial point x_1, each iteration computes a direction, and then makes a line-search. At the current iteration k, a number of gradients $g_1, \ldots, g_k$ and of directions $d_1, \ldots, d_{k-1}$ have been computed from the beginning. Then consider the subspace U_k spanned by all of these vectors:

$$U_k := \left\{ v = \sum_{i=1}^{k} \alpha_i g_i + \sum_{i=1}^{k-1} \beta_j d_j \ : \ \alpha \in \mathbb{R}^k, \ \beta \in \mathbb{R}^{k-1} \right\};$$

this definition is valid for $k > 1$ only, U_1 being the subspace spanned by g_1.

In any iterative method, it is normal to seek the next iterate with the help of the information collected about the problem to solve. Here, we seek the next iterate in the affine manifold V_k containing x_k and parallel to U_k. Said otherwise: x_{k+1} will have the form $x_k + u$ with $u \in U_k$, or equivalently the direction d_k is taken in U_k. The conjugate gradient method is then defined as follows.

Axiom We assume that f is quadratic elliptic and we take x_{k+1} minimizing f in $V_k = x_k + U_k$. □

Thus, x_{k+1} is the best possible iterate, given the information collected at the kth iteration.

Remark 5.1. A generic point in V_k is $x(\alpha) = x_k + \sum \alpha_i u_i$, where the u_i's form a basis of U_k. Then consider the function $h(\alpha) = f(x(\alpha))$; we have

$$\frac{\partial h}{\partial \alpha_i}(\alpha) = \sum_{j=1}^{n} \frac{\partial f}{\partial x^j}(x(\alpha)) \frac{\partial x^j}{\partial \alpha_i}(\alpha) = (f'(x(\alpha)), u_i).$$

To say that $x(\alpha)$ minimizes f in V_k is to say that α minimizes h, or that the gradient of f at $x(\alpha)$ is orthogonal to each u_i.   □

In the following, we will assume that $x_2, \ldots, x_k$ satisfy the above axiom; we will also assume that no g_i is 0 (otherwise we are done: the corresponding x_i is already optimal). The subspace spanned by a set B will be denoted by $[B]$; for example $U_k = [d_1, \ldots, d_{k-1}, g_1, \ldots, g_k]$.

Theorem 5.2. *With this system of notation, the three subspaces*

$$U_k' := [d_1, \ldots, d_{k-1}, g_k], \quad U_k'' := [g_1, \ldots, g_k], \quad U_k''' := [d_1, \ldots, d_k]$$

coincide with U_k (knowing that we set $U_1' = U_1 = [g_1]$).

Proof. By definition and by construction, $U_1 = U_1' = U_1'' = U_1'''$; suppose recursively that $U_{k-1} = U_{k-1}' = U_{k-1}'' = U_{k-1}'''$. The key is to show that both g_k and d_k lie outside this subspace.

Knowing that $g_k \neq 0$, we have $f(x_k) > f(x_k - tg_k) \geqslant f(x_{k+1})$ for $t > 0$ small enough (the second inequality comes from the axiom, using $g_k \in U_k$). This implies $x_{k+1} \notin V_{k-1}$ (otherwise x_k would not minimize f in V_{k-1}); hence $d_k \notin U_{k-1}$ and certainly has a nonzero component along g_k:

$$d_k = \alpha_k g_k + u \quad \text{with} \quad u \in U_{k-1} \text{ and } \alpha_k \neq 0$$

hence

$$g_k = \frac{d_k}{\alpha_k} - \frac{u}{\alpha_k} \quad \text{with} \quad \frac{-u}{\alpha_k} \in U_{k-1}.$$

We conclude that the following operations produce the same subspace:
- to construct U_k''' by appending d_k to $U_{k-1} = U_{k-1}'''$;
- to construct U_k'' by appending g_k to $U_{k-1} = U_{k-1}''$;
- to construct U_k by appending d_k and g_k to U_{k-1};
- to construct U_k' by appending g_k to $U_{k-1}'' = U_{k-1}'''$.   □

According to Remark 5.1, we already know that the gradients are mutually orthogonal: g_k is orthogonal to every preceding gradient (and to every preceding direction too). At each iteration k, it suffices to append to U_k either d_{k+1} or g_{k+1}; appending both is useless, the dimension of the subspace U_k will increase by one anyway. An immediate consequence is that the algorithm has to terminate after n iterations at most; this demonstrates well enough the interest of this strategy.

5.2 Developing the Method

Let us write our quadratic function in the form $f(x) = \frac{1}{2}(Ax, x) + (b, x) + a$, so that its gradient is $g(y) = Ay + b = g(x) + A(y - x)$. Our problem is now to find d_k issued from x_k and pointing to the minimum of f in V_k; afterwards, there will be no difficulty to find x_{k+1} by an exact minimization along d_k.

Theorem 5.3. *Let f be quadratic elliptic and suppose that x_k minimizes f in V_{k-1}. Then the following two statements are equivalent:*
– the direction d_k issued from x_k points to the minimum of f in V_k,
– $(d_k, Ad_i) = 0$ for $i = 1, \ldots, k - 1$.

Proof. Use again Remark 5.1: the first statement means that there is $t \in \mathbb{R}$ such that $g(x_k + td_k) = g_k + tAd_k$ is orthogonal to every generator of U_k. Using $U_k = U'_k$, this is to say that the following system in t is compatible:

$$(g_k, d_i) + t(Ad_k, d_i) = 0 \quad i = 1, \ldots, k - 1$$
$$(g_k, g_k) + t(Ad_k, g_k) = 0.$$

By the last equation, t cannot be 0 (because $g_k \neq 0$). On the other hand, every (g_k, d_i) is 0 by assumption (note that $d_{k-1} \in U_{k-1}$ by construction of d_{k-1}, the other d_i's are in U_{k-1} by construction of U_{k-1}). For the system to be compatible, it is therefore necessary to have

$$(Ad_k, d_i) = 0 \quad i = 1, \ldots, k - 1.$$

Conversely, suppose this last property is true, we have to show that the same system as before is compatible, i.e. $(g_k, Ad_k) \neq 0$ (all other equations are trivially satisfied). But d_k is a combination of $d_1, \ldots, d_{k-1}, g_k$. If we had simultaneously

$$(Ad_k, d_1) = \cdots = (Ad_k, d_{k-1}) = (Ad_k, g_k) = 0$$

we would obtain by linear combination $(Ad_k, d_k) = 0$, impossible. $\square$

Remark 5.4. By symmetry, we therefore have $(d_i, Ad_j) = 0$ for $i \neq j$. The d_i's are then said to be *conjugate* with respect to the matrix A, hence the name of the method. Remember that the gradients are mutually orthogonal, i.e. they are conjugate with respect to the identity matrix.

Besides, since $t_i \neq 0$ (see the proof of Theorem 5.3), we also have $(d_k, g_{i+1} - g_i) = 0$ for $i = 1, \ldots, k - 1$. Geometrically, d_k has the same scalar product with every gradient till the kth: d_k is orthogonal to the affine hull of these gradients. $\square$

5.3 Computing the Direction

There remains to compute explicitly $d_k \in U_k$. For this, it is convenient to generate U_k by the gradients ($U_k = U_k''$), expressing the direction in the form $\sum \alpha_j g_j$. The α_j's are computed by expressing that (d_k, g_i) is a constant, which gives the linear system

$$(d_k, g_i) = \sum_{j=1}^{k} \alpha_j (g_i, g_j) = b_k \quad i = 1, \ldots, k;$$

here, the scalar unknown b_k is a multiplicative factor which plays no role, since we just want a direction. This system is diagonal and its solution is straightforward: $\alpha_j = b_k/|g_j|^2$. Actually, the following recursive computation is more handy:

Theorem 5.5. *The sequence $\{d_k\}$ of directions is given by:*

$$d_1 = -g_1 \text{ for } k = 1, \quad \text{then}$$

$$d_{k+1} = -g_{k+1} + c_k d_k \quad \text{with} \quad c_k = \frac{|g_{k+1}|^2}{|g_k|^2}.$$

Proof. We case of d_1 is clear. For $k > 1$, we have just seen that $d_k = b_k \sum_{j=1}^{k} g_j/|g_j|^2$. Then choose $b_k = -|g_k|^2$, hence:

$$d_k = -g_k - |g_k|^2 \sum_{j=1}^{k-1} \frac{g_j}{|g_j|^2} = -g_k - |g_k|^2 \frac{d_{k-1}}{b_{k-1}} = -g_k + |g_k|^2 \frac{d_{k-1}}{|g_{k-1}|^2}. \quad \square$$

Now for the stepsize: we must minimize f along d_k, i.e. make an exact line-search. With a quadratic f, this is straightforward: solve for t

$$0 = (g(x_k + td_k), d_k) = (g_k + tAd_k, d_k) = (g_k, d_k) + t(Ad_k, d_k)$$

and obtain $t_k = -\dfrac{(g_k, d_k)}{(Ad_k, d_k)}$.

5.4 The Algorithm Seen as an Orthogonalization Process

Forget now the minimization aspect, and consider a problem from linear algebra. Let A be a given symmetric positive definite matrix, g_1 a given vector. We want to construct two sequences $\{g_k\}$ and $\{d_k\}$ by the formula

$$g_{k+1} = g_k + t_k Ad_k, \quad d_{k+1} = -g_{k+1} + c_k d_k$$

and we want to have, for $i \neq j$,

$$(g_i, g_j) = 0 \quad \text{and} \quad (d_i, Ad_j) = 0$$

(the connection with the previous sections is obvious).

Theorem 5.6. *Just take* $d_1 = -g_1$, *then*

$$t_k = -\frac{|g_k|^2}{(g_k, Ad_k)}, \qquad c_k = \frac{(Ad_k, g_{k+1})}{(Ad_k, d_k)}$$

and stop as soon as $g_{k+1} = 0$.

Proof. Suppose recursively that the desired properties hold until order k:

$$g_k \neq 0, \quad d_k \neq 0$$

and (if $k > 1$),

$$(g_k, g_i) = 0, \quad (d_k, Ad_i) = 0 \qquad i = 1, \ldots, k - 1.$$

We need to establish these properties at the order $k + 1$.

First observe that $(d_k, Ad_k) > 0$, hence c_k exists; t_k as well, because: first, $(g_1, Ad_1) = -(g_1, Ag_1) < 0$, and then

$$(g_k, Ad_k) = (-d_k + c_{k-1}d_{k-1}, Ad_k) = -(d_k, Ad_k) < 0.$$

Now g_{k+1} is orthogonal to the previous gradients. Indeed $(g_{k+1}, g_k) = 0$ by the choice of t_k. Then, for $i < k$ (if $k > 1$),

$$(g_{k+1}, g_i) = (g_k, g_i) + t_k(Ad_k, g_i) = t_k(Ad_k, g_i).$$

If $i = 1$ OK. Otherwise we obtain $(g_{k+1}, g_i) = t_k(Ad_k, -d_i + c_{i-1}d_{i-1}) = 0$ from the recursion assumption.

Then d_{k+1} is conjugate to the previous directions. Indeed $(d_{k+1}, Ad_k) = 0$ by the choice of c_k; and for $i < k$ (if $k > 1$),

$$(d_{k+1}, Ad_i) = -(g_{k+1}, Ad_i) + c_k(d_k, Ad_i)$$
$$= -(g_{k+1}, Ad_i) = -(g_{k+1}, \frac{g_{i+1} - g_i}{t_i})$$

(note that $t_i \neq 0$) which is 0 from the first part of the proof.

Finally $d_k \in U_k''' = U_k''$ by definition (Theorem 5.2). From the first part of the proof, $g_{k+1} \perp U_k''$. Therefore $(g_{k+1}, d_k) = 0$ and we conclude $(g_{k+1}, d_{k+1}) = -|g_{k+1}|^2 \neq 0$.   □

Remark 5.7. This proof reveals the key-elements of the algorithm:
- the choice of t_k and of c_k simply guarantees conjugacy of the last two pairs of vectors: $(g_{k+1}, g_k) = (d_{k+1}, Ad_k) = 0$;
- for the preceding pairs, the recursion is transmitted by itself from k to $k+1$ without any other assumption;
- nevertheless, taking $d_1 = -g_1$ is necessary to start the recursion $(g_{k+1}, g_1) = t_k(Ad_k, g_1) = 0$.   □

Corollary 5.8. *With the notation of §5.1, suppose that g_1 has the form $g_1 = Ax_1 + b$. In the algorithm above, construct $x_{k+1} = x_k + t_k d_k$. The conjugate gradient method is obtained; in particular, $g_{k+1} \perp U_k$.*

Proof. Everything is rather clear: the strictly convex function f has a unique minimum point x_{k+1} in V_k. The sequence $\{x_k\}$ constructed by conjugate gradient is unambiguously defined. In view of Theorem 5.3, this sequence can also be constructed as in Theorem 5.6.   □

Remark 5.9. Uniqueness of the sequence $\{x_k\}$ implies in particular that c_k and t_k are in turn defined unambiguously; yet, Theorem 5.6 gives seemingly different values for t_k and c_k (see Theorem 5.5 and the expression $-(g_k, d_k)/(d_k, Ad_k)$ of the optimal stepsize). We leave it as an exercise to check that the various orthogonality relations reconcile all of these different values.   □

5.5 Application to Non-Quadratic Functions

In addition to the initial motivation of conjugate gradient, Theorems 5.5, 5.6 give an interesting way of computing the direction, in a general descent algorithm. In view of the preceding developments, the following descent scheme can be applied to a general (non-quadratic) function:

Algorithm 5.10 (Nonlinear conjugate gradient).
Step 0 (Initialization). $x_1 \in \mathbb{R}^n$ and $\varepsilon > 0$ are given; set $k = 1$.
Step 1 (Stopping test). Compute $g_k = g(x_k)$; if $|g_k| \leqslant \varepsilon$ stop.
Step 2 (Computing the direction). If $k = 1$ set $d_k = -g_k$;
  otherwise compute c_{k-1} and $d_k = -g_k + c_{k-1} d_{k-1}$.
    If $(d_k, g_k) < 0$ go to Step 3. Otherwise set $d_k = -g_k$.
Step 3. Line-search along d_k to obtain $t_k > 0$.
Step 4 (Loop). Set $x_{k+1} = x_k + t_k d_k$; increase k by 1 and go to Step 1.   □

At Step 2, f is pretended to be quadratic and the direction is computed accordingly. This direction is then less "Markovian" than the pure gradient: it depends on the preceding information, mimicking what is done in quasi-Newton methods. The direction thus obtained is "restarted" on the gradient if it appears to be uphill; note, however, that this would never happen if the line-searches were exact, since in this case

$$(d_k, g_k) = (-g_k, g_k) + c_{k-1}(d_{k-1}, g_k) = -|g_k|^2 .$$

Remark 5.11. Of course, exact line-searches are impossible in the non-quadratic case. In a (nonlinear) conjugate-gradient context, some comments can be made about the line-search.

– First, it can be considered as being a particular instance of the general problem studied in Chap. 3;
– but one can consider that the one-dimensional minimization of f is an important thing in the development of conjugate gradients (we used continually the property $(d, g_+) = 0$);
– however, this argument is not completely clear, since conjugacy has so little meaning in the non-quadratic case.

Indeed the question is still controversial... and probably pointless, as will be seen later in §§5.6, 6.3.   □

Consider now the question of computing c in Algorithm 5.10. Two possible values were given in Theorems 5.5 and 5.6. For example, we can simply choose $c_{k-1} = |g_k|^2/|g_{k-1}|^2$, which results in the so-called *Fletcher-Reeves* method (F-R). However, this technique is too heavily based on the quadratic character of the objective function and on the starting axiom: not even speaking of exact line-searches, minimizing f (non-quadratic) in the manifold V_k is now out of question. Nevertheless, the mean-value formula says that

$$g(x_{k+1}) = g(x_k) + t_k A_k d_k$$

where A_k is the average Hessian between x_k and x_{k+1}:

$$A_k = \int_0^1 f''(x_k + st_k d_k)\mathrm{d}s.$$

Extrapolating the proof of Theorem 5.6, it can be thought (a dary thought, indeed) that the most important thing is to conjugate at least d_k and d_{k+1} with respect to A_k, which will occur if we take $c_k = (A_k d_k, g_{k+1})/(A_k d_k, d_k)$. But A_k is unknown! The following result is therefore used:

Theorem 5.12. *With A_k as above, $(d_k, A_k d_{k+1}) = 0$ providing that*

(i) $c_k = \dfrac{(g_{k+1} - g_k, g_{k+1})}{|g_k|^2}$ *and*

(ii) the $(k-1)^{\text{st}}$ and k^{th} line-searches are exact.

Proof. It suffices to show that the value of c_k given in Theorem 5.6 gives the present expression. By definition of A_k,

$$\frac{(A_k d_k, g_{k+1})}{(A_k d_k, d_k)} = \frac{(g_{k+1} - g_k, g_{k+1})}{(g_{k+1} - g_k, d_k)}.$$

If t_k is optimal, then $(g_{k+1}, d_k) = 0$ and the denominator is $-(g_k, d_k)$; but if t_{k-1} is optimal, then $(g_k, d_k) = (g_k, -g_k + c_{k-1}d_{k-1}) = -|g_k|^2$.   □

The value $c_k = (g_{k+1} - g_k, g_{k+1})/|g_k|^2$ is called the *Polak-Ribière* formula (P-R). The comparative merits of both formulae (F-R and P-R) give birth to a paradox, which is not uncommon in applied mathematics, and which illustrates some of the difficulties encountered in this science.

(i) It can be proved that Fletcher-Reeves converges globally (the argument is that of Theorem 3.9: the cosine of the angle between d and g does not tend to 0 too fast).

(ii) A counter-example exists, where Polak-Ribière does not converge, in the sense that no cluster point of the sequence $\{x_k\}$ is stationary (however we mention that the global convergence of P-R can be proved if f is locally elliptic: the cosine is bounded from below by the condition number of f'').

(iii) Yet, it is well-known that P-R behaves much better (i.e. converges perceivably faster) than F-R, which is no longer used by anybody. Altogether, P-R is much better in practice than F-R, while the contrary prevails theoretically.

Luckily, the next section suggests (and §6.3 confirms) that this difficulty has little importance anyway: indeed there exists a much better motivated formula, which in addition lends itself to fruitful generalizations.

5.6 Relation with Quasi-Newton

Consider again quasi-Newton methods of Chap. 4, where the directions were given by $d_k = -W_k g_k$. It can be shown that most conceivable quasi-Newton formulae produce directions mutually conjugate in the "perfect" case, i.e. when f is quadratic and the line-searches are exact. These methods therefore also consist in minimizing a quadratic function in the manifold V_k cumulating all past information; a connection is thus established between Chaps. 4 and 5. Notwithstanding, the real motivation of quasi-Newton (to approximate the Hessian matrix) is much richer than minimizing a quadratic function. Actually, a much more interesting connection exists between the two chapters, which has important numerical consequences.

The next result uses notation from Chap. 4: $s = x_{k+1} - x_k$, $y = g_{k+1} - g_k$ and the subscript "+" denotes the $(k+1)^{\text{st}}$ iterate.

Theorem 5.13. *At iteration* k, *suppose the line-search is exact:* $(g_+, d) = 0$, *and define the matrix*

$$W_+ = I - \frac{sy^T + ys^T}{(y,s)} + \left[1 + \frac{|y|^2}{(y,s)}\right]\frac{ss^T}{(y,s)}.$$

Then the direction $d_+ := -W_+ g_+$ *is* $d_+ = -g_+ + \dfrac{(y, g_+)}{(y, d)} d.$

Proof. Trivial: $(g_+, s) = t(g_+, d) = 0$, hence

$$d_+ = -g_+ + (y, g_+)\frac{s}{(y,s)} + 0 - 0 = -g_+ + (y, g_+)\frac{d}{(y,d)}. \qquad \square$$

This establishes a close link with Chap. 4:

- The matrix W_+ defined in this theorem is nothing other than BFGS after a reinitialization of W to the identity. Said otherwise: W_+ is the result of the BFGS formula (§4.4) applied *one time* to the identity, with the most recent pair $\{y, s\}$; while a standard quasi-Newton matrix is the result of the formula applied k *times* to the identity, with the sequence of pairs $\{s_i, y_i\}$, $i = 1, \ldots, k$. Remember the end of Remark 4.6.
- The "poor man" quasi-Newton direction of Theorem 5.13 is just a conjugate-gradient direction, namely that of Theorem 5.6.

Note that these relations are valid even when f is not quadratic: it suffices that the line-search be exact at the present k^{th} iteration. Now, an exact line-search cannot be tolerated in the non-quadratic case, but after all, it can perfectly be avoided: considering that the formulae from Theorems 5.5 and 5.6 are hardly justified in the non-quadratic case, they can simply be forgotten, to the advantage of the little more complicated formula of Theorem 5.13. It is an appropriate time to remember an important property of quasi-Newton formulae:

Theorem 5.14. *If $(g_+, d) > (g, d)$, the direction d_+ of Theorem 5.13 is a descent direction.*

Proof. Immediate consequence of Theorem 4.5: here $M = I$ is positive definite, and the assumption guarantees $(y, s) > 0$. It follows that M_+ and W_+ are positive definite. □

Then a straightforward idea is to define a variant of Algorithm 5.10: to compute the direction of Step 2 as in Theorem 5.13 (no reinitialization is necessary), and to use in Step 3 Wolfe's line-search (§3.4) which, we recall, guarantees $(g_+, d) > (g, d)$.

Thus, conjugate gradient is helped by quasi-Newton theory; but the converse is also true. This will be seen in §6.3, where the results above will even be enhanced, to define methods combining the two advantages: moderate memory requirement of conjugate gradient, and fast convergence of quasi-Newton. We will also see in §6.4 another aspect of the help brought by conjugate gradient to Newtonian methods: to allow a cheap resolution of the linear system $f''d = -g$.

Bibliographical Comments

Conjugate gradient is due to Hestenes et Stiefel in the paper [193] which, despite its age, still contains a lot of material worth meditating. The more recent book of Hestenes [192] deserves the same comment. Remembering that it is really a method belonging to the realm of linear algebra, [170] is a must. In particular, this last reference contains some material on *preconditioned conjugate gradient*, an important technique to solve large linear systems of equations.

For the non-quadratic case, the implementation considered best is [329]. The convergence proof of Fletcher-Reeves can be found in [376], and [294] contains a counter-example for which Polak-Ribière does not converge. For these questions, we mention [153], where a convergent variant of Polak-Ribière is proposed; its basic idea is to restart on the gradient when $c_k \leqslant 0$; besides, [2] proposes a variant of Wolfe's line-search, well suited to Fletcher-Reeves: global convergence is conserved and restarts are not necessary.

The formula given in Theorem 5.13 is commonly attributed to Perry, see [328].

6 Special Methods

This chapter is mostly devoted to methods which, although less "universal" than the preceding, are useful in a good number of cases. The first one (trust-region) is actually extremely important, and might supersede line-searches, sooner or later. The other methods deal with the direction; they are either classical (Gauss-Newton) or recent (limited-memory quasi-Newton, truncated Newton) and apply only in some well-defined subclasses of problems. The chapter also outlines the standard approach to solve quadratic optimization problems (item 2.2.2 in the classification of §1.1.2).

Up to now, we have extensively used the notation d, a minimization iteration being done according to the formula $x_+ = x + td$. In the present chapter, the concept of direction tends to disappear; most of the time, we will denote by h a step between two points such as x and x_+: $x_+ = x + h$.

6.1 Trust-Regions

Our first method is a variant of line-searches, particularly well-suited in a Newtonian context; it appeared in the context of nonlinearly constrained optimization. Take again the situation of Chap. 3 in a Newtonian context (Chap. 4). Starting from the current iterate $x = x_k$, one wishes to go to a supposedly excellent estimate x^N, obtained by the minimization of a model representing f in a neighborhood of x. Call $\tilde{f}$ this model; we have seen so far quadratic models, say

$$\tilde{f}(x + h) := f(x) + (g, h) + \frac{1}{2}(Mh, h);\qquad(6.1)$$

here $M = M_k$ can be the Hessian $f''(x_k)$, or a quasi-Newton approximation, or the Gauss-Newton approximation as in §6.2; in the next chapters, we will see situations where $\tilde{f}$ is not quadratic. Anyhow, we set $x^N = x + h^N$, where h^N minimizes $\tilde{f}(x + \cdot)$.

The fact of interest here is that the desired iterate x^N may not be convenient, and must perhaps be corrected; in all preceding chapters, this was the task of the line-search. Most often, x^N is not convenient because $x^N - x = h^N$ is too large; in a Newtonian method, for example, this could occur when

– the model is valid only for small h, and we have $f(x^N) \geqslant f(x)$,
– and/or the Hessian $M = f''(x)$ is not positive definite, which destroys any meaning of h^N.

It may also happen that h^N is too small: for example, in a quasi-Newton method, the initial matrix M_1 may be too large; since M_+ and M are close together (see §4.4), M_k will remain too large for a number of iterations.

In such cases, the idea of a line-search was to seek x_+ along the half-line $x + \mathbb{R}_+ h^N$; what is magic in this half-line? Answer; nothing. This idea was in fact inherited from first-order methods, where the direction was defined up to a positive constant (compare Remark 2.7 and the end of §4.2). Indeed, rather than shortening the stepsize, it is much more natural to modify the model; after all, the reason x^N is not convenient is that $\tilde{f}$ itself is not convenient. Based on this idea, one can proceed as follows:

– Perturb the model $\tilde{f}$ in a certain way (see below), the perturbation depending on a certain parameter Δ.
– Solve (6.1) and obtain a perturbed solution h_Δ.
– Adjust Δ so that this solution h_Δ is "convenient".
– This adjustment is made according to the principles developed in Chap. 3.

For a demonstration, let us show that line-searches of Chap. 3 can be recovered from these premises. The model being given by (6.1) with M positive definite, suppose that we perturb $\tilde{f}$ to

$$\tilde{f}_\Delta(x + h) := f(x) + (g, h) + \frac{1}{2\Delta}(Mh, h),$$

which gives $x_\Delta = x - \Delta M^{-1} g$. To adjust Δ (the stepsize! setting $\Delta = t$ and $h = d$ makes the allusion more blatant) so as to obtain an $x_+ = x_\Delta$ satisfying Wolfe's rule gives back the standard line-search of §3.4.

6.1.1 The Elementary Problem

There are several ways of perturbing the model, all more or less equivalent anyway. We limit ourselves to the one that is now universally adopted: the next iterate is forced into the ball centered at x and of radius Δ (the trust-region). In other words, h_Δ solves

$$\min_h \tilde{f}(x + h), \quad |h| \leqslant \Delta. \tag{6.2}$$

Here $\tilde{f}$ is the unperturbed model, for example the quadratic function (6.1); instead of perturbing the model, we rather restrict it.

It so happens that (6.2) can be efficiently solved, even when M is not positive definite; let us give an idea of how this is done. For $\mu \geqslant 0$ such that $M + \mu I$ is positive semi-definite (i.e. $\mu \geqslant \max\{0, -\lambda\}$, if λ is the smallest eigenvalue of M), consider the Lagrangian

$$\ell(h, \mu) = \tilde{f}(x + h) + \frac{\mu}{2}(|h|^2 - \Delta^2).$$

For fixed μ large enough, ℓ has a minimum point h, given by

$$h = h(\mu) := -(M + \mu I)^{-1}g. \tag{6.3}$$

Then the computation of h_Δ calls for duality theory (see §8.2 and §16.2). This theory predicts the value μ^* giving, via (6.3), the $h(\mu)$ solving (6.2). The following result is slightly informal, in that it neglects the (exceptional) case $\mu^* = -\lambda$.

Theorem 6.1. *The solutions to* (6.2) *are those* $h(\mu^*)$ *of* (6.3) *satisfying* $|h(\mu^*)| \leqslant \Delta$, *where* $\mu^* \geqslant \max\{0, -\lambda\}$ *is such that* $\mu^*(|h(\mu^*)| - \Delta) = 0$. *Besides, the function* $\mu \mapsto \ell(h(\mu), \mu)$ *is concave, infinitely differentiable, and maximal at* μ^*.

Proof. Omitted: it uses results from Chap. 8.2. Essentially, equation (6.3) (supposed to have a solution) expresses the stationarity of the Lagrangian ℓ, and the condition on μ is transversality.   □

It must be mentioned that the normalization cannot be arbitrary in this result. Luckily, the Euclidean normalization plays a privileged role, independently of the particular form of the model:

Proposition 6.2. *Suppose that* f *and* $\tilde{f}$ *coincide to first order in a neighborhood of* x: $\tilde{f}(x + h) = f(x + h) + o(|h|)$. *When* $\Delta \to 0$, *the direction* h_Δ *tends to* $-g$ *(up to the normalization).*

Proof. Remember §2.4: when $\Delta \to 0$, (6.2) defines the steepest-descent direction of $\tilde{f}$ (hence of f) associated with $|\cdot|$; and by assumption, $|\cdot|$ is the Euclidean norm defining the gradient.   □

Taking Theorem 6.1 into account, the actual computation of h_Δ reduces to a search on $\mu \geqslant 0$. This search can be viewed either as the resolution of the equation $|h(\mu)| = \Delta$, or as the maximization of the concave function $\ell(h(\mu), \mu)$. In both cases, Newton does the trick, via a differentiation of (6.3). Normally, one solves rather $1/|h(\mu)| - 1/\Delta = 0$, which seems a better conditioned problem.

6.1.2 The Elementary Mechanism: Curvilinear Search

Considering now the computation of h_Δ as solved, let us see how Δ can be adjusted, following the general principles of Chap. 3. The merit function called there $t \to q(t) = f(x + td)$, with d fixed, is now $\Delta \to f(x + h_\Delta)$, which implies a few differences with respect to the "line"-search:

(i) The trajectory $\{x + td\}_{t>0}$ becomes $\{x + h_\Delta\}_{\Delta>0}$, which is no longer a half-line but rather a curve in the space $\mathbb{R}^n$; we will rather speak of a *curvilinear* search.

(ii) If $\tilde{f}$ has a minimum point at finite distance, this trajectory has in turn a finite length. In case of extrapolations, there is then a difficulty: h_Δ eventually stops at a point minimizing $\tilde{f}(x + \cdot)$, even if $\Delta \to +\infty$.

(iii) The initial derivative $q'(0)$ becomes impossible to compute; the test $q(t) \leqslant q(0) + m_1 t q'(0)$, central in §§3.4, 3.5, is now impossible to implement (incidentally: it may even happen that $g = 0$ hence $q'(0) = 0$, while $h = 0$ does not solve (6.2); one of the strong points of trust-region, precisely, is the ability of escaping from stationary points which are not local minima). Here comes the time to remember the remarks at the end of §3.5.3: $tq'(0)$ and $\frac{1}{2}tq'(0)$ were interpreted as nominal decreases of f; here, a nominal decrease can be simply taken as the decrease of the model:

$$\delta(\Delta) := \tilde{f}(x) - \tilde{f}(x + h_\Delta) \left[= f(x) - \tilde{f}(x + h_\Delta)\right]. \tag{6.4}$$

(iv) In addition to $q'(0)$ being unknown, $q'(t)$ becomes $\frac{d}{d\Delta}f(x+h_\Delta)$, a number which can be computed but which has little meaning. As a result, Wolfe's rule (§3.4) is no longer so natural.

In view of all this, Armijo's rule (§3.5.2) makes life simpler and is commonly adopted. In particular, this rule excludes any extrapolation and accommodates easily a trajectory $\{x + h_\Delta\}$ of finite length. However, Remark 3.11 is still valid: a mechanism is needed to force an increase of Δ in case of necessity; the algorithm loses some purity. Using another rule (Wolfe, Goldstein and Price) would be possible, but to the price of an elaborate study of the mapping $\Delta \mapsto h_\Delta$; we skip this here, for lack of space.

In summary, the following algorithm is a very simple implementation of the trust-region technique. Armijo's test uses (6.4) and becomes

$$f(x + h_\Delta) \leqslant f(x) - m\delta(\Delta).$$

When it is not satisfied, h_Δ is too large, Δ must be decreased.

Algorithm 6.3. Are given: a starting point x, a model $\tilde{f}$, an Armijo coefficient $m > 0$, an initial size Δ of the trust-region, and a simulator computing f-values.
  – Compute h_Δ solving (6.2).
  – If $f(x + h_\Delta) \leqslant f(x) - m\delta(\Delta)$ terminate.
  – Else decrease Δ and loop.   □

We leave it to the reader to show that this algorithm terminates if the following properties hold: the decrease of Δ is significant (PROPERTY E of §3.2), $\tilde{f}(x) = f(x)$, $\tilde{f}'(x) = f'(x)$, and $m < 1$.

Note that Algorithm 6.3 considers the static aspect only: its calls must be chained within an algorithm updating $x = x_k$, to finally minimize f. Recall here Remark 3.11 again: a systematic initialization $\Delta = 1$ is dangerous. It is therefore recommended to allow an increase of Δ when the search is finished, to initialize the next search. We omit these details here.

6.1.3 Incidence on the Sequence x_k

It remains to check that this method does have the qualities ensuring convergence (global and possibly superlinear) of the sequence $\{x_k\}$ that it generates. The rudimentary Algorithm 6.3 gives only a vague idea of an actual implementation; in particular, it does not specify the management of $\Delta = \Delta_k$ from one iteration k to the next. We will therefore content ourselves with some indications.

Apart from the strategy to manage Δ, a trust-region method is essentially characterized by the model $\tilde{f}$ (just as any optimization method, actually). Limiting our convergence study to one particular $\tilde{f}$, the quadratic model (6.1), everything relies upon the behaviour of $M = M_k$. To study the method, one must of course express that $x_+ = x + h_\Delta$ is not arbitrary, but really a minimum point of $\tilde{f}$ in the trust region. This is the substance of the following result, comparing $\tilde{f}(x + h_\Delta)$ to what would be obtained with the steepest-descent direction.

Lemma 6.4. *Set $\tilde{L} := \max\{1, \Lambda\}$, where Λ is the largest eigenvalue of M in (6.1). Then we have in (6.4)*

$$\delta(\Delta) \geqslant \frac{1}{2}|g| \min\left(\Delta, \frac{|g|}{\tilde{L}}\right).$$

Proof. Let t_Δ be the optimal solution to the problem

$$\min_t \tilde{f}(x - tg) = f(x) - t|g|^2 + \frac{1}{2}t^2(g, Mg) \quad \text{subject to} \quad 0 \leqslant t|g| \leqslant \Delta.$$

Of course, $\tilde{f}(x + h_\Delta) \leqslant \tilde{f}(x - t_\Delta g)$, and we proceed to bound $\tilde{f}(x - t_\Delta g)$ from above. In the computations below, we use $\tilde{f}(x) = f(x)$.

If the trust region does constrain the solution t_Δ, the function $t \mapsto \tilde{f}(x - tg)$ has a nonpositive derivative at $t_\Delta = \Delta/|g|$; in other words $-|g|^2 + \frac{\Delta}{|g|}(Mg, g) \leqslant 0$. This results in a bound on (Mg, g) which, inserted in $\tilde{f}$, gives $\tilde{f}(x - t_\Delta g) \leqslant f(x) - \frac{1}{2}\Delta|g|$.

If t_Δ is an unconstrained minimum, its value is $t_\Delta = |g|^2/(Mg, g)$; the denominator is certainly positive. Since $(Mg, g) \leqslant \Lambda|g|^2 \leqslant \tilde{L}|g|^2$, we can write

$$\tilde{f}(x - t_\Delta g) = f(x) - \frac{1}{2}\frac{|g|^4}{(Mg, g)} \leqslant f(x) - \frac{1}{2\tilde{L}}|g|^2.$$

The formula follows from piecing together the two cases. □

With the help of this result, Armijo's rule gives directly

$$[f(x_+) =] f(x + h_\Delta) \leqslant f(x) - \frac{m}{2}|g| \min\left(\Delta, \frac{|g|}{\bar{L}}\right),$$

which is the necessary bound of $f(x) - f(x_+)$ for Argument 1 in a convergence proof (see Theorems 2.11 and 3.9). We therefore see that, to obtain this bound, an upper bound of the largest eigenvalue of M is needed. The situation is no longer as in line-searches, where a bound on the ratio of extreme eigenvalues was needed (see for example Lemma 3.8). By contrast, trust-regions would still work with $M_k \equiv 0$ (and it is normal since this would produce steepest-descent!)

To obtain Argument 2 (bounding $|x_+ - x|$ from below), a lower bound of Δ is needed. This means that the model should not be too "optimistic": more precisely, the smallest eigenvalue of M should not be too negative; everything will go right if, for example, M stays positive semi-definite.

We stop technicalities here and we just mention the main convergence results allowed by trust-regions. With an appropriate (but rudimentary) management of $\Delta = \Delta_k$, and assuming that $|M_k|$ stays bounded, it can be shown that:

(i) either $f(x_k) \to -\infty$ or $\liminf |g_k| = 0$,

(ii) and the criterion of Dennis-Moré (Theorem 4.11) implies superlinear convergence.

Nothing particularly amazing so far; but trust-regions enjoy two "pluses":

(iii) indeed $g_k \to 0$; and more importantly:

(iv) take $M_k = f''(x_k)$; if $\{x_k\}$ is bounded, it has a cluster point x^* such that $(f'(x^*) = 0$ and) $f''(x^*)$ is positive semi-definite. Thus, trust-region methods have every reason to avoid critical points which are not local minima. Needless to say, this "miracle" has its roots in Theorem 6.1.

6.2 Least-Squares Problems: Gauss-Newton

It should be clear from Chap. 4 that the name of the game in the present Part I is to approximate the Newton step $-(f'')^{-1}g$, i.e. after all the Hessian f''; this was the essence of §§4.4 to 4.7. There exists a rather favourable situation, where an idea of this Hessian is available, without having to compute second derivatives. This is *parameter identification* – the vast majority of optimization problems. In this type of problems, the objective function is the deviation between predictions and observations, a deviation which is often expressed in the ℓ_2-norm.

In a word, the following *least-square problem* is continually encountered in optimization:

$$\min f(x), \quad \text{where} \quad f(x) = \frac{1}{2}\sum_{j=1}^{p} f_j^2(x). \tag{6.5}$$

The f_j's generally have the form

$$f_j(x) = \varphi(x, e_j) - z_j,$$

where φ is the answer of a theoretical model to an input e, this answer being compared to a set of experimental answers z_j. The theoretical model depends (possibly linearly) on unknown parameters $x \in \mathbb{R}^n$ to be identified.

Elementary calculations give the gradient $f' = \sum f_j f'_j$ and the Hessian

$$f''(x) = \sum_{j=1}^{p} f'_j(x) f'_j(x)^\top + \sum_{j=1}^{p} f_j(x) f''_j(x)$$

which is thus a sum of two terms; the first one depends on the gradients, only the second one involves second derivatives. Remembering that one of the drawbacks of Newton's method is the need to compute such second derivatives, it is tempting to neglect this second term.

In least-square problems, one is therefore led to considering

$$G(x) := \sum_{j=1}^{p} f'_j(x) f'_j(x)^\top = J(x) J(x)^\top, \qquad (6.6)$$

where $J(x)$ denotes the matrix whose columns are the p gradients f'_j; $G(x)$ is called the *Gauss-Newton* matrix. A question is then: to what extent is $G(x)$ a good approximation of $f''(x)$? Actually, this will happen in two cases.

– When the f_j's are mildly non-affine (each $\varphi(\cdot, e_j)$ is mildly nonlinear), the f''_j's are quasi-constant, the f'''_j's are small. Then $G(x) \simeq f''(x)$. Indeed, the whole idea of Gauss-Newton consists in saying: suppose the f_j's are affine, i.e. suppose the initial problem is a mere linear least-square problem $\min \frac{1}{2}|J^\top x - z|^2$; then it is solved by the Newton iterate $x - (JJ^\top)^{-1} g(x)$ (note that $g(x) = J(J^\top x - z)$). This amounts to assuming that the matrix $G(x) = JJ^\top$ does not depend on x – and is invertible, of course.

– When the least-square solution gives a good fit, the optimal value of f is close to 0. Then all the $f_j(x)$'s are small within convergence and, asymptotically, the Gauss-Newton matrix becomes close to the Hessian. This is all what is needed, since the qualities of Newton's method come into play within convergence only.

Anyhow, the Gauss-Newton method consists in computing the direction $-G_k^{-1} g_k$ at each iteration k. A (Wolfe) line-search then allows the computation of the next iterate; the resulting algorithm is well-defined, provided that each G_k is positive definite.

Actually, the difficulty precisely lies at this point.

– When a matrix G_k is "hardly" positive definite, i.e. ill-conditioned, the method is in trouble; convergence may even be impaired (see §4.5).

– For this same reason, the line-search itself can often be in trouble: the Gauss-Newton direction can be orthogonal to the gradient, at least within roundoff errors.

Indeed, the designers of the method (neither Gauss nor Newton but Levenberg in 1944 and Marquardt in 1963) were aware of these difficulties due to ill-conditioning. They imagined, independently of each other, a stabilizing process: add to G_k a "suitably chosen" multiple $\lambda > 0$ of the identity. In view of §6.1.1 (and setting $\lambda = \mu$), we see that this is nothing other than a trust-region device. In summary: the Gauss-Newton method is dangerous when used with a line-search, and calls for trust-region. This gives an algorithm of the following type:

Algorithm 6.5. To solve (6.5), start from x_1; are given: the Armijo coefficient $m \in]0, 1[$ and the stopping tolerance $\varepsilon > 0$. Set $k = 1$.

Step 1. Compute the gradient g_k of f at x_k. If $|g_k| \leqslant \varepsilon$ stop.

Step 2. Compute the Gauss-Newton matrix $G_k = G(x_k)$ from (6.6) and define the model

$$\tilde{f}(x_k + h) = f(x_k) + (g_k, h) + \frac{1}{2}(G_k h, h).$$

Step 3. Use a trust-region algorithm as in §6.1 to obtain x_{k+1} satisfying

$$f(x_{k+1}) \leqslant f(x_k) - m(f(x_k) - \tilde{f}(x_{k+1})).$$

Step 4. Increase k by 1 and loop to Step 1.   □

Convergence of such an algorithm is not difficult to establish: through some smoothness of f, the matrices G_k, which are positive semi-definite, will be bounded, and results from §6.1.3 will be applicable.

6.3 Large-Scale Problems: Limited-Memory Quasi-Newton

Except possibly for least-square problems, the idea of quasi-Newton is basically the only way of computing really efficient directions. For a number of variables n exceeding 10^5, say, storing and managing the corresponding matrix W exceeds the abilities of present computers. In this situation, we have only seen conjugate gradient as a possible alternative, while §5.5 has somehow ended up in a deadlock. Yet, Theorem 5.13 suggested a fruitful link: conjugate gradient can be viewed as a "poor man" quasi-Newton method, with systematic re-initializations of W_k to the identity.

Let us push the idea further: a quasi-Newton matrix W_k is completely defined by (the initial matrix W_1 and) the $2(k-1)$ vectors $s_1, y_1, \ldots, s_{k-1}, y_{k-1}$. To be sure, a means therefore exists to compute the direction $d_k = -W_k g_k$ via an explicit use of only these $2(k-1)$ vectors, which makes $2(k-1)n$ numbers, instead of $n(n+1)/2$. Formulae do exist, it is easy to check that the following algorithm does the trick.

Algorithm 6.6. Are given: an initial matrix W^0, a vector g and m vector pairs $\{s_i, y_i\}$ for $i = 1, \ldots, m$.

Problem: compute $d = -Wg$, where W is the result of m BFGS updates of W^0.

Answer: $d = -h^m$, obtained as follows.

– Set $q^m = g$; for $i = m, \ldots, 1$ do

$$\alpha^i = \frac{(q^i, s_i)}{(y_i, s_i)} \quad \text{and} \quad q^{i-1} = q^i - \alpha^i y_i \,.$$

– Set $h^0 = W^0 q^0$; for $i = 1, \ldots, m$ do

$$\beta^i = \frac{(y_i, h^{i-1})}{(y_i, s_i)} \quad \text{and} \quad h^i = h^{i-1} + (\alpha^i - \beta^i) s_i \,. \qquad \Box$$

Remark 6.7. We use superscripts (q^i, α^i, etc.) to suggest that the above iterations need not be related to the iterations minimizing f. For example, if Algorithm 6.6 is used to compute the kth direction d_k of such a minimization algorithm, then we will set $g = g_k$ and $d_k = -h^m$. Said otherwise, the m iterations of Algorithm 6.6 will be systematically executed at each iteration k of the minimization algorithm. $\qquad \Box$

Equipped with these formulae, take a computer allowing the storage of $2m$ vectors of dimension n; if $n = 10^5$ for example, a standard personal computer will easily accept $m = 10$. On this computer, start a quasi-Newton algorithm without explicit storage of the matrices W_k, but with a direct computation of $W_k g_k$ via Algorithm 6.6. The iterations $k = 1, \ldots, m$ can be performed normally. Afterwards, the "standard" matrix W_k can no longer be used: it needs $k > m$ pairs $\{s_i, y_i\}$; among these k pairs, m must be chosen. To respect the essence of quasi-Newton, which constructs a local model of f in a neighborhood of x_k, one chooses of course the last m computed pairs: normally, they were computed at corresponding x's close to x_k. This results in an algorithm intermediate between 4.3 and 5.10; its first m iterations are identical to those of Chap. 4, let us describe the current iteration after the mth:

Algorithm 6.8.

Step 1 (initializing iteration k). We have the m pairs of vectors

$$s_1 = x_{k-m+1} - x_{k-m}, \ldots, s_m = x_k - x_{k-1},$$
$$y_1 = g_{k-m+1} - g_{k-m}, \ldots, y_m = g_k - g_{k-1}.$$

Choose a "simple" matrix W^0, for example the identity.

Step 2 (poor-man qN formulae). Compute the direction $d_k = -W_k g_k$ by an application of Algorithm 6.6 to the matrix W^0, starting from $g = g_k$.

Step 3. Obtain the next iterate x_{k+1} and its gradient g_{k+1} by a Wolfe line-search along d_k, or alternatively by a curvilinear search of the type 6.3.

Step 4 (refreshing the information). For $i = 1, \ldots, m-1$, replace each pair $\{s_i, y_i\}$ by $\{s_{i+1}, y_{i+1}\}$. Store $s_m = x_{k+1} - x_k$ and $y_m = g_{k+1} - g_k$ and loop to Step 1.   □

An important numerical remark is the following. In standard quasi-Newton methods, the role of the initial matrix W_1 is not too crucial: in view of the successive updates, this role fades away along the iterations. Nevertheless, it is experimentally observed that a bad initial W_1 entails bad W_k's for many iterations. In a limited memory method such as Algorithm 6.8, we have an initial matrix $W^0 = W_k^0$ at each iteration k. According to the above experimental observation, W^0 must be chosen with great care; to content oneself with $W^0 = I$ at each iteration results in a bad algorithm. We do not elaborate here on appropriate initializations of W^0 in Algorithm 6.6.

Algorithm 6.8 is excellent; it is at present the best choice, often the only possible one, for large-scale problems. However, it raises a rather frustrating question. One could think that its convergence is faster when the number m of stored pairs is larger; yet, one empirically observes that its speed of convergence stalls beyond a modest value of m; for larger m, the total computing time worsens, since the cost of one iteration is proportional to m. Rather frequently, one sees Algorithm 6.8 with $m = 20$ terminating faster than the standard BFGS algorithm. Even worse: the "critical" value of m, beyond which the above stalling phenomenon occurs, seems relatively stable, say $m \simeq 20$, independently of n. No satisfactory explanation has been suggested so far.

From a theoretical point of view, Algorithm 6.8 has no reason to converge superlinearly. Global convergence has received little attention.

6.4 Truncated Newton

Let us stay in the Newtonian framework. On several occasions, it has been said that the pure Newton method has two practical drawbacks (among others):
- to compute second derivatives,
- to solve a linear system at each iteration.

In some cases, the first drawback may disappear: for example in least-square problems (§6.2), or also when second derivatives can easily be computed. In these cases, how can we cope with the second drawback? Here comes a usefulness of conjugate gradient.

Let M be the matrix of the system to be solved: M can be f'', or the Gauss-Newton matrix G_k, or any other symmetric matrix; M can even be a quasi-Newton matrix, possibly not positive definite because Wolfe's condition was not satisfied (see Theorem 4.5). We want to solve $Mh = -g$ economically.

Suppose for the moment that M is positive definite; then we can equally minimize the quadratic function $\tilde{f}$ of (6.1). Now make an important remark:

since a Newtonian method reaches full efficiency only asymptotically, only an *approximate* minimization of $\tilde{f}$ is necessary. As a result, conjugate gradient is fully justified, since it decreases the model $\tilde{f}$ at each of its own iterations: it will produce a "reasonable" direction even if its convergence is stopped "manually". In fact, call $h^i, i = 0, 1, \ldots$ the iterates produced by a descent method – conjugate gradient or any other – applied to the minimization of $\tilde{f}(x + h)$ (with respect to h). If this algorithm is initialized on $h^0 = 0$, there holds $\tilde{f}(x + h^i) < \tilde{f}(x + h^0) = f(x)$ at each iteration. From positive definiteness of M, there even holds

$$(g, h^i) < \tilde{f}(x + h^i) - f(x) < \tilde{f}(x + h^0) - f(x) = 0 \,.$$

Thus, each h^i is a descent direction, along which a (Wolfe) line-search can be performed. In case of conjugate gradient, h^i is even an excellent direction, minimizing $\tilde{f}$ in a certain hyperplane: remember the axiom of §5.1. In case $M = f''(x_k)$, one can also anticipate that superlinear convergence will be preserved, providing that the direction $d_k = h^{i_k}$ is close enough to the "ideal" Newton step $x^N - x_k$.

When M is not positive definite, minimizing $\tilde{f}$ becomes meaningless. In addition, while the hypothesis M positive definite is not crucial theoretically for conjugate gradient (exercise: redo Chap. 5 under the mere hypothesis A invertible), it is another story in practice: the method becomes unstable, with a tendency to produce isotropic directions (i.e. q's such that $(Aq, q) = 0$). This situation places us again in the field of application of trust-region.

Altogether, the situation is as follows: we want to solve (6.1), (6.2) approximately; but we do not want Theorem 6.1, which implies to factor matrices such as $M + \mu I$ (by assumption we do not want to solve $Mh = -g$, i.e. to factor M). Here comes a piece of luck: it suffices to apply blindly conjugate gradient, with an appropriate stopping test.

First of all, let us adopt notation adapted to the present situation. To solve the system $Mh + g = 0$, we write conjugate gradient in the form

$$h^{i+1} = h^i + t^i q^i \,, \quad q^{i+1} = -Mh^{i+1} - g + c^i q^i \,, \quad \text{with}$$

$$h^0 = 0, q^0 = -g \,, \quad t^i = -\frac{(Mh^i + g, q^i)}{(Mq^i, q^i)} \,, \quad c^i = \frac{|Mh^{i+1} + g|^2}{|Mh^i + g|^2} \,; \qquad (6.7)$$

q^i thus denotes the "direction" issued from h^i (but has nothing to do with the forthcoming direction d_k issued from x_k); remark that the residual $Mh^i + g$ is the gradient $\tilde{f}'(x + h^i)$.

Proposition 6.9. *As long as* $(Mq^i, q^i) > 0$*, the iterations of the above algorithm satisfy* $|h^{i+1}| > |h^i|$*.*

Proof. It is best to see the argument geometrically, remembering the axiom of §5.1: each line $h^i + tq^i$ is orthogonal at h^i to a subspace containing h^i (with notation as in §5.1, $U^i = V^i$ since $h^0 = 0$). Hence h^i is the projection of 0 onto such a line, therefore $|h^i| < |h^i + tq^i|$ for any $t \neq 0$, in particular for $t = t^i$. □

Then start solving (6.1), (6.2), via formulae (6.7). At the current iteration i, the following events can occur:

(i) The curvature (Mq^i, q^i) of $\tilde{f}$ in the direction q^i is negative: the constraint $|h| \leqslant \Delta$ is certainly active in (6.2). For example, h^i can then be extrapolated up to the edge of the trust region: we take $d_k = \Delta h^i / |h^i|$.

(ii) The current iterate h^i has popped out of the trust region: here again, the constraint $|h| \leqslant \Delta$ is certainly active in (6.2). For example, d_k can be interpolated between h^i and h^{i-1}.

(iii) None of the above cases: conjugate gradient proceeds normally, its iterations can be continued *ad libitum*, for example until

(iv) the residual $Mh^i + g$ is deemed small enough; then stop: $d_k = h^i$ estimates well enough the Newton step.

Remark 6.10. The initialization $h^0 = 0$ is important. Not only does it guarantee the downhill character of each h^i, but it also implies that h^1 minimizes $t \mapsto \tilde{f}(x - tg)$ in the trust region. This is important for convergence of $\{x_k\}$, see the proof of Lemma 6.4. □

We do not describe more the "complete" truncated Newton algorithm; first, there is no consensus yet on the best version, and the details are rather technical anyway. We just make two comments:

– The general qualities of trust-region are forwarded to the present method: global convergence via a rudimentary management of Δ, and convergence to a local minimum of f.

– The relevant property for superlinear convergence is that the residual go to 0 infinitely faster than the gradient: $f''(x_k)d_k + g_k = o(|g_k|)$.

6.5 Quadratic Programming

This section is devoted to quadratic programs (item 2.2.2 in the classification of §1.1.2): we demonstrate the *active set* mechanism used by standard quadratic solvers. To this aim, we consider the "simplest" (convex) quadratic program

$$\begin{cases} \min f(x), \\ x \geqslant 0, \end{cases} \quad \text{with } f(x) := \frac{1}{2}x^\top M x - b^\top x. \tag{6.8}$$

Here the symmetric $n \times n$ matrix M is assumed positive definite; this avoids a good deal of technicalities. Note at this point that the scalar product in $\mathbb{R}^n$ has to be the ordinary dot-product; otherwise the constraints $x \geqslant 0$ would have nothing special: they should be viewed just as general constraints $Ax \geqslant a$.

Following the general principles of optimization (descent property, line-searches etc.), the algorithm constructs a sequence of feasible points x_k such that $f(x_k)$ decreases. To do so, direct advantage is of course taken of the simple structure of the problem.

Denote by v^i the ith component of a vector $v \in \mathbb{R}^n$ and by

$$I(x) := \{i \in \{1,\ldots,n\} : x^i = 0\} \tag{6.9}$$

the set of components of $x \geqslant 0$ that cannot be decreased. The optimal solution of (6.8) is then *characterized* by the equations

$$\begin{cases} (Mx - b)^i = 0 & \text{if } i \notin I(x) \text{ (i.e. } x^i > 0) \text{,} \\ (Mx - b)^i \geqslant 0 & \text{if } i \in I(x) \text{,} \end{cases} \tag{6.10}$$

illustrated by Fig. 6.1. These equations are indeed necessary and sufficient for optimality. To obtain them,

– either form the gradient $\nabla f(x) = Mx - b$ of the objective function and apply standard optimality conditions (see (13.1); they are necessary and sufficient because of convexity);

– or establish by direct calculations the formula

$$f(x + h) - f(x) = (Mx - b)^\top h + \frac{1}{2} h^\top M h \,; \tag{6.11}$$

observe that the quadratic term is nonnegative and negligible for small h.

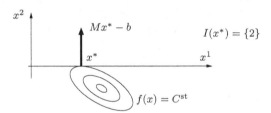

Fig. 6.1. The minimum of a quadratic function over the first orthant

The optimality conditions (6.10) can also be viewed as a *complementarity problem*, which is often symbolized in the condensed form

$$0 \leqslant (Mx - b) \perp x \geqslant 0 \,. \tag{6.12}$$

This latter problem is thus *equivalent* to the optimization problem (6.8).

6.5.1 The basic mechanism

Even though (6.8) is completely determined by the (finitely many) entries of M, no explicit formula can solve it directly. In fact, take a set $I \subset \{1,\ldots,n\}$ (which will stand for the set of active constraints) and consider the problem

$$\min \{f(x) : x \in H^I\}, \quad \text{where} \quad H^I := \{x \in \mathbb{R}^n : x^i = 0, \ i \in I\}. \quad (6.13)$$

In other words, impose arbitrarily some constraints of (6.8) as equalities, and neglect the others. The solution of (6.13) is explicitly given by the linear system

$$\begin{cases} (Mx - b)^i = 0, & i \notin I, \\ x^i = 0, & i \in I. \end{cases} \quad (6.14)$$

There are 2^n such sets I and the whole question is to find $I(x^*)$ of (6.9), for x^* optimal in (6.8). In fact, solving (6.13) or (6.14) with $I = I(x^*)$ will explicitly produce the optimal solution x^*. The idea is then to iterate over I, and one iteration of the solution algorithm works schematically as follows:

Algorithm 6.11 (Schematic QP iteration). The set $I \subset \{1, \ldots, n\}$ is given.

Step 1. Call $\hat{x}$ the solution of (6.13) or (6.14).

Step 2. If $\hat{x}$ satisfies (6.10) then stop.

Step 3. If $\hat{x} \not\geq 0$ append some appropriate index to I and loop to Step 1.

Step 4. If $\hat{x} \geq 0$ but $M\hat{x} - b \not\geq 0$, remove some appropriate index from I and loop to Step 1. □

Together with the index set I, the algorithm maintains a feasible point, whose management in Step 3 will be seen in §6.5.2 below.

Remark 6.12. The matrix of the linear system in (6.14) is some principal submatrix of the positive definite M (the one with rows and columns indexed out of I). As such, it is positive definite, so (6.14) is appropriately solved by a Cholesky factorization, see Remark 4.2. The change in I at each iteration allows the use of economic formulae for a quick update of the Cholesky factors.

Reaching (6.10) may require to explore all possible sets I, so tolerances can be inserted to help the algorithm stop earlier. Actually one stops when the two following properties hold:

– negative components of x are small,

– the *projected gradient* $g \in \mathbb{R}^n$ defined by

$$g^i = \begin{cases} (Mx - b)^i & \text{if } x^i > 0, \\ \min \{0, (Mx - b)^i\} & \text{otherwise} \end{cases}$$

has small components as well. These properties guarantee that the complementarity property (6.12) holds approximately (0 begin replaced by some small negative number in the extreme left- and righthand sides). □

6.5.2 The solution algorithm

The algorithm starts from some x_1 and sets $I_1 = I(x_1)$; for example $x_1 = 0$ and $I_1 = \{1, \ldots, n\}$. Then it generates a sequence of feasible points x_k and a sequence of sets I_k of "activated constraints". At the current iteration k, (6.14) is solved with $I = I_k$ and produces $\hat{x} = \hat{x}_k \in \mathbb{R}^n$. If (6.10) does not hold, there are two cases.

Case $\overline{F}$: $\hat{x}_k \not\geqslant 0$ There is some r ($\notin I_k$) such that $\hat{x}_k^r < 0$. To obtain the next (feasible) iterate x_{k+1}, we make a line-search, to minimize f starting from x_k and moving toward $\hat{x}_k$: we solve

$$\min \{ f(x(t)) : x(t) \geqslant 0 \}, \quad \text{where} \quad x(t) := x_k + t(\hat{x}_k - x_k).$$

By convexity of f (which decreases all the way from x_k to $\hat{x}_k$), this also means to take the largest possible t, and the optimal t is therefore:

$$\bar{t}_k = \min_{r \in \{1,\ldots,n\}} \left\{ \frac{x_k^r}{(x_k - \hat{x}_k)^r} : (x_k - \hat{x}_k)^r > 0 \right\} < 1. \tag{6.15}$$

See Fig. 6.2, where the above $\bar{t}_k$ is obtained at $r = 1$. The left part represents the space $\mathbb{R}^n$ and the right part is the restricted function $t \mapsto f(x(t))$.

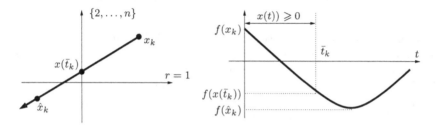

Fig. 6.2. Minimizing the convex function f over the segment $x(t) \geqslant 0$

In this Case $\overline{F}$, the algorithm sets

$$x_{k+1} = x(\bar{t}_k) \quad \text{and} \quad I_{k+1} = I(x_{k+1}). \tag{6.16}$$

Note that I_{k+1} contains at least one r which was not in I_k.

Case F : $\hat{x}_k \geqslant 0$ For notational convenience, we describe the operations as if the current iteration were the $(k-1)^{\text{st}}$.

In this case, some constraint was unduly imposed as an equality in (6.13): there is s ($\in I_{k-1}$) such that $(M\hat{x}_{k-1} - b)^s < 0$. Choose such an s and remove it from I_{k-1} (i.e. free the component x^s). For example we can take for s the most negative $(M\hat{x}_{k-1} - b)^i$ but this is not essential.

In this case, the algorithm sets

$$x_k = \hat{x}_{k-1} \quad \text{and} \quad I_k = I(x_k)\backslash\{s\}. \tag{6.17}$$

Remark 6.13. The aim in relaxing s is to let the next iterate x_{k+1} enter the feasible domain. This is guaranteed only if *one* index is removed from $I(x_k)$. If several indices are removed, the crucial Lemma 6.16 below does not apply.

$\square$

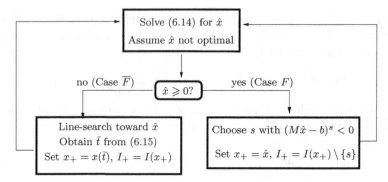

Fig. 6.3. Flow-chart of QP algorithm

Figure 6.3 (in which the iteration index is dropped) summarizes the operations described in this section.

Remark 6.14. If only because of roundoff errors, the linear system (6.14) is not likely to produce any x^i (with $i \notin I$) exactly null. In other words, we normally have $I(\hat{x}) = I$ at each iteration, so that we will have $I_+ = I \cup \{r\}$ or $I_+ = I \setminus \{s\}$, depending on which case occurred. We see that this algorithm is very similar to the simplex algorithm for linear programming: an iteration essentially consists in managing the matrix of a linear system by pivoting operations. The main difference is that the dimension of that matrix fluctuates along the iterations.   □

6.5.3 Convergence

The argument for convergence is that f decreases at each iteration, and there are only finitely many possible I_k's. Some subtleties in the proofs are due to degenerate situations, in which an extra component of $\hat{x}$ in (6.14) vanishes (despite Remark 6.14), or x_k is already optimal in H^{I_k}. We start with a few elementary results.

Lemma 6.15. *At each iteration k,*

(i) $I(\hat{x}_k) \supset I_k$,
(ii) $\hat{x}_k$ *minimizes f in* $H^{I(\hat{x}_k)}$,
(iii) $f(\hat{x}_k) \leqslant f(x_k)$, *with strict inequality if* $\hat{x}_k \neq x_k$.

Proof. Compare (6.14) with (6.9) to establish (i), which implies (ii) because $\hat{x}_k \in H^{I(\hat{x}_k)} \subset H^{I_k}$. Now realize from (6.16), (6.17) that x_k always lies in H^{I_k}; so (iii) is clear by definition of $\hat{x}_k$, the strict inequality coming from the strict convexity of f (which has a unique minimum on any convex set).   □

Now we make sure that relaxing s in Case F does let the next iterate enter the feasible domain.

Lemma 6.16. *Suppose iteration $k - 1$ was in Case F, so that x_k has been set to $\hat{x}_{k-1}$ and $\hat{x}_k$ is the solution of (6.13) or (6.14) with I replaced by $I_k = I(x_k)\backslash\{s\}$. Then $f(\hat{x}_k) < f(x_k)$ and $\hat{x}_k^s > 0$.*

Proof. We will use the notation

$$g^i := \frac{\partial f}{\partial x^i}(x_k) = (Mx_k - b)^i \quad \text{for } i = 1,\dots,n .$$

Let e_s be the sth basis vector. Then:
– the choice of s gives $\nabla f(x_k)^\top e_s = g^s < 0$, so that $f(x_k + te_s) < f(x_k)$ for $t > 0$ small enough;
– the definition (6.17) of I_k gives $x_k + te_s \in H^{I_k}$ for all $t \in \mathbb{R}$.
We conclude $f(\hat{x}_k) \leqslant f(x_k + te_s) < f(x_k)$.
   Now, from Lemma 6.15(ii), $x_k = \hat{x}_{k-1}$ minimizes f in $H^{I(\hat{x}_{k-1})} = H^{I(x_k)}$, so that $g^i = 0$ for $i \notin I(x_k)$ (look at Fig. 6.1 again). Then

$$\nabla f(x_k)^\top(\hat{x}_k - x_k) = \sum_{i \in I(x_k)} g^i(\hat{x}_k - x_k)^i = g^s(\hat{x}_k - x_k)^s = g^s\hat{x}_k^s$$

(note: $\hat{x}_k^i = x_k^i = 0$ for all $i \in I(x_k)$ except $i = s$).
   Piecing together, we obtain from convexity or from (6.11)

$$g^s\hat{x}_k^s = \nabla f(x_k)^\top(\hat{x}_k - x_k) \leqslant f(\hat{x}_k) - f(x_k) < 0 .$$

The result follows since $g^s < 0$.   □

   This result does not imply that $\hat{x}_k$ is feasible: some other component may become negative, in which case the next iteration will be in Case $\overline{F}$.

Lemma 6.17. *Suppose iteration k is in Case $\overline{F}$. Then $f(x_{k+1}) < f(x_k)$.*

Proof. Consider an index r reaching the minimum in (6.15), so that $\hat{x}_k^r < 0$. From (6.14), $r \notin I_k$, and we claim that $x_k^r > 0$, i.e. that $r \notin I(x_k)$ – see (6.9). In fact:
– if iteration $k - 1$ was in Case $\overline{F}$, this holds from (6.16);
– if iteration $k - 1$ was in Case F, $I(x_k) = I_k \cup \{s\}$ from (6.17). If $r \notin I_k$ were in $I(x_k)$, we would have $r = s$; but this is impossible since $\hat{x}_k^s > 0$ (Lemma 6.16).
Our claim is proved.
   As a result, we can write

$$t_k \geqslant \frac{x_k^r}{(x_k - \hat{x}_k)^r} > 0 .$$

Furthermore x_k is feasible and $\hat{x}_k$ is not: from Lemma 6.15(iii), $f(\hat{x}_k) < f(x_k)$. Then the result is easy to visualize on the right part of Fig. 6.2.   □

We can now prove the main result:

Theorem 6.18. *The property* $f(x_k) \leqslant f(x_{k-1})$ *holds at each iteration, and equality implies* $f(x_{k+1}) < f(x_k)$.
It follows that the algorithm stops after finitely many iterations.

Proof. Lemmas 6.15(iii) and 6.17 state that f can never increase. Now the property $f(x_k) = f(x_{k-1})$ implies that iteration $k - 1$ is in Case F (Lemma 6.17). Then $\hat{x}_k^s > 0$ (Lemma 6.16), while $x_k^s = 0$; hence $\hat{x}_k \neq x_k$ and $f(\hat{x}_k) < f(x_k)$, which implies $f(x_{k+1}) < f(x_k)$ (invoke Lemma 6.17 if iteration k is in Case $\overline{F}$).

Now assume iteration $k-1$ is in Case F, so that x_k minimizes f in $H^{I_{k-1}}$. From the first part of the proof, no subsequent $\hat{x}_{k'}$ can lie in $H^{I_{k-1}}$, which implies that no subsequent $I_{k'}$ can be equal to I_{k-1}: there can be at most 2^n iterations in Case F.

Finally, if an iteration k is in Case $\overline{F}$, then I_{k+1} properly contains I_k (Lemma 6.15(i) is important for this): there can be at most n iterations between two successive occurrences of Case F. Altogether the algorithm is finite. □

To finish, let us mention some delicate points in this approach.

(i) Classical algorithms simply define the next working set I_+ by adding or subtracting *one* index (r or s) from the current I, disregarding the set $I(x_+)$. Such implementations are simpler than ours but Lemma 6.17 disappears; so some precautions must be taken to prevent cycling, in case x is not changed by an $\overline{F}$-iteration.

(ii) A difficulty is the definition of $I(x)$ (remember Remark 6.14): under what threshold should an x^i be considered as 0? Even though the progress $f(x_k) - f(x_{k+1})$ is positive "on paper" in Case $\overline{F}$, a very small x_k^i may make this progress insignificant in practice. Then zigzags may occur, which are a weakened form of the cycling phenomenon mentioned in (i).

   The cure is to append the corresponding i into I_k but then, what means "very small"?

(iii) Likewise, a tolerance should be used to test positivity of $M\hat{x} - b$; but again, the concept of a "negative but sufficiently small" $(M\hat{x} - b)^s$ may be delicate to quantify – see also Remark 6.12.

(iv) Along the same lines, substantial difficulties appear in Case F when M is only positive semidefinite. Removing s from I may result in a degenerate system (6.14). In this situation, a number of decisions have to be made:

   – When (6.14) should it be considered as degenerate?
   – In this case, when (6.13) should it be considered as having a subspace of solutions (the alternative being no solution at all)?

– In this case, some index not in I is redundant and should be removed. Which one?

Again the difficulty in these decisions is numerical rather than theoretical.

(v) A general quadratic problem has constraints of the type $Ax \geqslant a$. There is not much difference with (6.8) – where A was the identity matrix and a was 0; I indices activated constraints and (6.8) becomes

$$\min f(x), \quad A^i x = a^i, \text{ for } i \in I . \tag{6.18}$$

In Case $\overline{F}$, the properties

$$A^i(\hat{x}_k - x_k) = 0 \text{ for all } i \in I_k \quad \text{and} \quad A^r(\hat{x}_k - x_k) < 0$$

imply that the new row A^r is linearly independent of the set $\{A^i\}_{i \in I_k}$ – an important property for an easy resolution of (6.18).

Bibliographical Comments

A conclusion of this chapter is that the trust-region approach, which a priori seems nothing more than a natural variant of line-searches, actually opens a vast field of applications. The main reason is that, beyond the motivation we gave (to argue against the value of the half-line $\{x + td\}_{t>0}$), it is particularly well-suited to treat a model $\tilde{f}$ ill-conditioned (case of Gauss-Newton), or even nonconvex (pure Newton, quasi-Newton without the property $(y, s) > 0$).

This method has progressively appeared during the 70's, and imposes itself more and more, due to its robustness and its ease of implementation. We just cite the synthesis [264], a reference paper.

Despite its name, Gauss-Newton is traditionally associated with the names of Levenberg and Marquardt [237, 248]. In its classical version, it is coupled with a stabilization of the type $G_k + \lambda I$ (the *Levenberg-Marquardt parameter*); the modern tendency goes toward trust-region instead: see [264] again.

Limited-memory BFGS formulae were given by J. Nocedal in [275], and [238, 152] give implementations of the resulting algorithms. If f is elliptic, it is observed in [238] that the proof of Theorem 4.9 simplifies and applies to the limited-memory version.

The truncated Newton method appeared in [102]; our development of §6.4 is essentially due to [345]. It should be mentioned here that the techniques for automatic or computational differentiation, alluded to at the end of §1.6, can be used to compute second derivatives; or at least to compute Mh, the Hessian times a vector, without an explicit use of the second derivatives appearing in M. This provokes a revival of the interest for Newton methods. Even if such computational differentiation holds its promises, however (such is not the case yet), the result of a competition with quasi-Newton is still unsure.

Active set methods for quadratic programming are described in most textbooks on optimization. As mentioned earlier, their implementation is delicate, we can cite [165]. It is worth mentioning that they may become impractical for large problems. Just for an illustration, suppose for example that the algorithm is initialized on $x_1 = 0$ but that the optimal solution x^* is entirely positive. In view of Remark 6.13, at least n iterations will be necessary, a possibly very large number. New ideas are then necessary. Two such ideas can be found for example in [98, 266]; one can also use the interior-point paradigm, developed in Part IV of this book.

Note that convexity of the quadratic function f (i.e. the property $M \succcurlyeq 0$) is absolutely necessary in our framework: the nonconvex case contains combinatorial optimization, a field which we definitely leave out of this book – see Remark 1.1.

7 A Case Study: Seismic Reflection Tomography

We conclude this first part of the book with the presentation of a problem which can be implemented in a short course on numerical unconstrained optimization, using high-level languages like MATLAB [249] or SCILAB [327]. It can be used to test the algorithms we have described[1]. Perhaps more importantly, its implementation involves various operations, typical when dealing with nontrivial applications. In a way, the problem resembles the real-world examples reviewed in §1.2 – although widely simplified. As such, it is well-suited to train oneself in the writing of a significant simulator in the sense of §1.3. The present chapter describes the necessary material to write the corresponding simulator.

We propose a schematic version of an engineering technique which has been employed for many years to explore the geological structure of the subsurface [100, 101, 162] and has been improved recently [99]. The idea is to measure the time taken by waves to travel from various sources to various receivers, after their refraction and reflection on the interfaces separating the layers of the subsurface. These measurements/observations serve to determine the positions of these interfaces and the wave velocities in the layers. From an optimization viewpoint, the problem consists in minimizing a nonlinear least-squares function expressing the mismatch between the observed traveltimes and those calculated by ray tracing in a subsoil model.

7.1 Modelling

Seismic tomography is based on the simplified model of *geometrical optics*, in which waves are supposed to propagate along rays. Here we consider a 2D model with a single reflector. Then the waves are issued from sources, reflect on this single interface, and are captured by receivers. The traveltimes between sources and receivers (all of which are placed on the ground surface, assumed horizontal) are measured and used to recover the position of the reflector. We take the additional assumption that the wave velocity is constant and known in the considered subsoil layer between the ground surface and

---

[1] Elementary algorithm testing can also use the famous Rosenbrock banana function $\mathbb{R}^2 \ni (x, y) \mapsto (1 - x)^2 + 100(y - x^2)^2$.

the unknown interface. When the velocity is constant, a ray follows a trajectory which is a straight line between a source/receiver and the reflector. The model is shown in Fig. 7.1, in which one can see a source at a given position

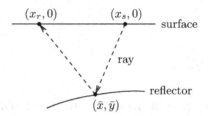

Fig. 7.1. A single reflection on the interface

$(x_s, 0)$, a receiver at a given position $(x_r, 0)$, and the reflection point of the wave at the position $(\bar{x}, \bar{y})$; the latter has to be computed.

It is assumed that the interface to be retrieved has its x-coordinate between 0 and 1 and is located at a depth less than 1. Since the rays have a horizontal component, it is a good idea to consider a region with x-coordinates between `xmin` $:= -0.5$ and `xmax` $:= 1.5$, so that the domain of interest is [`xmin`, `xmax`] $\times [-1, 0]$. The unknown interface is supposed to be a function of x; overhanging structures are thus discarded. To have a problem in finite dimension, the interface has to be discretized. Taking a representation by a cubic spline is attractive since the interface is then of class C^2: a property allowing the computation of the reflection points, see §7.2 below.

A *cubic spline* on the interval [`xmin`, `xmax`] is a piecewise polynomial function. The polynomials are cubic on each of the subintervals of [`xmin`, `xmax`], which can be assumed to have the same length `dx` $=$ (`xmax` $-$ `xmin`)/`ndx`, where `ndx` is typically around 100. Transition conditions between polynomials defined on adjacent subintervals make the full cubic spline a C^2 function.

Cubic splines can be represented and stored in various ways. The one given in the previous paragraph roughly corresponds to the "pp" form in the MATLAB toolbox `splines`. However, the most useful representation in our context is the one that makes use of the decomposition of the splines in a basis of cubic B-splines (the "sp" form in the toolbox `splines`); this eases computations to be seen §7.3 below. A cubic spline $x \mapsto y(x)$ can indeed be decomposed as a finite sum

$$y(x) = \sum_i a_i B_i(x), \tag{7.1}$$

where the coefficients a_i are real numbers and the functions B_i are basic cubic splines, the B-splines. These are cubic splines, positive on 4 adjacent subintervals and null elsewhere. They are associated with the discretization. Therefore, the cubic spline (7.1) above is fully described by the $n := $ `ndx` $- 3$

coefficients a_i. We use the MATLAB functions spmak, fnval, and fnplot to construct, evaluate, and plot these splines:

```
cs = spmak([...], [...]);  y = fnval(cs,x);  fnplt(cs);
```

7.2 Computation of the Reflection Points

The main task to make our simulator is to write a MATLAB function, say

```
function [time,time_der,x] = ttime (src,rcv,cs)
```

computing the traveltime time of a wave from a given source $(x_s, 0)$ (src = x_s) to a given receiver $(x_r, 0)$ (rcv = x_r), via a reflection on the interface at a point $(\bar{x}, \bar{y})$ ($x = \bar{x}$) to be computed; see Fig. 7.1. This function also computes the derivative time_der of the traveltime with respect to the interface – a computation to be explained in §7.3 below. Finally, cs is the MATLAB structure describing the cubic spline of the interface (see the use of spmak in the previous section).

Consider a piecewise linear path from $(x_s, 0)$ to $(x_r, 0)$, which encounters the interface at point (x, y). Assuming unit velocity of the wave, the time spent along this path is its length

$$\ell(x, y) = \sqrt{(x - x_s)^2 + y^2} + \sqrt{(x - x_r)^2 + y^2}, \qquad (7.2)$$

where the y-coordinate is actually given by (7.1). To determine the actual reflection point $(\bar{x}, y(\bar{x}))$, we apply the *Fermat law*, according to which the traveltime is stationary. In our simplified setting, we just consider the minimization of the function $\varphi(x) := \ell(x, y(x))$ with respect to x (note from (7.1) that y, hence φ, also depends on a); when there are several reflection points, we thus compute one with smallest traveltime (which is actually the point of view adopted in practice). This is a one-dimensional optimization problem, for which we must solve the equation

$$\varphi'(x) = \ell'_x(x, y(x)) + \ell'_y(x, y(x))y'(x) = 0. \qquad (7.3)$$

This formula shows that differentiating φ involves the derivatives of the cubic spline $x \mapsto y(x)$ in (7.1), which can be obtained by the MATLAB function fnder and evaluated by fnval.

Newton's method can be used to solve (7.3): at its current iteration k, let (see §4.1):

$$x_k^N := x_k - \frac{\varphi'(x_k)}{\varphi''(x_k)} \qquad (7.4)$$

be the Newton estimate. Taking $x_{k+1} = x_k^N$ is the pure, local, Newton algorithm. A robust implementation should globalize it, by way of a line-search (§4.2), which requires the local decrease of φ when moving from x_k toward x_k^N.

To guarantee this property, $\varphi''(x_k)$ (supposed to be positive) can be replaced by 1 if it becomes less than some small $\varepsilon > 0$. The iterations can be started at $x_1 = (x_s + x_r)/2$.

Figure 7.2 shows a reflector having the shape of a step-pyramid and the

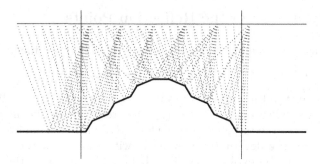

Fig. 7.2. Ray tracing for the step-pyramid interface (the y-axis is stretched)

result of the ray tracing operation, which computes the reflection points for various source-receiver pairs. These results are obtained by repeated execution of the `ttime` function, for each source-receiver pair. In the present case, we have $m = 960$ such pairs, but only one 16th of them are plotted for clarity.

7.3 Gradient of the Traveltime

In the previous section, the interface was assumed to be known. In the actual problem, however, the interface – characterized by a in (7.1) – is unknown: we need to identify it, with the help of the measured traveltimes; remember the meteorological problem of §1.2.2. In `ttime` (the essential part of the simulator), `time` will give objective values but the optimization algorithm will also need the derivatives `time_der` with respect to the a_i's.

Now, besides the simple functions $x \mapsto y$ from (7.1) (which should rather be written $(x, a) \mapsto y$ to highlight the dependence on the interface) and $(x, y) \mapsto \ell$ from (7.2), we have the implicit mapping $a \mapsto (\bar{x}, \bar{y})$ defined by the minimization of φ, i.e. by (7.3); $\bar{y}$ stands for $y(\bar{x})$ – a correct writing should be $y(\bar{x}(a), a)$. The result is a composite mapping $a \mapsto \tau(a) := \ell(\bar{x}(a), y(\bar{x}(a), a))$, which we have to differentiate.

To compute $\nabla \tau(a)$, assume that the necessary differentiability properties hold and make a formal calculation, using (7.1) and (7.3):

$$
\begin{aligned}
\nabla \tau(a) &= \ell_x'(\bar{x}, \bar{y}) \nabla \bar{x}(a) + \ell_y'(\bar{x}, \bar{y}) \left[y_x'(\bar{x}, a) \nabla \bar{x}(a) + \nabla_a y(\bar{x}, a) \right] \\
&= \varphi'(\bar{x}) \nabla \bar{x}(a) + \ell_y'(\bar{x}, \bar{y}) \nabla_a y(\bar{x}, a) \\
&= \ell_y'(\bar{x}, \bar{y}) \nabla_a y(\bar{x}, a) \, ,
\end{aligned}
$$

which does not depend on $\nabla \bar{x}(a)$ (luckily!). Thus, computing $\nabla \tau(a)$ involves elmentary calculus only: to differentiate (7.2) (with respect to y) and (7.1) (with respect to a)[2]. Incidentally, (7.1) shows that $\nabla_a y(\bar{x}, a)$ has at most 4 nonzero components; the above computation can be performed in a time independent of n.

7.4 The Least-Squares Problem to Solve

Given an interface, described by the parameters $a \in \mathbb{R}^n$, and an acquisition survey (locations of m source-receiver pairs), we have shown how a vector of traveltimes $T(a) \in \mathbb{R}^m$ can be computed (§7.2) and differentiated (§7.3) by ray tracing. Seismic reflection tomography is the corresponding *inverse problem*. Its purpose is to adjust the interface a so that $T(a)$ best matches a vector of traveltimes $T^{\text{obs}} \in \mathbb{R}^m$ (the observed traveltimes) picked on seismic data. Since Gauss [139; 1809], it is both classical and natural to formulate this problem as a least-squares one:

$$\min_{a \in \mathbb{R}^n} \frac{1}{2} |r(a)|_2^2, \quad \text{with} \quad r(a) := T(a) - T^{\text{obs}}, \tag{7.5}$$

in which one minimizes the norm of the mismatch or residual between $T(a)$ and T^{obs}.

The fact that problem (7.5) may be ill-posed has been pointed out by many (the problem may have no solution or its solutions may not depend continuously on the data), which yields oscillations in the interface. To ensure well-posedness, a curvature regularization is often introduced: one adds to the cost function in (7.5) a term penalizing the L^2-norm of y'' in (7.1). A short computation shows that such a term is $a^\top R a$; here R is the symmetric positive semidefinite matrix whose element (i, j) is

$$R_{ij} = \int_{\texttt{xmin+3dx}}^{\texttt{xmax-3dx}} B_i''(\xi) B_j''(\xi) \, d\xi \,.$$

Instead of (7.5), the regularized least-squares problem is then

$$\min_{a \in \mathbb{R}^n} \left(\frac{1}{2} |r(a)|_2^2 + \frac{\varepsilon}{2} \, a^\top R a \right) \,. \tag{7.6}$$

The choice of the regularization parameter $\varepsilon > 0$ is a difficult task; L-curve technique [187] is sometimes used.

The necessary material is now available for an actual implementation. To construct a synthetic dataset, proceed for example as follows:

---

[2] A rigorous justification of the above calculation is far beyond the scope of this book – even of Part II, which deals with nonsmooth optimization. Observe here and now that existence of $\nabla \tau(a)$ implies a crucial property: (7.3) must have a *unique* solution $\bar{x}(a)$. Checking $\nabla \tau$ by finite differences is certainly a good idea.

– Choose an interface, characterized by a function $x \mapsto y(x)$ to be inserted in (7.2) – the spline representation (7.1) is useless at this point! The interface may look like the step-pyramid of Fig. 7.2 but something more handy may be advisable; say a polynomial or a trigonometric function, behaving mildly in $[0, 1]$. Note from (7.4) that $y(\cdot)$ must be twice differentiable.

– Place a reasonable number of sources and receivers on the ground surface.

– Apply the method defined in §7.2 to compute the traveltimes between each pair (r, s) of source-receiver.

– Call T_{rs}^{obs} the value thus obtained, to be used in (7.5).

This procedure is unlikely to allow an exact fitting: the residuals will be nonzero at the optimal solution. For a safer debugging, it may be advisable to choose a discretization of $[\texttt{xmin}, \texttt{xmax}]$ and to invent coefficients a_i, wich will give the exact interface via (7.1).

7.5 Solving the Seismic Reflection Tomography Problem

The unconstrained optimization problem (7.6) can be solved by a number of algorithms described and analyzed in this first part of the book. We quote the following two candidates:

– The quasi-Newton algorithm (§4.4) only needs the gradient $J(a)^{\top} r(a) + \varepsilon R a$ of the cost function in (7.6), where $J(a) := r'(a)$ is the Jacobian of the residual (each of its lines is the gradient of a traveltime).

– The Gauss-Newton algorithm (§6.2) uses the gradient of the cost function, as well as the approximation $J(x)^{\top} J(x) + \varepsilon R$ of its Hessian.

Recall that the Jacobian $J(a)$ can be computed in a time similar to $T(a)$ (see §7.3). Therefore, the Gauss-Newton algorithm should be favoured in this problem, since it is able to use a part of the Hessian with the same computational effort as the one to get the gradient. Both quasi-Newton and Gauss-Newton algorithms can be globalized by a line-search (chapter 3); Gauss-Newton should preferably be globalized by trust regions (§6.1).

Figure 7.3 shows the retrieved interfaces corresponding to the unper-

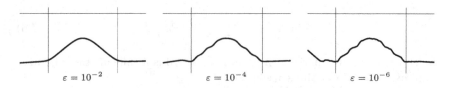

$\varepsilon = 10^{-2}$ $\varepsilon = 10^{-4}$ $\varepsilon = 10^{-6}$

Fig. 7.3. Retrieving the step-pyramid interface

turbed traveltimes issued from the step-pyramid interface shown in Fig. 7.2, for various values of the regularization parameter ε in (7.6) and a discretization of $[\texttt{xmin}, \texttt{xmax}]$ in 100 subintervals (500 subintervals were used for the ray tracing operation). The details of the interface appear progressively as ε decreases. Observe on Fig. 7.2 that the extreme parts of the interface are not lit by the waves emanating from the sources, so that their recovered positions are only determined by the regularization. This explains why these parts of the interface have a position which does not correspond to the original interface, in particular when ε is small.

Needless to say, the above model is only a schematic representation of the real problem. Among others, we can mention: more complex geometry (10 reflectors is common and they can cross each other), sources and reflectors that can be situated in wells, noisy data, large scale (10^4 unknown parameters, 10^6 data), possible constraints to take into account a priori information, etc.

General Conclusion

This study of unconstrained optimization conveys two important messages. One is that optimization algorithms can be given a "global" character, thanks to stabilization devices, such as line-search or trust-region. Second, the whole business for efficient algorithms is a proper use of second-order information. Concerning the latter, we have seen a variety of possibilities:

(i) The "standard" one, in which nothing is known beyond first order. This is unambiguously the realm of quasi-Newton methods, possibly with limited memory. It can be used in conjunction with line-search or with trust-region.

(ii) The case of least-squares problems, where an attractive approximation G_k (the matrix of Gauss-Newton) is directly available. Even though G_k is certainly positive semi-definite, it may be ill-conditioned. An appropriate parameter λ (of Levenberg-Marquardt) is needed for efficiency, as well as successive solutions to the system $(G_k + \lambda I)d = -g_k$, for successive values of λ.

(iii) In some situations, second derivatives can be computed, or conveniently approximated. The resulting matrix M_k need not be positive semi-definite. As a result, line-searches are definitely inappropriate: one needs either truncated conjugate gradient (§6.4) or trust-region (§6.1). The latter approach results in computations similar to those of case (ii): an appropriate parameter Δ (the size of the trust-region) is needed, as well as successive factorizations of $M_k + \mu I$, for successive values of the Lagrange multiplier μ (§6.1.1); and each iteration on μ implies the factorization of the perturbed matrix $M_k + \mu I$.

Even when several of these possibilities are available, the best strategy is not necessarily obvious. Indeed consider a least-squares problem, where the exact Hessian can be computed, for example via automatic differentiation of §1.6. Should one choose (i), (ii) or (iii)? and in case (i), should one choose

a line-search? Actually the answer depends largely on *computing times*: how long does it take
- to compute f and g?
- to compute f''?
- to obtain an appropriate solution to $f''h = -g$ via truncated conjugate gradient?
- to obtain an appropriate λ in Gauss-Newton?
- to obtain an appropriate Δ in trust-region?

One sees that the question cannot be given a general answer *in abstracto*.

Part II

Nonsmooth Optimization

Claudia A. SAGASTIZÁBAL

Nonsmooth optimization (NSO) is devoted to optimization problems in which the objective function f (and possibly the constraints) is not continuously differentiable.

In these problems, the task is to find a point $\bar{x}$ such that $f(\bar{x}) \leq f(x)$ for all x in the domain of interest. We will focus most of our study in unconstrained problems, with f convex and finite everywhere (hence locally Lipschitzian), but not continuously differentiable. To solve such problems, it is necessary to consider a special object: the *subdifferential* of a convex function f, denoted by $\partial f(x)$, which is a fundamental concept in convex analysis.

This second part is organized as follows: Chapter 8 contains some basic results of subdifferential calculus and duality theory, and two examples of nonsmooth functions that appear often in practice, such as pointwise maxima and dual functions. In Chapter 9 we study the NSO methods of steepest descent, subgradients and cutting-planes. We present in detail bundle methods in Chapter 10. Chapter 11 is devoted to explaining some important applications of NSO, including the decomposition of large-scale or complex problems using Lagrangian duality, as well as extensions of bundle methods for handling varying dimensions, for solving constrained problems, and for solving generalized equations. We finish in Chapter 12 with some computational exercises that help getting a better understanding of NSO methods.

8 Introduction to Nonsmooth Optimization

The aim of this chapter is to review some general theoretical issues concerning the problem

$$\min_{x \in \mathbb{R}^n} f(x),\tag{8.1}$$

where differentiability assumptions are replaced by convexity. We recall the necessary theory for solving (8.1), and give two important examples of non-differentiable functions, namely max-functions and dual functions.

8.1 First Elements of Convex Analysis

We make here a brief review of basic concepts in subdifferential calculus. For proofs and more details, we refer to [195], especially its Chapter VI; see also the abridged version [196]. The book of Rockafellar [309] is a classical reference in the domain. A more general subdifferential theory, that includes the nonconvex setting, can be found in [80] and [314]. We refer also to [52] for a concise treatment of the area of nonsmooth analysis underlying computational optimization techniques.

Let f be a convex function as in Definition 2.2, defined on the whole of $\mathbb{R}^n$. Then f is continuous and locally Lipschitzian. Moreover, the directional derivative $f'(x; d) := \lim_{t \searrow 0} \frac{f(x+td)-f(x)}{t}$ exists for each fixed x and d in $\mathbb{R}^n$. As a result, f has a gradient almost everywhere. When $\nabla f(x)$ does not exist, the point x is called a *kink*.

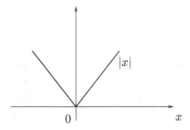

Fig. 8.1. Minimizers are often kinks

Even though kinks form a set of zero measure, in practice minimizers are often kinks. The simplest example of nondifferentiable convex function, the absolute value function, confirms this observation. Namely, its only kink is precisely the unique minimizer, $x = 0$; see Figure 8.1.

To describe the behavior of the function near a kink x, the concept of gradient must be generalized. Instead of just a vector, a certain *set* will be associated with x. This is the subdifferential of f at x:

$$\partial f(x) := \{ s \in \mathbb{R}^n : f(y) \geq f(x) + \langle s, y - x \rangle \text{ for all } y \in \mathbb{R}^n \}. \qquad (8.2)$$

This set is nonempty, closed, convex, and locally bounded as a (set-valued) function of x; it reduces to the singleton $\{\nabla f(x)\}$ if and only if f is differentiable at x. Each element of $\partial f(x)$ is called a *subgradient*. An equivalent definition, expressed in terms of the directional derivative, is:

$$\partial f(x) := \{ s \in \mathbb{R}^n : \langle s, d \rangle \leq f'(x; d) \text{ for all } d \in \mathbb{R}^n \}. \qquad (8.3)$$

Also, letting "conv S" denote the convex hull of the set S, defined as the smallest convex set containing S, i.e., as

$$\text{conv } S := \left\{ \sum_i \alpha_i s^i : s^i \in S, \sum_i \alpha_i = 1, \alpha_i \in [0, 1] \right\};$$

the relation

$$\partial f(x) = \text{conv} \left\{ \lim_{x^i \to x} \nabla f(x^i) \text{ for all } x^i \text{ for which } \nabla f(x^i) \text{ and the limit exist} \right\}$$

holds. This last definition allows a generalization of subdifferential calculus to nonconvex functions, known as *Clarke subdifferential*, [80].

By using (8.3), it can be shown that

$$f'(x; d) = \max\{ \langle s, d \rangle : s \in \partial f(x) \}. \qquad (8.4)$$

Therefore, $f'(x; \cdot)$ is Lipschitz continuous, with the same constant as f.

When comparing all the expressions above for the subdifferential, we see that (8.2) gives a geometric interpretation. It says that $\partial f(x)$ is made up of the slopes of the hyperplanes supporting the epigraph of f at $(x, f(x)) \in \mathbb{R}^n \times \mathbb{R}$. This characterization is called the *subgradient inequality*:

$$s \in \partial f(x) \text{ if and only if } f(y) \geq f(x) + \langle s, y - x \rangle \text{ for all } y \in \mathbb{R}^n. \qquad (8.5)$$

Even though the concept of "gradient" is extended from a vector-function to a set-valued function, a good part of the classical differential calculus remains valid for nondifferentiable functions. For an illustration, we include here some results, concerning primarily the first-order setting.

– First order expansion:

$$\forall \varepsilon > 0 \quad \exists \delta > 0 : \|h\| \leq \delta \implies |f(x + h) - f(x) - f'(x; h)| \leq \varepsilon \|h\|.$$

– Mean-value theorem: For $x \neq y$ in $\mathbb{R}^n$, there exist

$$t \in \,]0,1[\text{ and } s \in \partial f(ty + (1-t)x)$$

such that

$$f(y) - f(x) = \langle s, y - x \rangle \,.$$

– Optimality in (8.1): it can be characterized by a generalization of Fermat condition in the differentiable case, $\nabla f(\bar{x}) = 0$. A direct application of definitions (8.2), (8.3) gives the following result.

For $f : \mathbb{R}^n \to \mathbb{R}$ convex, the following three statements are equivalent:
  (i) f is minimized at $\bar{x}$, i.e., $f(y) \geq f(\bar{x})$ for all $y \in \mathbb{R}^n$;
  (ii) $0 \in \partial f(\bar{x})$;
  (iii) $f'(\bar{x}; d) \geq 0$ for all $d \in \mathbb{R}^n$.

Thus, a minimum point $\bar{x}$ is characterized by $0 \in \partial f(\bar{x})$, or $f'(\bar{x}; d) \geq 0$ for all $d \in \mathbb{R}^n$. Conversely, a non-optimal x is characterized by the existence of at least one direction d such that $f'(x; d) < 0$.

A direction satisfying $f'(x; d) < 0$ is called a *descent direction* of f at the point x (or downhill direction). As in the differentiable case, descent directions are fundamental in numerical nonsmooth optimization, since they allow the generation of iterates with decreasing objective values.

From the definition of directional derivative and by convexity, d is a direction of descent of f at x when there exists $t > 0$ for which $f(x + td) < f(x)$. Also, from (8.4), this is equivalent to having $\langle s, d \rangle < 0$ for all $s \in \partial f(x)$. Furthermore, see [195; Theorem VIII.1.1.3], from a geometrical point of view, using again (8.3) and (8.4), we see that to find a descent direction corresponds to finding a hyperplane separating strictly the two (closed convex) sets $\partial f(x)$ and $\{0\}$:

A descent direction d is such that, for $\alpha \in [f'(x;d), 0[$, the hyperplane $\{z \in \mathbb{R}^n : \langle z, d \rangle = \alpha\}$ separates $\partial f(x)$ and $\{0\}$ strictly:

$$\langle s, d \rangle \leq \alpha < 0 \quad \text{for all } s \in \partial f(x) \,.$$

8.2 Lagrangian Relaxation and Duality

Duality is an important source of nondifferentiable problems; see Chapter 11.1. We now give some elements of Lagrangian duality.

8.2.1 Primal-Dual Relations

Consider the primal problem

$$\begin{cases} \min_{p \in P} f_o(p) \\ c_j(p) = 0, & j \in E := \{1, \dots, n_E\} \\ c_j(p) \leq 0, & j \in I := \{n_E + 1, \dots, n := n_E + n_I\}, \end{cases} \tag{8.6}$$

with $f_o, c_j : P \to \mathbb{R}$ and $P \subset \mathbb{R}^N$. With the primal problem (8.6), we associate

- a closed convex set X, the space of "multipliers" or dual variables,
- a function (the "Lagrangian") $L : P \times X \to \mathbb{R} \cup \{+\infty\}$ having the property that, for all $p \in P$, $\sup_{x \in X} L(p, x) = f_o(p)$ if p satisfies the constraints in (8.6), and $+\infty$ otherwise.

Then (8.6) is equivalent to solving

$$\inf_{p \in P} \sup_{x \in X} L(p, x). \tag{8.7}$$

The idea is to invert the order of "sup" and "inf" in (8.7). This inversion is practically interesting only if the *dual function*

$$\theta(x) := \inf_{p \in P} L(p, x) \tag{8.8}$$

is easy to compute, and it is not identically $-\infty$, a particularly nasty case. Instead of solving (8.6) or (8.7), one works with its *dual*:

$$\sup_{x \in X} \inf_{p \in P} L(p, x) = \sup_{x \in X} \theta(x), \tag{8.9}$$

which can be simpler to solve, due to the particular structure of the problem (this somewhat vague claim will become obvious in the examples of § 11.1.1, 11.1.2 below).

An abstract duality theory can be developed from these premises. Here, we limit ourselves to a particular dual of (8.6), which is the most popular. Specifically, we shall consider the dual coming from the Lagrangian function classical in optimization. It is characterized by:

- $X := \mathbb{R}^{n_E} \times \mathbb{R}_+^{n_I}$, whose elements will be denoted by $x = (x_E, x_I)$. Constraints will be likewise condensed into $c(p) = (c_E(p), c_I(p))$.
- $L(p, x) := f_o(p) + \langle x, c(p) \rangle$, where $\langle \cdot, \cdot \rangle$ is the Euclidean scalar product in $\mathbb{R}^{n_E + n_I}$.

The method which finds solutions to (8.6) by solving (8.9) is called *Lagrangian relaxation*. In this method, the dual problem (8.9) -the maximization of the dual function θ defined by (8.8)- can be put in the framework of (8.1) -the minimization of a convex function- by defining $f := -\theta$; we analyze such functions in § 8.3.2 below.

Suppose that both the primal and dual problems ((8.7) and (8.9), respectively) have solutions. Admitting that (8.9) is simpler to solve than (8.7) (and hence than (8.6), our initial problem), we shall be interested in finding primal-dual pairs which solve at the same time both problems.

In Linear Programming, having a primal optimal value that is finite amounts to having dual feasibility and, hence, a solvable dual problem. For general data as in (8.6), it is known that the key object for (8.9) to have dual solutions is the image of the constraints:

$$Im\,\mathcal{C} := \{(c_E(p), c_I(p)) : p \in P\} + \{0 \in \mathbb{R}^{n_E}\} \times \mathbb{R}^{n_I}_+$$
$$= \{(c_E(p), c_I(p) + r) : p \in P, r \in \mathbb{R}^{n_I}_+\}\,.$$

Let "ri S" denote the relative interior of a convex set S, defined as the interior of S in the smallest affine space in which S is contained, endowed with the induced topology. Proposition XII.2.4.1 in [195] states the following result:

Suppose θ is not identically $-\infty$. Then
$$0 \in \mathrm{ri}\,\mathrm{conv}\,Im\,\mathcal{C} \implies \text{(8.9) has at least one solution.}$$

The assumption that $0 \in \mathrm{ri}\,\mathrm{conv}\,\mathcal{C}$ is a condition of qualification of constraints (cf. the end of §13.3). Suppose (8.6) is a convex problem, i.e., P, f_o and the inequality constraints c_I are convex, while the equality constraints c_E are affine. A classical constraint qualification in this setting is Slater condition:

$$\text{there exists } p^0 \in \mathrm{ri}\,P \text{ such that } c_E(p^0) = 0 \text{ and } c_I(p^0) < 0\,. \qquad (8.10)$$

If this condition holds, there are dual solutions (in fact, (8.10) just says that $0 \in \mathrm{ri}\,Im\,\mathcal{C}$ and, with our assumptions, $Im\,\mathcal{C} = \mathrm{conv}\,Im\,\mathcal{C}$).

For general data, we shall only assume that the primal functions in (8.6) are such that the infimum defining the dual function (8.8) is attained:

$$\text{for all } x \in X \text{ such that } \theta(x) > -\infty, \text{ there exists } p_x \in P : \theta(x) = L(p_x, x)\,.$$

8.2.2 Back to the Primal. Recovering Primal Solutions

The dual approach will be totally successful if a primal solution can be recovered from a dual solution (8.9). The *duality gap* measures the difference between the dual and primal optimal values:

$$\min_{p \in P} \{f_o(p) : c_E(p) = 0\,, c_I(p) \le 0\} - \sup_{x \in X} \theta(x)\,. \qquad (8.11)$$

This quantity is always nonnegative, see (8.14) below, and measures how good is the dualization scheme, in the sense of approximating well the primal problem. Along these lines, we now establish a first fundamental result, stating that each computation of the dual function has associated a minimizer p_x which solves a *perturbation* of (8.6).

Theorem 8.1 (Everett). *For fixed $x \in X$, suppose the dual function (8.8) has a solution $p_x \in P$:*

$$p_x \quad \text{solution to } \inf_{p \in P} L(p, x) = \theta(x)\,. \qquad (8.12)$$

Then p_x also solves

$$\begin{cases} \min_{p \in P} f_o(p) \\ c_E(p) = c_E(p_x) \\ c_j(p) \le c_j(p_x), & j \in I \text{ such that } x_j > 0 \\ c_j(p) \text{ arbitrary}, & j \in I \text{ such that } x_j = 0. \end{cases} \quad (8.13)$$

Proof. Let $p \in P$ satisfy the constraints in (8.13). Because p_x solves (8.8), we have $L(p_x, x) \le L(p, x)$, which implies that $f_o(p_x) \le f_o(p)$. Since p_x trivially satisfies the constraints in (8.13), the result is immediate. □

We now give further relations between the primal and dual problems. In particular, relation (8.14), known as the "weak duality" inequality. The wording "p feasible for (8.6)" below means that p satisfies the constraints given by c_E and c_I in (8.6).

Lemma 8.2. *Consider the primal and dual problems (8.6) and (8.7). Then*

(i) The duality gap (8.11) is never negative, i.e.,

$$f_o(p) \ge \theta(x) \quad (8.14)$$

for all $p \in P$ feasible in (8.6) and for all $x \in X$.

(ii) Any $p \in P$ feasible in (8.6) for which $f_o(p) = \theta(x)$ is a solution to (8.8).

(iii) If the maximal value of (8.9) is $+\infty$, then (8.6) has no feasible solution. Conversely, if (8.6) has the minimal value $-\infty$, then $\theta \equiv -\infty$ in (8.9).

(iv) If there exist $\bar{p} \in P$ feasible in (8.6) and $\bar{x} \in X$ for which $\theta(\bar{x}) = f_o(\bar{p})$, then $\bar{p}$ solves (8.6) and $\bar{x}$ solves (8.9).

Proof. (i) Because $p \in P$ is feasible for (8.6), we have

$$f_o(p) \ge f_o(p) + \langle x, c(p) \rangle = L(p, x) \ge \theta(x)$$

for all $x \in X$, from which (8.14) follows immediately.

(ii) Any feasible $p \in P$ for which $f_o(p) = \theta(x)$ is, using (8.14), a minimum point in (8.8). The remaining items are straightforward. □

A consequence of weak duality is that solving the dual problem (8.9) gives the *best lower bound* for the primal optimal value that can be obtained by Lagrangian relaxation. This property is often used in combinatorial optimization.

From Theorem 8.1, we see that to solve the primal problem (8.6) via its dual, it is enough to find x such that p_x from (8.12) is also primal feasible. This matter is not simple at all, since nothing guarantees that the process computing $\theta(x)$ will manage to find a solution p_x of (8.12) that is primal feasible. In order to recover primal solutions from dual problems two assumptions are required. First,

$$\exists (p_x, x) \in P \times X \text{ for } p_x \text{ given by } (8.12) \text{ and feasible for } (8.6). \qquad (8.15)$$

Second, a property for some primal-dual pairs (p, x):

$$(p, x) \in P \times X \quad \text{satisfying} \quad c(p) \leq 0 \quad \text{and} \quad \langle x, c(p) \rangle = 0. \qquad (8.16)$$

In other words, such p satisfies the constraints of (8.6), with complementarity for the multiplier x:

$$c_E(p) = 0, \quad c_I(p) \leq 0 \quad \text{and} \quad c_j(p) = 0 \text{ for all } j \in I \text{ such that } x_j > 0;$$

see also §13.3.

Lemma 8.3. *Suppose* (8.15) *holds and let* $(p_{x'}, x')$ *be a pair satisfying* (8.16). *The following primal-dual relations hold:*

(i) $f_o(p_{x'}) = \theta(x')$.

(ii) x' *solves the dual problem* (8.9).

(iii) *Any solution* $\bar{x}$ *of* (8.9) *is such that the relations in* (8.15) *hold for the pair* $(p_{x'}, \bar{x})$.

(iv) $\bar{p}$ *is a solution to the primal problem* (8.6) *if and only if* $\bar{p}$ *solves* (8.12) *with* x *replaced by* $\bar{x}$, *a solution to the dual problem* (8.9), *and the primal-dual pair* $(\bar{p}, \bar{x})$ *satisfies* (8.16).

Proof. (i) By (8.16), $L(p_{x'}, x') = f_o(p_{x'})$, together with (8.12), item (i) holds. (ii) Combine item (i) with (8.14) written for $p = p_{x'}$ and $x \in X$ arbitrary, to obtain the optimality of x' in (8.9). (iii) Take $\bar{x}$ solving (8.9): $\theta(\bar{x}) \geq \theta(x') = f_o(p_{x'})$, by item (i). Inequality (8.14) applied to $p_{x'}$ and $\bar{x}$ gives $f_o(p_{x'}) \geq \theta(\bar{x})$, so $\theta(\bar{x}) = f_o(p_{x'})$. Since the pair $(p_{x'}, x')$ satisfies (8.16), $c(p_{x'}) = 0$ and $L(p_{x'}, x) = f_o(p_{x'})$ for any $x \in X$. In particular, this means that $x_{x'}$ solves (8.12) when $x = \bar{x}$. (iv) Suppose $\bar{p}$ solves (8.6). Then, by item (iii) and by item (i) written for $\bar{x}$ and $\bar{p}$, $f_o(\bar{p}) \leq f_o(p_{x'}) = \theta(\bar{x}) \leq f_o(\bar{p})$. As a result, by definition of the dual function, $f_o(\bar{p}) = \theta(\bar{x}) \leq L(\bar{p}, \bar{x}) = f_o(\bar{p}) + \langle \bar{x}, c(\bar{p}) \rangle \leq f_o(\bar{p})$. Thus, $\langle \bar{x}, c(\bar{p}) \rangle = 0$ and $\bar{p}$ solves (8.12) for $\bar{x}$.
The converse assertion follows from item (i). $\qquad \square$

According to Lemma 8.3, to cancel the duality gap and to obtain $p_{\bar{x}}$ solving (8.6), it "suffices" to find an admissible primal-dual pair $(p_{\bar{x}}, \bar{x})$, with $\bar{x}$ solving (8.9). The next result shows the importance of convexity for these facts to hold.

Theorem 8.4. *Consider problem* (8.6) *and suppose that*

(i) *the set P is the whole space $\mathbb{R}^n$;*

(ii) *the equality constraint functions are affine and the inequality constraints are convex and differentiable; and*

(iii) *a constraint-qualification condition is satisfied, for example* (8.10).

Then $\bar{p}$ solves (8.6) *if and only if there exists a multiplier $\bar{x} \in X$ such that the Karush Kuhn-Tucker conditions are satisfied:*

$$\nabla_p L(\bar{p}, \bar{x}) = 0 \quad and \quad \bar{x}_i c_i(\bar{p}) = 0, \ with \ c_E(\bar{p}) = 0 \ and \ c_I(\bar{p}) \le 0.$$

Said otherwise, the primal solutions are exactly the points $p_{\bar{x}}$ as defined in (8.12) *such that $(p_{\bar{x}}, \bar{x})$ is admissible and $\bar{x}$ solves the dual problem* (8.9).

Thus, the duality gap vanishes if (8.6) has convex data and a constraint qualification is satisfied (for example, of Slater-type). In this case, a dual solution $\bar{x}$ is a Lagrange multiplier given by the Karush-Kuhn-Tucker Theorem.

8.3 Two Convex Nondifferentiable Functions

We finish this chapter with two important examples of NSO arising from finite minimax problems and dual problems.

8.3.1 Finite Minimax Problems

Functions defined as the pointwise maximum of a finite collection of smooth functions appear frequently as objective functions in (8.1). Suppose that

$$f(x) := \max\{f_j(x) : j = 1, \ldots, np\} \tag{8.17}$$

where each f_j is a convex smooth function, and let $J(x) := \{j : f_j(x) = f(x)\}$ denote the set of *active indices* at x. Then

$$f'(x; d) = \max\{\langle \nabla f_j(x), d \rangle : j \in J(x)\}$$

and

$$\partial f(x) = \operatorname{conv}\{\nabla f_j(x) : j \in J(x)\}.$$

Therefore, f has a gradient at x if there is a **unique** f_j giving the maximum, i.e., if there is only one active index (or if, by chance, the active gradients $\nabla f_j(x)$ have the same value for all $j \in J(x)$).

We shall often consider the particular case of affine functions, i.e., when $f_j(x) = \langle s^j, x \rangle + b_j$, for given $s^j \in \mathbb{R}^n$ and $b_j \in \mathbb{R}$. In this case,

$$\partial f(x) = \operatorname{conv}\{s^j : j \text{ such that } f(x) = \langle s^j, x \rangle + b_j\}. \tag{8.18}$$

Note that, given a fixed x, knowing just one active index $j \in J(x)$ amounts to knowing $f(x) = f_j(x)$ and one subgradient, $\nabla f_j(x) \in \partial f(x)$. The so-called *black-box methods* are developed on the basis of this minimal information of the objective function in (8.1); see § 9.3 below.

8.3.2 Dual Functions in Lagrangian Duality

When comparing the objective function in the minimax problem in § 8.3.1, and the dual function (8.8), we see that the latter is only an extension to an *infinite* collection of functions;

$$-\theta(x) = \sup_{p \in P} -L(p, x) = \sup_{p \in P}\{-\theta_p(x) := -f_o(x) - \langle x, c(p) \rangle\},$$

where the finite set of indices $\{1, \ldots, np\}$ in (8.17) is replaced by the infinite set $\{p \in P\}$. Likewise, the role of $\{f_j, j \le np\}$ in (8.17) is played by the collection of functions $\{-\theta_p, p \in P\}$. The following scheme summarizes these relations:

	finite minimax (8.17)	(negative of) dual function (8.8)
objective function	$f(x) = \max_j f_j(x)$	$-\theta(x) = \sup_p -\theta_p(x)$
collection of functions	$\{f_j(x)\}$	$\{-\theta_p(x) = -f_o(p) - \langle x, c(p) \rangle\}$
gradient of functions	$\nabla f_j(x)$	$-\nabla_x \theta_p(x) = -c(p)$
index set	$j \in \{1, \ldots, np\}$	$p \in P$
active indices	$j \in J(x)$	p_x solving (8.8)

The negative of the dual function θ is given by the supremum of functions that are affine with respect to x. In order to fully characterize its subdifferential, and obtain a version of (8.18) in this infinite setting, some additional assumptions are required. More precisely, the application of [195; Theorem VI.4.4.2] to the function $-\theta$ gives the following result.

Suppose that in (8.6) P is compact, f_o and the inequality constraints c_I are lower semicontinuous functions, and the equality constraints c_E are continuous. Then

$$\partial\left(-\theta\right)(x) = \operatorname{conv}\left\{-c(p_x) = -(c_E(p_x), c_I(p_x)) \text{ for all } p_x \text{ solving (8.8)}\right\}.$$

In particular,

$$-c(p_x) \in \partial\left(-\theta\right)(x). \tag{8.19}$$

We mention that the compactness assumption on P can be replaced by additional conditions on the functions. For example, to require the constraints to be bounded from below and the objective function f_o to be coercive on P:

$$\frac{f_o(p)}{\|p\|} \to +\infty \text{ if } p \in P \text{ and } \|p\| \to +\infty.$$

With these assumptions, it is possible to replace P in (8.8) by the (compact) set $P \cap B(0; \rho)$, where $B(0; \rho)$ is the ball in $\mathbb{R}^n$ with radius ρ, for ρ large enough. Note that this does not change the value of θ.

Most of the numerical NSO methods we consider here are of the *black-box* type. This means that they only need the knowledge of the function and **one** subgradient. Accordingly, for any given x, it is enough to compute just one minimizer p_x in (8.8): $-\theta(x) = -L(p_x, x)$ and, by (8.19) $-c(p_x)$ is a subgradient.

Remark 8.5. As in the finite minimax case, differentiability of θ depends on uniqueness of "active indices". For the dual function, this means uniqueness of the minimizer p_x in (8.8). When the Lagrangian is strictly convex as a function of p (i.e., for strictly convex f_o, affine c_E , and convex c_I), there is a unique minimum point p_x and $\nabla\theta(x) = (c_E(p_x), c_I(p_x))$. Otherwise, to define a strictly convex Lagrangian, one should add a suitable penalty term, forcing uniqueness. The resulting function is the *augmented Lagrangian*. For example, when (8.6) has only equality constraints, the augmented Lagrangian has the form

$$L_\pi(p, x) = L(p, x) + \pi\psi(\|c_E(p)\|).$$

For $L(\cdot, x) + \pi\psi(\|c_E(\cdot)\|)$ to be strictly convex on P, the positive penalization parameter π should be large enough and $L(\cdot, x)$ should be strictly convex on the kernel of the Jacobian $Jc_E(\cdot)$. The penalty function ψ is increasing and satisfies $\psi(t) > 0$ for all $t > 0$ with $\psi(0) = 0$. A typical example is $\psi(t) = t^2$; more complicated expressions for ψ appear when both equality and inequality constraints are present in (8.6); see cf. § 16.3. Note that the price of gaining differentiability of the dual function is in the delicate problem of properly setting π, by finding an adequate balance between optimality and feasibility.

9 Some Methods in Nonsmooth Optimization

Once a characterization of an optimal point for (8.1) has been derived, we are interested in the problem of *computing* a minimizer. We consider in this chapter the main difficulties arising when f is not differentiable, and give a first group of numerical methods.

9.1 Why Special Methods?

As a general rule, and for the sake of consistency, NSO methods strive to mimic as much as possible those of differentiable optimization. In general, the algorithms presented in the following sections generate iterates x^k by first finding a direction d^k (a descent direction in favorable cases) and then a scalar stepsize $t_k > 0$. The update of the iterate is given by $x^{k+1} = x^k + t_k d^k$.

Thus, at first glance, analyzing separately the nonsmooth case could seem useless. Yet, NSO has some traps in which a non-acquainted reader might easily fall:

- Trap of the stopping test: this issue is extremely delicate, because the condition "$g \in \partial f(x^k)$ with $\|g\| \leq \varepsilon$", directly translated from "$\|\nabla f(x^k)\| \leq \varepsilon$", may never happen. This situation occurs even in very simple cases. For example, the absolute value function of Figure 8.1 has $|g| = 1$ for all $x^k \neq 0$, the optimum point.

- Trap of approximate subgradients: often in practical situations, the subgradient (and even the function) is not computed exactly. Instead, it is obtained from f-values by finite differences. This approach is valid only in the smooth case, because $f'(x; d) = \langle \nabla f(x), d \rangle$ is **linear** in d and can be approximated by difference quotients. When f is not differentiable, the mapping $x \mapsto \partial f(x)$ is not continuous (cf. § 9.2.2 below). As a result, difference quotients do not necessarily belong to the subdifferential, not even in the limit. Consider in $\mathbb{R}^3$ the function $f(x) = \max\{x_1, x_2, x_3\}$. As shown in (8.18) above, $\partial f(x) = \{\alpha \in \mathbb{R}^3 : \alpha_i \geq 0, \sum_i \alpha_i = 1\}$ for all "diagonal" x, i.e., such that $x_i = \xi$ for all i. However, when $\xi = 0$ the forward, backward and central finite difference approximations are $(1, 1, 1)$, $(0, 0, 0)$ and $(1/2, 1/2, 1/2)$, respectively. None of these "approximated gradients" is in the unit simplex $\partial f(0, 0, 0)$.

– Curse of nondifferentiability: since the (set-valued) function $x \mapsto \partial f(x)$ is not continuous, a small variation on x^k may produce large variations on $\partial f(x^k)$. Directions d^k are computed on the available information on the subdifferential, so their computation may vary drastically and produce very different iterates x^{k+1}. This phenomenon occurs also when running the *same* program on different processors: roundoff errors are such that the sequences generated on different computers are no longer comparable! This unavoidable problem makes extremely difficult any numerical comparison.

9.2 Descent Methods

The philosophy of so-called descent methods is to generate a sequence $\{x^k\}$ such that each iteration guarantees a decrease of f. Different characterizations of descent directions were given at the end of § 8.1. Essentially, for all non-optimal x^k ($0 \notin \partial f(x^k)$) there exists a descent direction d^k, which corresponds to the strict separation of the sets $\{0\}$ and $\partial f(x^k)$. Figure 9.1 displays directions that are downhill for f at x^k. Note that each of them makes an obtuse angle with **every** element of the set $\partial f(x^k)$.

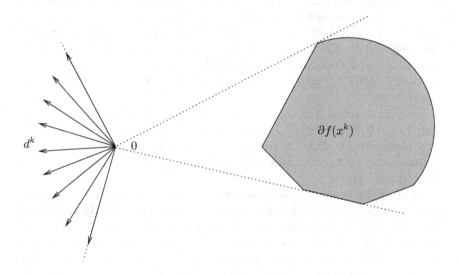

Fig. 9.1. Descent directions

The algorithmic scheme underlying most descent methods is the following.

Descent pattern. Take $x^1 \in \mathbb{R}^n$ and set $k = 1$.

STEP 1 (formal stopping test). If $0 \in \partial f(x^k)$, stop.

STEP 2 (descent). Find a descent direction d^k of f at x^k.

STEP 3 (line-search). Find a stepsize $t_k > 0$ such that $f(x^k + t_k d^k) < f(x^k)$.
STEP 4 (loop). Define $x^{k+1} := x^k + t_k d^k$. Change k to $k + 1$, go to 1.

The stopping test of STEP 1 is purely formal. The delicate issue of defining implementable stopping tests will be treated in more detail for two particular algorithms, cutting-planes and bundle methods, in § 9.3 and Chapter 10, respectively.

We shall consider different possibilities in STEP 2, showing how the non-differentiability of f limits the choice of directions. In particular, we shall see that some methods (subgradients, cutting-planes) do not follow at all the descent scheme in STEPS 2 and 3 above. The reason for such modification is that in NSO descent directions cannot always be generated; see Figure 9.6 below.

With respect to STEP 3, for simplicity we shall often set $t_k = 1$ for all k (even if $f(x^k + d^k) > f(x^k)$). For line-searches in NSO we refer to [269], [229]; see also the stepsize strategy in [195; Chapters XIV.3. and XV.3.3] and the curve-search algorithm in [236].

9.2.1 Steepest-Descent Method

A first idea (very natural but very unfortunate) to find a descent direction d^k consists in looking for the best possible descent at each iteration. This is the steepest-descent direction:

$$d^k \in \underset{\|d\|=1}{\operatorname{Argmin}} f'(x^k; d),$$

or equivalently, via (8.4),

$$d^k \in \underset{\|d\|=1}{\operatorname{Argmin}} \ \underset{s \in \partial f(x^k)}{\max} \ \langle s, d \rangle.$$

In this problem, the norm chosen to bound the feasible set could be any one (for differentiable functions, Euclidean normalizations have already been considered in § 2.5).

From a geometric point of view, $\{0\}$ and $\partial f(x)$ are separated by the hyperplane orthogonal to the *projection* of 0 onto $\partial f(x)$. The *steepest-descent direction*, boldfaced in Figure 9.1, is just opposite to this particular subgradient:

$$d^k = -\frac{\gamma^k}{\|\gamma^k\|} \text{ where } \gamma^k := P_{\partial f(x^k)}(0) \text{ belongs to } \partial f(x^k). \qquad (9.1)$$

Steepest-descent algorithms suffer from two important drawbacks which make them inefficient:

– The computation of a descent direction requires the knowledge of the *whole* subdifferential, and this at each iteration. This requirement is excessive, if not impossible, in most practical applications. For the finite minimax problem corresponding to (8.17), it would mean to identify all the active indices at every point. Likewise, for the dual function (8.8), one would need to compute all the minimizers p_x, which can form an infinite set. The so-called *black-box* methods, see § 9.3 and Chapter 10 below, are algorithms defined on the knowledge of the objective function value and just one subgradient, which is a more reasonable requirement.

– The sequence $\{x^k\}$ may oscillate and converge to a non-optimal point. This *zigzagging phenomenon* is demonstrated in Example 9.1 below; see also § VII.2.2, [195]. Zig-zags of steepest-descent methods for smooth functions were already mentioned in § 2.5, nondifferentiability is bound to only amplify the phenomenon. In § 9.2.2 we shall explain how to stabilize the steepest descent algorithm in order to eliminate such oscillations.

Example 9.1 (Instability of steepest-descent). Consider in $\mathbb{R}^2$ the minimax problem with objective function defined as follows:

$$f(x) := \max\{f_0(x), f_{-1}(x), f_{-2}(x), f_1(x), f_2(x)\},$$

where $f_0(x) := -100; \quad f_{\pm 1}(x) := 3x_1 \pm 2x_2; \quad f_{\pm 2}(x) := 2x_1 \pm 5x_2$.

The optimal value is $\bar{f} = -100$, and the (infinite) set of minimizers is $\{(x_1, x_2) \in \mathbb{R}^2 : x_1 \leq -50 \text{ and } |x_2| \geq 0.4x_1 + 20\}$, i.e., the region where f_0 is active. Figure 9.2 displays the regions where the various functions f_j are active; the locus of the kinks, in boldface in the figure, is made up of the half-lines K_{02}, K_{0-2}, K_{12}, K_{-1-2}, $K_{\pm 1}$, and $K_{\pm 2}$, where

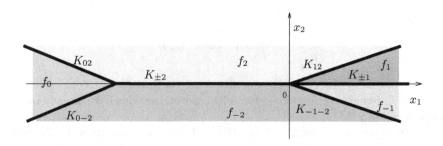

Fig. 9.2. Minimax function

$K_{02} := \{x : f(x) = f_0(x) = f_2(x)\} = \{(x_1, x_2) : x_2 = -20 - 0.4x_1\}$,
$K_{0-2} := \{x : f(x) = f_0(x) = f_{-2}(x)\} = \{(x_1, x_2) : x_2 = 20 + 0.4x_1\}$,
$K_{12} := \{x : f(x) = f_1(x) = f_2(x)\} = \{(x_1, x_2) : 3x_2 = x_1\}$,
$K_{-1-2} := \{x : f(x) = f_{-1}(x) = f_{-2}(x)\} = \{(x_1, x_2) : -3x_2 = x_1\}$,
$K_{\pm 1} := \{x : f(x) = f_1(x) = f_{-1}(x)\} = \{(x_1, x_2) : x_2 = 0, \, x_1 \geq 0\}$,
$K_{\pm 2} := \{x : f(x) = f_2(x) = f_{-2}(x)\} = \{(x_1, x_2) : x_2 = 0, -50 \leq x_1 < 0\}$.

Consider the region $x_1 \geq 0$. For any x on K_{-1-2}, by (8.18), the active gradients are ∇f_{-1} and ∇f_{-2}. Therefore

$$x \in K_{-1-2} \Rightarrow \partial f(x) = \mathrm{conv}\{(3,-2),(2,-5)\} \text{ so } \gamma = P_{\partial f(x)}(0) = (3,-2).$$

Similarly,

$$x \in K_{12} \Rightarrow \partial f(x) = \mathrm{conv}\{(3,2),(2,5)\} \text{ and } \gamma = P_{\partial f(x)}(0) = (3,2).$$

Start the steepest-descent algorithm at $x^1 := (9,-3) \in K_{-1-2}$. Omitting normalization, the steepest-descent direction is $d^1 = (-3,2)$. The line-search function $0 \leq t \mapsto q(t) := f(x^1 + td^1)$ has two kinks, $x^1 + \frac{3}{2}d^1 \in K_{\pm 1}$ and $x^1 + 2d^1 \in K_{12}$:

$$q(t) = \begin{cases} 33 - 13t, & \text{if } 0 \leq t \leq 3/2, \\ 21 - 5t, & \text{if } 3/2 \leq t \leq 2, \\ 3 + 4t, & \text{if } 2 \leq t \leq 3. \end{cases} \quad [\text{since } x_1^1 + td_1^1 \geq 0 \iff t \leq 3]$$

The exact line-search at STEP 3 would give the optimal stepsize $t_1 = t^* = 2$, for which $x^1 + t_1 d^1 \in K_{12}$. As for Wolfe's line-search (§ 3.4), starting from the data $q(0) = 33$ and $q'(0) = -13$, it would reject any $t \leq 3/2$ as "too small" (the test $q'(t) \geq m_2 q'(0)$ is not satisfied). In fact, when seeking a stepsize $t \geq 3/2$ not "too large", any adjustment of t adapted to piecewise linear functions, should find $t_1 = t^*$. The next iterate is $x^2 = (3,1) \in K_{12}$, which in turn produces $x^3 = (1,-1/3) \in K_{-1-2}$, and the zigzagging phenomenon between the two half-lines K_{12} and K_{-1-2} becomes blatant. As shown in Figure 9.3, the sequence $\{x^k = (3^{3-k}, (-1)^k 3^{2-k})\}_{k \geq 1}$ converges (very slowly!) to 0, which is a **non-optimal** kink.

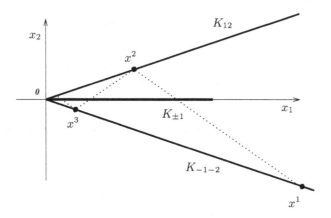

Fig. 9.3. Zigzagging trajectory

9.2.2 Stabilization. A Dual Approach. The ε-subdifferential

For a descent algorithm to be convergent, it should generate a sequence $\{x^k\}$ satisfying the following conditions:
- the sequence $\{f(x^k)\}$ is strictly decreasing,
- the sequence $\{x^k\}$ is a *minimizing* for problem (8.1), i.e.,
  $\liminf f(x^k) = f(\bar{x}) \leq f(y)$ for all $y \in \mathbb{R}^n$. In other words,
  - $\{x^k\}$ has a cluster point $\bar{x}$,
  - $\bar{x}$ is a minimizer of f.

In Example 9.1 the sequence $\{x^k\}$ does have decreasing objective values ($f(x^k) = 11 * 3^{2-k}$) and does have a cluster point, the zero vector. However, this cluster point is not a minimum of f. To understand what is wrong with the steepest descent method, consider the sequence of distances

$$\{\mathcal{D}_k := \text{dist}(0, \partial f(x^k)) = \|\gamma^k\|\},$$

where γ^k is as in (9.1). The multifunction $x \mapsto \partial f(x)$ has a closed graph, i.e.,

$$\text{if } \left\{\left(x^k, g^k \in \partial f(x^k)\right)\right\} \to (\bar{x}, \bar{g}) \text{ then } \bar{g} \in \partial f(\bar{x}). \tag{9.2}$$

Therefore, if $\{\mathcal{D}_k\}$ tends to 0, then $0 \in \partial f(\bar{x})$ and the cluster point is a minimizer. But for our example, the method generates a sequence of gradients $g^k = \gamma^k = (3, (-1)^k 2)$, for which $\mathcal{D}_k = \sqrt{3^2 + |2|^2}$ for all k.

In order to ensure that $\bar{x}$ is a minimizer of f (i.e., to ensure that $\{\mathcal{D}_k\} \to 0$), we would need the subdifferential to be continuous as a multifunction. This means that $\partial f(\cdot)$ has to be both outer and inner semicontinuous[1]. Outer semicontinuity holds because of the closedness property (9.2). However, the following property, of inner-continuity, **does not** hold:

if $\{x^k\} \to \bar{x}$ and $\bar{g} \in \partial f(\bar{x})$ then there exists $\{g^k \in \partial f(x^k)\} \to \bar{g}$.

This phenomenon occurs in Example 9.1, but it also occurs for the absolute value function (take the sequence $\{x^k = 1/k\}$, which minimizes the function $f(x) = |x|$ and whose (sub)gradients are constantly equal to 1).

The important continuity property of subgradients can be enforced by the introduction of a tolerance, or viscosity parameter, in the definition of the subdifferential. More precisely, for a given $\varepsilon \geq 0$, the *ε-subdifferential of f at x* is defined as

$$\partial_\varepsilon f(x) := \{s \in \mathbb{R}^n : f(y) \geq f(x) + \langle s, y - x \rangle - \varepsilon \text{ for all } y \in \mathbb{R}^n\}. \tag{9.3}$$

This set approximates $\partial f(x)$, with $\partial f(x) \subseteq \partial_\varepsilon f(x)$ for all $\varepsilon \geq 0$. Moreover, it is a continuous multifunction of x:

---

[1] We prefer here the terminology of [195] to that of an upper semicontinuous *multi*function. The reader will find an explanation of this terminology in the appendix of the cited book.

– it is outer semicontinuous because its graph is closed (Proposition XI.4.1.1 in [195]):

$$\left\{ \left(x^k, \varepsilon_k, s_k \in \partial_{\varepsilon_k} f(x^k) \right) \right\}_k \to (\bar{x}, \bar{\varepsilon}, \bar{s}) \implies \bar{s} \in \partial_{\bar{\varepsilon}} f(\bar{x}). \tag{9.4}$$

– Since f is Lipschitz-continuous, we have for fixed $\varepsilon > 0$ that

$$\forall \rho > 0 \ \exists \delta > 0 \ : \ \|x^k - \bar{x}\| \leq \delta \implies \partial_\varepsilon f(\bar{x}) \subset \partial_\varepsilon f(x^k) + B(0; \rho),$$

which means that $\partial_\varepsilon f(\cdot)$ is also inner semicontinuous (at $\bar{x}$).

The smearing parameter ε is also introduced in other objects related to the subdifferential. For instance, the ε-optimality condition is $0 \in \partial_\varepsilon f(\bar{x})$, and it means that $f(y) \geq f(\bar{x}) - \varepsilon$ for all $y \in \mathbb{R}^n$. Likewise, for any point x^k that is not ε-optimal there are directions d for which the approximate directional derivative $f'_\varepsilon(x^k; d) := \lim_{t \searrow 0} \frac{f(x+td) - f(x) + \varepsilon}{t}$ is negative. In such directions, called of ε-descent at x^k, there exists a positive scalar t such that $f(x^k + td) < f(x^k) - \varepsilon$.

The ε-subdifferential calculus is rather involved (see Chapter XI in [195]). When f has the form (8.17), for the particular case of a maximum of affine functions f_j, it holds that

$$\partial_\varepsilon f(x) = \left\{ \sum_{i=1}^{np_x} \alpha_i s^i : \alpha_i \geq 0, \sum_{i=1}^{np_x} \alpha_i = 1 \text{ and } f(x) \leq \sum_{i=1}^{np_x} \alpha_i f_i(x) + \varepsilon \right\},$$

where $np_x := \min(np, n+1)$ depends on x. Unlike the subdifferential formula (8.18), which only makes use of active gradients, the smeared subgradients may employ all the gradients ∇f_j for $j = 1, \ldots, np_x$. In particular, for Example 9.1, after some calculations we obtain that, for $\varepsilon > 0$ small enough,

$$x \in K_{-1-2} \implies \partial_\varepsilon f(x) = \text{conv}\{(3, -2), (2, -5), (3, 2)\}$$

and $\gamma_\varepsilon := P_{\partial_\varepsilon f(x)}(0) = \left(\frac{140}{47}, -\frac{46}{47} \right)$. The associated ε-steepest descent direction d^1_ε now points out of the region defined by K_{-1-2} and K_{12}. The corresponding line-search function $q(t)$ no longer crosses $K_{\pm 1}$ but crosses $K_{\pm 2}$ at the only one kink $t = \frac{141}{46}$:

$$q(t) = \begin{cases} 33 - \frac{510}{47} t, & \text{if } 0 \leq t \leq \frac{141}{46}, \\ 3 - \frac{50}{47} t, & \text{otherwise}. \end{cases}$$

The exact line-search at STEP 3 would give the optimal stepsize $t_1 = t^* = \frac{141}{46}$, for which $x^1 + t_1 d^1_\varepsilon = (-\frac{3}{23}, 0)$. The dangerous kink $x = (0,0)$ is left behind, and the algorithm proceeds; see Figure 9.4.

Methods that use the ε-subdifferential to compute directions are called of ε-descent. They generate sequences $\{x^k\}$ and $\{\varepsilon_k\}$ in such a way that

– $\{f(x^k)\}$ is decreasing;

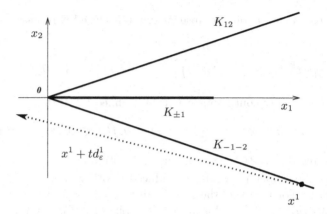

Fig. 9.4. Escaping non-optimal kinks by ε-steepest descent

– $\{\varepsilon_k\}$ tends to 0;
– 0 is a cluster point of the sequence $\{\mathcal{D}_{\varepsilon_k} = \mathrm{dist}(0, \partial_{\varepsilon_k} f(x^k)) = \|\gamma_\varepsilon^k\|\}$.
These methods can be interpreted as a stabilization in the dual space (i.e., in the space of subgradients) of the steepest descent algorithm.

Note that ε-steepest descent methods are still not implementable since, to compute γ_ε^k, the *whole* ε-subdifferential must be known. We shall see in Chapter 10 that bundle methods construct smeared subgradients $\hat{s}^k$ *without* projecting onto $\partial_{\varepsilon_k} f(x^k)$. Instead, they define a polyhedral approximation for the subdifferential that can be built along iterations using only the black-box information. At the same time, bundle methods approximate the objective function f by a polyhedral model. This results in a primal stabilization (in the x- space); see § 10.1 below.

9.3 Two Black-Box Methods

Knowing that the *whole* subdifferential $\partial f(x)$ is usually not available, we shall deal with a much weaker requirement. For any given point, the user computes the value of the function and only *one* arbitrary subgradient.

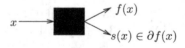

Fig. 9.5. Black box

The *black box* routine displayed in Figure 9.5 is also sometimes called oracle or simulator. Algorithms will be constructed based on this sole information, given by the user, and they will aim at being descent schemes.

However, due to the little information available, algorithms may not always be able to produce (or even recognize) descent directions. As a result, STEP 2 of the Descent Pattern only produces *candidates* to become a direction of descent (rather than certain descent directions). The performance of the various algorithms will heavily depend on their capacity to generate and recognize which candidates are "good enough", in the sense of decreasing the objective value.

The term "black box" comes from the fact that optimization methods generate a minimizing sequence *independently* of the way the calculations are organized to obtain $f(x)$ and $s(x)$. This part of the program is left to the user's responsibility, who is supposed to know the real nature of the problem.

9.3.1 Subgradient Methods

In the differentiable case, the direction $-\nabla f(x)$ is downhill, as well as any direction in the half-space opposite to the gradient, since they make an obtuse angle with $\nabla f(x)$. Starting from this remark, it is conceivable to mimic this behavior with the information obtained from the black box. Similarly to the implementable scheme in § 3.6, we define the following algorithmic pattern:

Algorithm 9.2 (of subgradients). Take $x^1 \in \mathbb{R}^n$ and set $k = 1$.

STEP 1 (Calling the black box – formal stopping test). Call the black box of Figure 9.5 with $x = x^k$. If $0 \in \partial f(x^k)$, stop.

STEP 2 (Candidate descent-direction). Set $d^k := -\dfrac{s(x^k)}{\|s(x^k)\|}$.

STEP 3 (line-search). Find a stepsize $t_k > 0$ satisfying, if possible, the condition $f(x^k + t_k d^k) < f(x^k)$.

STEP 4 (loop). Define $x^{k+1} := x^k + t_k d^k$. Change k to $k + 1$, go to 1.

STEP 3 is a relaxed form of the usual line-search requirement in smooth optimization (for example, Armijo's rule in § 3.5.2), in the sense that descent is imposed whenever possible. This is because in the nondifferentiable case a direction opposite to a subgradient need not always be a descent direction. As shown in Figure 9.1, to be downhill, a direction must make an obtuse angle with the *whole* subdifferential, and not with the sole element $s(x)$. Therefore, having as an output of the black box just one subgradient, it is not possible to check if a given direction is downhill.

Figure 9.6 displays the level-lines for two functions minimized at $0 \in \mathbb{R}^2$: the differentiable function $f_L(x_1, x_2) = x_1^2 + 2x_2^2$ on the left and the nondifferentiable function $f_R(x_1, x_2) := |x_1| + 2|x_2|$ on the right. The shadowed area shows all descent-directions in the two cases. Observe that for this simple case, the direction opposite to $s = (1, 2) \in \partial f_R(0, x_2)$, for $x_2 > 0$, is **not** a direction of descent.

Even though Algorithm 9.2 is not a descent pattern, an adequate choice of stepsizes t_k does produce a convergent method. To see what can be done,

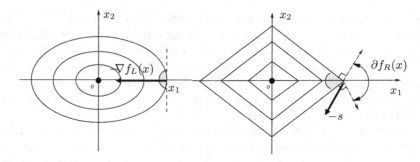

Fig. 9.6. Differentiable and nondifferentiable cases

given any point $y \in \mathbb{R}^n$, consider the sequence of square distances $\|x^{k+1} - y\|^2$. Expanding the squares, and using the definitions of x^{k+1} and d^k, we obtain

$$\|x^{k+1} - y\|^2 = \|x^k - y - t_k \frac{s(x^k)}{\|s(x^k)\|}\|^2$$
$$= \|x^k - y\|^2 + t_k^2 + 2 \frac{t_k}{\|s(x^k)\|} \langle y - x^k, s(x^k) \rangle .$$

The subgradient inequality (8.5) written at $x = x^k$ gives the relation

$$\|x^{k+1} - y\|^2 \le \|x^k - y\|^2 + t_k^2 + 2 \frac{t_k}{\|s(x^k)\|} (f(y) - f(x^k)) . \qquad (9.5)$$

The following convergence result makes repeated use of this relation. We assume that the stopping test in Algorithm 9.2 never holds and, hence, an infinite sequence of iterates $\{x^k\}$ is generated.

Theorem 9.3. *Assume* (8.1) *has minimizers. Suppose Algorithm 9.2 loops forever* $(k \to \infty)$. *If at* STEP 3 *stepsizes are chosen so that*

$$(a) \sum_k t_k = +\infty \quad and \quad (b) \sum_k t_k^2 < \infty ,$$

then $\{x^k\}$ *converges to a minimum point of* (8.1), *at sub-linear rate.*

Proof. Write (9.5) for $y = \bar{x}$, a minimizer in (8.1), which exists by the assumption. The corresponding relation has a third right hand side term that is negative, so $0 \le \|x^{k+1} - \bar{x}\|^2 \le \|x^k - \bar{x}\|^2 + t_k^2$. Then, by condition (b), the sequence of square errors $\{\|x^k - \bar{x}\|^2\}$ converges (for now, not necessarily to 0). Hence, the sequence $\{x^k\}$ is bounded and, by the local boundedness property of $\partial f(\cdot)$, so is the sequence $\{s(x^k)\}$, say by a constant M. Suppose, for contradiction purposes, that $\{x^k\}$ is not a minimizing sequence for (8.1). Then $\liminf f(x^k) > f(\bar{x})$ and there exist a positive constant C, a vector $\bar{y}$ and an index K such that

$$f(\bar{y}) \leq f(x^k) - C \text{ for all } k \geq K.$$

Choosing $y = \bar{y}$ in (9.5), this means that

$$\|x^{k+1} - \bar{y}\|^2 \leq \|x^k - \bar{y}\|^2 + t_k(t_k - 2C')$$

for some positive constant $C' = C/M$. Since $t_k \to 0$ by condition (b), the factor $(t_k - 2C')$ above is eventually smaller than $-C'$ and

$$\|x^{k+1} - \bar{y}\|^2 \leq \|x^k - \bar{y}\|^2 - t_k C' \text{ for } k \text{ big enough.}$$

Summing up the terms in the last inequality, and using again the fact that the square errors converge, yields a contradiction of condition (a). Thus, it must hold that $\liminf f(x^k) = f(\bar{x})$. By continuity of f, the (bounded) sequence $\{x^k\}$ has a cluster point $\hat{x}$ which minimizes f. For this minimizer, reasoning as for $\bar{x}$, the sequence of errors $\|x^k - \hat{x}\|$ converges, now to 0, and the conclusion follows.

To prove the rate of convergence result, we use again a contradiction argument. Suppose that the rate is linear, and recall Theorem 1.6. There exist $C > 0$ and $\theta \in]0, 1[$ such that $\|x^k - \bar{x}\| \leq C\theta^k$ for all k. Since $t_k = \|x^{k+1} - x^k\| \leq \|x^{k+1} - \bar{x}\| + \|\bar{x} - x^k\| < 2C\theta^k$, the series $\sum_k t_k$ would be convergent, contradicting condition (a). $\qquad\square$

A general description of subgradient methods can be found in Shor's monograph [332]. In view of condition (a) of Theorem 9.3, subgradient methods are also called of "divergent series". A major difficulty of these methods is to define practical rules for t_k which ensure convergence. For example, choosing t_k by means of a classical line-search may generate a zigzagging sequence as in Example 9.1.

With rules other than (a), Algorithm 9.2 can have better convergence rates. It is shown in [163] that if $t_k = t_0\theta^k$, for $t_0 > 0$ and $\theta \in]0, 1[$, the rate of convergence becomes geometric. However, the sequence converges to a point $\bar{x}$ that need not be optimal. Global convergence holds when t_0 and θ satisfy certain hypotheses, that cannot be checked in practice.

Another variant is possible when the optimal value $\bar{f} = f(\bar{x})$ is known. In this case, it suffices to take $t_k = 2m(f(x^k) - \bar{f})/\|s(x^k)\|$, with $m \in]0, 1[$ to improve the rate of convergence. Note that, with this choice, (9.5) written for $z = \bar{x}$ implies that the error decreases by $2m|m - 1|(|f(x^k) - \bar{f}|/\|s(x^k)\|)^2$ at each iteration.

When the value $\bar{f}$ is not available, *dilation* methods propose a rule that uses information from previous iterations to choose the stepsize. These "accelerated" methods avoid the generation of consecutive directions producing zig-zags. We can cite in this family the ellipsoid algorithm [206], a particular case of dilation along a subgradient, [331]. Along the same lines, subgradient level methods approximate f's level sets in order to obtain an estimate of $\bar{f}$, [55], [215], [164]. In these methods, there is a subsequence of "records", generated by remembering the points with smallest objective value, that is monotonically decreasing.

In general, subgradient methods suffer from important drawbacks (lack of implementable stopping test, lack of descent, possible poor rate of convergence). Nevertheless, they are extremely popular among practitioners, because of their simplicity of implementation. The methods we present in the following sections overcome some of these drawbacks. As a price to pay, subproblems defining directions at STEP 2 will have an increasing complexity.

9.3.2 Cutting-Planes Method

Essentially, subgradient methods use the information given by the black box only once at a time, without a memory of past iterations. If, instead, past information is kept, it is possible to define a model of the objective function. Cutting-planes methods, [78], [205], use the values

$$f_i := f(x^i) \text{ and } s^i := s(x^i), \text{ for } i = 1, \dots, k$$

obtained so far, to construct the following piecewise-affine model for f:

$$\check{f}_k(y) := \max_{i=1,\dots,k} \{f_i + \langle s^i, y - x^i \rangle\}. \tag{9.6}$$

The minimization of the model $\check{f}_k$, on a convex compact set S, to be determined before starting the method, gives the new iterate x^{k+1}. Figure 9.7 shows different cutting-planes for a (smooth) function f.

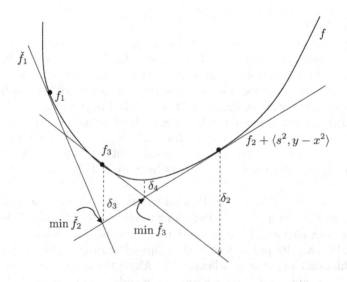

Fig. 9.7. Cutting-planes iterations

Note that by definition of the model,

$$\text{for all } k \quad \check{f}_k \leq \check{f}_{k+1} \quad \text{and} \quad \check{f}_k \leq f, \tag{9.7}$$

where the last relation follows from the subgradient inequality (8.5). So, by the convexity of f, the graph of the cutting-planes model $\check{f}_k$ approaches the graph of f from outside, with increasing accuracy as k grows. This is why the method is convergent; see Theorem 9.6 below. In addition, there is now an implementable stopping test (note the decreasing values of the distances δ_k in Figure 9.7).

Algorithm 9.4 (cutting-planes). Let $\text{tol} \geq 0$ be a given stopping tolerance and let $S \neq \emptyset$ be a compact convex set containing a minimum point of f. Choose $x^1 \in S$ and set $k = 1$. Define $\check{f}_0 \equiv -\infty$.

STEP 1 (Calling the black box – implementable stopping test).
Call the black box from Figure 9.5 with $x = x^k$. Compute

$$\delta_k := f(x^k) - \check{f}_{k-1}(x^k) . \tag{9.8}$$

If $\delta_k \leq \text{tol}$, stop.
STEP 2 (Candidate descent direction). Find

$$d^k \in \text{Argmin}_{x^k + d \in S} \check{f}_k(x^k + d) .$$

STEP 3 (Line-search – constant stepsize). Take $t_k = 1$.
STEP 4 (loop). Define $x^{k+1} := x^k + t_k d^k$. Change k to $k + 1$, go to 1.

Remark 9.5.
– For simplicity, we present here a version without line-search. STEP 2 can be replaced by

$$x^{k+1} \in \text{Argmin}_{y \in S} \check{f}_k(y) , \tag{9.9}$$

and STEP 3 can be omitted.

– We shall not specify the choice of the set S, introduced to guarantee the existence of a direction d^k (otherwise, at the initial iterations the minimization subproblem in STEP 2 may be unbounded from below). Choosing this set is a key element to overcome the intrinsic instability of cutting-planes (cf. the final comments in Example 9.7).

– Having on hand $\check{f}_k$, a *model* of f, allows us to quantify in STEP 1 a "nominal decrease" $\delta_k > 0$, predicted by $\check{f}_k$. If the model is good (i.e., if k is large enough), the stopping test will eventually be activated; see Figure 9.7 and Theorem 9.6.

– Finally, note that a price was paid for obtaining a stopping test. Suppose S is polyhedral, for example, a box. Then, to compute d^k in STEP 2, instead of the straightforward calculation in the Subgradients Algorithm 9.2, now it is necessary to solve a linear programming (LP) problem:

$$
\begin{cases}
\min_{(d,r)} r \\
r \geq f_i + \langle s^i, x^k - x^i \rangle + \langle s^i, d \rangle, i = 1, \ldots, k \\
x^k + d \in S \text{ and } r \in \mathbb{R}.
\end{cases}
\tag{9.10}
$$

Like in subgradient methods, the sequence generated by Algorithm 9.4 does not necessarily have decreasing objective values $f(x^k)$. A simple example of this situation appears in Figure 9.8, where we made a zoom of Figure 9.7. Basically, the closer to an optimum is x^k (x^4 in the figure), the worse will be the next iterate. This is due to the introduction in the model $\check{f}_k$ of an almost "horizontal" affine function ($f_4 + \langle s^4, y - x^4 \rangle$ in the figure). As a result, the minimizer x^{k+1} of the model is "pushed away" from the minimum of f (cf. x^5 in the figure).

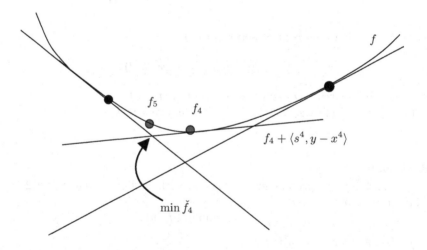

Fig. 9.8. Cutting-planes iterations near a minimizer

Bundle methods, described in Chapter 10, propose variants to force the decrease of f, while keeping the model idea, important to define stopping tests.

We now prove convergence for the sequence of records $\{\bar{f}_k := \min_{1 \leq i \leq k} f_i\}$, i.e., the best values obtained for f along the iterations of Algorithm 9.4. Note that the assumption (at the initialization step of Algorithm 9.4) that S is a compact set containing a minimizer of (8.1) implies that the optimal value $\bar{f}$ is finite and that the sequence $\{x^k\}$ is bounded.

Theorem 9.6. *Consider the sequence $\{x^k\}$ generated by Algorithm 9.4.*

(i) If the algorithm loops forever ($k \to \infty$), then

$$\lim_{k\to\infty} \check{f}_{k-1}(x^k) = \bar{f} = \lim_{k\to\infty} \bar{f}_k = \liminf_{k\to\infty} f(x^k).$$

Therefore, Algorithm 9.4 is guaranteed to stop at some iteration k_{last} whenever tol > 0.

(ii) If the sequence is finite, its last element, $x^{k_{last}}$ is tol$-$*optimal:*

$$f(x^{k_{last}}) \le \bar{f} + \text{tol}.$$

Proof. (i) If $k \to \infty$, by (9.9) and the left inequality in (9.7), the sequence $\{\check{f}_{k-1}(x^k)\}$ is nondecreasing and bounded above by $\bar{f}$, because $\check{f}_k \le f$ for all k. Therefore, $\check{f}_k(x^{k+1}) \nearrow \bar{f} - C$, where $C \ge 0$ is some constant. So, it holds that

$$\text{for all } k \quad \check{f}_{k-1}(x^k) \le \bar{f} - C. \tag{9.11}$$

We now show that assuming $C > 0$ yields a contradiction. Take a convergent subsequence $\{x^{k_l}\}_{k_l}$ of the (bounded) sequence $\{x^k\} \subset S$, and let M be an upper bound of diam($\partial f(x^{k_l-1})$), for k_l big enough. In addition, take k_l large enough to satisfy $\|x^{k_l-1} - x^{k_l}\| \le C/(2M)$. The desired contradiction results from the following chain of inequalities:

$$
\begin{aligned}
\bar{f} - C &\ge \check{f}_{k_l-1}(x^{k_l}) & \text{[by (9.11) with } k = k_l] \\
&\ge \check{f}_{k_{l-1}}(x^{k_l}) & \text{[by (9.7), since } k_l - 1 \ge k_{l-1}] \\
&\ge f_{k_{l-1}} + \langle s^{k_{l-1}}, x^{k_l} - x^{k_{l-1}} \rangle & \text{[by (9.6)]} \\
&\ge f_{k_{l-1}} - M\|x^{k_{l-1}} - x^{k_l}\| & \text{[by Cauchy-Schwarz]} \\
&\ge \bar{f} - MC/(2M) & \text{[for } k_l \text{ large enough]} \\
&= \bar{f} - \frac{C}{2}.
\end{aligned}
$$

We now show that $\liminf_{k\to\infty} f(x^k) = \bar{f}$ (note that this automatically implies that the decreasing sequence of records $\{\bar{f}_k\}$ converges to $\bar{f}$). Suppose there exists $\tilde{C} > 0$ such that $f(x^k) \ge \bar{f} + \tilde{C}$ for all k. By the boundedness of $\{x^k\}$, there exists a convergent subsequence $\{x^{k_l}\}_{k_l}$. Let L be a common Lipschitz constant for f and $\check{f}_k$ and take k_l large enough to satisfy $\|x^{k_{l+1}} - x^{k_l}\| \le \tilde{C}/(2L)$. The following chain of inequalities shows that $\tilde{C}$ cannot be positive:

$$
\begin{aligned}
\bar{f} + \tilde{C} &\le f(x^{k_l}) \\
&= \check{f}_{k_l}(x^{k_l}) & \text{[by (9.6)]} \\
&= \check{f}_{k_l}(x^{k_l}) - \check{f}_{k_l}(x^{k_l+1}) + \check{f}_{k_l}(x^{k_l+1}) \\
&\le \check{f}_{k_l}(x^{k_l}) - \check{f}_{k_l}(x^{k_l+1}) + \check{f}_{k_{l+1}-1}(x^{k_l+1}) & \text{[by (9.7), since } k_l \le k_{l+1} - 1] \\
&\le \check{f}_{k_l}(x^{k_l}) - \check{f}_{k_l}(x^{k_l+1}) + \bar{f} & \text{[by (9.11) with } k = k_{l+1}] \\
&\le L\|x^{k_{l+1}} - x^{k_l}\| + \bar{f} & [\check{f}_{k_l} \text{ Lipschitz}] \\
&\le L\tilde{C}/(2L) + \bar{f} & \text{[for } k_l \text{ large enough]} \\
&= \bar{f} + \frac{\tilde{C}}{2}.
\end{aligned}
$$

Finally, note that when `tol` is positive, the existence of a subsequence $\{f(x^{k_i})\} \to \bar{f}$ ensures that the stopping test will eventually be activated.

(*ii*) Since the sequence is finite, $\delta_{k_{last}} \leq$ `tol`, and

$$
\begin{aligned}
f(x^{k_{last}}) &\leq \check{f}_{k_{last}-1}(x^{k_{last}}) + \mathtt{tol} && \text{[by (9.8)]}\\
&\leq \check{f}_{k_{last}-1}(y) + \mathtt{tol} && \text{[for all } y \in S \text{, by (9.9)]}\\
&\leq f(y) + \mathtt{tol} && \text{[by the right inequality in (9.7).]}
\end{aligned}
$$

Since, by assumption, S contains a minimum of f, the conclusion follows.

$\square$

Cutting-planes methods have good convergence properties for functions whose graph is V-shaped, or *sharp* functions; see [313], [286]. In particular, for polyhedral functions the method has finite convergence: this is a consequence of Theorem 9.6(*i*), since polyhedral functions coincide with their cutting-planes model $\check{f}_k$ for some k, say k_{last}. Then $\delta_{k_{last}} = 0$, the sequence $\{x^k\}$ is finite and the stopping test holds for any tolerance `tol`, in particular `tol` $= 0$.

For general functions, however, the cutting-planes algorithm may present instabilities and bad numerical behavior. As shown graphically in Figure 9.8, when "horizontal" affine functions are added to the model, iterates can move far away from the set of minima and increase the current function value. The following function, known as BADGUY in the literature, is due to A. Nemirovskii (exercise 3, § 4.3.6, [272]; see also § XV.1.1.2, [195]). It gives an analytical extreme example of the catastrophic consequences of instability.

Example 9.7 (Instability of cutting-planes). Take in $\mathbb{R}^n$ the function

$$
f(x) := \max\{0, -1 + 2\lambda + \|x\|\},
$$

where $\lambda \in]0, 1/2[$ is an arbitrary parameter. The optimal value is $\bar{f} = 0$, and the optimal set is the ball $B(0; 1 - 2\lambda)$, centered at the origin and with radius $1 - 2\lambda$. The function f is nondifferentiable on the sphere of kinks $\bar{B}(0; 1 - 2\lambda) := \{x \in \mathbb{R}^n : \|x\| = 1 - 2\lambda\}$. From § 8.3.1, the subdifferential has the following expression:

$$
\partial f(x) = \begin{cases} \{0\}, & \text{if } \|x\| < 1 - 2\lambda; \\ \{x/\|x\|\}, & \text{if } \|x\| > 1 - 2\lambda; \\ \mathrm{conv}\{0, x/\|x\|\}, & \text{if } \|x\| = 1 - 2\lambda. \end{cases}
$$

To start Algorithm 9.4, we assume that whenever (9.9) has more than one minimum, we choose a solution such that x^{k+1} has norm equal to 1. With this convention, for all k

$$
f(x^{k+1}) = 2\lambda, \quad \partial f(x^{k+1}) = \{\nabla f(x^{k+1})\} = x^{k+1}, \text{ and}
$$
$$
f_{k+1} + \langle s^{k+1}, y - x^{k+1} \rangle = 2\lambda + \langle x^{k+1}, y - x^{k+1} \rangle = -1 + 2\lambda + \langle x^{k+1}, y \rangle.
$$

Take $S = B(0;1) \subset \mathbb{R}^n$ as the unit-ball and start with $x^1 = 0 \in \mathbb{R}^n$. We have $\delta_1 = +\infty$, $\check{f}_1 \equiv 0$ and $x^2 \in \mathrm{Argmin}_S 0 = S$. So

$$\check{f}_2(y) = \max\{0, -1 + 2\lambda + \langle x^2, y \rangle\}.$$

This function is 0 on the region $S_2 := \{y \in \mathbb{R}^n : \langle x^2, y \rangle \leq 1 - 2\lambda\}$. The minimization (9.9) eliminates this portion of S:

$$x^3 \text{ solution to } \min_{y \in S} \check{f}_2(y) \equiv -1 + 2\lambda + \min_{y \in (S \setminus S_2)} \langle x^2, y \rangle.$$

Since both x^2 and x^3 are unit norm vectors, the minimum above is attained when the angle between x^3 and x^2 is π. Along iterations, the same phenomenon occurs: by the fact that $\check{f}_1 \equiv 0$, a section of S (up to a π-rotation) is eliminated at each iteration:

$$x^{k+1} \text{ solution to } -1 + 2\lambda + \min_{y \in (S \setminus \cup_2^k S_j)} \max_{j=2,\ldots,k} \langle x^j, y \rangle.$$

Again, the angle between x^{k+1} and x^k is π. Therefore, as long as there remain vectors $y \in (S \setminus \cup_2^k S_j)$ with $\|y\| = 1$, there will be a unit norm x^{k+1} and the black box will invariably answer the value $f(x^{k+1}) = 2\lambda$. The number of iterations necessary to obtain a value $f(x^{k+1}) < 2\lambda$ depends of the number of cuts S_k necessary to eliminate the unit-sphere $\bar{B}(0;1)$ from the unit ball. To estimate this number, one should compare the areas of $\bar{B}(0;1)$ and of an arbitrary cut $S_k \cap \bar{B}(0;1)$, denoted by A_S and A_λ, respectively (the area of a cut does not depend on k, but on λ).

Let θ be the angle for which $\cos \theta = 1 - 2\lambda$, we have $\theta \simeq 2\lambda^{\frac{1}{2}}$. Some (pages of) computations give the relations $A_S = 2S_{n-1}/n$ and $A_\lambda = \theta^n S_{n-1}/n$, where S_{n-1} is the area of the unit-sphere in $\mathbb{R}^{n-1}$.

Therefore, Algorithm 9.4 will need $k = A_S/A_\lambda$ iterations until all unit-vectors are eliminated from the unit ball, the feasible set in (9.9), i.e.,

$$k \geq \frac{2}{\theta^n} \simeq 2(4\lambda)^{-n/2}.$$

Since $f(x^2) - \bar{f} = \cdots = f(x^k) - \bar{f} = 2\lambda$ as long as there remain such unit norm vectors, if, for example, $\lambda = 0.025$ and $n = 20$, a number $k := 2 * 10^{10}$ of iterations is necessary before reducing the initial error of 0.05!

The situation can change dramatically if the feasible set S in (9.9) is allowed to vary along iterations. Suppose that, for the same function f, we now generate iterates satisfying

$$\tilde{x}^{k+1} \in \mathrm{Argmin}_{y \in \tilde{S}_k} \check{f}_k(y),$$

where $\tilde{S}_k := B(0; 1 - 2\lambda + \kappa_k)$, for $\kappa_k = \frac{2\lambda}{k+1}$. Given $\tilde{x}^1 = 0 \in \mathbb{R}^n$, we have that $\tilde{S}_1 = B(0; 1 - \lambda)$ and

$$\tilde{x}^2 \in \text{Argmin}_{y \in \tilde{S}_1} 0 \,.$$

Adopting again the convention that whenever the LP problem has more than one minimum, it chooses a solution on the border of $\tilde{S}_k$, i.e., such that $\|\tilde{x}^{k+1}\| = 1 - 2(1 - \frac{1}{k+1})\lambda$, we would have that $\|\tilde{x}^2\| = 1 - \lambda$, so $f(\tilde{x}^2) = \lambda$. At every iteration, the point $\tilde{x}^{k+1} \in \tilde{S}_k$ has norm smaller or equal than $1 - 2\lambda + \frac{2}{k+1}\lambda$. Since the feasible set $\tilde{S}_k$ shrinks with k, each iteration reduces the error:

$$f(\tilde{x}^2) - \bar{f} = \lambda, f(\tilde{x}^3) - \bar{f} \le \frac{2}{3}\lambda, \cdots, f(\tilde{x}^{k+1}) - \bar{f} \le \frac{2}{k+1}\lambda.$$

Taking $\lambda = 0.025$ as before, we see that already at the third iteration the error would be reduced from 0.05 to 0.05/3.

Our choice of feasible sets $\tilde{S}_k$, with $\kappa_k \to 0$ as $k \to \infty$, was made in order to make $\{\tilde{S}_k\}$ converge to the solution set $B(0; 1-2\lambda)$. Without the knowledge of the solution set (or, at least, of a "tight" set containing a minimizer), such choice is not possible a priori. Instead, one could define a ball centered at the best point obtained so far (with smallest function value, for example), with a radius that would vary along iterations, for instance $\tilde{S}_k = B(x_{best}^k; \kappa_k)$. This is the basic idea of bundle methods, in their trust region variant presented in Example 10.4.

Remark 9.8. Another serious drawback of the cutting-planes method is the infinite accumulation of affine functions defining the model. This phenomenon is amplified by the instability problem mentioned above. More precisely, as k grows, the linear program (9.10) defining iterates has more and more constraints. Because of instability, many constraints are similar. So the linear program (9.10) becomes extremely difficult to solve; not only due to its size, but also due to its bad conditioning. A way out of this tailing-off effect is to *clean the model*, and eliminate the *less active* constraints in (9.9), following some criteria. By the nature of cutting-planes methods, the selection of "inactive" constraints can only be done using heuristics, which are not supported by any convergence result. Bundle methods, by contrast, use the so-called *aggregation* technique to control the size (and, possibly, conditioning) of subproblems defining directions *without impairing* the original properties of global convergence.

10 Bundle Methods. The Quest for Descent

None of the black-box methods considered so far are descent schemes. Only the steepest-descent method guarantees a decrease of the objective function at each iteration. However, it requires the complete knowledge of ∂f, and can be trapped at non-optimal kinks.

We now focus our attention on a family of black-box methods that combine both *descent* and *stability* properties, called bundle methods [226], [362], [227], because they keep memory of past iterations in a *bundle* of information:

$$\mathcal{B} = \{f_i, y^i, s^i, i = 1, \ldots, k\} \text{ and } x^k, \text{ the point with "best" objective value}.$$

10.1 Stabilization. A Primal Approach

With the information collected along iterations, bundle methods construct both a model $(\check{f}_k)$ for the objective function f and a polyhedral approximation of its subdifferential $(\partial \check{f}_k)$. Keeping in mind Example 9.7 and § 9.2.2, bundle methods can be considered stabilized variants of cutting-planes (primal stabilization) as well as of steepest-descent methods (dual stabilization).

We explain now the stabilization scheme in the primal space. Recall the intrinsic instability of cutting-planes methods observed in Figure 9.8. To prevent the objective function from increasing, it would be desirable for the algorithm to "remember" the best point obtained so far (i.e., x^4 giving f_4 in Figure 9.8). With this extra information kept along iterations, the algorithm can generate two sequences of points. One is the sequence of sample points used to define the model $\check{f}_k$. We call those points *candidate* points and denote them by y^k. A second sequence consists of those sample points that decreased sufficiently the objective function f, in the sense of (10.1) below. We call these points *stability centers* and denote them by x^k. Note that $\{x^k\}$ is a subsequence of $\{y^k\}$.

In order to generate candidate points and select stability centers, we define subproblems by modifying (9.9) in Algorithm 9.4, following some *stabilization principles* determined by:

(i) the choice of a model φ_k, which approximates f (for instance, $\varphi_k = \check{f}_k$). The fact of having a model gives a "nominal decrease", as in (9.8);

(ii) the choice of a stability center x^k, for which $f(x^k)$ is the "best" value obtained so far and from which the decrease of f will be measured;

(iii) the choice of a normalization $|\cdot|_k$ to prevent big oscillations.

We use these elements to define a stabilized subproblem, whose solution y^{k+1} is considered just a **candidate** to make f decrease. The quality of candidates is measured using the nominal decrease δ_{k+1}. Only "good" candidates, i.e., those satisfying the relation

$$f(y^{k+1}) \leq f(x^k) - m\delta_{k+1} \tag{10.1}$$

for $m \in]0, 1[$, will become stability centers.

There are several possibilities to define stabilized subproblems and their associated nominal decrease, we describe some of them in § 10.2 below. Essentially, all the variants modify the cutting-planes subproblem (9.9) to prevent a move "too far away" from x^k.

The norm $|\cdot|_k$, measuring distance to x^k, can be made "less tight" in the beginning of the iterations (to progress fast) and become more stringent as k grows (to prevent instability when approaching a solution).

Algorithm 10.1 (general bundle). Let $\texttt{tol} \geq 0$ and $m \in]0, 1[$ be given parameters. Choose x^1, call the black box from Figure 9.5 with $x = x^1$, construct the model φ_1, and set the algorithm parameters, such as $|\cdot|_1$. Set $k = 1$ and $\delta_1 = \infty$.

STEP 1 (implementable stopping test). If $\delta_k \leq \texttt{tol}$, stop.

STEP 2 (candidate). Solve

$$y^{k+1} \in \text{Argmin stabilized pbm}(\varphi_k, x^k, |\cdot|_k), \quad \text{and} \tag{10.2}$$

Define $\delta_{k+1} = \delta(\varphi_k, x^k, |\cdot|_k, y^{k+1}) \geq 0$.

STEP 3 (calling the black box – assessing the candidate). Call the black box from Figure 9.5 with $x = y^{k+1}$.

  Descent test:

$$f(x^k) - f(y^{k+1}) \geq m\delta_{k+1}? \begin{cases} \text{Yes: } x^{k+1} := y^{k+1} \text{ (descent-step)} \\ \\ \text{No: } x^{k+1} := x^k \quad \text{(null-step)}. \end{cases}$$

STEP 4 (improving the model – loop). Append y^{k+1} to the model, i.e., construct φ_{k+1}. Define the algorithm parameters for the next iteration, such as $|\cdot|_{k+1}$. Change k to $k + 1$, go to STEP 1.

Remark 10.2.

– Since the nominal decrease is nonnegative, the descent test of STEP 3 decides if the candidate provides a sufficient descent for f. If f decreases by at least a fraction m of the decrease predicted by the model, the stability center will be moved to y^{k+1}. This Armijo-like test is very close to the trust region philosophy of § 6.1; see again the end of § 3.5.3. When y^{k+1} is a good candidate, we shall say that a "serious" step, or descent-step is made.

– If the candidate brings no significant descent for f, or no descent at all, the stability center is not changed (y^{k+1} is not "good enough"). In this case, we hope to obtain a better candidate after the model is enriched at STEP 4. Since nothing is done in terms of updating centers, this case is called a "null" step.

– Note that since the relations

$$f(x^k) - f(y^{k+1}) > 0 \quad \text{and} \quad f(x^k) - f(y^{k+1}) < m\delta_{k+1}$$

can hold simultaneously, the algorithm may generate null steps with smaller objective function value than $f(x^k)$, but that were not considered "good enough" in terms of the nominal decrease. This is the reason why we refer to x^k as the point with "best" objective value (instead of smallest objective value).

– Typically, φ_k is the cutting-planes approximation $\check{f}_k$, i.e., (9.6) written with $x^i = y^i$. By "improving" the model in STEP 4 we mean the fact of introducing the affine function defined by the black-box information obtained at y^{k+1}:

$$\varphi_{k+1}(y) = \max\{\varphi_k(y), f(y^{k+1}) + \langle s(y^{k+1}), y - y^{k+1} \rangle\}.$$

The more general notation φ_k (instead of just $\check{f}_k$) reflects the possibility of *cleaning the model*, already mentioned in Remark 9.8. This is the aggregation technique described in § 10.3.2 below.

– In general, since at null steps there is no new reliable information, the metric $|\cdot|_k$ is only updated when there is a descent-step. The update of the metric is an extremely complex topic, yet crucial for good numerical results. In § 10.3.3 we give a rule that proves good for implementation.

– Line-search is not considered here. This concept becomes somewhat fuzzy in a serious-null steps framework; see the final comments in Example 10.6 below.

The set of indices k for which a *new* descent-step is done is denoted by K_s. In particular, when there are infinitely many descent-steps, the infinite sequence $\{\delta_k\}_{k \in K_s}$ is convergent, whenever (8.1) has minimizers.

Lemma 10.3. *Consider Algorithm 10.1 and suppose it loops forever ($k \to \infty$). Use the notation $f_* := \lim_{k \in K_s} f(x^k)$ and assume $f_* > -\infty$. Then*

$$(0 \le) \sum_{k \in K_s} \delta_k \le \frac{f_1 - f_*}{m}.$$

Proof. Note first that, since $\texttt{tol} \ge 0$, for the algorithm to loop forever the nominal decrease must satisfy $\delta_k > 0$ for all k. Take an arbitrary $k \in K_s$. Since the descent test is satisfied, $x^{k+1} = y^{k+1}$ and

$$f(x^k) - f(x^{k+1}) = f(x^k) - f(y^{k+1}) \geq m\delta_{k+1}.$$

Let k' be the index following k in K_s. Between k and k' the algorithm makes null-steps only, without moving the stability center: $x^{k+1} = x^{k+j}$, for all $j = 2, \ldots, k' - k$. The descent test at k' gives

$$f(x^{k+1}) - f(x^{k'+1}) \geq m\delta_{k'+1}.$$

Hence, for any $k'' \in K_s$,

$$m \sum_{k \in K_s}^{k''} \delta_k \leq \sum_{k \in K_s}^{k''} f(x^k) - f(x^{k+1}) = f_1 - f_{k''} \leq f_1 - f_*.$$

Now letting $k'' \to \infty$ gives the desired result.   □

This simple result is useful for proving convergence whenever the algorithm generates an infinite number of serious-steps. Namely, since the sequence $\{f(x^k)\}_{k \in K_s}$ is strictly decreasing, either $\{f(x^k)\} \searrow -\infty$ (in which case f is unbounded from below and $\{x^k\}$ is trivially a minimizing sequence), or by Lemma 10.3, $\{f(x^k)\} \searrow f_*$ and $\{\delta_k\} \to 0$ (a crucial relation for showing convergence, cf. items (ii) and (iii) in Lemma 10.8 below). Finally, note that when $\{\delta_k\} \to 0$, if tol is positive then Algorithm 10.1 will stop at some iteration k_{last}.

10.2 Some Examples of Stabilized Problems

The short examples that follow are particular forms of subproblems (10.2). Each variant is characterized by a *parameter*, to be updated at each iteration in STEP 4. For the moment, we do not explain how to choose neither the model φ_k nor the normalization $|\cdot|_k$. We address these topics in more detail for one of the variants, namely the penalized bundle method described in § 10.3.

Example 10.4 (Trust region). Consider subproblem (9.9) in the cutting-planes Algorithm 9.4. The model is minimized over a fixed set S, possibly big (and, thus, the way is open to uninvited oscillations as in Example 9.7). Instead, one could define a feasible set that varies along iterations, where the model is considered reliable, i.e., a *trust region*.

Having the parameter $\kappa_k > 0$, define the ball centered at $x = x^k$ with radius κ_k. The stabilized subproblem (10.2) to be solved at STEP 2 is:

$$y^{k+1} \text{ solution to } \begin{cases} \min \varphi_k(y) \\ |y - x^k|_k^2 \leq \kappa_k. \end{cases} \tag{10.3}$$

The corresponding nominal decrease is defined by $\delta_{k+1} := f(x^k) - \varphi_k(y^{k+1})$. The trust region is managed in such a way that the parameter $\kappa_k \to 0$ when $k \to \infty$.

A curvilinear search adapted to this variant is outlined in § XV.1.3.1 of [195]. Essentially, the parameter κ_k varies in the interval $]\kappa_L, \kappa_R[$, according to an Armijo-like rule; as in § 3.5.2.

Example 10.5 (Levels). This approach is somewhat dual to Example 10.4. It seeks to minimize the radius of the ball centered at the stability center x^k, while reaching a prescribed decrease for the model function, or *level*.

Having the parameter ℓ_k, subproblem (10.2) is the following:

$$y^{k+1} \text{ solution to } \begin{cases} \min \frac{1}{2}|y - x^k|_k^2 \\ \varphi_k(y) \leq \ell_k \, . \end{cases} \tag{10.4}$$

The nominal decrease δ_{k+1} is the same as in the previous example.

If the optimal value $\bar{f}$ is known, the management of ℓ_k is easy. Otherwise, the update of ℓ_k is more delicate, since the feasible set in (10.4) could be empty. Nevertheless, the method presents good numerical performances; see [230].

10.3 Penalized Bundle Methods

In this variant, stabilization is the result of introducing a quadratic term in the model used for subproblems.

Example 10.6 (Penalization of the model). For a parameter $\mu_k > 0$, subproblem (10.2) is:

$$y^{k+1} \text{ solution to } \begin{cases} \min \varphi_k(y) + \frac{1}{2}\mu_k|y - x^k|_k^2 \\ y \in \mathbb{R}^n \, . \end{cases} \tag{10.5}$$

The nominal decrease is

$$\delta_{k+1} := f(x^k) - \left(\varphi_k(y^{k+1}) + \frac{1}{2}\mu_k|y^{k+1} - x^k|_k^2 \right) \, . \tag{10.6}$$

When compared to the nominal decrease in previous examples, we see that, $\delta_{k+1} = \delta_{k+1}^{prev} - \frac{1}{2}\mu_k|y^{k+1} - x^k|_k^2$. Therefore, with this nominal decrease, penalized bundle methods will be less strict for accepting a candidate as the next stability center. If desired, the same nominal decrease, i.e., δ_{k+1}^{prev}, could be used with this variant.

The line-search [236] corresponds to introducing a parameter $t > 0$ in the quadratic term of (10.5). The corresponding solution depends on this parameter, i.e., $y^{k+1} = y^{k+1}(t)$. Then STEP 3 is modified in order to allow interpolations and extrapolations of t during the assessment of the candidate. Note that a change in the stepsize results in a different stabilized subproblem. In this sense, the t-adjustment is closer in spirit to a curvilinear search ; see also [213], [326].

Figure 10.1 shows one iteration of the penalized bundle method.

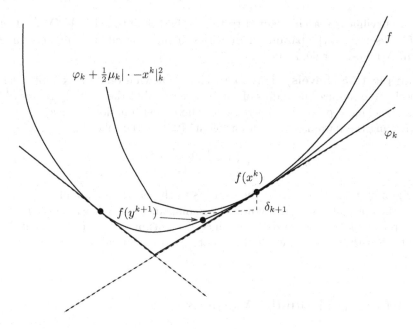

$\varphi_k + \frac{1}{2}\mu_k |\cdot - x^k|_k^2$

f

φ_k

$f(x^k)$

$f(y^{k+1})$

δ_{k+1}

Fig. 10.1. One iteration of the penalized bundle method

Some correspondences can be established between the examples presented so far.

Theorem 10.7. *Consider Examples 10.4, 10.5 and 10.6, with same metric $|\cdot|_k$. The following holds:*

- (i) *For given $\kappa_k > 0$, let y^{k+1} solve the stabilized problem of Example 10.4. Then there exists $\mu_k > 0$ such that y^{k+1} solves the stabilized problem of Example 10.6.*
- (ii) *For given $\mu_k > 0$, let y^{k+1} solve the stabilized problem of Example 10.6. Then there exists $\ell_k \geq 0$ such that y^{k+1} solves the stabilized problem of Example 10.5.*
- (iii) *For given $\ell_k \geq 0$, let y^{k+1} solve the stabilized problem of Example 10.5. Then there exists $\kappa_k \geq 0$ such that y^{k+1} solves the stabilized problem of Example 10.4.*

Proof. (i) The stabilized problem (10.3) of Example 10.4 is convex, and a constraint qualification condition holds (see for instance, Theorem 8.4 or Chapter 13, § 13.3). Then there exists an associated multiplier $0 \leq \lambda \in \mathbb{R}$ for which solutions to (10.2) are obtained by minimizing the Lagrangian, i.e.,

$$y^{k+1} = \operatorname{argmin}_{y \in \mathbb{R}^n} L(y, \lambda) \quad \text{where } L(y, \lambda) := \varphi_k(y) + \frac{1}{2}\lambda\left(|y - x^k|_k^2 - \kappa_k\right).$$

In other words, y^{k+1} solves (10.5) in Example 10.6, for $\mu_k := \lambda$.

(ii) Let y^{k+1} solve (10.5) in Example 10.6. For all $y \in \mathbb{R}^n$ we have

$$\varphi_k(y^{k+1}) + \frac{1}{2}\mu_k|y^{k+1} - x^k|_k^2 \leq \varphi_k(y) + \frac{1}{2}\mu_k|y - x^k|_k^2.$$

In particular, let $\ell_k := \varphi_k(y^{k+1})$; then, for any y such that $\varphi_k(y) \leq \ell_k$, we have

$$\frac{1}{2}\mu_k|y^{k+1} - x^k|_k^2 \leq \frac{1}{2}\mu_k|y - x^k|_k^2.$$

Since the parameter μ_k is positive, the conclusion follows.

(iii) Let y^{k+1} solve (10.4) of Example 10.5. Then $\varphi_k(y^{k+1}) = \ell_k$, so letting $\kappa_k := |y^{k+1} - x^k|_k$ it holds that y^{k+1} also minimizes $\varphi_k(y)$ in $\{y \in \mathbb{R}^n : |y - x^k|_k^2 \leq \kappa_k\}$. $\qquad\square$

This result establishes the *formal* equivalence of all these variants. In practice, they differ in the heuristics used to update their respective parameters at STEP 4 of Algorithm 10.1. Their numerical performances will vary according to these heuristics. In § 10.3.3 we explain how to update the parameter for the penalized bundle method, using a rule related to quasi-Newton methods.

For all the examples, stabilized subproblems are (convex) quadratic programming problems (QP). Cutting-planes methods have subproblems (9.9) that are LPs, recall (9.10). The additional complication of solving a QP is the price to pay to obtain stability. As a thumb rule, the more robust and sophisticated is the black-box method, the more complex is the corresponding subproblem. Nevertheless, convex QPs are not much more difficult to solve than LPs. The underlying resolution technique is essentially the same (active sets strategies, interior points, etc). In actual implementations a QP such as (10.5) is solved through its dual. This dual QP has a very special structure, namely, a simplicial feasible set (see Lemma 10.8 below). Good QP solvers exploit such structure, as well as the fact of knowing how QP-data changes along (serious step or null step) iterations in order to define warm starts. As a result, solving (10.5) does not require much more computational work than solving an LP like (9.10). Finally, a pay-off compensating the extra computational burden is that having QP subproblems allows to compress the bundle of information (i.e., to *clean the model*) by using the aggregation techniques described in § 10.3.2.

Other variants of stabilized subproblem were considered more recently in [241], [240], [135]. In [241], [240], the idea is to use a model with affine functions appended with a quadratic term centered at x^k:

$$\varphi_k(y) = \max_{i \leq k}\left\{f_i + \langle s^i, y - y^i\rangle + \frac{1}{2}\langle x^k - y^i, H_k(x^k - y^i)\rangle\right\},$$

where H_k is a certain quasi-Newton matrix. As for [135], it develops a comprehensive theory for generalized bundle methods, for which the stabilizing term can be any closed convex function. However, it should be noted that, since the work of solving a general convex program has to be repeated at each iteration, the resulting subproblems may become too difficult to solve.

10.3.1 A Trip to the Dual Space

In order to obtain a general formulation for the model φ_k, we first rewrite the cutting-planes model $\check{f}_k$ by referring it to the stability center:

$$\check{f}_k(y) = f(x^k) + \max_{i=1,\ldots,k} \{-e_i + \langle s^i, y - x^k \rangle\},$$

where the terms e_i are the *linearization errors* at x^k

$$(0 \leq) \quad e_i := f(x^k) - f_i - \langle s^i, x^k - y^i \rangle, i = 1, \ldots, k.$$

Note that, by (9.3), $s^i \in \partial_{e_i} f(x^k)$. With this notation, the bundle of past information is

$$\mathcal{B} = \{(s^i, e_i); \text{ with } s^i \in \partial_{e_i} f(x^k), i = 1, \ldots, k\} \text{ and } (x^k, f(x^k), s(x^k)).$$

When the bundle becomes too big (k large), we shall compress it to keep only np_k elements, with np_k possibly much smaller than k. With the new notation, the piecewise-linear models have the form

$$\varphi_k(y) = f(x^k) + \max_{i=1,\ldots,np_k} \{-e_i + \langle s^i, y - x^k \rangle\}. \tag{10.7}$$

At the beginning of the iterative process, while k is not too large, $np_k = k$ and $\varphi_k = \check{f}_k$. Later on, even though $np_k << k$, the compressed bundle is still formed by couples (s^i, e_i) satisfying $s^i \in \partial_{e_i} f(x^k)$; see § 10.3.2 below. However, for some of the couples there may no longer be a point y^i for which $s^i \in \partial f(y^i)$.

Before writing a dual of (10.5), we further specify the metric therein by choosing a square positive definite matrix M_k of order n, and letting

$$|\cdot|_k^2 := \langle M_k \cdot, \cdot \rangle.$$

The corresponding dual norm is denoted by $\|\cdot\|_k^2 := \langle \cdot, M_k^{-1} \cdot \rangle$.

Lemma 10.8. *Let y^{k+1} be the unique solution to (10.5) and assume $\mu_k > 0$. Then*

$$y^{k+1} = x^k - \frac{1}{\mu_k} M_k^{-1} \hat{s}^k \quad where \quad \hat{s}^k := \sum_{i=1}^{np_k} \bar{\alpha}_i s^i \tag{10.8}$$

and $\bar{\alpha} = (\bar{\alpha}_1, \ldots, \bar{\alpha}_{np_k})$ is a solution to

$$\begin{cases} \min_{\alpha_i} \frac{1}{2} \| \sum_{i=1}^{np_k} \alpha_i s^i \|_k^2 + \mu_k \sum_{i=1}^{np_k} \alpha_i e_i \\ \alpha \in \Delta_k := \{ z \in [0,1]^{np_k} : \sum_{i=1}^{np_k} z_i = 1 \} . \end{cases} \tag{10.9}$$

In addition, the following relations hold:

(i) $\hat{s}^k \in \partial \varphi_k(y^{k+1})$,

(ii) $\delta_{k+1} = \varepsilon_k + \dfrac{1}{2\mu_k} \| \hat{s}^k \|_k^2$, *where we defined* $\varepsilon_k := \sum_{i=1}^{np_k} \bar{\alpha}_i e_i$.

(iii) $\hat{s}^k \in \partial_{\varepsilon_k} f(x^k)$.

Proof. Write (10.5) as a QP with an extra scalar variable r as follows

$$\begin{cases} \min_{(y,r) \in \mathbb{R}^n \times \mathbb{R}} r + \frac{1}{2}\mu_k |y - x^k|_k^2 \\ r \geq f(x^k) - e_i + \langle s^i, y - x^k \rangle , i = 1, \dots, np_k . \end{cases} \tag{10.10}$$

The corresponding Lagrangian is, for $\alpha \in \mathbb{R}_+^{np_k}$,

$$L(y, r, \alpha) = r + \frac{1}{2}\mu_k |y - x^k|_k^2 + \sum_{i=1}^{np_k} \alpha_i (f(x^k) - e_i + \langle s^i, y - x^k \rangle - r) ,$$

i.e., rearranging terms,

$$= (1 - \sum_{i=1}^{np_k} \alpha_i) r + \frac{1}{2}\mu_k |y - x^k|_k^2 + \sum_{i=1}^{np_k} \alpha_i (f(x^k) - e_i + \langle s^i, y - x^k \rangle) .$$

In view of strong convexity, (10.5) has the unique solution y^{k+1}. Furthermore, the equivalent problem (10.10) has affine constraints; hence, there exists an optimal multiplier $\bar{\alpha}$ associated with y^{k+1}. Since there is no duality gap (recall § 8.2.2), $(y^{k+1}, \bar{\alpha})$ can be obtained either by solving the primal problem derived from (10.10) or by solving its dual.

$$(10.10) \equiv \min_{(y,r) \in \mathbb{R}^n \times \mathbb{R}} \max_{\alpha \in \mathbb{R}_+^{np_k}} L(y, r, \alpha) \equiv \max_{\alpha \in \mathbb{R}_+^{np_k}} \min_{(y,r) \in \mathbb{R}^n \times \mathbb{R}} L(y, r, \alpha) .$$

All the problems above have the same finite optimal value. However, the dual (rightmost) problem involves the unconstrained minimization of L with respect to r. For the dual value to be finite, the term multiplying r in L has to vanish, i.e. , α must lie in the unit-simplex Δ_k. As a result, y^{k+1} and $\bar{\alpha}$ solve the primal and dual problems

$$\min_{y \in \mathbb{R}^n} \max_{\alpha \in \mathbb{R}_+^{np_k}} L(y, \alpha) \equiv \max_{\alpha \in \mathbb{R}_+^{np_k}} \min_{y \in \mathbb{R}^n} L(y, \alpha) ,$$

where $L(y, \alpha) := f(x^k) + \frac{1}{2}\mu_k |y - x^k|_k^2 + \sum_{i=1}^{np_k} \alpha_i (-e_i + \langle s^i, y - x^k \rangle)$. Consider the last dual problem. For each $\alpha \in \Delta_k$ fixed, the optimality condition defining $y(\alpha) = \operatorname{argmin}_y L(y, \alpha)$ is $0 = \nabla_y L(\alpha, y(\alpha))$, i.e.,

$$0 = \mu_k M_k(y(\alpha) - x^k) + \sum_{i=1}^{np_k} \alpha_i s^i \,. \qquad (10.11)$$

In particular, when $\alpha = \bar{\alpha}$, because $y(\bar{\alpha})$ is y^{k+1}, (10.8) holds.

To see that $\bar{\alpha}$ also solves (10.9), multiply (10.11) by $y(\alpha) - x^k$ and by $\frac{1}{\mu_k} \sum_{i=1}^{np_k} \alpha_i s^i$ to write

$$0 = \mu_k |y(\alpha) - x^k|_k^2 + \sum_{i=1}^{np_k} \alpha_i \langle s^i, y(\alpha) - x^k \rangle$$
$$= \sum_{i=1}^{np_k} \alpha_i \langle s^i, y(\alpha) - x^k \rangle + \frac{1}{\mu_k} \left\| \sum_{i=1}^{np_k} \alpha_i s^i \right\|_k^2 \,,$$

which in turn implies that

$$\mu_k |y(\alpha) - x^k|_k^2 = \frac{1}{\mu_k} \left\| \sum_{i=1}^{np_k} \alpha_i s^i \right\|_k^2$$

and

$$L(y(\alpha), \alpha) = f(x^k) - \frac{1}{2\mu_k} \left\| \sum_{i=1}^{np_k} \alpha_i s^i \right\|_k^2 - \sum_{i=1}^{np_k} \alpha_i e_i \,.$$

Altogether,

$$\bar{\alpha} \text{ solves } \max_{\alpha \in \Delta_k} L(y(\alpha), \alpha)$$
$$= f(x^k) - \min_{\alpha \in \Delta_k} \left\{ \frac{1}{2\mu_k} \left\| \sum_{i=1}^{np_k} \alpha_i s^i \right\|_k^2 + \sum_{i=1}^{np_k} \alpha_i e_i \right\} \,. \qquad (10.12)$$

To show (i), write the optimality condition for (10.5) and use the definition of $\hat{s}^k$ in (10.8): $0 \in \partial\varphi_k(y^{k+1}) + \mu_k M_k(y^{k+1} - x^k) = \partial\varphi_k(y^{k+1}) - \sum_{i=1}^{np_k} \bar{\alpha}_i s^i = \partial\varphi_k(y^{k+1}) - \hat{s}^k$.

To show (ii), note first that, since there is no duality gap, the primal optimal value in (10.5) is equal to the dual optimal value in (10.12):

$$\varphi_k(y^{k+1}) + \frac{1}{2}\mu_k |y^{k+1} - x^k|_k^2 = f(x^k) - \frac{1}{2\mu_k} \left\| \sum_{i=1}^{np_k} \bar{\alpha}_i s^i \right\|_k^2 - \sum_{i=1}^{np_k} \bar{\alpha}_i e_i \,.$$

Together with (10.6) and (10.5), the relation follows.

Finally, to show (iii), use that $f \geq \varphi_k$ and item (i) to write for any $y \in \mathbb{R}^n$

$$f(y) \geq \varphi_k(y) \geq \varphi_k(y^{k+1}) + \langle \hat{s}^k, y - y^{k+1} \rangle \,.$$

Using (10.8) this inequality can be re-written as

$$f(y) \geq \varphi_k(y^{k+1}) + \langle \hat{s}^k, y \pm x^k - y^{k+1} \rangle$$
$$= \varphi_k(y^{k+1}) + \langle \hat{s}^k, y - x^k \rangle - \langle \hat{s}^k, y^{k+1} - x^k \rangle$$
$$= \varphi_k(y^{k+1}) + \langle \hat{s}^k, y - x^k \rangle + \frac{1}{\mu_k} \|\hat{s}^k\|_k^2$$
$$= f(x^k) + \langle \hat{s}^k, y - x^k \rangle - \left(f(x^k) - \varphi_k(y^{k+1}) - \frac{1}{\mu_k} \|\hat{s}^k\|_k^2 \right) \,.$$

The relations in (10.6) and item (ii) give the desired result. $\qquad\square$

The multiplier $\bar{\alpha}$ solving (10.9) is unique only when the bundle subgradients $\{s^i\}_{i \leq np_k}$ are linearly independent (in this case, the (LI-CQ) condition in Chapter 13 holds for (10.10)). In the absence of uniqueness, QP solvers usually find the minimum-norm solution. Note, however, that $\hat{s}^k$, the subgradients convex combination, is the same for any multiplier $\bar{\alpha}$ solving (10.9), because y^{k+1} is unique.

The last result established in Lemma 10.8 gives a formal meaning to the dual stabilization effect of bundle methods, already announced in § 9.2.2. More precisely, the subgradient $\hat{s}^k$ is also a smeared subgradient of the objective function f at the center x^k.

We are now in a position to "dualize" Example 10.6 and obtain another form of stabilized subproblem (10.2).

Example 10.9 (Dual stabilization). The parameter μ_k in (10.9) can be considered as a multiplier associated with the linear term $\sum_k \alpha_i e_i$. Given a parameter $\epsilon_k \geq 0$, the stabilized subproblem (10.2) is given by y^{k+1} and $\hat{s}^k$ as in (10.8), where now

$$\bar{\alpha} = (\bar{\alpha}_1, \ldots, \bar{\alpha}_{np_k}) \text{ is a solution to } \begin{cases} \min_{\alpha_i} \frac{1}{2} \| \sum_{i=1}^{np_k} \alpha_i s^i \|_k^2 \\ \alpha \in \Delta_k \\ \sum_{i=1}^{np_k} \alpha_i e_i \leq \epsilon_k . \end{cases} \quad (10.13)$$

The nominal decrease δ_{k+1} is the same as in (10.6). Since at a solution we have that $\sum_{i=1}^{np_k} \bar{\alpha}_i e_i = \epsilon_k$, it follows that the relation $\delta_{k+1} = \epsilon_k + \frac{1}{2\mu_k} \|\hat{s}^k\|_k^2$, similar to Lemma 10.8 (ii), holds.

We finish with a word on the stopping test. The expression given for δ_{k+1} in Lemma 10.8 (ii), shows two different terms to measure approximate optimality of x^k. Keeping in mind (9.4), by Lemma 10.8 (iii) we see that for $\{x^k\}$ to be a minimizing sequence, it must hold that $\{\varepsilon_k\} \to 0$ and $\{\hat{s}^k\} \to 0$. For this reason, it is preferable to check, instead of $\delta_{k+1} \leq \mathtt{tol}$, the following split stopping test at STEP 1

$$\varepsilon_k \leq \mathtt{tol}_\varepsilon \quad \text{and} \quad \|\hat{s}^k\|_k \leq \mathtt{tol}_s . \quad (10.14)$$

A harmonious choice of these bounds is important to obtain good numerical performances.

10.3.2 Managing the Bundle. Aggregation

We already mentioned that models φ_k from (10.7) essentially follow the cutting-planes approximation $\check{f}_k$. As iterations go along, the number of elements in the bundle increases. When the size of the bundle becomes too big, it is necessary to compress it and clean the model. Algorithm 10.1 must be appended with a *selection and compression* mechanism introduced at STEP 4 that keeps control of the bundle size.

Suppose that at iteration k the bundle has np_k couples, which define the model (10.7). Each couple has the form (s^i, e_i), with $s^i \in \partial_{e_i} f(x^k)$, but in the expression defining $\hat{s}^k$ in (10.8) only count subgradients s^i for which $\bar{\alpha}_i > 0$ (likewise, for the errors e_i in the convex sum defining ε_k in Lemma 10.9(ii)). For this reason, we call *indispensable* couples (respectively, *dispensable*) the pairs (s^i, e_i) in the bundle corresponding to *active* (resp. *inactive*) indices, i.e., to i such that $\bar{\alpha}_i > 0$ (resp. $\bar{\alpha}_i = 0$). When the algorithm reaches an iteration where the number np_k becomes too big, the following steps are executed:

– *selection* of dispensable couples, that can be discarded;
– if the remaining, indispensable, couples are still too many, *compression* of the indispensable information into a single couple, called aggregate.

Aggregation is the synthesis mechanism that condenses the essential information of the bundle into one single couple, given by $(\hat{s}^k, \varepsilon_k)$ as defined in Lemma 10.8. The corresponding affine function, inserted in the model when there is compression, is called *aggregate linearization*:

$$f_a(y) := f(x^k) - \varepsilon_k + \langle \hat{s}^k, y - x^k \rangle. \tag{10.15}$$

This function, which has the same form as any other affine function of (10.7), summarizes all the information generated up to iteration k. Note, however, that unlike other elements in the bundle, there is no previous y^i for which $\hat{s}^k \in \partial f(y^i)$. The aggregate linearization has the following properties.

Lemma 10.10. *For f_a defined by (10.15), it holds that*

(i) $f_a(y) = \varphi_k(y^{k+1}) + \langle \hat{s}^k, y - y^{k+1} \rangle$, *for all* $y \in \mathbb{R}^n$.
(ii) $f_a(y) \le \varphi_k(y)$, *for all* $y \in \mathbb{R}^n$.
(iii) *Let* $\psi : \mathbb{R}^n \to \mathbb{R}$ *be a convex function such that* $\psi(y) \ge f_a(y)$ *for all* $y \in \mathbb{R}^n$ *and* $\psi(y^{k+1}) = f_a(y^{k+1})(= \varphi_k(y^{k+1}))$. *Then*

$$y^{k+1} = \operatorname*{argmin}_{y \in \mathbb{R}^n} \left\{ \psi(y) + \frac{1}{2}\mu_k|y - x^k|_k^2 \right\}. \tag{10.16}$$

Proof. Item (i) is immediate from (10.15) and the relations in Lemma 10.8, while item (ii) follows from (i) and Lemma 10.8(i).

(iii) Write the optimality condition of (10.16) using (10.8):

$$0 \in \partial\psi(y^{k+1}) + \mu_k M_k(y^{k+1} - x^k) = \partial\psi(y^{k+1}) - \hat{s}^k,$$

where $\hat{s}^k$ is defined in (10.8). It suffices to prove $\hat{s}^k \in \partial\psi(y^{k+1})$. For arbitrary $y \in \mathbb{R}^n$, the assumption on ψ and item (i) give $\psi(y) \ge f_a(y) = \psi(y^{k+1}) + \langle \hat{s}^k, y - y^{k+1} \rangle$, i.e., the desired result. $\square$

The last item in Lemma 10.10 shows that, a posteriori, nothing is changed if instead of φ_k one uses f_a or a function ψ as above. The aggregate linearization synthesizes indispensable information of active bundle elements, while the model

$$\psi(y) = f(x^k) + \max_{i:\,\bar{\alpha}_i > 0}\{-e_i + \langle s^i, y - x^k\rangle\}$$

expresses the same information in a disaggregate form. In this sense, the function f_a is "minimal", in the class of functions sandwiched between f_a and φ_k as in (iii) that leave invariant y^{k+1}.

Let $np_{\max}$ be the maximal size of the bundle, and np_k be its current size. The compression sub-algorithm to be appended at STEP 4 of Algorithm 10.1 is the following:

Algorithm 10.11 (compression).

STEP 4 (Algorithm 10.1)

Let $np_{\max}$ be given. We have np_k elements (s^i, e_i) in the bundle. The algorithm has solved $(10.5)=(10.8)\&(10.9)$ to compute y^{k+1} and has called the black box.

Compression test: If $np_k \geq np_{\max}$ then

    Selection: Let $n_{act} := \{i \leq np_k : \bar{\alpha}_i > 0\}$ be the cardinality of active indices.

    If $n_{act} \leq np_{\max} - 1$, then delete all inactive couples from the bundle, set $n_{left} = n_{act}$, and define $np_{k+1} = n_{left} + 1$.

    Otherwise, discard two or more couples (s^i, e_i) from the bundle. Otherwise, discard two or more couples (s^i, e_i) from the bundle. The resulting bundle cardinality, n_{left}, should be smaller or equal to $np_{\max} - 2$. Define $np_{k+1} = n_{left} + 2$.

    Aggregation: If $n_{left} \neq n_{act}$, append $(s^{np_{k+1}-1}, e_{np_{k+1}-1}) := (\hat{s}^k, \varepsilon_k)$ to the bundle.

Improving/updating the bundle: Set $s^{np_{k+1}} := s(y^{k+1})$.

    – Append (s^{np_k+1}, e_{np_k+1}) to the bundle, with

$$e_{np_k+1} = \begin{cases} 0, & \text{if serious step,} \\ f(x^k) - \left(f(y^{k+1}) + \langle s^{k+1}, x^k - y^{k+1}\rangle\right), & \text{if null-step.} \end{cases}$$

    – In case of serious step, update the linearization errors:
$$e_i := e_i + f(y^{k+1}) - f(x^k) - \langle s^i, y^{k+1} - x^k\rangle, \ i = 1, \ldots, np_{k+1}-1. \quad (10.17)$$

Model – loop Define the algorithm parameters for the next iteration, such as $|\cdot|_{k+1}$. Construct φ_{k+1} as in (10.7), written with k replaced by $k+1$. Change k to $k+1$, go to STEP 1.

We now comment on some features of the compression sub-algorithm.

Remark 10.12.

- Until $np_k \geq np_{max}$ for the first time, the procedure above is just the old STEP 4 in Algorithm 10.1. The model is $\varphi_k = \check{f}_k$, and $np_k = k$. If there is no compression, the relation (10.17) above just gives a smart implementation for updating the bundle information, by dynamically redefining linearization errors which, although not explicit in the notation, do depend on x^k.
- The necessity of appending the aggregate couple in the Compression test above depends on which information is discarded at the Selection step. More precisely, if all active couples are kept, the aggregate couple is not needed, because all the indispensable information condensed in the aggregate couple is already present (in disaggregate form).
- When the maximum capacity is reached, for instance when $k = np_{max}$, suppose we decide to discard the elements 1 and 2 from the bundle, and to append the aggregate couple. The resulting model will be

$$\varphi_{k+1}(y) = \max \left\{ \max_{\{3 \leq i \leq k+1\}} \{f_i + \langle s^i, y - y^i \rangle\}, \varphi_k(y^{k+1}) + \langle \hat{s}^k, y - y^{k+1} \rangle \right\}$$

$$= \max \left\{ f(x^{k+1}) + \max_{\{3 \leq i \leq k+1\}} \{-e_i + \langle s^i, y - x^{k+1} \rangle\}, f_a(y) \right\}.$$

In view of Lemma 10.10, the last affine function can be replaced by any function $\psi \geq f_a$. Note that in any case, by construction, for all k and for all $y \in \mathbb{R}^n$, and somewhat similarly to (9.7),

$$f_a(y) \leq \varphi_{k+1}(y) \leq f(y) \text{ and } \varphi_{k+1}(y) \geq f_{k+1} + \langle s^{k+1}, y - y^{k+1} \rangle. \quad (10.18)$$

- The parameter np_{max} determines the maximum size of each stabilized subproblem (10.5) and, hence, the dimension of the dual variable in (10.9).
- There are many possibilities to choose which couples to discard. For example, all those for which $\bar{\alpha}_i = 0$ in (10.9). Or those with smallest e_i. Or the oldest ones. Or just all of them. As long as the aggregate couple is introduced in the bundle when some active element has been discarded, the algorithm will remain convergent. Different selections of discarded couples may result in different speeds of convergence, though.

10.3.3 Updating the Penalization Parameter. Reversal Forms

We now give a rule for updating μ_k for $k \in K_s$, i.e., at serious steps. For simplicity, we use $M_k = I$ for all k, so that the metric is $| \cdot |_k^2 = \| \cdot \|_k^2 = \| \cdot \|^2 = \langle \cdot, \cdot \rangle$, i.e., the Euclidean norm. A special feature of this rule is that it is not completely heuristic, since it is supported by convex analysis and quasi-Newton theory.

For a convex function f the *Moreau-Yosida regularization*, of f at a given point $x \in \mathbb{R}^n$ is denoted by $F_\mu(x)$ and defined as the optimal value of

$$\min_{y \in \mathbb{R}^n} \left\{ f(y) + \frac{1}{2}\mu\|y - x\|^2 \right\}, \tag{10.19}$$

where μ is a positive parameter. The corresponding unique minimizer is the *proximal point* of f at x, denoted by $p_\mu(x)$. The optimality condition for (10.19) is

$$p_\mu(x) = x - \frac{1}{\mu}s(p_\mu(x)), \text{ where } s(p_\mu(x)) \in \partial f(p_\mu(x)), \tag{10.20}$$

and

$$\nabla F_\mu(x) = \mu(x - p_\mu(x)) \tag{10.21}$$

is a Lipschitz continuous function of x, [268],[313]. It also well known that minimizing f is equivalent to minimizing F_μ for any $\mu \geq 0$.

Consider Algorithm 10.1 with subproblem (10.5), i.e., a penalized bundle method, and suppose that $\varphi_k = \check{f}_k$. A sequence of null steps between two stability centers merely improves the model, leaving unchanged the center x^k in (10.5). When y^{k+1} satisfies the descent condition (10.1), it becomes x^{k+1} and, by Lemma 10.8,

$$\text{for all } k \in K_s \quad x^{k+1} = x^k - \frac{1}{\mu_k}\hat{s}^k, \quad \text{where } \hat{s}^k \in \partial \check{f}_k(x^{k+1}). \tag{10.22}$$

As observed in [15], [136], [89], the process of making null steps until satisfaction of the descent condition can be interpreted as an implementable procedure to compute the proximal point of f at x^k. More precisely, consider (10.20) written with $x = x^k$. By comparing the resulting relations to (10.22), we observe that the parameter μ now varies with k and that

$$p_{\mu_k}(x^k) \approx x^{k+1} \text{ and } s(p_{\mu_k}(x^k)) \approx \hat{s}^k = \mu_k(x^k - x^{k+1}) \in \partial \check{f}_k(x^{k+1}). \tag{10.23}$$

From this point of view, the descent test (10.1) is nothing but an assessment of how good y^{k+1} is as an approximation of $p_{\mu_k}(x^k)$ (the same interpretation is used in a more general setting in § 11.2.3 below).

Suppose the sequence $\{\mu_k\}_{k \in K_s}$ has a limit $\tilde{\mu}$. By (10.21) written with $(\mu, x) = (\mu_k, x^k)$, we have that $\hat{s}^k \approx \nabla F_{\mu_k}(x^k)$. Since minimizing f is equivalent to minimizing $F_{\tilde{\mu}}$, the update in (10.22) can be seen as a step of a preconditioned gradient method applied to the minimization of $F_{\tilde{\mu}}$, similarly to § 2.6. In order to increase the rate of convergence, the preconditioner should vary along iterations, trying to approximate second-order information for $F_{\tilde{\mu}}$ at a minimizer $\bar{x}$, in the spirit of quasi-Newton framework (see Chapter 4).

The update of μ_k is based on this remark. More precisely, we use a variable metric preconditioner of the form $\frac{1}{\mu_k}I$, updated using the so-called symmetric rank-one formula; see [105], which gives the following quasi-Newton formula, called *poor man's* in [236] :

$$\mu_{k+1} = \frac{\|v\|^2}{\langle v, u \rangle} .$$

(10.24)

As usual in quasi-Newton methods, the pair (u, v) is formed by some difference of points and gradients. A classical choice would be

$$(u, v) = \left(x^{k+1} - x^k, \nabla F_{\mu_k}(x^{k+1}) - \nabla F_{\mu_k}(x^k)\right)$$
$$\approx \left(x^{k+1} - x^k, \nabla F_{\mu_k}(x^{k+1}) - \hat{s}^k\right) .$$

Note that, since μ in (10.19) is now varying with k, we shall be working with a varying Moreau-Yosida regularization, F_{μ_k}. Moreover, from (10.21), we see that, to compute $\nabla F_{\mu_k}(x^{k+1})$, an additional calculation is required. For this reason, to define the pair (u, v) we use, instead of the classical choice, the *reversal form* introduced in [236], that we describe next.

After x^{k+1} has been defined, the black box gives a subgradient $s^{k+1} \in \partial f(x^{k+1})$. The idea is to use information already available to define a point z^{k+1} such that $x^{k+1} = p_{\mu_k}(z^{k+1})$, by using (10.20)-(10.21). More precisely, writing (10.20) with $(\mu, p_\mu(x), x, s(p_\mu(x)))$ replaced, respectively, by $(\mu_k, x^{k+1}, z^{k+1}, s^{k+1})$ yields

$$x^{k+1} = z^{k+1} - \frac{1}{\mu_k} s^{k+1} \quad \Leftrightarrow \quad \begin{cases} z^{k+1} = x^{k+1} + \frac{1}{\mu_k} s^{k+1} \\ \nabla F_{\mu_k}(z^{k+1}) = \mu_k(z^{k+1} - x^{k+1}) = s^{k+1} . \end{cases}$$

Similar relations hold for (μ_k, x^k, z^k, s^k), so

$$(v, u) = (\nabla F_{\mu_k}(z^{k+1}) - \nabla F_{\mu_k}(z^k), \qquad z^{k+1} - z^k)$$
$$= (s^{k+1} - s^k \qquad , x^{k+1} - x^k + \frac{1}{\mu_k}(s^{k+1} - s^k)) .$$

Using this expression for the quasi-Newton pair (u, v) in the formula (10.24) gives the following (reciprocal of the) update

$$\frac{1}{\mu_{k+1}} = \frac{1}{\mu_k} + \frac{\langle x^{k+1} - x^k, s^{k+1} - s^k \rangle}{\|s^{k+1} - s^k\|^2} .$$

(10.25)

This reversal formula is used in STEP 3 of the penalized bundle method to update μ_k when there is a descent step.

Although, in general, symmetric rank-one updates do not preserve positive definiteness (i.e., do not guarantee that $\mu_{k+1} > 0$), when using the reversal form for (u, v), the numerator of the second term in (10.25) can be made positive by incorporating a Wolfe-like condition in the curvilinear search mentioned at the end of Remark 10.2. Alternative formulæ, based on similar ideas were developed in [305].

The use of variable metrics allows an acceleration of penalized bundle methods. Research on this subject is not much developed yet. "More than first order" nonsmooth optimization methods are still an open ground; see for example [254], [42], [234], [235], [236], [258], [77]. In particular, it is shown in [233] that under appropriate assumptions (related to f not being "sharp", cf. comments after Theorem 9.6), the sequence of serious steps generated by penalized bundle methods using the reversal update has superlinear rate of convergence. A related subject is to find adequate second-order developments of convex functions; see [234], [235], [231], [255], [256].

A recent breakthrough in the area is the $\mathcal{VU}$-bundle method in [257], a fully implementable algorithm with superlinear convergence of serious iterates. The price to pay for gaining in speed of convergence is that each iteration requires the solution of two QP problems, one similar to (10.5), followed by another QP yielding a Newton-like direction that speeds up the method.

A sophisticated code based on [236], with curvilinear search and reversal quasi-Newton update of μ_k, is available upon request at INRIA. It is called N1CV2 and is free for an academic use; see

    http://www-rocq.inria.fr/estime/modulopt/optimization-routines/
    n1cv2.html

Other efficient codes, are BT and NOA, based on [326], and [213], respectively. We should also mention the proximal analytic center cutting-planes method [18].

The figure displayed on the cover of this book shows iterates obtained by N1CV2 for a test problem called MAXQUAD. This academic example is due to (the somewhat random typing of) Claude Lemaréchal; a related function can be found in § VIII.3.3.3 of [195]. MAXQUAD is a finite minimax problem, as in § 8.3.1, defined by $np = 5$ functions f_j in (8.17). Each of these functions is quadratic and has the form $f_j(x) := \langle x, A_j x \rangle + \langle x, b_j \rangle$ for $x \in \mathbb{R}^{10}$. Letting $j = 1, \ldots, np$, the corresponding vectors and matrices are, for all $i = 1, \ldots, 10$, $b_j(i) := e^{i/j} \sin(ij)$, and

$$A_j(i,k) := \begin{cases} e^{i/k} \cos(ik) \sin(j) & \text{for } k = i+1, \ldots, 10 \\ A_j(k,i) & \text{for } k = 1, \ldots, i-1 \\ |\sin j| i/n + \sum_{l \neq i} |A_j(i,l)| & \text{for all } k = i. \end{cases}$$

At the (rounded) optimal solution

$$\bar{x} = (-.1263, -.0344, -.0069, .0264, .0673, -.2784, .0742, .1385, .0840, .0386)^\top$$

the first four quadratic functions f_j $(j = 1, \ldots, 4)$ are active and equal $\bar{f} = -0.8414080$. The cover figure shows a 2D-view of MAXQUAD level sets about $\bar{x}$, as well as the corresponding components of a few serious steps computed by N1CV2; see the computational exercises in Chapter 12, § 12.1.1.

10.3.4 Convergence Analysis

When the parameter `tol` in Algorithm 10.1 is taken strictly positive, by Lemma 10.3 there is an index k_{last} for which $\delta_{k_{last}} \leq$ `tol` if (8.1) has minimizers. By Lemma 10.8 (ii), both $\varepsilon_{k_{last}}$ and $\|\hat{s}^{k_{last}}\|_{k_{last}}/\mu_{k_{last}}$ are small. Therefore, by Lemma 10.8 (iii), the last serious step satisfies the following approximate optimality condition:

$$\forall y \in \mathbb{R}^n \quad f(y) \geq f(x^{k_{last}}) - \|\hat{s}^{k_{last}}\|_{k_{last}} |y - x^{k_{last}}|_{k_{last}} - \varepsilon_{k_{last}}.$$

When `tol` $= 0$, the algorithm either stops having found a solution to (8.1) (the inclusion $0 \in \partial_0 f(x^{k_{last}})$ holds), or it loops indefinitely. In this case, to analyze the global convergence properties of Algorithm 10.1, we consider the following particular instance.

Algorithm 10.13 (penalized with aggregation).

– Matrices M_k are scalar multiples of a symmetric positive definite matrix M, chosen at the initialization step. The scalar factors are updated only at serious steps, i.e., $M_k := \eta_k M$, with η_k varying for $k \in K_s$. With this choice, both the primal and dual norms can be bounded by the Euclidean norm $\| \cdot \|^2$, using the extreme eigenvalues λ and Λ of M:

$$\eta_k \lambda \| \cdot \|^2 \leq |\cdot|_k^2 \leq \eta_k \Lambda \| \cdot \|^2 \text{ and } \frac{1}{\eta_k \Lambda} \| \cdot \|^2 \leq \| \cdot \|_k \leq \frac{1}{\eta_k \lambda} \| \cdot \|^2,$$
$$(10.26)$$

– y^{k+1} of (10.5) is computed by the dual (10.8)&(10.9), so the nominal decrease if δ_{k+1} of (10.6) while ε_k is as in Lemma 10.8 (ii),
– the model φ_k of (10.7) is constructed according to the Compression subalgorithm 10.11, and
– the stopping tolerance is `tol` $= 0$.

Our choice of the metric corresponds to the separation between two different effects of the quadratic term in subproblem (10.5). Namely, one related to the *strength* of the penalization (corresponding to μ_k), and another related to the *shape* of the penalization (corresponding to $|\cdot|_k$). This separation brings more flexibility into the way parameters are updated. For example, the update of μ_k could follow some (curved) line-search ideas, while the update of η_k could be based in the variable metric ideas from § 10.3.3. Or one could define $M = I$, $\eta_k = 1$ for all $k \in K_s$, and only change μ_k for $k \in K_s$ according to § 10.3.3. As long as the resulting parameters satisfy the conditions (10.27) and (10.31) given below, Algorithm 10.13 will always generate a minimizing sequence for (8.1). If condition (10.28) is also satisfied, the whole sequence will converge to a minimizer.

To show convergence, we suppose that the algorithm never stops. In this case, there are two possibilities for the sequence of descent steps $\{x^k\}_{k \in K_s}$. Either it has infinitely many elements, or there is an iteration k_{last} where a last serious step is done, i.e., $x^k = x^{k_{last}}$ for all $k \geq k_{last}$. We consider these two situations separately.

Theorem 10.14. *Suppose Algorithm 10.13 generates infinitely many descent-steps x^k for $k \in K_s$. Then either (8.1) has an empty solution set and $\{f(x^k)\} \searrow -\infty$, or the following holds:*

(i) Both $\{\delta_k\} \to 0$ and $\{\varepsilon_k\} \to 0$ as $k \to \infty$ in K_s.

(ii) If

$$\text{for all } k \in K_s \quad \eta_{k+1} \leq \eta_k \quad \text{and} \quad \sum_{k \in K_s} \frac{\eta_{k+1}}{\mu_k \eta_k} = +\infty, \quad (10.27)$$

the sequence $\{x^k\}$ is minimizing for (8.1).

(iii) If, in addition to (10.27), there exist positive constants $\eta_{\min}$ and B satisfying, for all $k \in K_s$,

$$\eta_{\min} \leq \eta_k \quad \text{and} \quad \frac{\eta_{k+1}}{\mu_k \eta_k} \leq B, \quad (10.28)$$

then the sequence $\{x^k\}$ is bounded and converges to a minimizer of (8.1).

Proof. Note first that, since $\texttt{tol} = 0$ and the algorithm does not stop, it holds that $\delta_{k+1} > 0$ for all $k \in K_s$. As a result, the infinite sequence of objective values $\{f(x^k)\}$ is strictly decreasing. If (8.1) has no solution, the sequence goes to $-\infty$. Otherwise, by Lemma 10.3, the series $\sum_{k \in K_s} \delta_{k+1}$ converges and item (i) follows by Lemma 10.8 (ii).

To see (ii), we apply a reasoning similar to the one yielding (9.5) in Chapter 9. More precisely, given an arbitrary $x \in \mathbb{R}^n$, use the definition of M_{k+1} and of x^{k+1} ($= y^{k+1}$ in (10.8)) to expand the square below as follows:

$$
\begin{aligned}
|x^{k+1} - x|^2_{k+1} &= \left\langle x^{k+1} - x, \eta_{k+1} M(x^{k+1} - x) \right\rangle \\
&= \eta_{k+1} \left\langle x^k - x - \frac{1}{\mu_k \eta_k} M^{-1} \hat{s}^k, M(x^k - x) - \frac{1}{\mu_k \eta_k} \hat{s}^k \right\rangle \\
&= \frac{\eta_{k+1}}{\eta_k} \left\langle x^k - x, \eta_k M(x^k - x) \right\rangle - 2 \frac{\eta_{k+1}}{\mu_k \eta_k} \left\langle x^k - x, \hat{s}^k \right\rangle \\
&\quad + \frac{\eta_{k+1}}{\mu_k \eta_k} \frac{1}{\mu_k} \left\langle \frac{1}{\eta_k} M^{-1} \hat{s}^k, \hat{s}^k \right\rangle \\
&= \frac{\eta_{k+1}}{\eta_k} |x^k - x|^2_k + \frac{\eta_{k+1}}{\mu_k \eta_k} \left(2 \left\langle x - x^k, \hat{s}^k \right\rangle + \frac{1}{\mu_k} \|\hat{s}^k\|^2_k \right).
\end{aligned}
$$

We bound the second right hand side term by using Lemma 10.8 (iii):

$$2 \left\langle x - x^k, \hat{s}^k \right\rangle \leq 2(f(x) - f(x^k) + \varepsilon_k),$$

while a bound for the third right hand side term is given by Lemma 10.8 (ii):

$$\frac{1}{\mu_k} \|\hat{s}^k\|^2_k \leq 2(\delta_{k+1} - \varepsilon_k).$$

As a result, using the left hand side condition in (10.27), we obtain the relation

$$|x^{k+1} - x|^2_{k+1} \leq |x^k - x|^2_k + 2\frac{\eta_{k+1}}{\mu_k \eta_k}(f(x) - f(x^k) + \delta_{k+1}). \tag{10.29}$$

To show item (ii), suppose for contradiction purposes that there exist $\tilde{x} \in \mathbb{R}^n$ and $\rho > 0$ such that $f(\tilde{x}) \leq f(x^k) - \rho$ for all $k \in K_s$. Since, by item (i), $\{\delta_k\} \to 0$, there exists k_ρ such that $\delta_{k+1} \leq \rho/2$ for all $k \geq k_\rho$, and, hence, writing relation (10.29) for $x = \tilde{x}$ we obtain

$$0 \leq |x^{k+1} - \tilde{x}|^2_{k+1} \leq |x^k - \tilde{x}|^2_k + 2\frac{\eta_{k+1}}{\mu_k \eta_k}(-\rho + \rho/2) \quad \text{for all } k \in K_s, k \geq k_\rho.$$

Summing the inequalities over $k_\rho \leq k \in K_s$ yields

$$0 \leq |x^{k_\rho} - \tilde{x}|^2_{k_\rho} - \rho \sum_{k \in K_s, k \geq k_\rho} \frac{\eta_{k+1}}{\mu_k \eta_k}.$$

Letting $k_\rho \to \infty$, we obtain a contradiction for the divergence assumption in (10.27).

To see item (iii), take in (10.29) $x = \bar{x}$, a solution to (8.1), and sum over $k \in K_s$. Since $f(\bar{x}) \leq f(x^k)$ for all k, we see that $\lim_{k \to \infty} |x^k - \bar{x}|^2_k < +\infty$. By (10.26) and (10.28), $\eta_{\min}\lambda\|x^k - \bar{x}\|^2 \leq |x^k - \bar{x}|^2_k$, so the sequence $\{x^k\}$ is bounded. Extract a subsequence $\{x^{k_i}\}_{k_i \in K_s}$ converging to $\bar{x}$ as $i \to \infty$. To show that the whole sequence converges to $\bar{x}$, given any $\rho > 0$, take i big enough to ensure that

$$\|x^{k_i} - \bar{x}\|^2 \leq \frac{\rho\eta_{\min}\lambda}{2\eta_{k_i}\Lambda} \quad \text{and} \quad \sum_{k \in K_s, k \geq k_i} \delta_{k+1} \leq \frac{\rho\eta_{\min}\lambda}{2}. \tag{10.30}$$

The sum of (10.29), written with $x = \bar{x}$, over $k \in K_s$ and going from k_i to an arbitrary $\tilde{k} > k_i$ yields

$$|x^{\tilde{k}+1} - \bar{x}|^2_{\tilde{k}+1} \leq |x^{k_i} - \bar{x}|^2_{k_i} + \sum_{k \in K_s, k=k_i}^{\tilde{k}} \delta_{k+1}.$$

By (10.26) and (10.28), this means that

$$\eta_{\min}\lambda\|x^{\tilde{k}+1} - \bar{x}\|^2 \leq |x^{\tilde{k}+1} - \bar{x}|^2_{\tilde{k}+1} \leq |x^{k_i} - \bar{x}|^2_{k_i} + \textstyle\sum_{k \in K_s, k=k_i}^{\tilde{k}} \delta_{k+1}$$
$$\leq \eta_{k_i}\Lambda\|x^{k_i} - \bar{x}\|^2 + \textstyle\sum_{k \in K_s, k \geq k_i} \delta_{k+1},$$

which, together with (10.30), yields that $\|x^{\tilde{k}+1} - \bar{x}\|^2 \leq \rho$, as desired. □

We now address the case of finitely many descent-steps.

Theorem 10.15. *Suppose Algorithm 10.13 generates a last descent-iterate $x^{k_{last}}$, followed by infinitely many null-steps. If*

$$\text{for all } k \geq k_{last} \quad \mu_{k+1} \geq \mu_k, \quad \text{and} \quad \mu_k \leq \mu_{\max} \tag{10.31}$$

for some positive constant $\mu_{\max}$, then the sequence $\{y^k\}$ converges to $x^{k_{last}}$ and $x^{k_{last}}$ minimizes f.

Proof. For simplicity, in the proof that follows we drop some (sub/supra) indices k_{last}, and denote $x^{k_{last}}$ and $\eta_{k_{last}}$ by x and η, respectively. Since matrices M_k only change at serious steps, for all $k \geq k_{last}$,

$$| \cdot |_k^2 = \eta \langle \cdot, M \cdot \rangle =: \eta | \cdot |_M^2 \quad \text{and} \quad \| \cdot \|_k^2 = \frac{1}{\eta} \langle \cdot, M^{-1} \cdot \rangle =: \frac{1}{\eta} | \cdot |_{M^{-1}}^2.$$

For any $y \in \mathbb{R}^n$, consider the function

$$L_k(y) := \varphi_k(y^{k+1}) + \frac{1}{2} \mu_k \eta |y^{k+1} - x|_M^2 + \frac{1}{2} \mu_k \eta |y^{k+1} - y|_M^2.$$

By definition of y^{k+1} in (10.5), $\varphi_k(y^{k+1}) + \frac{1}{2} \mu_k \eta |y^{k+1} - x|_M^2 \leq \varphi_k(x)$ and, by (10.18), $\varphi_k(x) \leq f(x)$, so

$$L_k(y^{k+1}) \leq f(x) \quad \text{for all } k \geq k_{last}. \tag{10.32}$$

Furthermore, the right hand side inequality in (10.18) and the identity $\mu_k \eta M(x - y^{k+1}) = \hat{s}^k$ from (10.8) give the relations

$$\varphi_{k+1}(y) \geq \varphi_k(y^{k+1}) + \mu_k \eta \langle M(x - y^{k+1}), y - y^{k+1} \rangle = f_a(y). \tag{10.33}$$

Using condition (10.31), inequality (10.33) written for $y = y^{k+2}$, and the definition of L_k, we obtain that

$$\begin{aligned}
L_{k+1}(y^{k+2}) &= \varphi_{k+1}(y^{k+2}) + \frac{1}{2} \mu_{k+1} \eta |y^{k+2} - x|_M^2 \\
&\geq \varphi_{k+1}(y^{k+2}) + \frac{1}{2} \mu_k \eta |y^{k+2} - x|_M^2 \\
&\geq \varphi_k(y^{k+1}) + \langle \mu_k \eta M(x - y^{k+1}), y^{k+2} - y^{k+1} \rangle \\
&\quad + \frac{1}{2} \mu_k \eta |y^{k+2} - x|_M^2 \\
&= L_k(y^{k+1}) - \frac{1}{2} \mu_k \eta |y^{k+1} - x|_M^2 \\
&\quad + \langle \mu_k \eta M(x - y^{k+1}), y^{k+2} - y^{k+1} \rangle + \frac{1}{2} \mu_k \eta |y^{k+2} - x|_M^2.
\end{aligned}$$

By expanding the difference of squares we see that

$$\begin{aligned}
|y^{k+2} - x|_M^2 - |y^{k+1} - x|_M^2 &= \langle y^{k+2} - x + y^{k+1} - x, M(y^{k+2} - y^{k+1}) \rangle \\
&= \langle y^{k+2} - y^{k+1} + 2(y^{k+1} - x), M(y^{k+2} - y^{k+1}) \rangle \\
&= \langle y^{k+2} - y^{k+1}, M(y^{k+2} - y^{k+1}) \rangle \\
&\quad + 2 \langle y^{k+1} - x, M(y^{k+2} - y^{k+1}) \rangle \\
&= |y^{k+2} - y^{k+1}|_M^2 \\
&\quad + 2 \langle M(y^{k+1} - x), y^{k+2} - y^{k+1} \rangle,
\end{aligned}$$

yielding the inequality

$$L_{k+1}(y^{k+2}) \geq L_k(y^{k+1}) + \frac{1}{2} \mu_k \eta |y^{k+2} - y^{k+1}|_M^2. \tag{10.34}$$

Since the increasing sequence $\{L_k(y^{k+1})\}$ is bounded from above by (10.32), it must converge.

We now show that the sequence $\{y^{k+1}\}$ is bounded, with $\{y^{k+1} - y^k\} \to 0$.

Using once more the identity $\mu_k \eta M(x - y^{k+1}) = \hat{s}^k$ and the relation $f_a \leq f$ in (10.18), we see that

$$L_k(y^{k+1}) + \frac{1}{2}\mu_k \eta |y^{k+1} - x|_M^2 = \varphi_k(y^{k+1}) + \langle \hat{s}^k, y^{k+2} - y^{k+1} \rangle = f_a(x) \leq f(x).$$

Since the first term in the left hand side converges as $k \to \infty$ and, by (10.31), $\mu_k \geq \mu_{k_{last}}$, we obtain that the sequence $\{y^{k+1}\}$ must be bounded. In addition, using again that $\mu_k \geq \mu_{k_{last}}$ by (10.34), and passing to the limit, we conclude that $\{y^{k+2} - y^{k+1}\} \to 0$.

Being a convex function, f is locally Lipschitzian with subdifferential locally bounded. Let L and M denote the respective constants on a bounded set containing the sequence $\{y^k\}$. Using the right hand side inequality in (10.18), we see that

$$-M|y^{k+1} - y^k| \leq \langle s^k, y^{k+1} - y^k \rangle \leq \varphi_k(y^{k+1}) - f(y^k),$$

while, using the inequality $\varphi_k \leq f$ in (10.18), we have that

$$\varphi_k(y^{k+1}) - f(y^k) \leq f(y^{k+1}) - f(y^k) \leq L|y^{k+1} - y^k|.$$

Therefore, $\{\varphi_k(y^{k+1}) - f(y^k)\} \to 0$, because $\{y^{k+1} - y^k\} \to 0$ as $k \to \infty$. From the bounded sequence $\{y^k\}$ extract a subsequence $\{y^{k_i}\} \to \bar{y}$ as $i \to \infty$ and note that $\{y^{k_i+1}\} \to \bar{y}$ because $\{y^{k+1} - y^k\} \to 0$. Write

$$f(y^{k_i+1}) - \varphi_{k_i}(y^{k_i+1}) = f(y^{k_i+1}) - f(y^{k_i}) + f(y^{k_i}) - \varphi_{k_i}(y^{k_i+1})$$

to see that

$$f(y^{k_i+1}) - \varphi_{k_i}(y^{k_i+1}) \to 0 \quad \text{and} \quad \varphi_{k_i}(y^{k_i+1}) \to f(\bar{y}) \quad \text{as } i \to \infty. \quad (10.35)$$

To show that x minimizes (8.1), recall that for all $k \geq k_{last}$, the descent-test (STEP 3, Algorithm 10.1) is never satisfied. This means that $f(y^{k_i+1}) - f(x) > -m\delta_{k_i+1}$, so, adding δ_{k_i+1} to both sides of the inequality, and using (10.6), we obtain

$$\begin{aligned}
0 \leq (1 - m)\delta_{k_i+1} &\leq f(y^{k_i+1}) - f(x) + \delta_{k_i+1} \\
&= f(y^{k_i+1}) - \varphi_{k_i}(y^{k_i+1}) - \tfrac{1}{2}\mu_{k_i}\eta |x - y^{k_i+1}|_M^2 \\
&\leq f(y^{k_i+1}) - \varphi_{k_i}(y^{k_i+1}).
\end{aligned}$$

Passing to the limit as $i \to \infty$ and using (10.35) we conclude that $\delta_{k_i+1} \to 0$. By Lemma 10.9(ii), it follows that $\varepsilon_{k_i} + \frac{1}{2\mu_{k_i}\eta}|\hat{s}^{k_i}|_{M^{-1}}^2$ converges to 0. Then both ε_{k_i} and $\hat{s}^{k_i}$ converge to 0 as $i \to \infty$, because $\mu_{k_i} \leq \mu_{\max}$, by (10.31). Since from Lemma 10.8 (iii), $\hat{s}^{k_i} \in \partial_{\varepsilon_{k_i}} f(x)$, we have that $f(y) \geq f(x) + \langle \hat{s}^{k_i}, y - x \rangle - \varepsilon_{k_i}$ for all $y \in \mathbb{R}^n$. Passing to the limit as $i \to \infty$, the inequality shows that x minimizes f on $\mathbb{R}^n$.

Finally, we show that any cluster point $\bar{y}$ is equal to x. Use the facts that

x minimizes f and that $f \geq \varphi_{k_i}$ by (10.18), together with the definition of y^{k_i+1} from (10.5) and the inequality $\mu_{k_i} \geq \mu_{k_{last}}$ from (10.31), to write the relations

$$f(\bar{y}) \geq f(x) \geq \varphi_{k_i}(x) \geq \varphi_{k_i}(y^{k_i+1}) + \frac{1}{2}\mu_{k_i}\eta|y^{k_i+1} - x|_M^2$$
$$\geq \varphi_{k_i}(y^{k_i+1}) + \frac{1}{2}\mu_{k_{last}}\eta|y^{k_i+1} - x|_M^2.$$

By (10.35), we obtain in the limit that

$$f(\bar{y}) \geq \lim_{i \to \infty}\left(\varphi_{k_i}(y^{k_i+1}) + \frac{1}{2}\mu_{k_{last}}\eta|y^{k_i+1} - x|_M^2\right)$$
$$= f(\bar{y}) + \frac{1}{2}\mu_{k_{last}}\eta|\bar{y} - x|_M^2,$$

an inequality that is possible only if $\bar{y} = x$. Since the relations above hold for any cluster point of the sequence $\{y^{k+1}\}$, there can only be one of such points, namely x. □

Recall that, for the Subgradients Algorithm 9.2 to converge, conditions (a) and (b) are required to hold in Theorem 9.3. In practice, no choice of t_k can satisfy both conditions simultaneously. By contrast, conditions (10.27), (10.28), and (10.31), ensuring convergence of the Penalized Bundle Algorithm 10.13, are rather mild to satisfy. For example, it is enough to set $\eta_k = \eta_{\min}$ for all $k \in K_s$ and $\mu_k = \mu_{\max}$ for all k. Of course, the method will be faster with "smarter" choices of parameters, such as the reversal update described in § 10.3.3.

We finish with a summary of the advantages of bundle methods, especially when (10.2) is solved via its dual ((10.8)&(10.9)= (10.5)).
– Construction of a model φ_k and of a nominal decrease δ_k, which result in *implementable* stopping tests.
– Better adjustment of the stopping tolerance, thanks to (10.14).
– Possibility of aggregating the bundle to avoid memory overflows, due to an infinite accumulation of the affine functions defining the model.
– Quick resolution of (10.2), whose dual has dimension np_k at most. This aspect is crucial when n is large.

11 Applications of Nonsmooth Optimization

This final chapter is devoted to particular variants and extensions of NSO methods, that can be developed when the problem to be solved has some special structure. For example, such is the case

- when (8.1) corresponds to a certain dual problem, arising in large-scale or combinatorial optimization; or
- when (8.1) is a *constrained* problem; or
- when (8.1) is solved via its optimality condition, i.e., by finding $\bar{x}$ such that $0 \in \partial f(\bar{x})$.

11.1 Divide to conquer. Decomposition methods

One of the most important applications of nonsmooth optimization methods is decomposition. Decomposition techniques are used to solve large-scale (or complex) problems, replacing them by a sequence of reduced-dimensional (or easier) *local* problems linked by a *master* program, as shown schematically in Figure 11.1.

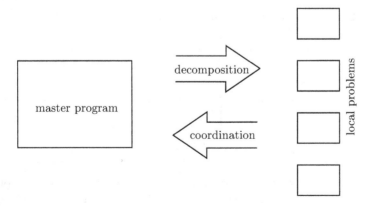

Fig. 11.1. Decomposition/coordination

These methods prove efficient when the structure of the problem is naturally separable. For example, the optimal management of a set of power plants; see § 1.2.4, and also [232], [19] and [21].

Two basic decomposition schemes give birth to all the methods of this class. These are resource- and price-decomposition, dual to each other.

– In *resource decomposition*, the central manager sends directives to each unit (decomposition). Each plant minimizes its own cost while respecting the directives, and answers in return a price according to these directives. With this new data, the master program adjusts the directives (coordination).
– *Price decomposition* works the opposite way. The central manager sends a set of prices to the subsidiary companies; each of them minimizes its own cost, driven by these prices. Depending on the local answers, the master program adjusts the prices (coordination).

Nonsmoothness comes into play in the master program, which is a dual problem, as in § 11.1.1, 11.1.2 below. The solution of a primal problem via its dual is interesting only if the calculation of the dual function is much easier than solving the primal problem directly. This is the case when the primal problem

$$\begin{cases} \min_p f_o(p) \\ p \in P \\ c_E(p) = 0 \\ c_I(p) \leq 0 \end{cases} \tag{11.1}$$

has some structure suitable for decomposition:

– For instance, the data can be block-separable: $f_o(x) = \sum_{i=1}^{K} f_o^i(p^i)$, with $p^i \in P^i \subset \mathbb{R}^{N_i}$, and each $f_o^i : \mathbb{R}^{N_i} \to \mathbb{R}$; the constraints being decomposable as well, as in (11.2) below. After relaxation of the constraints, there are K local problems each one of dimension N_i; see § 11.1.1.
– When the data is not decomposable, separability can be induced by *duplicating variables*. More precisely, suppose P in (11.1) is discrete, and that it is difficult to check simultaneously that $p \in P$ and that the remaining constraints, $c(\cdot)$, are also satisfied. One can induce separability by rewriting the problem as follows

$$\begin{cases} \min_{p,\tilde{p}} f_o(p) \\ \tilde{p} \in P \\ c_E(p) = 0 \\ c_I(p) \leq 0 \\ \tilde{p} = p. \end{cases}$$

Relaxing the last constraint yields a dual with two local problems, one in the continuous variable p (checking satisfaction of constraints given by $c(\cdot)$), the other in the discrete variable $\tilde{p}$ (checking that $\tilde{p} \in P$). The principle "divide to conquer" applies.

In general, decomposing amounts to solving some dual problem. As shown by Theorem 8.4, only when the Karush-Kuhn-Tucker conditions are necessary and sufficient for optimality, the dual solution gives a solution to the primal problem. Nevertheless, even in the presence of a positive duality gap, the numerical resolution of the dual problem with a bundle method will generate a sequence $\{x^k\}$ converging to a maximizer $\bar{x}$, with subgradients $\hat{s}^k \in \partial_{\varepsilon_k}(-\theta)(x^k)$, where both $\hat{s}^k$ and ε_k are close to 0 for k large enough. Since $\hat{s}^k$ is a convex combination of gradients $s^i = -c(p_{y^i})$ for some primal points p_{y^i}, the primal convex combination $\sum_i \bar{\alpha}_i p_{y^i}$ is a reasonable approximation of a primal optimum by Theorem 8.1, specially when the relaxed constraints are affine or convex.

In addition, we stress the fact that, even without convexity, passing to the dual problem can always be used to *bound* the optimal value $f_o(\bar{p})$ from below by $\theta(\bar{x})$. This property is useful in combinatorial optimization, to define an evaluation function in branch-and-bound methods (see [188] for an early example).

If one really wants to eliminate the duality gap (to know exactly the optimal primal value, for example), augmented Lagrangians can still be used. The augmented Lagrangian of Remark 8.5 is a basis for *multiplier methods*, [191],[288], [311], [25], [27]. The augmented Lagrangian gives an augmented dual function that is differentiable. However, it may no longer be decomposable (cf. § 11.1.4 below).

11.1.1 Price Decomposition

Duality shows all its potential when (11.1) is block-separable. Price decomposition schemes appear when using Lagrangian relaxation to exploit separability.

General Algorithm Suppose that the data can be split into K blocks, as follows. Let the dimension of the primal space be $N = \sum_{i=1}^{K} N_i$, with

$$p = (p^1, \ldots, p^K), \qquad p^i \in P^i \subset \mathbb{R}^{N_i}, \qquad P = \prod_{i=1}^{K} P^i,$$

$$f_o(p) = \sum_{i=1}^{K} f_o^i(p^i), \, c_E(p) = \sum_{i=1}^{K} c_E^i(p^i), \text{and } c_I(p) = \sum_{i=1}^{K} c_I^i(p^i). \tag{11.2}$$

The decomposed functions only operate on each block:

$$f_o^i : P^i \to \mathbb{R}, \quad c_E^i : P^i \to \mathbb{R}^{n_E}, \text{ and } \quad c_I^i : P^i \to \mathbb{R}^{n_I}.$$

In this setting, the Lagrangian of (11.1) has a separable structure

$$L(p, x) = \sum_{i=1}^{K} \left(f_o^i(p^i) + \langle x, c^i(p^i) \rangle \right),$$

with $x \in \mathbb{R}^{n_E + n_I}$ and the decomposed vector of constraints is $c^i(p^i) = (c^i_E(p^i), c^i_I(p^i))$. The associated dual function is also separable:

$$\theta(x) = \sum_{i=1}^{K} \theta^i(x) := \sum_{i=1}^{K} \min_{p^i \in P^i} \left\{ f^i_o(p^i) + \langle x, c^i(p^i) \rangle \right\} . \qquad (11.3)$$

When compared to (8.8), we see that one evaluation of $\theta(x)$ now requires the resolution of K **independent** local problems, each one of dimension N_i, instead of one (big) problem of size $N = \sum_{i=1}^{K} N_i$. Thus, computations will be dramatically simplified for large-scale primal problems, like in § 1.2.4.

Typically, the master program will choose a multiplier x ("price") sent to the K local solvers. Once solutions $\{p^i_x, i = 1, \ldots, K\}$ to (11.3) are obtained, the master program computes $\theta(x)$ and one subgradient, and defines a new x, by making one step of some NSO method to maximize the dual function. Figure 11.2 shows the particular form of Figure 11.1 for this decomposition method.

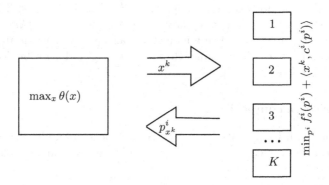

Fig. 11.2. Price decomposition

Algorithm 11.1 (price decomposition). Choose $x^1 \in X$ and set $k = 1$.

STEP 1 (Calling the local problems – decomposition). For $i = 1, \ldots, K$, compute:

$$p^i_{x^k} := \operatorname{argmin}_{p^i \in P^i} \left\{ f^i_o(p^i) + \langle x^k, c^i(p^i) \rangle \right\} .$$

STEP 2 (Evaluating the dual function). We have

$$\theta(x^k) = \sum_{i=1}^{K} \left(f^i_o(p^i_{x^k}) + \langle x^k, c^i(p^i_{x^k}) \rangle \right) .$$

From (8.19), a subgradient is available: $-\sum_{i=1}^{K} c^i(p^i_{x^k}) \in \partial\left(-\theta\right)(x^k)$.

STEP 3 (Master program – coordination). Compute x^{k+1} in order to maximize $\theta(x)$. For example, by using any one of the black-box methods in § 9.3 or in Chapter 10.

STEP 4 (loop). Change k to $k+1$, go to 1.

We now show that the well-known decomposition method of Dantzig-Wolfe [95] is a particular case of price decomposition.

Relation with Dantzig-Wolfe Decomposition Suppose that at STEP 3 of Algorithm 11.1 the cutting-planes method from § 9.3 is used. Following Algorithm 9.4, at iteration k a piecewise linear (concave) model $\check{\theta}_k(x)$ is constructed using the past information

$$\left(\theta(x^j), s^j\right) = \left(L(p_{x^j}, x^j), c(p_{x^j})\right) = \left(\sum_{i=1}^{K} f_o^i(p_{x^j}^i), \sum_{i=1}^{K} c^i(p_{x^j}^i)\right)$$

for $j = 1, \ldots, k$. Each affine function defining the cutting-planes model takes the form:

$$\theta(x^j) + \langle s^j, y - x^j \rangle = \sum_{i=1}^{K} \left(f_o^i(p_{x^j}^i) + \langle c^i(p_{x^j}^i), y \rangle\right),$$

for all $y \in X$. Therefore, the model (9.6) for θ (written with $j, -\theta(x^j), -s^j$ instead of i, f_i, s^i) becomes

$$\check{\theta}_k(y) = \min_{j=1,\ldots,k} \sum_{i=1}^{K} (f_o^i(p_{x^j}^i) + \langle c^i(p_{x^j}^i), y \rangle).$$

The cutting-planes Algorithm 9.4 (applied to the maximization of the concave function θ) computes the next iterate as follows:

$$x^{k+1} \quad \text{a solution to} \quad \max_{y \in S} \check{\theta}_k(y) = \min_{y \in S} \left(-\check{\theta}_k\right)(y).$$

If the feasible set S is polyhedral, for example a box, this subproblem is a linear program, so a solution can be found by solving its dual. To write this dual problem, we proceed as in the proof of Lemma 10.8. Without going into details, the cutting-planes subproblem (9.9) can be considered an instance of the penalized bundle subproblem (10.5) without quadratic term ($\mu_k = 0$). Starting from the equivalent primal LP problem (recall (9.10)),

$$\begin{cases} \min_{\{r \in \mathbb{R}, y \in X\}} -r \\ r \leq \sum_{i=1}^{K} (f_o^i(p_{x^j}^i) + \langle c^i(p_{x^j}^i), y \rangle), j = 1, \ldots, k, \end{cases} \tag{11.4}$$

the analogous of (10.9) is the following dual problem:

$$\begin{cases} \min_{\alpha \in \Delta_k} \sum_{j=1}^{k} \alpha_j \sum_{i=1}^{K} f_o^i(p_{x^j}^i) \\ \sum_{j=1}^{k} \alpha_j \sum_{i=1}^{K} c_E^i(p_{x^j}^i) = 0 \\ \sum_{j=1}^{k} \alpha_j \sum_{i=1}^{K} c_I^i(p_{x^j}^i) \leq 0, \end{cases} \tag{11.5}$$

where $\Delta_k := \{z \in [0,1]^k : \sum_{j=1}^{k} z_j = 1\}$ is the unit-simplex in $\mathbb{R}^k$. Theorem 8.4 gives the relation between the primal and dual solutions: x^{k+1} is an optimal Lagrange multiplier for the constraints in (11.5).

The well known *Dantzig-Wolfe* method generates iterates precisely by means of (11.5). At each iteration, local problems ("slave"-programs) return to the master program the solutions $\{p_{x^k}^i, i = 1, \ldots, K\}$ of STEP 1. With this information, the master defines at STEP 3 the next x^{k+1} as the multiplier associated with the constraints of (11.5).

From the master's point of view, Dantzig-Wolfe decomposition **is a cutting-planes method**. The well known tailing-off effect, that reduces speed of convergence as k increases, is nothing but the intrinsic instability of cutting-planes method that was mentioned in Remark 9.8.

Remark 11.2.

– Because the master program iterates using a cutting-planes algorithm, for nonlinear problems convergence of the Dantzig-Wolfe decomposition method is guaranteed only when the number of affine functions defining the model goes to infinity. Thus, it can be preferable to solve the master by using instead a bundle method, which allows aggregation and keeps bounded the size of quadratic subproblems.

– Another possibility for an acceleration is the *disaggregation* of the master objective function; see [19]. Denote by $\theta^i(x^k)$ each optimal value at STEP 1 of Algorithm 11.1; in view of (11.3), $\theta(x^k) = \sum_{i=1}^{K} \theta^i(x^k)$. Disaggregation then consists in maximizing the dual function at STEP 3 via the *separate* modeling of each function $\theta^i(\cdot)$. The cutting-planes model $\check{\theta}_k$ is replaced by the (more accurate) model $\sum_{i=1}^{K} \check{\theta}_k^i$. However, note that this will require to store past information in a disaggregate form:

$$\left(f_o^i(p_{x^j}^i), c^i(p_{x^j}^i) \right), \text{ for } i = 1, \ldots, K \text{ and } j = 1, \ldots, k.$$

Depending on the problem, this extra computational burden may have a negative effect in the overall performance of the method.

– Since each iteration of the cutting-planes method appends to (11.4) one constraint (i.e. the model is enriched with one affine function), this process is called *row generation*. By contrast, the quantity increasing along iterations in (11.5) is the dimension of the variables α; one then speaks of *column generation*.

11.1.2 Resource Decomposition

Consider the following reformulation of the primal problem (11.1), obtained by introducing a right hand side perturbation $u \in \mathbb{R}^{n_E} \times \mathbb{R}^{n_I}$, corresponding to directives or "resources":

$$
\begin{cases} \min v(u) \\ u = (u_E, u_I) \\ u_E = 0 \\ u_I \leq 0 \end{cases}
\quad \text{where} \quad v(u) := \begin{cases} \min_{p \in P} f_o(p) \\ c_E(p) = u_E \\ c_I(p) \leq u_I. \end{cases}
$$

With respect to the decomposition scheme in Figure 11.1, the left hand side problem above is the master program, while the right hand side problem corresponds to local problems. More precisely, using the separable structure described in (11.2), the master program is

$$
\begin{cases} \min_{u^i \in U^i} \sum_{i=1}^K v^i(u^i) \\ \sum_{i=1}^K u_E^i = 0 \\ \sum_{i=1}^K u_I^i \leq 0, \end{cases}
\tag{11.6}
$$

where for each $i = 1, \ldots, K$ we defined the feasible set

$$
U^i := \left\{ (u_E^i, u_I^i) \in \mathbb{R}^{n_E} \times \mathbb{R}^{n_I} : \exists p^i \in P^i : c_E^i(p^i) = u_E^i, c_I^i(p^i) \leq u_I^i \right\};
$$

while the local problems are

$$
v^i(u^i) := \begin{cases} \min_{p^i \in P^i} f_o^i(p^i) \\ c_E^i(p^i) = u_E^i \\ c_I^i(p^i) \leq u_I^i. \end{cases}
\tag{11.7}
$$

Problem (11.6) has no reason to be differentiable, because the individual value functions $v^i(\cdot)$ themselves need not be differentiable. Under appropriate continuity and convexity assumptions, however, for each local problem Theorem 8.4 holds. In this case, there is no duality gap, and (11.7) has a minimizer $\bar{p}_{u^i}^i$ and an optimal Lagrange multiplier $\bar{x}_{u^i}^i$, both depending on u^i, for which

$$
v^i(u^i) = f_o^i(\bar{p}_{u^i}^i) = \theta_{u^i}(\bar{x}_{u^i}^i),
\tag{11.8}
$$

where θ_{u^i} is the dual function obtained from (11.7) after relaxation of the constraints. Furthermore,

$$-\bar{x}_{u^i}^i \in \partial v^i(u^i).$$

To see that the multiplier $\bar{x}_{u^i}^i$ gives a subgradient, consider (11.7) written with u^i replaced by $\tilde{u}^i$, another perturbation. The weak duality relation (8.14) applies to such problem:

$$f_o^i(p^i) \geq \theta_{\tilde{u}^i}(x) := \inf_{p^i \in P^i} \{f_o^i(p^i) + \langle x, c^i(p^i) - \tilde{u}^i \rangle\},$$

for arbitrary $x \in X = \mathbb{R}^{n_E} \times \mathbb{R}_+^{n_I}$ and $p^i \in P^i$. In particular, for $p^i = \bar{p}_{\tilde{u}^i}^i$, i.e., a point in P^i such that $v^i(\tilde{u}^i) = f_o^i(\bar{p}_{\tilde{u}^i}^i)$,

$$v^i(\tilde{u}^i) = f_o^i(\bar{p}_{\tilde{u}^i}^i) \geq \theta_{\tilde{u}^i}(x)$$
$$= \inf_{p^i \in P^i} \{f_o^i(p^i) + \langle x, c^i(p^i) - u^i \rangle\} + \langle x, u^i - \tilde{u}^i \rangle$$
$$= \theta_{u^i}(x) + \langle x, u^i - \tilde{u}^i \rangle$$

for arbitrary $x \in X$. In particular, taking $x = \bar{x}_{u^i}^i$, and using the relation (11.8),

$$v^i(\tilde{u}^i) \geq v^i(u^i) + \langle -\bar{x}_{u^i}^i, \tilde{u}^i - u^i \rangle.$$

Therefore, at the coordination step, both $v(u) = \sum_{i=1}^K f_o^i(\bar{p}_{u^i}^i)$ and the subgradient $-\sum_{i=1}^K \bar{x}_{u^i}^i$ are available after solving the K local problems (11.7). Thus, a black-box method can be applied to solve the master program (11.6) (note that this is a *constrained* NSO problem, though).

Figure 11.1 specializes for this method as shown in Figure 11.3.

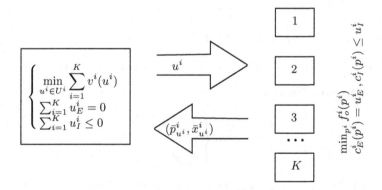

Fig. 11.3. Resource decomposition

Observe that the information sent by the master to the local solvers lies in the constraint space. From an economic point of view, the quantities u^i correspond to resource allocations sent from an upper level. The local units return values $-\bar{x}_{u^i}^i$, which are the marginal utilities associated with the allocated resource u^i, used to produce $\bar{p}_{u^i}^i$ units of product.

11.1.3 Variable Partitioning or Benders Decomposition

Benders decomposition is useful when the variables are separable into "hard" and "easy" ones. An example of this situation is multistage stochastic programming [316]: easy/hard variables correspond to first/second stages of decision. The primal problem (11.1) has the form

$$\begin{cases} \min f_o(p) + \tilde{f}_o(\tilde{p}) \\ (p, \tilde{p}) \in P \times \tilde{P} \\ c_E(p) + \tilde{c}_E(\tilde{p}) = 0 \\ c_I(p) + \tilde{c}_I(\tilde{p}) \leq 0, \end{cases}$$

where hard data is denoted by using a superscript $\sim$. The equivalent bilevel form is

$$\begin{cases} \min f_o(p) + v^{opt}(p) \\ p \in P \end{cases}$$

where we defined the *optimality function*

$$v^{opt}(p) := \begin{cases} \min \tilde{f}_o(\tilde{p}) \\ \tilde{p} \in \tilde{P} \\ \tilde{c}_E(\tilde{p}) = -c_E(p) \\ \tilde{c}_I(\tilde{p}) \leq -c_I(p). \end{cases} \tag{11.9}$$

In many cases this value function has a separable block structure, like (11.2), but similarly to (11.7), the function v^{opt} may not be differentiable. Depending on the data, v^{opt} can be lower semicontinuous and even convex. In this case, a subgradient can be computed using a Lagrange multiplier associated with the constraints. More precisely, writing (11.9) with $p = p^k$, and letting $\tilde{p}_{p^k}$ be a solution with associated Lagrange multiplier $\tilde{x}_{p^k}$, we have that

$$v^{opt}(p^k) = \tilde{f}_o(\tilde{p}_{p^k}) \quad \text{and} \quad s(p^k) := Jc(p^k)^T \tilde{x}_{p^k} \in \partial v^{opt}(p^k),$$

where we use the (transposed) Jacobian matrix $Jc(p^k) = (Jc_E(p^k), Jc_I(p^k))$.

Once again, the master program approximates $v^{opt}(\cdot)$ by using a cutting-planes model $\check{v}_k^{opt}$. The second level problem (which may be separable) plays the role of the local problem, and serves to evaluate $v^{opt}(p^k)$ and a subgradient, for every p^k sent by the master.

Suppose, for simplicity, that we use a cutting-planes method, that $f_o(p) = \langle a, p \rangle$ for some vector a, and that P is polyhedral. A straightforward application of the cutting-planes method gives for the master program an LP problem of the form:

$$\begin{cases} \min \langle a, p \rangle + \check{v}_k^{opt}(p) \\ p \in P \end{cases} \Leftrightarrow \begin{cases} \min_{p \in P, r \in \mathbb{R}} \langle a, p \rangle + r \\ r \geq \tilde{f}_o(\tilde{p}_{p^j}) + \langle Jc(p^j)^T \tilde{x}_{p^j}, p - p^j \rangle \text{ for } j \leq k. \end{cases}$$

At this point, it is important to notice that, depending on p^k, the feasible set in (11.9) corresponding to the local problems, i.e.

$$\left\{ \tilde{p} \in \tilde{P} : \tilde{c}_E(\tilde{p}) = -c_E(p^k) \text{ and } \tilde{c}_I(\tilde{p}) \leq -c_I(p^k) \right\},$$

may be empty. In this case, the value of $v^{opt}(p^k)$ and a subgradient cannot be computed. Instead, a new p^{k+1}, for which problem (11.9) written with $p = p^{k+1}$ is feasible, should be generated by the master program.

The new iterate is computed by introducing in the master LP problem a *feasibility* cut. The idea is, before trying to compute $v^{opt}(p^k)$, to define the *feasibility function*

$$v^{feas}(p^k) := \begin{cases} \min \|z\|_1 \\ \tilde{p} \in \tilde{P}, z_E^+, z_E^-, z_I^+ \geq 0 \\ \tilde{c}_E(\tilde{p}) + z_E^+ - z_E^- = -c_E(p^k) \\ \tilde{c}_I(\tilde{p}) - z_I^+ \leq -c_I(p^k) \\ z = (z_E^+ - z_E^-, z_I^+) \in \mathbb{R}^{n_E + n_I}, \end{cases} \quad (11.10)$$

where $\|z\|_1 := \sum_{i=1}^n |z_i|$ is the 1-norm in $\mathbb{R}^n$. Note that the feasibility function is nonnegative, with $v^{feas}(p) = 0$ if and only if problem (11.9) is feasible. In order to avoid infeasible local problems, the master program should rather be written

$$\begin{cases} \min f_o(p) + v^{opt}(p) \\ p \in P \\ v^{feas}(p) = 0. \end{cases}$$

Denote by $(\tilde{p}'_{p^k}, \tilde{z}_{p^k})$ a solution to (11.10), and by $\tilde{x}'_{p^k}$ an optimal multiplier corresponding to the constraints involving $c_E(p^k)$ and $c_I(p^k)$. If $\tilde{z}_{p^k} = 0$, the feasible set in (11.9), written with $p = p^k$, is nonempty. Otherwise, if $\tilde{z}_{p^k} \neq 0$, instead of trying (and failing) to compute $v^{opt}(p^k)$ and $s(p^k) \in \partial v^{opt}(p^k)$, we define black-box information for $v^{feas}(p^k)$, knowing that

$$v^{feas}(p^k) = \|\tilde{z}_{p^k}\|_1 \quad \text{and} \quad s'(p^k) := Jc(p^k)^T \tilde{x}'_{p^k} \in \partial v^{feas}(p^k).$$

With the introduction of feasibility cuts (corresponding to a cutting-planes model of v^{feas}), a typical master program iteration in Benders decomposition has the form:

$$\begin{cases} \min_{p \in P} \langle a, p \rangle + \check{v}_k^{opt}(p) \\ \check{v}_k^{feas}(p) \leq 0 \end{cases} \Leftrightarrow \begin{cases} \min_{p \in P, r \in \mathbb{R}} \langle a, p \rangle + r \\ r \geq \tilde{f}_o(\tilde{p}_{p^j}) + \langle Jc(p^j)^T \tilde{x}_{p^j}, p - p^j \rangle \text{ for } j \in O_k \\ 0 \geq \|\tilde{z}_{p^l}\|_1 + \langle Jc(p^l)^T \tilde{x}'_{p^l}, p - p^l \rangle \text{ for } l \in F_k, \end{cases}$$

where O_k and F_k correspond to past iterations where, respectively, optimality or feasibility cuts were generated.

When (11.9) is a linear programming problem, in general there is no need to make the extra computations required to introduce feasibility cuts. Namely, when required to solve an infeasible problem (11.9), any good LP solver detects infeasibility along the iterative process. In this case, there is an exit mode that can also provide a *recession direction* (of dual unboundedness), corresponding to the feasibility cuts above.

The general decomposition scheme for this method is shown in Figure 11.4.

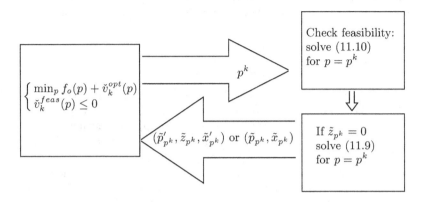

Fig. 11.4. Benders decomposition

In contrast with resource decomposition, Benders decomposition sends to the local units values in the (primal) variable-space. In [221] it is shown that in the linear case, this method (primal for the master program) is dual to the Dantzig-Wolfe decomposition applied to the dual.

Note that the resource decomposition problem (11.6) was formulated using the feasible sets U^i, which may not be known apriori, except for some particular applications. Such sets can be eliminated from (11.6) by introducing feasibility cuts, similarly to the procedure just shown for Benders decomposition.

Since Benders decomposition is a cutting-planes method from the master's point of view, it also presents tailing-off effects when the number of iterations becomes large. For this reason, a bundle method for solving the (constrained NSO) master problem might be preferable in some applications.

11.1.4 Other Decomposition Methods

We finish this section with a quick review of other decomposition techniques.

–Decomposition methods based on the principle of *auxiliary problem*. For a problem with hard constraints, the idea is to approximate the starting problem by a sequence of problems in which the hard functions are linearized; see for instance [93]. A penalty term is added if necessary, so that local problems, called auxiliary problems, are coercive and, hence, solvable. This decomposition is useful when using augmented Lagrangians with quadratic terms. Crossed terms kill separability, the auxiliary problem principle allows to introduce approximations in the augmented term, in order to obtain separable local problems.

–*Proximal decomposition methods* are suitable for problems where the feasible set is a subspace (such is the case when separability was induced by duplicating variables); see [343], [245], [76]. For the convex problem

$$\begin{cases} \min f_1(p_1) + f_2(p_2) \\ p_1 \in P_1 , p_2 \in P_2 \\ Ap_1 - p_2 = 0, \end{cases}$$

the method in [76] applies proximal point iterations to the subdifferential of the Lagrangian function $L(p_1, p_2, x) = f_1(p_1) + f_2(p_2) + \langle x, Ap_1 - p_2 \rangle$, by alternately fixing the primal variables (p_1, p_2) or the multiplier x. A nice feature of this decomposition scheme is that the minimization for the local problems is carried out separately in the spaces P_1 and P_2, and the two minimization problems decompose further according to the separable structure of the functions f_1 and f_2. The "alternating projection-proximal method" in [352] extends the same decomposition scheme to more general problems. These methods, however, are only applicable if local subproblems can be solved *exactly*. Although this may be the case in some applications (linear multi-commodity flow problems, for example), such assumption can be too restrictive in general.

–*Inexact proximal decomposition methods* address the question of local problems that cannot be solved exactly. For the decomposition [76] mentioned above, when f_1 and f_2 are convex nondifferentiable functions, each local problem would require the application of a NSO method (a bundle method for example) until finding an optimum. Since such process is repeated for each iteration of the master program, the resulting decomposition scheme becomes inapplicable. To overcome this handicap, the method in [335] gives a decomposition method where local problems are not solved exactly. For the example mentioned in the previous item, it is enough to make bundle iterations on each local problem until a relaxed stopping test is satisfied. The decomposition approach proposed in [335] contains [76] and [352] as special cases and gives a unified insight into those schemes by relating them to the hybrid inexact proximal methods described in § 11.2.3 below. The same hybrid framework, applied in a different context, gives the ε-proximal decomposition in [279], and the parallel bundle method in [58].

11.2 Transpassing Frontiers

In some applications, the special structure of the problem to be solved leads to the development of a new NSO method. We describe some of such extensions in this section.

11.2.1 Dynamic Bundle Methods

Consider problem (11.1) with only inequality constraints

$$\begin{cases} \min_p f_o(p) \\ p \in P \\ c_j(p) \le 0 \text{ for } j = 1, \ldots, n_I, \end{cases} \tag{11.11}$$

where f_o is a convex function, the constraints c_j are affine, n_I is a large number, and $P \subset \mathbb{R}^N$ can be any set (including a discrete set).

For the Lagrangian relaxation approach to make sense in practice, two points are fundamental. First, the dual function should be much simpler to evaluate (at any given x) than solving the primal problem directly; we assume this is the case for (11.11). Second, the dual problem should not be too large: in NSO this means that the dimension of the dual variable is less than 10000. So the approach is simply not applicable in our setting, because in (11.11) n_I is too big. Instead of dualizing all the n_I constraints at once, an alternative approach is to choose *subsets* of constraints to be dualized at each iteration. In this dynamical relaxation, subsets J have cardinality $|J|$ much smaller than n_I. As a result, the corresponding dual function, defined on the nonnegative orthant $\mathbb{R}_+^{|J|}$, is manageable from the NSO point of view. However, since the dual function now varies at each iteration, to obtain convergent algorithms, primal and dual information should be combined adequately.

Dual Information A first important consequence of considering a subset J instead of the full set $\{1, \ldots, n_I\}$ is that complete knowledge of dual function is no longer available. The dual function for (11.11) is

$$\theta(x) = \min_{p \in P} L(p, x) = \min_{p \in P} \{ f_o(p) + \langle x, c(p) \rangle \}.$$

In the dynamic setting, for any given x only

$$\min_{p \in P} \left\{ f_o(p) + \sum_{j \in J} x_j c_j(p) \right\} \quad \textbf{and not} \quad \min_{p \in P} \left\{ f_o(p) + \sum_{j=1}^{n_I} x_j c_j(p) \right\}$$

is known. This means that rather than having θ, only its trace, $\theta \circ \mathcal{P}_J$, is available. Here, $\mathcal{P}_J$ is the linear operator in $\mathbb{R}^{n_I}$ defined as the orthogonal projection on the subspace $\mathbb{R}^{|J|} \times \{0 \in \mathbb{R}^{n_I - |J|}\} \subset \mathbb{R}^{n_I}$.

Similarly to p_x from (8.12) in § 8.3.2, it is easy to see that the point

$$p_{J,x} \in \mathrm{Argmin}_{p \in P} \left\{ f_o(p) + \sum_{j \in J} x_j c_j(p) \right\}$$

still gives a subgradient, i.e, $-c(p_{J,x}) \in \partial(-\theta)(x)$. With respect to Figure 9.5, the black-box information now depends on J, as shown in Figure 11.5.

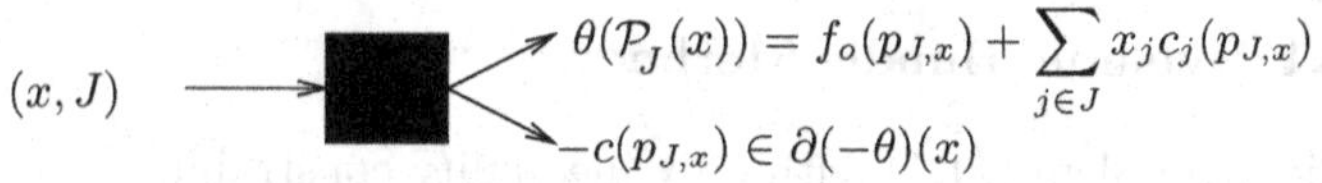

$$(x, J) \longrightarrow \blacksquare \begin{array}{l} \nearrow \theta(\mathcal{P}_J(x)) = f_o(p_{J,x}) + \sum_{j \in J} x_j c_j(p_{J,x}) \\ \\ \searrow -c(p_{J,x}) \in \partial(-\theta)(x) \end{array}$$

Fig. 11.5. Black box depending on a set J

Primal Information To choose which constraints are to be dualized at each iteration, we assume that a *Primal Oracle* is available. A *Primal Oracle* is a procedure $PrimOr$ that, given $p \in \mathbb{R}^N$ and $J \subseteq \{1, \ldots, n_I\}$, identifies constraints c_j with $j \notin J$ that are not satisfied by p. The output of the procedure is an index set I, i.e., $I = PrimOr(p, J)$, which can be empty.

We assume that, as long as there remain violated constraints with indices in $j \in \{1, \ldots, n_I\} \setminus J$, the primal procedure is able to identify one of such constraints. With this assumption,

$$PrimOr(p, J) = \emptyset \Leftrightarrow \{j \in \{1, \ldots, n_I\} : c_j(p) > 0\} \subseteq J$$
$$\Leftrightarrow \{j \in \{1, \ldots, n_I\} : c_j(p) \leq 0\} \supseteq \{1, \ldots, n_I\} \setminus J.$$

In particular, $PrimOr(p, \{1, \ldots, n_I\})$ is always the empty set. In combinatorial optimization, this assumption corresponds to an *exact separation* algorithm for the family of inequalities defined by the constraints.

Note that the primal oracle can only add indices to the set J, thus increasing the dual dimension. In the absence of some criterion to drop indices from the set J, we might soon be working with a dual dimension that is too large to handle. For this reason, every time a candidate is declared a serious step, we eliminate indices corresponding to zero components of the new stability center. The idea is that, since the dual variable is a multiplier for (11.11), in view of the complementarity relation (8.16), zero dual components should correspond to feasible constraints.

The Algorithm We now give the dynamic version of the penalized bundle method with $|\cdot|_k = \|\cdot\|$, i.e., using the Euclidean norm for all iterations; we refer to [22] for full details.

Recall that, instead of (8.1) we are now minimizing the negative of the dual function, $-\theta$ in the nonnegative orthant $\mathbb{R}_+^{n_I}$. The set $\hat{J}_k$ below corresponds to the index set used when the current serious step, x^k, was generated. Note that, by (11.12) below, each iterate is defined so that $y^k = \mathcal{P}_{\hat{J}_k}(y^k)$ with $\hat{J}_k \subset J_k$.

Since all the black-box information can be recovered from $p_{J,x}$, the bundle of information will now be

$$\mathcal{B} = \left\{ (e_i, p_{J_i, y^i}), i = 1, \ldots, np_k \right\} \text{ and } \hat{J}_k, x^k,$$

where linearization errors are defined for the convex function $-\theta$:

$$e_i = -\theta(p_{\hat{J}_k, x^k}) - f_o(p_{J_i, y^i}) - \langle c(p_{J_i, y^i}), x^k \rangle.$$

Algorithm 11.3 (dynamic bundle). Let $\texttt{tol} \geq 0$, $m \in]0, 1[$, and $k_{comp} > 0$ be given parameters. Choose a nonnegative $x^1 \in \mathbb{R}_+^{n_I}$ such that $J_1 := \{j \leq n_I : x_j^1 > 0\} \neq \emptyset$, and call the black box from Figure 11.5 with $(x, J) = (x^1, J_1)$. Construct the model φ_1 for the convex function $-\theta$, and set the algorithm parameters, such as μ_1. Set $k = 1$, $\delta_1 = \infty$, and $I_1 = \emptyset$. Define $\hat{J}_1 = J_1$.

STEP 1 (dynamic stopping test). If $\delta_k \leq \texttt{tol}$ and $I_k = \emptyset$, stop.

STEP 2 (candidate). Solve

$$y^{k+1} \text{ solution to } \begin{cases} \min \varphi_k(y) + \frac{1}{2}\mu_k \|y - x^k\|^2 \\ y \in \mathbb{R}^{J_k} \times \{0 \in \mathbb{R}^{n_I - |J_k|}\} \end{cases}, y \geq 0. \tag{11.12}$$

Define $\delta_{k+1} = -\theta(x^k) - \varphi_k(y^{k+1}) - \frac{1}{2}\mu_k \|y^{k+1} - x^k\|^2$.

Define the primal convex point to be used by the Primal Oracle:

$$\hat{p}^{k+1} = \sum_{i=1}^{np_k} \bar{\alpha}_i p_{J_i, y^i},$$

where $\bar{\alpha}$ solves the dual problem of (11.12), similar to (10.9).

STEP 3 (calling the black box and the Primal Oracle).   Call the black box from Figure 11.5 with $(x, J) = (y^{k+1}, J_k)$.

Call the Primal Oracle to compute $I_{k+1} = PrimOr(\hat{p}^{k+1}, J_k)$

STEP 4 (assessing the candidate and choosing constraints). Descent test:

$$-\theta(x^k) + \theta(y^{k+1}) \geq m\delta_{k+1}? \begin{cases} \text{Yes:} & \text{(descent-step)} \\ \text{No:} & \text{(null-step)}. \end{cases}$$

If y^{k+1} was declared a null step, define

$$x^{k+1} = x^k, \hat{J}_{k+1} = \hat{J}_k, J_{k+1} = J_k \cup I_{k+1}, \text{ and go to Step 5.}$$

If y^{k+1} was declared a serious step, then

– If $k \geq kcomp$ and $I_{k+1} \neq \emptyset$, define $O_{k+1} = \emptyset$.

– Otherwise, compute $O_{k+1} = \{j \in J_k : y_j^{k+1} = 0\}$.

Define

$$x^{k+1} = y^{k+1}, \hat{J}_{k+1} = J_k \backslash O_{k+1}, J_{k+1} = \hat{J}_{k+1} \cup I_{k+1}, \text{ and go to Step 5.}$$

STEP 5 (improving the model – loop). Append y^{k+1} to the model, i.e., construct φ_{k+1}. Define the algorithm parameters for the next iteration, such as μ_{k+1}. Change k to $k + 1$, go to 1.

When compared to Algorithm 10.1, we see that the new extra calculations correspond essentially to the management of index sets J_k. More precisely, as long as null steps are done, we keep on adding constraints to the dual function (the cardinality of J_k increases). By contrast, when a candidate is declared a serious step, we eliminate those constraints corresponding to zero components of the new serious step x^{k+1}. The parameter k_{comp} ensures that eventually (as $k \to \infty$) indices will be removed only when the convex primal point $\hat{p}^{k+1}$ is feasible for all the n_I constraints. In order to keep a low dimensionality of J_k and make the algorithm faster, such condition is not required at early iterations ($k < k_{comp}$).

In a schematic way, we now have the decomposition scheme described in Figure 11.6.

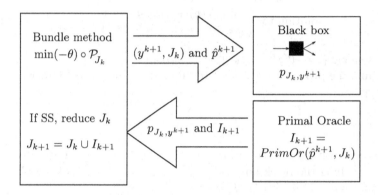

Fig. 11.6. Dynamic bundle method

Along the lines of Algorithm 10.11, it is possible to introduce a compression sub-algorithm in Algorithm 11.3. Namely, primal bundle points p_{J_i, y^i} are aggregated into the convex primal point $\hat{p}^{k+1}$, while linearization errors are aggregated as usual. However, such aggregation results in no loss of information **only if** the dualized constraints are affine.

Similarly to Theorems 10.14 and 10.15 in Chapter 10, in [22] the dynamic bundle method is shown to generate a sequence $\{x^k\}$ that is minimizing for $-\theta$ on the nonnegative orthant $\mathbb{R}^{n_I}_+$. Moreover, when in addition to c, f_o in (11.1) is also affine, it is shown that if Algorithm 11.3 stops at some iteration k_{last} with $\delta_{k_{last}} = 0$ and $I_{k_{last}} = \emptyset$, then the primal convex point $\hat{p}^{k_{last}}$ solves the primal problem (11.1), written with P replaced by conv P. When, in addition, P is a finite set and, hence, θ is a polyhedral function, the algorithm is shown to have finite termination for a suitable choice of parameters.

This dynamic methodology can also be grooved together with the so-called *maximum violation* primal oracle, i.e., a procedure giving the index of the constraint with higher value. Maximum violation oracles require to explore the whole constraint set, and may not be well adapted for some particular problems. By constrast, the primal oracle presented here, depending on the set J, allows for the use of heuristic methods. The choice of suitable primal oracles to identify violated inequalities depends on the particular application. For the dynamic method to work, the primal oracle should essentially ensure that eventually all violated constraints are included in the working set J.

11.2.2 Constrained Bundle Methods

Consider a general convex constrained problem of the form

$$\begin{cases} \min_{x \in \mathbb{R}^n} f(x) \\ c(x) \leq 0, \end{cases} \tag{11.13}$$

where $f, c : \mathbb{R}^n \to \mathbb{R}$ are convex functions, in general nondifferentiable. There is no loss of generality in formulating (11.13) with only one constraint: if necessary, c can be defined as the pointwise maximum of finitely many convex functions (like (8.17) in § 8.3.1), since the resulting function would still be convex.

Associated with problem (11.13) there is the *improvement function*, depending on a fixed $x \in \mathbb{R}^n$, defined by

$$h_x(y) := \max\{f(y) - f(x), c(y)\}, \quad \text{for all } y \in \mathbb{R}^n. \tag{11.14}$$

When a Slater condition holds for (11.13), $\bar{x}$ is a solution to (11.13) if and only if $\bar{x}$ solves the unconstrained problem of minimizing $h_{\bar{x}}$.

Ideally, one could apply an unconstrained bundle method to minimize the function $h_{\bar{x}}(\cdot)$ directly. In practice, since $f(\bar{x})$ is not known, the idea is to take $x = x^k$, the last serious step, and proceed until a new serious step is generated. Then the base point in the improvement function moves, i.e., we set $x = x^{k+1}$ in (11.14). The resulting method, introduced in [318], works on an objective function that varies along the iterations. Given a stability center x^k, the usual penalized bundle technique is applied to the improvement function $h_k := h_{x^k}$, until a descent condition for this function with respect to the value $h_k(x^k) = \max\{c(x^k), 0\} =: c^+(x^k)$ is achieved. At this time, the corresponding candidate y^{k+1} is accepted as the next stability center and the algorithm proceeds, working with the new improvement function, h_{k+1}.

When written for $f = h_k$, the descent condition (10.1), i.e, $h_k(y^{k+1}) \leq h_k(x^k) - m\delta_{k+1}$, becomes

$$f(y^{k+1}) - f(x^k) \leq c^+(x^k) - m\delta_{k+1}, \tag{11.15}$$

and

$$c(y^{k+1}) \leq c^+(x^k) - m\delta_{k+1}, \tag{11.16}$$

where $\delta_{k+1} = c^+(x^k) - \varphi_k(y^{k+1}) - \frac{1}{2}\mu_k|y^{k+1} - x^k|_k^2$.

Suppose both (11.15) and (11.16) hold, so that $x^{k+1} = y^{k+1}$. If x^k is infeasible, then $f(x^{k+1}) > f(x^k)$ is possible (since $c^+(x^k) > 0$). Therefore, the method is not monotone with respect to f when outside of the feasible region. However, outside of the feasible region the method is monotone with respect to c, because $c(x^{k+1}) < c^+(x^k) = c(x^k)$ for x^k infeasible. This seems intuitively reasonable: while it is natural to accept the increase in the objective function value in order to decrease infeasibility, it is not so clear why one would want to decrease the objective function at the expense of moving away from the feasible region.

The situation reverses when x^k is feasible. In that case, $c^+(x^k) = 0$, so that $f(x^{k+1}) < f(x^k)$. But although (11.16) implies that x^{k+1} is feasible too, it is possible that $c(x^{k+1}) > c(x^k)$ (except when $c(x^k)$ is exactly zero). This also appears completely reasonable: while preserving feasibility, we allow c to increase (so that the boundary of the feasible set can be approached), at the same time obtaining a decrease in the objective function. In particular, if the starting point x^0 is feasible, the method will operate in a feasible mode (all serious steps are feasible).

Conditions (11.15) and (11.16) result in an important difference with respect to the unconstrained penalized bundle method in Chapter 10. Every time a serious step is declared, the cutting-planes model φ_k has to be properly revised to make sure that the relation $\varphi_{k+1} \le h_{k+1}$ holds (the important relation $\varphi_{k+1} \le f$ in (10.18), written with f replaced by h_{k+1} should be satisfied). Since $f(x^{k+1}) > f(x^k)$ is perfectly possible, we may have $h_{k+1} \le h_k$, where the inequality can be strict for some points. This means that a lower approximation for h_k (i.e., φ_k satisfying $\varphi_k \le h_k$) may no longer be valid for h_{k+1} (i.e., $\varphi_k \not\le h_{k+1}$). Note that this adjustment is independent of compressing the bundle, which will also require additional care.

Suppose for the moment that the bundle contains only black-box information, i.e., that no aggregation has been done. Since the unconstrained bundle method now works with an objective function $h_k(\cdot)$ which varies with $k \in K_s$, past information relevant for constructing the model is no longer just the bundle

$$\mathcal{B} = \left\{ (e_i, s_h^i \in \partial_{e_i} h_k(x^k)), i = 1, \ldots, np_k \right\}.$$

Instead, separate information about the objective and constraint functions needs to be kept. For (11.13), black-box information has the form ($f_i = f(y^i), c_i = c(y^i)$) and ($s_f^i \in \partial f(y^i), s_c^i \in \partial c(y^i)$). Or, equivalently,

$$\mathcal{B}^{bb} = \left\{ (e_{f_i}, e_{c_i}, s_f^i \in \partial_{e_{f_i}} f(x^k), s_c^i \in \partial_{e_{c_i}} c(x^k)) \right\}, i = 1, \ldots, np_k,$$

where the linearization errors for f and c, are the usual ones:

$$e_{f_i} := f(x^k) - f_i - \langle s_f^i, x^k - y^i \rangle, \text{ and}$$

$$e_{c_i} := c(x^k) - c_i - \langle s_c^i, x^k - y^i \rangle, \text{ for } i \in \mathcal{B}^{bb}.$$

The purpose of keeping the bundle information separated is to allow the computation of the function and subgradient values for different functions h_k. As shown in [318; Lemma 3.1], for each $i \in \mathcal{B}^{bb}$, setting

$$\begin{cases} e_i := e_{f_i} + c^+(x^k) & \text{and } s^i_{h_k} := s^i_f, & \text{if } f_i - f(x^k) \geq c_i, \\ e_i := e_{c_i} + c^+(x^k) - c(x^k) \text{ and } s^i_{h_k} := s^i_c, & \text{if } f_i - f(x^k) < c_i, \end{cases}$$

guarantees that $e_i \geq 0$ and $s^i_{h_k} \in \partial_{e_i} h_k(x^k)$ for every $k \in K_s$.

Similarly to (10.17), separate linearization errors can be updated by a simple formula, even when h_k changes. Suppose now the algorithm has performed some compression and aggregation, and that we split the bundle into two subsets, corresponding to black box and aggregate information:

$$\mathcal{B} = \mathcal{B}^{bb} \cup \mathcal{B}^{agg}.$$

For example, with respect to Algorithm 10.11, for the first iteration k when there is aggregation, $\mathcal{B}^{bb}$ contains the n_{left} couples kept, while $\mathcal{B}^{agg} = \{(\varepsilon_k, \hat{s}^k)\}$. Every time a new stability center is defined ($x^{k+1} = y^{k+1}$), linearization errors for $i \in \mathcal{B}^{bb}$ are updated according to the relations

$$e_{f_i} = e_{f_i} + f(y^{k+1}) - f(x^k) + \langle s^i_f, x^k - y^{k+1} \rangle, \text{ and}$$
$$e_{c_i} = e_{c_i} + c(y^{k+1}) - c(x^k) + \langle s^i_c, x^k - y^{k+1} \rangle;$$

while aggregate errors for $i \in \mathcal{B}^{agg}$ are updated as follows:

$$\varepsilon_i = \varepsilon_i + c^+(x^{k+1}) - c^+(x^k) + \left(f(y^{k+1}) - f(x^k)\right)^+ + \langle \hat{s}^i, x^k - y^{k+1} \rangle;$$

see [318; Lemma 3.2]. In the same work, the algorithm is shown to be convergent, similarly to Theorems 10.14 and 10.15.

Bibliographical Comments Constrained nonsmooth problems are very complex, only a few practical methods can be found in the literature.

Convex problems with "easy" constraints, such as bound or linear constraints, can be solved either by inserting the constraints directly into each stabilized subproblem, or by projecting iterates onto the feasible set; see for instance [135] and [211, 212]. For more general constraints, one possibility is to solve an equivalent unconstrained problem with an exact penalty objective function; see [209, 214]. This approach, however, presents some drawbacks which are typical whenever a penalty function is employed. Specifically, estimating a suitable value of the penalty parameter is sometimes a delicate task. Furthermore, if a large value of the parameter is required to guarantee the exactness of a given penalty function, then numerical difficulties arise.

Other bundle-type methods for (11.13) that do not use penalization are [252, 253] and [210; Chapter 5]. In these works, infeasible candidates are automatically declared "null steps". The resulting sequence of serious steps is then both feasible and monotone in f. Since serious steps include the starting point, such methods require to compute a feasible point to start the algorithm. This "phase I" general (nonsmooth) convex feasibility problem may be as difficult to solve as (11.13) itself. As a result, the overall computational burden of solving the problem may increase considerably. On the other hand, feasible methods can be useful in applications where problem function(s) may not be defined everywhere outside of the feasible region.

Infeasible bundle methods are very rare. We can mention the "phase I-phase II" modification of the feasible method in [210; Chapter 5.7] and the constrained level bundle methods of [230]. In [230], successive approximations of the exact improvement function $h_{\bar{x}}$ are used in the algorithm. Specifically, in the expression

$$h_{\bar{x}}(y) = \lambda f(y) + (1 - \lambda)c(y) - \lambda f(\bar{x}) \text{ for some } \lambda \in [0, 1],$$

the values of λ and of $f(\bar{x})$ are estimated at each iteration. Those estimates are used to define a certain gap function and an associated level parameter ℓ_k for the QP. Like the unconstrained variant described in Example 10.5, such methods are especially suitable for those problems where the optimal value $f(\bar{x})$ is either known or is easy to estimate. Otherwise, estimating the optimal value is a delicate issue and inappropriately chosen values may lead to infeasible quadratic programming subproblems.

Finally, the filter strategy [130] was recently proposed in [131] and [203] as an alternative to the use of a penalty function in the framework of bundle methods for solving (11.13). The method of [131] is based on solving linear programming subproblems obtained by replacing the objective and constraint functions by their respective cutting-planes models, subject to a box-constrained trust region, like in Example 10.4. However, as stated, the method of [131] does not allow for a compression mechanism. By contrast, the method in [203] does use bundle compression, by applying the penalized bundle method to the improvement function h_k and using a filter criterion to declare a candidate point a new stability center. Like smooth filter methods, both [131] and [203] are infeasible methods, but require a (feasibility) *restoration* step to ensure convergence when the procedure fails to produce a point acceptable by the filter criterion.

11.2.3 Bundle Methods for Generalized Equations

We have seen in Chapter 10, equations (10.22) and (10.23), that serious-step iterations of the bundle method can be interpreted as an implementation of the proximal point method. Similar ideas can be used to solve generalized equations of the form $0 \in T(x)$, where T is a maximal monotone operator. The relation comes from the fact that the subdifferential of a convex function is a particular case of maximal monotone operator. Thus, problem (8.1) is equivalent, via the optimality condition, to finding a zero of the maximal monotone operator $T = \partial f$.

In order to make the link with bundle methods, recall that the generalized equation $0 \in T(x)$ can be solved applying the classical proximal point method, [312]. This algorithm consists of solving a sequence of subproblems of the following structure:

Given x^k, find x^{k+1} and $s^{k+1} \in T(x^{k+1})$ such that $0 = s^{k+1} + \mu_k(x^{k+1} - x^k)$.

For implementations, error terms need to be considered in subproblems. The corresponding analysis was first done in [312]. However, to be convergent, this inexact algorithm requires that asymptotically the subproblems are solved exactly, in the sense that *absolute* errors should tend to zero (in fact, form a convergent series). More recently, other forms of inexact proximal methods, called "hybrid", were considered in [337] ; see also [336] and [338]. In these methods, subproblems also accept errors in the maximal monotone operator, by means of an appropriate enlargement, introduced in [57] and denoted by T^ε. This enlargement plays the role of the smeared subdifferential, $\partial_\varepsilon f$, noting that for $T = \partial f$ such enlargement is "big": in [57] it is shown that $\partial_\varepsilon f \subseteq \partial^\varepsilon f$, where the inclusion can be strict for some functions.

The *hybrid approximate extra-gradient method* in [337] defines subproblems and iterates as follows:

Given x^k, find y^{k+1} and $\hat{s}^k \in T^{\hat{\varepsilon}_k}(y^{k+1})$ such that $\qquad$ (11.17)

$$\left\| \frac{1}{\mu_k} \hat{s}^k + y^{k+1} - x^k \right\|^2 + 2\frac{1}{\mu_k}\hat{\varepsilon}_k \leq \sigma^2 \left(\left\| \frac{1}{\mu_k} \hat{s}^k \right\|^2 + \|y^{k+1} - x^k\|^2 \right) \quad (11.18)$$

$$x^{k+1} = x^k - \frac{1}{\mu_k}\hat{s}^k, \qquad\qquad (11.19)$$

where $\sigma \in [0, 1[$ is a given parameter. In [338; Theorems 6 and 8] it is shown that this method is globally convergent with linear rate, under the same assumptions needed for exact proximal methods. Moreover, there is no need of solving subproblems with more and more accuracy, in the sense that *relative* errors can be bounded away from zero (in fact, they can be fixed along iterations).

When compared to a penalized bundle method applied to minimizing f, note that the update for x^{k+1} in (11.19) is the same as in (10.8), whenever y^{k+1} gives a descent step and, hence, $p^k = y^{k+1}$. Furthermore, using Lemma 10.8 (*iii*) and defining

$$\hat{\varepsilon}_k := \varepsilon_k + f(y^{k+1}) - f(x^k) + \frac{1}{\mu_k}\|\hat{s}^k\|^2 \quad (= f(y^{k+1}) - \varphi_k(y^{k+1})),$$

we see that $\hat{s}^k \in \partial_{\hat{\varepsilon}_k} f(y^{k+1}) \subseteq \partial^{\hat{\varepsilon}_k} f(y^{k+1})$, so (11.17) holds.

We now relate condition (11.18) to the descent condition (10.1). By (10.8), the first left hand side term in (11.18) is null, while the two right hand side terms are both equal to $\frac{\sigma^2}{\mu_k^2}\|\hat{s}^k\|^2$. Altogether, (11.18) can be re-written as

$$\hat{\varepsilon}_k \leq \sigma^2 \frac{1}{\mu_k}\|\hat{s}^k\|^2 \text{ i.e., } \varepsilon_k + f(y^{k+1}) - f(x^k) + \frac{1}{\mu_k}\|\hat{s}^k\|^2 \leq \sigma^2 \frac{1}{\mu_k}\|\hat{s}^k\|^2.$$

Rearranging terms, the right hand side inequality becomes

$$f(x^k) - f(y^{k+1}) \geq (1 - \sigma^2)\frac{1}{\mu_k}\|\hat{s}^k\|^2 + \varepsilon_k$$
$$= 2(1 - \sigma^2)\left(\frac{1}{2\mu_k}\|\hat{s}^k\|^2 + \varepsilon_k\right) + (2\sigma^2 - 1)\varepsilon_k$$
$$= 2(1 - \sigma^2)\delta_{k+1} + (2\sigma^2 - 1)\varepsilon_k \,.$$

Therefore, choosing $\sigma \in]\frac{1}{\sqrt{2}}, 1[$ and letting $m := 2(1 - \sigma^2)$, the test (11.18) would declare y^{k+1} a serious step when

$$f(x^k) - f(y^{k+1}) \geq m\delta_{k+1} + (1 - m)\varepsilon_k \,,$$

a stronger condition than (10.1), used in penalized bundle methods. Such relation is natural, since by working with the maximal monotone operator $T = \partial f$ instead of with f itself, important structural information is not actually being used.

Bundle methods for maximal monotone operators are considered in [59] and [317]. Since there is no longer a function to minimize, the concept of model function disappears. Accordingly, the only black-box output is $s \in T(x)$ for any given x. Hence, the bundle of past information consists of pairs of the form $(y^i, s^i \in T(y^i))$ that are used to solve a QP, somewhat similar to (10.9), of the form

$$\begin{cases} \min \|\sum_{i \in B_k} \alpha_i s^i\|^2 \\ \alpha \in \Delta_k \\ B_k = \{i \leq k : \|y^i - x^k\| \leq \rho_k\}, \end{cases}$$

where ρ_k is a parameter of the algorithm. Candidates y^{k+1} are generated by making a line-search along $\hat{s}^k := \sum_{i \in B_k} \bar{\alpha}_i s^i$, where $\bar{\alpha}$ is a solution to the QP above. The output of the line-search is either a serious-step satisfying $\langle s^{k+1}, \hat{s}^k \rangle > m \min\{\|s^{k+1}\|, \|\hat{s}^k\|\}$, or a null-step. Note that the QP subproblem is also close to the dual stabilization subproblem (10.13). We refer to [317] for more details.

12 Computational Exercises

In order to get a better understanding of the different methods presented in this NSO Part, we now give some computational exercises.

For initial experiments with optimization programs, it is a good idea to use a numerical computing environment, such as SCILAB [327], which is freely downloadable, or MATLAB [249]. Both scientific software packages are well adapted for numerical computations: their ready-to-use functions can be used as modules to easily build more complex algorithms. In particular, SCILAB optimization routines linpro and quapro, for linear and quadratic programming respectively, are robust and reliable for small-scale problems.

Following the *General Principles of Resolution* stated in § 1.3 in the General Introduction of this book, the exercises below consider a separate coding for the simulator and for the optimization algorithm.

12.1 Building Prototypical NSO Black Boxes

We start with some elementary calculations involving the test function MAXQUAD, to be used to test and compare various NSO algorithms.

12.1.1 The Function MAXQUAD

Consider the function MAXQUAD defined in page 153, at the end of § 10.3.3, and denoted by f.

Exercise: Write the general form of its subdifferential, at an arbitrary point $x \in \mathbb{R}^{10}$. Check in particular the form of $\partial f(\bar{x})$, where

$$
\bar{x} = \begin{pmatrix}
-0.12635719028771 \\
-0.03441290527305 \\
-0.00686417677098 \\
0.02631958976703 \\
0.06733200924718 \\
-0.27845349219516 \\
0.07425434369432 \\
0.13858219818536 \\
0.08409149779995 \\
0.03861740126733
\end{pmatrix}
$$

is the minimizer of f. To eliminate rounding errors, consider that $f(\bar{x}) = -0.84140$ to determine the set of active indices $J(\bar{x})$, as introduced in § 8.3.1.

Exercise: Write a black box simulator for MAXQUAD, following the general abstract structure

$$[\texttt{f,s,ind}] = \texttt{name\_of\_the\_function(x)}, \qquad (12.1)$$

where
- x is an input parameter, having the given point $x \in \mathbb{R}^{10}$, and
- f, s, ind are output parameters, containing, respectively, the value of the function at x; a subgradient; and a flag that is 0 if calculations were performed correctly. For MAXQUAD, ind is always set to 0.

For some functions, it may be useful to add an additional input parameter, acting as a flag that makes the simulator produce different output (for example, only compute the function value, but not a subgradient, or visceversa).

To check correctness of your coding, see if $f(\bar{x})$ is approximately equal to -0.8414080.

Exercise: Write a program generating the data needed to compute MAXQUAD (i.e., (A_j, b_j), for $j = 1, \ldots, 5$) that calls the simulator for a given value of x and makes a 2D graph for f near x. To make a 2D graph for this 10-dimensional function fix, for example, variables 3 to 10 and display the level curves of f when varying the first 2 components in a neighbourhood of x.

Compare your graph with the cover figure when $x = \bar{x}$.

12.1.2 The Function MAXANAL

A possibility to deal with the nondifferentiability of MAXQUAD is to *smooth* the function, for example by using a logarithmic barrier (cf. Chapter 20 in Part IV). More precisely, start by rewriting the maximum operation defining f as a convex sum:

$$f(x) = \sum_{j=1}^{5} \alpha_j f_j(x), \sum_{j=1}^{5} \alpha_j = 1, \text{ with } \alpha_j \geq 0 \text{ for all } j,$$

and recall that nonsmoothness for f comes from multiple active indices at x, i.e., from having nonzero α_j for more than one index j. A logarithmic penalization of these convex factors gives the regularized function

$$F_\mu(x) := \max \left\{ \sum_{j=1}^{5} \alpha_j f_j(x) + \mu \sum_{j=1}^{5} log(\alpha_j) : \sum \alpha_j = 1 \right\}$$

where μ is a given positive parameter. As $\mu \downarrow 0$, the regularization satisfies that $F_\mu(x) \uparrow f(x)$ for any given x. However, small values of μ, say $\mu \le 0.001$, give stiff yet differentiable problems, with highly non symmetric level sets near $\bar{x}$.

The function F_μ, called MAXANAL in the NSO literature, is a simple example of a common fact in NSO, already present in § 11.1. Namely, to evaluate the function f and a subgradient at some point x, the black box needs to solve an *optimization problem* parameterized by x.

Exercise: Write a black box simulator for MAXANAL, following the general abstract structure given in (12.1). There are various possibilities for solving the optimization problem involved in the definition of $F_\mu(x)$, including using optimization routines from MATLAB and SCILAB. Beware that, since the optimization problem becomes ill conditioned as μ goes to 0, a straightforward application of a built-in function may not always be successful. For this reason, MAXANAL output values of the flag ind may be non null, to indicate failure in the computation of f or s.

Exercise: Write a program generating the data needed to compute MAXANAL that calls the simulator for a given value of x and makes a 2D graph for f near x. Compare the level curves of MAXQUAD and MAXANAL for $x = \bar{x}$ for different values of μ.

12.2 Implementation of Some NSO Methods

We now pass to the implementation of some of the algorithms presented in Part II. The implementation will be general purpose, in the sense that it can be used to minimize any function with black box simulator coded with a calling list like in (12.1).

The following general structure gives a list of typical parameters of an unconstrained minimization algorithm:

$$[\texttt{xf,ff,sf,nbb,reason}]=\texttt{name\_of\_NSO\_algorithm(simul,}$$
$$\texttt{x0,f0,s0,maxbb,tolopt, printlev)}. \quad (12.2)$$

Like in (12.1), right hand side arguments are input parameters while left hand side ones is output data, produced by the algorithm.

More precisely,

- simul is a string with the name of the function to minimize, coded in a black box form, according to (12.1).
- x0, f0, s0 are the initial black box values; maxbb the maximum allowed number of calls to the black box; tolopt the stopping tolerance; and printlev is a flag used to determine how much information is to be printed along the iterative process (ranging from no print out to printing information at every iteration).

– **xf , ff , sf** correspond to the final black box values, found after having called the black box **nbb** times. The flag **reason** indicates different exit modes (success, failure, maximum of iterations reached, etc).

The list above is just meant to serve as an example, and is far from being exhaustive.

Exercise: Write a code for the subgradient method (Algorithm 9.2) without using a line-search, but including instead the possibility of using different stepsizes, according to some input parameter. For example, try different values of constant stepsize, then take the stepsize $t_k = 1/k$, and $t_k = 2m(f(x^k) - \bar{f})/\|s(x^k)\|$ with $m \in]0, 1[$, to be used with those simulators for which either the optimal value $\bar{f}$ is known or it can be well estimated.

Exercise: Write a code for the cutting planes method (Algorithm 9.4). The set S therein can be defined by a box:

$$ S := \left\{ x \in \mathbb{R}^n : x_{low} \leq x \leq x_{up} \right\} $$

where the vectors x_{low} and x_{up} are additional input parameters in (12.2).

Exercise: Write a first code for the penalized bundle method (Algorithm 10.1 with subproblem (10.5)) according to the following specifications:
– matrices M_k are set to the identity matrix for all k;
– constant penalization parameter μ_k;
– y^{k+1} of (10.5) is computed by the dual problem (10.8)&(10.9);
– split stopping test (10.14);
– the model φ_k of (10.7) is just the cutting-planes model $\check{f}_k$ from (9.6) (there is no aggregation).

12.3 Running the Codes

Exercise: As a first try, write a black box simulator for the function $\|x\|^2$, where $x \in \mathbb{R}^n$ and n is arbitrary. Run your NSO codes playing with different values of the input parameters. In particular, for the bundle code, analyze how different values for the stopping tolerance and the penalization parameter μ_k modify the output in this very simple case.

Exercise: Use your three implemented NSO methods to minimize **maxquad**, taking as starting point **x** a vector with all components equal to one. Repeat your runs with starting point **x**$= 0 \in \mathbb{R}^{10}$.

Exercise: Repeat yours runs, using now MAXANAL, with different values for the barrier μ (try $\mu \in \{0.00001, 0.0001, 0.001, 0.01, 0.1, 1\}$). Comparing MAXQUAD optimal value to the optimal value in the smoothened problem gives an idea of how small μ is, i.e. how much MAXQUAD is perturbed.

12.4 Improving the Bundle Implementation

Exercise: Consider the following variants for your bundle code:
- Penalization parameter updated as suggested in § 10.3.3
- Bundle management that keeps only active indices at each iteration.
- Compression sub-Algorithm 10.11, allowing both for selection and aggregation of the bundle.

Compare the effectiveness of each variant on the different black boxes. In particular, analyze the quality of the output when the bundle is fully compressed at each iteration ($np_{\max} = 2$).

Exercise: Since MAXANAL is a smooth function, it can be minimized by using some smooth unconstrained algorithm of Part I. For example, try using BFGS method with Wolfe's line-search (§§ 4.4,3.4) for different values of μ. Compare the obtained output with the one given by the simple bundle method in the previous exercise, both in terms of precision and number of black box calls required to stop.

12.5 Decomposition Application

We consider an energy problem along the lines of § 1.2.4. More precisely, we aim at determining the optimal management of a set of I of thermal units (a power plant may be composed of several units, or generators) for the next 2 days. The planning horizon is discretized in half hours, i.e, the stepsize t ranges in $\{1, \ldots, T\}$, with $T = 48$ half-hours. Let $d_1, \ldots, d_T$ denote the (known) demand, and let p_t^i denote the energy produced by the production unit $i \in I$ during the period t. In addition, to determine when the ith plant is to be switched on or off, we define binary variables

$$u_t^i = \begin{cases} 1 \text{ if unit } i \text{ is on at time step } t \\ 0 \text{ otherwise.} \end{cases}$$

A simple formulation of the well known *unit-commitment* problem in energy optimization consists in solving the optimization problem

$$\begin{cases} \min_{u,p} \ \sum_{i \in I} c^i(p^i, u^i) \\ \quad \sum_{i \in I} p_t^i = d_t \qquad \text{for all } t \\ \quad p_{low}^i u_t^i \leq p_t^i \leq p_{up}^i u_t^i \text{ for all } i, t \\ \quad u_t^i \in \{0, 1\} \qquad \text{for all } i, t. \end{cases} \tag{12.3}$$

The two last constraints define the technological constraint set D^i defined in (1.8) for each unit. For simplicity, we consider energy bounds p_{low}^i and p_{up}^i

that do not vary with t; note that such capacity bounds will only be required if the unit is switched on ($u_t^i = 1$). As usual in this type of problems, production costs are separable along time:

$$c^i(p^i, u^i) = \sum_{t=1}^{T} c_t^i(p_t^i, u_t^i) \quad \text{for all } i \in I;$$

in the expression above, the dependence on u_t^i refers to eventual start-up or shut-down costs. In a general abstract form,

$$c_t^i(p_t^i, u_t^i) = \alpha_t^i u_t^i + \beta_t^i(1 - u_t^i) + \gamma_t^i p_t^i + \delta_t^i {p_t^i}^2$$

where the values $(\alpha_t^i, \beta_t^i, \gamma_t^i, \delta_t^i)$ are given apriori.

Exercise: Write the dual problem that results from dualizing demand constraints in (12.3). Show that the negative of the dual function has the form

$$f(\lambda) := \lambda^\top d + \sum_{t=1}^{T} \sum_{i \in I} f_t^i(\lambda_t)$$

where each function f_t^i is the optimal value of an optimization problem:

$$f_t^i(\lambda_t) := \begin{cases} \min_{u_t^i, p_t^i} -c^i(p^i, u^i) - \lambda_t p_t^i \\ p_{low}^i u_t^i \le p_t^i \le p_{up}^i u_t^i \\ u_t^i \in \{0, 1\}. \end{cases} \quad (12.4)$$

Exercise: Write a black box simulator for the function f above, according to the structure in (12.1). Even though each local problem (12.4) has mixed 0-1 variables, the simplicity of its formulation allows for a direct solution, by exploring the only two possibilities for the 0-1 variable, i.e., $u_t^i = 0$ and $u_t^i = 1$.

Exercise: Apply all the NSO methods implemented in § 12.2 to solve the dual problem. To compare performances, check the final dual and primal values obtained, as well as the total number of calls to the black box required by each method to stop. Check also to which extent the dualized constraint of demand is violated at the primal points $p_t^i(\lambda^{final})$. Try your code with different number of units (cardinality of I varying between 2 and 300, for example) and different values of the input data, such as capacity bounds p_{low} and p_{up}, as well as cost parameters $\alpha, \beta, \gamma, \delta$. In particular, consider setting $\delta_t^i = 0$ for all i, t. Then compare with the same configuration, but setting a small value for the quadratic cost, say $\delta_t^i = 0.0001$. What do you observe?

Part III

Newton's Methods
in Constrained Optimization

J. Charles GILBERT

In this part, we introduce and study numerical techniques based on Newton's method to solve nonlinear optimization problems: objective function and functional constraints can all be nonlinear, possibly nonconvex. Such methods, in the form called *sequential quadratic programming* (SQP), date back at least to R.B. Wilson's thesis in 1963 [359], but were mainly popularized in the mid-seventies with the appearance of their quasi-Newton versions and their globalization, see U.M. Garcia Palomares and O.L. Mangasarian [280], S.P. Han [184, 185], M.J.D. Powell [291, 292, 293], and the references therein; let us also mention the earlier contributions by B.N. Pshenichnyj [300] and S.M. Robinson [306, 307]. Ongoing research on SQP deals with the efficient use of second derivatives, particularly for nonconvex or large-scale problems, the use of trust regions [86], the treatment of singular or nearly singular situations and of equilibrium constraints [242], globalization by filters, *etc.* SQP also appears as an auxiliary tool in interior point methods for nonlinear programming [65].

Like Newton's algorithm in unconstrained optimization, SQP is more a methodology than a single algorithm. Here, the basic idea is to linearize the optimality conditions of the problem and to express the resulting linear system in a form suitable for calculation. The interest of linearization is that it provides algorithms with fast local convergence. The linear system is made up of equalities and inequalities, and is viewed as the optimality conditions of a quadratic program. Thus, SQP transforms a nonlinear optimization problem into a *sequence* of quadratic optimization problems (quadratic objective, linear equality and inequality constraints), which are simpler to solve. This process justifies the name of the SQP family of algorithms. The approach is attractive because efficient algorithms are available to solve quadratic problems: active-set methods [160, 128], augmented Lagrangian techniques [98], and interior-point methods (for the last, see part IV of the present volume).

The above-mentioned principle alone is not sufficient to derive an implementable algorithm. In fact, one must specify how to solve the quadratic program, how to deal with its possible inconsistency, how to cope with a first iterate that is far from a solution (globalization of the method), how the method can be used without computing second derivatives (quasi-Newton versions), how to take advantage of the negative curvature directions, *etc.* These questions have several answers, whose combinations result in various algorithms, more or less adapted to a particular situation. There is little to be gained from our describing each of these algorithms. Rather, our aim is to present the concepts that form the building blocks of these methods and to show why they are relevant. A good understanding of these tools should allow the reader to adapt the algorithm to a particular problem or to choose the right options of a solver, in order to make it more efficient.

The present review of Newton-like methods for constrained optimization is probably more analysis- than practice-oriented. The aim in this short account is to make an inventory of the main techniques that are continuously used to analyze these algorithms. In particular, we state and prove precise results on their properties. We also introduce and explain their structure in some detail. However, theory does not cover all aspects of an algorithm. We therefore strive to describe some heuristics that are important for efficient implementations. In fact, it is no exaggeration to say that a method is primarily judged good on the basis of its numerical efficiency. The analysis often comes afterwards to try to explain such a good behavior. Finally, let us mention that all the mathematical concepts used in the present text are simple. In particular, even though we use nonsmooth merit functions, very few notions of nonsmooth analysis are employed, so as to make the text accessible to many.

This part is organized as follows. We start in chapter 13 by recalling some theory on constrained optimization (optimality conditions, constraint qualification, projection onto a convex set, *etc.*) and Newton's method for nonlinear equations and unconstrained minimization. This chapter ends with the presentation of a numerical project that will go with us along the next chapters of this part (in §§ 14.7, 15.4, 17.4, and 18.4). This project will give us the opportunity to discuss fine points of the implementation of some of the proposed algorithms and to illustrate their behavior in various situations; it also shows, incidentally, that it is relatively easy to write one's own SQP code, provided a solver of quadratic optimization problems is available.

After these preliminaries come two chapters dealing with local methods, whose convergence is ensured if the first iterate is sufficiently close to a solution. Chapter 14 is devoted to problems with only equality constraints. Here we are in the familiar domain of Analysis, where the objects involved (functions and feasible sets) are smooth. The tools are classical as well: mainly linear algebra and differential calculus. A few concepts of differential geometry may be useful to interpret the algorithms. Chapter 15 considers the case where equalities and inequalities are present. Introducing inequality constraints results in an important additional difficulty, due to intrinsic combinatorics in the problem. This comes from the fact that one does not know a priori which inequality constraints are *active* at a solution, i.e., those that vanish at a solution. If they were known, the algorithms from chapter 14 would apply. The algorithms themselves must therefore determine the set of active constraints, among 2^{m_I} possibilities (m_I being the number of inequality constraints). Combinatorics is a serious difficulty for algorithms, but SQP copes with it by gracefully forwarding it to a quadratic subproblem, where it is easier to manage. This also implies a change of style in the analysis of the problem. Indeed, various sets of indices must be considered (active or inactive, weakly or strongly active), with an accuracy that is not obtained immediately.

The concept of exact penalty is central to force convergence of algorithms, independently of the initial iterate (a concept known as "globalization"); this is studied in chapter 16. First, the exactness properties of the Lagrangian and augmented Lagrangian can be analyzed thanks to their smoothness. These results are then used to obtain the exactness of a nondifferentiable merit function. In chapter 17, it is shown how this latter function can be used and how the local algorithms can be modified to obtain convergence of the generated iterates from a starting point that can be far from a solution. The transition from globally convergent algorithms to algorithms with rapid local convergence is also studied in that chapter.

In the quasi-Newton versions of the algorithms, the matrices containing second derivatives are replaced by matrices updated with adequate formulae; this is the subject of chapter 18.

The Problem to Solve

This text presents efficient algorithms for minimizing a real-valued function $f : \Omega \to \mathbb{R}$, defined on an *open* set Ω in $\mathbb{R}^n$, in the presence of functional constraints on the parameters $x = (x_1, \ldots, x_n)$ to optimize. Equality constraints $c_i(x) = 0$, for $i \in E$, as well as inequality constraints $c_i(x) \leq 0$, for $i \in I$, can be present. It is supposed that the index sets E (for equalities) and I (for inequalities) are *finite*, having respectively m_E and m_I elements. These constraints can also be written

$$c_E(x) = 0 \quad \text{and} \quad c_I(x) \leq 0.$$

Vector inequalities, such as $c_I(x) \leq 0$ above, are to be understood componentwise. Hence $c_I(x) \leq 0$ means that all the components of the vector $c_I(x) \in \mathbb{R}^{m_I}$ must be nonpositive. The functions f and c need not be convex.

We therefore look for a point $x_* \in \Omega$ that minimizes f on the *feasible set*

$$X = \{x \in \Omega : c_E(x) = 0, \ c_I(x) \leq 0\}.$$

A point in X is said to be *feasible*. The problem is written in a condensed way as follows:

$$(P_{EI}) \quad \begin{cases} \min_x f(x) \\ c_E(x) = 0 \\ c_I(x) \leq 0 \\ x \in \Omega. \end{cases}$$

The open set Ω appearing in (P_{EI}) cannot be used to express general constraints, since a solution cannot belong to its boundary. It is simply the domain of definition of the functions f, c_E, and c_I. It is also the set where some useful properties are satisfied. For example, we always suppose that c_E is a *submersion* on Ω, i.e., that its Jacobian matrix at $x \in \Omega$,

$$A_E(x) := \nabla c_E(x)^\top,$$

of dimension $m_E \times n$ (the rows of $A_E(x)$ contain the transposed gradients $\nabla c_i(x)^\top$, $i \in E$, for the Euclidean scalar product), is surjective (or onto), for any $x \in \Omega$. Also, f and c are assumed to be smooth on Ω, for example of class C^2 (twice continuously differentiable).

We recall from definition 2.2 that problem (P_{EI}) is said to be *convex* when Ω is convex, f and the components of c_I are convex and c_E is affine. In this case, the feasible set X is convex.

Notation

We denote by

$$m = m_E + m_I$$

the total number of functional constraints. It will be often convenient to assume that E and I form a partition of $\{1, \ldots, m\}$:

$$E \cup I = \{1, \ldots, m\} \quad \text{and} \quad E \cap I = \emptyset.$$

Then, for $v \in \mathbb{R}^m$, we denote by v_E the m_E-uple made up of the components v_i of v, with indices $i \in E$; likewise for v_I. The constraints c_E and c_I are then considered to be obtained from a single function $c : \Omega \to \mathbb{R}^m$, whose components indexed in E [resp. I] form c_E [resp. c_I].

With a vector $v \in \mathbb{R}^m$, one associates the vector $v^\# \in \mathbb{R}^m$, defined as follows:

$$(v^\#)_i = \begin{cases} v_i & \text{if } i \in E \\ v_i^+ & \text{if } i \in I, \end{cases}$$

where $v_i^+ = \max(0, v_i)$. With this notation, (P_{EI}) is concisely written as:

$$\begin{cases} \min_x f(x) \\ c(x)^\# = 0 \\ x \in \Omega. \end{cases}$$

Indeed, $c(x)^\# = 0$ if and only if $c_E(x) = 0$ and $c_I(x) \leq 0$.

Let $x \in \Omega$. If $c_i(x) = 0$, the constraint i is said to be *active* at x. We denote by

$$I^0(x) = \{i \in I : c_i(x) = 0\}$$

the set of indices of inequality constraints that are active at $x \in \Omega$.

The Euclidean or ℓ_2 norm is denoted by $\| \cdot \|_2$. We use the same notation for the associated matrix norm.

Codes

A number of pieces of software based on the algorithmic techniques presented in this part have been written. We give a few words on some of them with a vocabulary that will be clear only after having read part III of the book.

- VF02AD by Powell [293; 1978] is part of the Harwell library. It uses Fletcher's VE02AD code (also part of the Harwell library) for solving the osculating quadratic problems [125].

- NLPQL by Schittkowski [323; 1985-86] can be found in the IMSL library. The osculating quadratic problems are solved by the dual method of Goldfarb and Idnani [165] with the modification proposed by Powell [296] (QL code).

- NPSOL by Gill, Murray, Saunders, and Wright [158; 1986] is available in the NAG library.

- FSQP by Lawrence, Tits, and Zhou [282, 222, 223, 224; 1993-2001] uses an SQP algorithm that evaluates the objective function only at points satisfying the inequality constraints. This nice property can be important for certain classes of applications.

- SPRNLP by Betts and Frank [29; 1994] can use second derivatives (if not positive definite, the Hessian of the Lagrangian is modified using a Levenberg parameter) and exploits sparsity information. It has been used to solve many optimal control problems after a direct transcription discretization.

- FAIPA by Herskovits et al. [189, 190; 1995-1998] also forces the iterates to be strictly feasible with respect to the inequality constraints. Interestingly, the algorithm requires to solve only linear systems of equations, no quadratic optimization problems [283]. This approach is connected to interior point algorithms.

- DONLP2 by Spellucci [342; 1998] is available on Netlib. It uses an active set technique on the nonlinear problem, so that the osculating quadratic problems have only equality constraints.

- SNOPT by Gill, Murray, and Saunders [156; 2002] is designed for sparse large-scale problems. The Hessian of the Lagrangian is approximated by limited memory BFGS updates (§ 6.3). The quadratic programs are solved approximately by an active set method. The globalization is done by linesearch on an augmented Lagrangian merit function.

- SQPAL by Delbos, Gilbert, Glowinski, and Sinoquet [99; 2006] can solve large-scale problems since it uses an augmented Lagrangian approach for solving the quadratic problems [98], a method that has the property of identifying the active constraints in a finite number of iterations.

Notes

Surveys on Newton's method for constrained optimization have been written by Bertsekas [26; 1982], Powell [295; 1986], Fletcher [128; 1987], Gill, Murray, Saunders, and Wright [159; 1989], Spellucci [340; 1993], Boggs and Tolle [35;

1995], Polak [285; 1997], Sargent [320; 1997], Nocedal and Wright [277; 1999, Chapter 18], Conn, Gould, and Toint [86; 2000, Chapter 15], and Gould, Orban, and Toint [178; 2005]. See also [242] for problems with equilibrium constraints and [28, 325] for applications to optimal control problems.

Acknowledgement

This part of the book has benefited from the remarks and constructive comments by friends and colleagues, including Paul Armand, Laurent Chauvier, Frédéric Delbos, Sophie Jan-Jégou, Xavier Jonsson, Jean Roberts, Delphine Sinoquet, and the other authors of this book. Some students of ENSTA (École Nationale de Techniques Avancées, Paris) and DEA students of Paris I Panthéon-Sorbonne, to whom the subject of this part has been taught for several years, have also contributed to improve the accessibility of this text. Finally, I would like to thank Richard James, who kindly accepted to supervise the English phrasing of several chapters. His careful reading and erudite recommendations were greatly appreciated.

13 Background

Before entering the subject itself, we recall some basic concepts. For further details, the reader is referred to the books by Fletcher [128; 1987], Ciarlet [79; 1988], Gauvin [141, 142; 1992-95], Hiriart-Urruty and Lemaréchal [195; 1993], and Bonnans and Shapiro [50; 2000]. See also [150] for an online review (in French).

13.1 Differential Calculus

A function is said to be of class C^k integer $k \geq 0$, if it is k times continuously differentiable. A function f is said to be of class $C^{k,1}$ it is of class C^k and if its kth derivative $f^{(k)}$ is Lipschitz continuous: for some norm $\| \cdot \|$ and constant $L > 0$, and for any x and y, one has

$$\|f^{(k)}(x) - f^{(k)}(y)\| \leq L\|x - y\|.$$

We shall frequently expand functions, i.e., give a polynomial approximation of $h \mapsto f(x+h)$, valid for more or less small h. This is due to the nature of our algorithms, obtained by linearizing optimality conditions. We give below some useful formulae for controlling the precision of these expansions.

Consider first the case of a scalar-valued function $f : \Omega \to \mathbb{R}$ defined on an open set Ω of $\mathbb{R}^n$. Let $x \in \Omega$ and $h \in \mathbb{R}^n$ be such that the segment $[x, x+h] = \{(1-\alpha)x + \alpha(x+h) : \alpha \in [0,1]\}$ lies in Ω. Suppose that f is $(k-1)$ times differentiable on Ω ($k \geq 1$) and k times differentiable on the open segment $]x, x+h[= \{(1-\alpha)x + \alpha(x+h) : \alpha \in]0,1[\}$. Then there exists $\theta \in]0, 1[$ such that

$$f(x+h) = \sum_{i=0}^{k-1} \frac{1}{i!} f^{(i)}(x) \cdot h^i + \frac{1}{k!} f^{(k)}(x + \theta h) \cdot h^k.$$

If $f : \Omega \to \mathbb{R}^m$ is vector-valued ($m > 1$), this expansion may not hold. Under the same conditions as above, however, we can write

$$\left\| f(x+h) - \sum_{i=0}^{k-1} \frac{1}{i!} f^{(i)}(x) \cdot h^i \right\| \leq \left(\sup_{z \in]x, x+h[} \|f^{(k)}(z)\| \right) \frac{\|h\|^k}{k!}.$$

Now if $f : \Omega \to \mathbb{R}^m$ is only $(k-1)$ times differentiable on Ω and k times differentiable at x, then

$$f(x+h) = \sum_{i=0}^{k} \frac{1}{i!} f^{(i)}(x) \cdot h^i + o(\|h\|^k).$$

If some more smoothness is assumed on f, the remainder term can be expressed in integral form. More precisely, if $f : \Omega \to \mathbb{R}^m$ is of class C^k on Ω and if the segment $[x, x+h]$ lies entirely in Ω, we have the formula:

$$f(x+h) = \sum_{i=0}^{k-1} \frac{1}{i!} f^{(i)}(x) \cdot h^i + \int_0^1 \frac{(1-t)^{k-1}}{(k-1)!} f^{(k)}(x+th) \cdot h^k \, dt.$$

As a weakened form of differentiability, we say that $f : \Omega \to \mathbb{R}^m$ has a *directional derivative* at x in the direction h if the limit exists in the following expression:

$$f'(x; h) := \lim_{t \to 0+} \frac{f(x+th) - f(x)}{t}.$$

The next lemma shows that the composition of functions having directional derivatives has directional derivatives too, providing the second function is Lipschitz continuous.

Lemma 13.1 (directional differentiability of a composition). *Suppose that $\varphi : \mathbb{R}^n \to \mathbb{R}^m$ has a directional derivative at x in the direction $h \in \mathbb{R}^n$ and that $\psi : \mathbb{R}^m \to \mathbb{R}^p$ is Lipschitz continuous in a neighborhood of $\varphi(x)$ and has a directional derivative at $\varphi(x)$ in the direction $\varphi'(x; h)$. Then $(\psi \circ \varphi)$ has a directional derivative at x in the direction h and there holds*

$$(\psi \circ \varphi)'(x; h) = \psi'(\varphi(x); \varphi'(x; h)).$$

Proof. For $t \to 0+$, use successively the directional differentiability of φ, the Lipschitz continuity, and the directional differentiability of ψ:

$$\begin{aligned}
(\psi \circ \varphi)(x + th) &= \psi(\varphi(x) + t\varphi'(x; h) + o(t)) \\
&= \psi(\varphi(x) + t\varphi'(x; h)) + o(t) \\
&= (\psi \circ \varphi)(x) + t\psi'(\varphi(x); \varphi'(x; h)) + o(t).
\end{aligned}$$

The result follows.   □

We shall also use the following result.

Lemma 13.2 (differentiability of a product). *If $g : \mathbb{R}^n \to \mathbb{R}^m$ is continuous at $x \in \mathbb{R}^n$ and $\alpha : \mathbb{R}^n \to \mathbb{R}$ is Fréchet differentiable at x with $\alpha(x) = 0$, then $f : \mathbb{R}^n \to \mathbb{R}^m$ defined by $f(x) = \alpha(x)g(x)$ is Fréchet differentiable at x and for $h \in \mathbb{R}^n$: $f'(x) \cdot h = (\alpha'(x) \cdot h)g(x)$.*

Proof. Let $h \to 0$. By the assumptions $\alpha(x+h) = \alpha'(x) \cdot h + o(\|h\|) = O(\|h\|)$ and $g(x+h) = g(x) + o(1)$. Therefore:

$$
\begin{aligned}
f(x+h) - f(x) - (\alpha'(x) \cdot h)g(x) &= \alpha(x+h)g(x+h) - (\alpha'(x) \cdot h)g(x) \\
&= (\alpha(x+h) - \alpha'(x) \cdot h) \, g(x) + o(\|h\|) \\
&= o(\|h\|).
\end{aligned}
$$

$\square$

13.2 Existence and Uniqueness of Solutions

Consider the optimization problem

$$
(P) \quad \begin{cases} \min_x f(x) \\ x \in X, \end{cases}
$$

where X is a subset of $\mathbb{R}^n$ (here unspecified).

Recall that a *global solution* to problem (P) is a point $x_* \in X$ minimizing f on the feasible set X:

$$
f(x_*) \leq f(x), \quad \text{for all } x \in X.
$$

A *local solution* to (P) is a feasible point x_*, minimizing f locally on the feasible set X: there exists $\varepsilon > 0$ such that

$$
f(x_*) \leq f(x), \quad \text{for all } x \in B(x_*, \varepsilon) \cap X.
$$

Here $B(x_*, \varepsilon)$ is the open ball centered at x_*, with radius ε. We say that x_* is a *strict local solution*, if the above is a strict inequality when $x \neq x_*$. Observe that a global solution is also a local solution. The algorithms to come are not aimed at finding global solutions. Hence, the word *solution* will hereafter be used for local solutions to (P).

Problem (P) has a solution if f is continuous and X is compact (closed and bounded) nonempty. The assumption "X compact" can be replaced by "X closed" (nonempty) if f has the property of tending to infinity at infinity on X, i.e., if:

$$
\lim_{\substack{x \in X \\ \|x\| \to \infty}} f(x) = +\infty.
$$

Besides, this problem has at most one solution if f is strictly convex and X is convex.

13.3 First-Order Optimality Conditions

Let x_* be a local solution to (P_{EI}). It is known that, if f and c are Gâteaux differentiable at x_* and if the constraints are qualified at x_* (see below), then there exists $\lambda_* \in \mathbb{R}^m$ such that the following Karush, Kuhn, Tucker conditions (KKT) hold:

$$
(\text{KKT}) \quad
\begin{cases}
(a) \ \nabla f(x_*) + A(x_*)^\top \lambda_* = 0 \\
(b) \ c_E(x_*) = 0, \quad c_I(x_*) \le 0 \\
(c) \ (\lambda_*)_I \ge 0 \\
(d) \ (\lambda_*)_I^\top c_I(x_*) = 0.
\end{cases}
\tag{13.1}
$$

We use the notation ∇ for a gradient with respect to the Euclidean scalar product (vector of partial derivatives). The above optimality conditions are called "of first-order", for they only involve first-order derivatives of f and c. They are continually used. Some comments on this optimality system will help the reader to memorize it.

Identity (a) is the *optimality equation* itself. We have used the notation $A(x_*)$ for the $m \times n$ Jacobian of the constraints at x_*: $A(x) = \nabla c(x)^\top$, so that its (i, j)th element is the partial derivative $\partial c_i / \partial x_j$ evaluated at x. This equation can also be written

$$
\nabla_x \ell(x_*, \lambda_*) = 0,
$$

where ℓ is the *Lagrangian* of the problem:

$$
\ell(x, \lambda) = f(x) + \lambda^\top c(x). \tag{13.2}
$$

The vector λ_* is called the *Lagrange multiplier*. The name *multiplier* comes from the fact that it multiplies the constraint vector in the Lagrangian. The vector has as many components as there are constraints. In (b), we recognize feasibility of x_*. Conditions (c) and (d) are only related to inequality constraints. By (c), the corresponding multipliers have a definite sign, depending on how (P_{EI}) is formulated. Here we have a "min", constraints c_I are "negative", equation (a) has a "+" sign, as well as the Lagrangian ℓ. Identity (d) is called *complementarity conditions*. As $(\lambda_*)_I \ge 0$ and $c_I(x_*) \le 0$, this amounts to writing

$$
(\lambda_*)_i c_i(x_*) = 0, \quad \text{for all } i \in I.
$$

Said otherwise, the multipliers corresponding to inactive constraints are zero:

$$
c_i(x_*) < 0 \quad \Longrightarrow \quad (\lambda_*)_i = 0.
$$

This comes from the fact that (13.1) expresses stationarity of x_*, which is a local property: if $c_i(x_*) < 0$, the constraint c_i must not appear in (13.1) because a small perturbation of this constraint does not affect stationarity of x_*. In some cases, we have the equivalence

$$c_i(x_*) < 0 \iff (\lambda_*)_i = 0. \tag{13.3}$$

We then say that *strict complementarity* holds.

A pair (x_*, λ_*) satisfying (KKT) is called a *primal-dual solution* to (P_{EI}), and x_* is said to be *stationary*. Given a primal-dual solution (x_*, λ_*), we use the notation

$$\begin{aligned}
I_*^0 &= I^0(x_*) = \{i \in I : c_i(x_*) = 0\}, \\
I_*^{0+} &= \{i \in I_*^0 : (\lambda_*)_i > 0\}, \\
I_*^{00} &= \{i \in I_*^0 : (\lambda_*)_i = 0\}.
\end{aligned}$$

Constraints with indices $i \in I_*^{0+}$ are said to be *strongly active* and those with indices $i \in I_*^{00}$ are said to be *weakly active*. The latter, though active $(c_i(x_*) = 0)$, can be removed from the problem without affecting stationarity of x_* (since $(\lambda_*)_i = 0$).

Constraint Qualification

As mentioned above, existence of a multiplier $\lambda_* \in \mathbb{R}^m$ such that (KKT) holds, is ensured when the constraints are qualified at x_*. The aim of the present subsection is not to give this concept a precise meaning. We shall therefore just give sufficient qualification conditions, which we shall call somewhat abusively *constraint qualification*.

Thus, we say that the constraints are qualified at x when one of the following conditions is satisfied.

(A-CQ) $c_{E \cup I^0(x)}$ is affine in a neighborhood of x.

(S-CQ) *Slater's Qualification* [334; 1950]:
- c_E is *affine* with c'_E surjective,
- the components of $c_{I^0(x)}$ are *convex*,
- there exists a point $\hat{x} \in X$ such that $c_{I^0(x)}(\hat{x}) < 0$.

(LI-CQ) The gradients of the active constraints $\{\nabla c_i(x) : i \in E \cup I^0(x)\}$ are linearly independent.

(MF-CQ) *Mangasarian-Fromovitz Qualification* [246; 1967]: if

$$\sum_{i \in E \cup I^0(x)} \alpha_i \nabla c_i(x) = 0, \quad \text{with } \alpha_i \geq 0 \text{ for } i \in I^0(x),$$

then $\alpha_i = 0$ for all $i \in E \cup I^0(x)$.

Observe that, if the constraint qualification condition (LI-CQ) is satisfied, there is *at most* one multiplier λ_* satisfying the first-order optimality conditions (KKT) for a given primal solution x_*. Indeed, from (KKT)$_d$, we have $(\lambda_*)_i = 0$ if $c_i(x_*) < 0$. For $i \in E \cup I_*^0$, (KKT)$_a$ and (LI-CQ) give uniqueness.

Observe also that the Mangasarian-Fromovitz condition is weaker (i.e., satisfied more often) than (LI-CQ). It can be shown to be equivalent to

$$\forall v \in \mathbb{R}^m, \ \exists d \in \mathbb{R}^n : \ c_E'(x) \cdot d = v_E \ \text{and} \ c_{I^0(x)}'(x) \cdot d \leq v_{I^0(x)}. \qquad (13.4)$$

In plain words, it is a kind of "weak-surjectivity" of the Jacobian of the active constraints; while (LI-CQ) expresses that this same operator is surjective. Recall that if (x_*, λ_*) satisfies (KKT), the set of Lagrange multipliers λ_* associated with x_* is bounded if and only if (MF-CQ) holds [140, 141].

13.4 Second-Order Optimality Conditions

We recall that a subset C of $\mathbb{R}^n$ is a *cone* when $\alpha C \subset C$, for all $\alpha > 0$. Said otherwise, $\alpha x \in C$ whenever $x \in C$ and $\alpha > 0$.

Let x_* be a point in the feasible set $X = \{x \in \Omega : c_E(x) = 0, \ c_I(x) \leq 0\}$ of problem (P_{EI}). The *critical cone* C_* at x_* associated with problem (P_{EI}) is defined by

$$C_* = \{d \in \mathbb{R}^n : c_E'(x_*) \cdot d = 0, \ c_{I_*^0}'(x_*) \cdot d \leq 0, \ f'(x_*) \cdot d \leq 0\}. \qquad (13.5)$$

The elements of C_* are called the *critical directions*. If (x_*, λ_*) is a primal-dual solution to (P_{EI}), C_* can also be written

$$C_* = \{d \in \mathbb{R}^n : c_{E \cup I_*^{0+}}'(x_*) \cdot d = 0, \ c_{I_*^{00}}'(x_*) \cdot d \leq 0\}. \qquad (13.6)$$

When the strict complementarity conditions (13.3) hold, $I_*^{00} = \emptyset$ and the critical cone becomes the null space of the Jacobian of the active constraints:

$$\{d \in \mathbb{R}^n : c_{E \cup I_*^0}'(x_*) \cdot d = 0\} = N\left(A_{E \cup I_*^0}(x_*)\right).$$

We can now state second-order necessary conditions (theorem 13.3, NC2) and second-order sufficient conditions (theorem 13.4, SC2) of optimality.

Theorem 13.3 (NC2). *Let x_* be a local solution to (P_{EI}). Suppose that f and $c_{E \cup I_*^0}$ are of class C^2 in a neighborhood of x_* and that $c_{I \setminus I_*^0}$ is continuous at x_*. Suppose also that the Mangasarian-Fromovitz constraint qualification (MF-CQ) holds at x_*. Then the set Λ_* of Lagrange multipliers λ_* associated with x_* such that conditions (KKT) hold is nonempty and*

$$\forall d \in C_*, \ \exists \lambda_* \in \Lambda_* : \ d^\top \nabla_{xx}^2 \ell(x_*, \lambda_*) d \geq 0. \qquad (13.7)$$

When the Mangasarian-Fromovitz constraint qualification (MF-CQ) holds at x_*, the set Λ_* of Lagrange multipliers associated with x_* is nonempty and compact, but this set is not necessarily reduced to a singleton. Therefore, the multiplier λ_* that gives the nonnegativity of $d^\top \nabla^2_{xx}\ell(x_*, \lambda_*)d$ in (13.7) may depend on $d \in C_*$.

Sufficient conditions of optimality do not include a constraint qualification.

Theorem 13.4 (SC2). *Suppose f and $c_{E \cup I^0_*}$ are differentiable in a neighborhood of $x_* \in \Omega$ and twice differentiable at x_*. Suppose also that the set Λ_* of Lagrange multipliers λ_* such that conditions (KKT) hold is nonempty and that*

$$\forall d \in C_* \backslash \{0\}, \quad \exists \lambda_* \in \Lambda_* : \quad d^\top \nabla^2_{xx}\ell(x_*, \lambda_*)d > 0. \tag{13.8}$$

Then x_* is a strict local minimum of (P_{EI}).

Condition (13.8) in theorem 13.4 is sometimes called the *weak second order sufficient condition of optimality*. In some cases, a stronger form of (13.8) is satisfied, in which a fixed λ_* can be chosen independently of the critical direction:

$$\exists \lambda_* \in \Lambda_*, \quad \forall d \in C_* \backslash \{0\}, \quad d^\top \nabla^2_{xx}\ell(x_*, \lambda_*)d > 0. \tag{13.9}$$

Condition (13.9) will be called the *semi-strong second order sufficient condition of optimality*. Finally, it is said that a solution x_* satisfies the *strong second order sufficient condition of optimality* if λ_* can be arbitrary in Λ_*:

$$\forall \lambda_* \in \Lambda_*, \quad \forall d \in C_* \backslash \{0\}, \quad d^\top \nabla^2_{xx}\ell(x_*, \lambda_*)d > 0. \tag{13.10}$$

When there is a unique Lagrange multiplier λ_* associated with x_* (for example because (LI-CQ) or the Kyparisis condition [220] holds), then (13.8), (13.9), and (13.10) are all equivalent.

We call *strong solution* to (P_{EI}) a pair (x_*, λ_*) satisfying the sufficient conditions of optimality stated in theorem 13.4 with one of the conditions (13.8), (13.9), or (13.10).

13.5 Speed of Convergence

Let $\{u_k\}_{k \geq 1}$ be a sequence in a normed space, converging to 0; and let $\{\alpha_k\}_{k \geq 1}$ be a sequence of positive numbers converging to 0 when $k \to \infty$. We write $u_k = O(\alpha_k)$, and we say that u_k is *big O* of α_k, if there is a constant C such that $\|u_k\| \leq C\alpha_k$ for all $k \geq 1$. We write $u_k = o(\alpha_k)$, and we say that u_k is *little o* of α_k, if for all $\varepsilon > 0$ there exists an index k_ε such that $\|u_k\| \leq \varepsilon\alpha_k$ for all $k \geq k_\varepsilon$. If $\{u'_k\}_{k \geq 1}$ is another sequence of points converging 0, we write

$$u_k \sim u'_k, \qquad (13.11)$$

and we say that $\{u_k\}$ and $\{u'_k\}$ are two *equivalent sequences* if $u_k = O(\|u'_k\|)$ and $u'_k = O(\|u_k\|)$.

Let $\{x_k\}$ converge to a point x_* and take a norm $\|\cdot\|$. We say that the convergence is *linear* for that norm, if there exist $r \in \,]0,1[$ and an index k_r such that

$$\|x_{k+1} - x_*\| \le r\|x_k - x_*\|, \quad \text{for all } k \ge k_r.$$

The convergence is *superlinear* if

$$x_{k+1} - x_* = o(\|x_k - x_*\|).$$

The convergence is *quadratic* if

$$x_{k+1} - x_* = O(\|x_k - x_*\|^2).$$

One speaks of linear, superlinear and quadratic convergence *in p steps* (p integer ≥ 1), if there holds respectively $\|x_{k+p} - x_*\| \le r\|x_k - x_*\|$ (for some $r \in \,]0,1[$ and for all k large enough), $x_{k+p} - x_* = o(\|x_k - x_*\|)$, and $x_{k+p} - x_* = O(\|x_k - x_*\|^2)$.

These are obviously stronger and stronger concepts: quadratic convergence implies superlinear convergence, which in turn implies linear convergence. The property of linear convergence depends on the norm, while superlinear and quadratic convergence do not.

Lemma 13.5 (equivalent sequences). *If the sequence $\{x_k\}$ converges superlinearly to x_*, then there holds $(x_k - x_*) \sim (x_{k+1} - x_k)$.*

Proof. We have

$$x_{k+1} - x_k = (x_{k+1} - x_*) - (x_k - x_*) = -(x_k - x_*) + o(\|x_k - x_*\|).$$

The result follows.   □

Other Use of the Notation $O(\cdot)$ and $o(\cdot)$.

We shall also use Landau's notation $O(\cdot)$ and $o(\cdot)$ with a slightly different point of view, without reference to convergent sequences. This notation will be useful when studying the local convergence of algorithms, before establishing the convergence of the considered sequences, so that the preceding definitions do not apply.

Let ψ and ϕ be two functions defined in a neighborhood V of a point $x_* \in \mathbb{R}^n$, with values in normed spaces. We say that ψ is *dominated* by ϕ in a neighborhood of x_*, and we write $\psi = O(\phi)$, if there exist a positive number C and a neighborhood $V_0 \subset V$ of x_* such that $\|\psi(x)\| \le C\|\phi(x)\|$ for all $x \in V_0$. We say that ψ is *negligible* compared with ϕ in a neighborhood x_*, and we write $\psi = o(\phi)$, if for all $\varepsilon > 0$, there exists a neighborhood $V_\varepsilon \subset V$ of x_* such that $\|\psi(x)\| \le \varepsilon\|\phi(x)\|$ for all $x \in V_\varepsilon$.

13.6 Projection onto a Closed Convex Set

Let $\langle \cdot, \cdot \rangle$ be a scalar product on $\mathbb{R}^n$, $\| \cdot \|$ the *associated norm* (i.e., $\|x\| = \langle x, x \rangle^{1/2}$), C a nonempty closed convex set in $\mathbb{R}^n$ and x a point of $\mathbb{R}^n$. Then there exists a unique element $x_p \in C$, called the *projection* of x onto C, such that

$$\|x_p - x\| \leq \|y - x\|, \quad \text{for all } y \in C.$$

It is thus the point of C that minimizes the distance of x to the points of C. We shall also use the notation $x_p = P_C x$.

Here are some properties of the projection that will be useful.

First, the projection $x_p = P_C x$ is *characterized* by one of the following equivalent conditions:

$$\langle x_p - x, y - x_p \rangle \geq 0, \quad \text{for all } y \in C, \tag{13.12}$$

$$\langle y - x, y - x_p \rangle \geq 0, \quad \text{for all } y \in C. \tag{13.13}$$

Thus, if $x_p \in C$ satisfies one of the above properties, it is the projection of x onto C. If C is a nonempty closed convex cone of $\mathbb{R}^n$ and $x \in \mathbb{R}^n$, another characterization of $x_p = P_C x$ is easily deduced from (13.12):

$$\langle x_p - x, x_p \rangle = 0 \quad \text{and} \quad \langle x_p - x, y \rangle \geq 0, \quad \text{for all } y \in C. \tag{13.14}$$

Furthermore, the mapping $x \mapsto P_C x$ has the following property:

$$\langle P_C x_2 - P_C x_1, x_2 - x_1 \rangle \geq \|P_C x_2 - P_C x_1\|^2, \quad \text{for all } x_1, x_2 \in \mathbb{R}^n,$$

which implies in particular that P_C is a monotone mapping. Using the *Cauchy-Schwarz inequality* $\langle u, v \rangle \leq \|u\| \, \|v\|$, one deduces:

$$\|P_C x_1 - P_C x_2\| \leq \|x_1 - x_2\|, \quad \text{for all } x_1, x_2 \in \mathbb{R}^n. \tag{13.15}$$

The projector P_C is therefore Lipschitz continuous with modulus 1.

13.7 The Newton Method

Consider a mapping $F : \mathbb{R}^N \to \mathbb{R}^N$. We want to solve numerically for $z \in \mathbb{R}^N$ the system with N equations in the N unknowns

$$F(z) = 0. \tag{13.16}$$

The *Newton method* (see also chapter 4) generates a sequence $\{z_k\}$ by the recurrence formula

$$z_{k+1} = z_k + d_k, \tag{13.17}$$

where the step d_k solves the equation (13.16) linearized at z_k:

$$F(z_k) + F'(z_k)d_k = 0. \tag{13.18}$$

This equation has a unique solution if $F'(z_k)$ is nonsingular. In this case,

$$d_k = -F'(z_k)^{-1}F(z_k). \tag{13.19}$$

The next theorem analyses the convergence of a slightly more general method, in which the direction d_k is given by

$$d_k = -M_k^{-1}F(z_k), \tag{13.20}$$

where M_k is a nonsingular matrix. Quasi-Newton methods enter this framework. We recall that nonsingular matrices form an open set in the normed space of matrices.

Theorem 13.6 (convergence of Newton's algorithm). *Let z_* be a zero of a map $F : \Omega \to \mathbb{R}^N$ defined on some neighborhood $\Omega \subset \mathbb{R}^N$ of z_*. Suppose that F is of class C^1 on Ω and that $F'(z_*)$ is nonsingular.*
1) *Then there exist $\varepsilon_z > 0$ and $\varepsilon_M > 0$ such that*

$$\|z_1 - z_*\| \le \varepsilon_z \qquad and \qquad \|M_k - F'(z_k)\| \le \varepsilon_M, \quad \forall k \ge 1, \tag{13.21}$$

imply that the recursion (13.17) with d_k given by (13.20) is well defined and generates a sequence $\{z_k\}$ converging linearly to z_*.
2) *If, in addition,*

$$\bigl(M_k - F'(z_*)\bigr)(z_k - z_*) = o(\|z_k - z_*\|),$$

then the convergence is superlinear.
3) *If, in addition, F' is Lipschitz continuous on Ω and*

$$\bigl(M_k - F'(z_*)\bigr)(z_k - z_*) = O(\|z_k - z_*\|^2),$$

then the convergence is quadratic.

Proof. Note $\beta := \|F'(z_*)^{-1}\|$ and choose $\varepsilon_M > 0$ such that $\beta\varepsilon_M < 1$ and

$$r := \frac{3\beta\varepsilon_M}{1 - \beta\varepsilon_M} < 1.$$

Now determine $\varepsilon_z > 0$ such that $\bar{B}(z_*, \varepsilon_z) \subset \Omega$ and $\|F'(z) - F'(z_*)\| \le \varepsilon_M$ for all $z \in \bar{B}(z_*, \varepsilon_z)$ (possible by the continuity of F').

Let M be an $N \times N$ matrix verifying $\|M - F'(z_*)\| \le \varepsilon_M$. Then $\|F'(z_*)^{-1}(M - F'(z_*))\| \le \beta\varepsilon_M < 1$ and, by Banach's perturbation lemma, the matrix M in nonsingular and satisfies $\|M^{-1}\| \le \beta/(1 - \beta\varepsilon_M)$. Applying this to $M = M_k$ and $M = F'(z)$, one finds that, for all $k \ge 1$ and all $z \in \bar{B}(z_*, \varepsilon_z)$, M_k and $F'(z)$ are nonsingular and

$$\|M_k^{-1}\| \quad \text{and} \quad \|F'(z)^{-1}\| \le \frac{\beta}{1 - \beta \varepsilon_M}.$$

In this case, d_k is well defined by formula (13.20).

Using $F(z_*) = 0$ and the fact that F is C^1 on $\bar{B}(z_*, \varepsilon_z)$, which allow us to use a Taylor expansion in integral form, there holds when $z_k \in \bar{B}(z_*, \varepsilon_z)$

$$\begin{aligned}
z_{k+1} - z_* &= z_k - z_* + d_k \\
&= M_k^{-1}\big(M_k(z_k - z_*) - F(z_k)\big) \\
&= M_k^{-1}\big(M_k - F'(z_k)\big)(z_k - z_*) \\
&\quad + M_k^{-1}\int_0^1 \Big(F'(z_k) - F'(z_* + t(z_k - z_*))\Big)(z_k - z_*)\,\mathrm{d}t.
\end{aligned}$$

Taking norms, also permuting norm and integral operator, we deduce that $\|z_{k+1} - z_*\| \le r\|z_k - z_*\|$. Therefore, by induction, all the sequence $\{z_k\} \subset \bar{B}(z_*, \varepsilon_z)$ if $z_1 \in \bar{B}(z_*, \varepsilon_z)$. Furthermore $z_k \to z_*$ (since $r < 1$). This concludes the proof of the first claim of the theorem.

With the additional assumption of the second point of the theorem, the estimate of the error $z_{k+1} - z_*$ above shows that $z_{k+1} - z_* = o(\|z_k - z_*\|)$, hence the convergence of $\{z_k\}$ is superlinear. Finally, with the additional assumptions of the third point of the theorem, the same error estimate provides $\|z_{k+1} - z_*\| \le C\|z_k - z_*\|^2$, for some constant $C > 0$, implying the quadratic convergence of $\{z_k\}$.. $\qquad\square$

Applying this theorem to the Newton algorithm ($M_k = F'(z_k)$ for all k) shows that, if F is C^1 in a neighborhood of z_*, if $F'(z_*)$ is nonsingular, and if the first iterate is close enough to z_*, the method converges superlinearly. It converges quadratically if, in addition, F' is Lipschitz continuous in a neighborhood of z_*.

The Osculating Quadratic Problem

To solve the unconstrained minimization problem

$$\min_{x \in \mathbb{R}^n} f(x), \tag{13.22}$$

a possibility is to solve its optimality condition (see § 2.2 or use (KKT))

$$\nabla f(x) = 0.$$

Taking $F = \nabla f$ in (13.16), the Newton equation (13.18) can be written

$$\nabla f(x_k) + \nabla^2 f(x_k)d_k = 0. \tag{13.23}$$

This equation is the first-order optimality condition of the problem

$$\min_d \nabla f(x_k)^\top d + \frac{1}{2} d^\top \nabla^2 f(x_k) d, \tag{13.24}$$

which is called the *osculating quadratic problem* to (13.22) at x_k.

When $\nabla^2 f(x_k)$ is positive definite (so is the case if x_k is close to a strong solution to (13.22)), it is equivalent to solve the linear system (13.23) or the quadratic problem (13.24). With (13.24), we have an optimization problem to solve, which may guide intuition when designing algorithms. This return to optimization after linearization of the optimality conditions will also be done for constrained problems.

Because Newton's algorithm is basically a method to solve nonlinear equations, it makes no distinction between the types of stationary points x_*, provided $\nabla^2 f(x_*)$ is nonsingular. According to theorem 13.6, the iterates are indeed attracted by such a regular point, even though it is not a local minimum of f; in particular it can be a maximum. When one tries to find a minimizer, this is not a nice property. There are techniques, however, such as truncated conjugate gradient iterations or the use of trust regions, which tend to overcome this undesirable feature. Some of them will be described in chapter 17, in the framework of constrained optimization problems.

13.8 The Hanging Chain Project I

We introduce here a test problem that will go with us along the chapters of this third part of the book. This problem will be used to implement and to test some of the optimization algorithms that will be presented and analyzed in the chapters 14 to 18. The model is simple enough to be implemented in a course on numerical constrained optimization, using languages like MAT-LAB [249] or SCILAB [327]. It is also rich enough to present all the difficult situations that a nonlinear optimization software has to deal with. The large amount of details given in this section and others have the goal to make an implementation easy.

The problem we propose to look at consists in finding the static equilibrium position of a chain made of rigid bars. Its extreme joints are fixed at two hooks and it is maintained above a given tilted flat floor. As we shall see, this problem can be modeled as a minimization problem with equality and inequality constraints, hence having the form of problem (P_{EI}) on page 193.

The hanging chain problem has been considered by several authors, with more or less generality and without the floor constraint. Luenberger [239] introduces it "to illustrate a wide assortment of theoretical principles and practical techniques". Veselić [356] provides a precise analysis of the problem. Bonnans and Shapiro [49] use it as an example for a perturbation analysis.

Modeling

We suppose that the chain has perfectly flexible joints and is subject to gravity, so that it lies in the vertical plane containing the hooks. Let it be the (x, y)-plane (see figure 13.1). We assume that the chain has **nb** rigid bars of

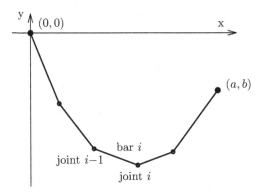

Fig. 13.1. Model of the hanging chain

given lengths L_i, $i = 1, \ldots, \mathbf{nb}$. Therefore, there are $\mathbf{nj} = \mathbf{nb} - 1$ free joints. The coordinates of the joints are denoted by (x_i, y_i), for $i = 1, \ldots, \mathbf{nj}$. These ones are the $n := 2(\mathbf{nj})$ variables to determine, since the position of the hooks is supposed to be given: at positions of coordinates $(x_0, y_0) = (0, 0)$ and $(x_{\mathbf{nb}}, y_{\mathbf{nb}}) = (a, b)$ say. There is no restriction on the value of (a, b). In particular, one can set $(a, b) = (0, 0)$, in which case the vertical plane containing the chain is supposed to be a problem data (it has not to be specified, but in this case the solution found in the (x, y)-plane is still a solution to the problem after a rotation around the vertical axis going through the point $(0, 0)$). We start with the premise that the position of the chain can be obtained by minimizing its potential energy. Let us give its analytic expression.

Consider a chain whose bars have not necessarily the given lengths and that is specified by the position of its free joints. We gather them in the vector of unknowns

$$x := (x_1, x_2, \ldots, x_{\mathbf{nj}}, y_1, y_2, \ldots, y_{\mathbf{nj}}).$$

The adopted order in the components of x is more convenient than taking $(x_1, y_1, x_2, y_2, \ldots)$. Since the ith bar is between the joints of coordinates (x_{i-1}, y_{i-1}) and (x_i, y_i), its length is given by

$$l_i(x) := \sqrt{(x_i - x_{i-1})^2 + (y_i - y_{i-1})^2}, \quad i = 1, \ldots, \mathbf{nb},$$

with $(x_0, y_0) := (0, 0)$ and $(x_{\mathbf{nb}}, y_{\mathbf{nb}}) = (a, b)$. Therefore, assuming unit weight per unit of length, the potential energy of the considered chain, which is defined up to a constant, can be written

$$E(x) = \sum_{i=1}^{nb} l_i(x) \frac{y_i + y_{i-1}}{2}.$$

The model must also specify the actual length of the bars: $l_i(x) = L_i$. Squaring $l_i(x)$ to have differentiability leads to the constraints $c(x) = 0 \in \mathbb{R}^{nb}$, where

$$c_i(x) = l_i(x)^2 - L_i^2, \quad \text{for } i = 1, \ldots, nb.$$

Without the floor constraint, the model consists in

$$\begin{cases} \min_x E(x) \\ c_i(x) = 0, \quad i = 1, \ldots, nb. \end{cases}$$

Note that on the feasible set, $l_i(x) = L_i$, so that the problem is not modified if $l_i(x)$ is substituted by L_i in the objective. The interest of this substitution is that the objective becomes linear:

$$\begin{cases} \min_x e(x) \\ c_i(x) = 0, \quad i = 1, \ldots, nb, \end{cases}$$

where

$$e(x) = \sum_{i=1}^{nb} L_i \frac{y_i + y_{i-1}}{2}.$$

We still have to model the floor constraint.

It is assumed that the floor is smooth, without roughness, so that the joints and the chain can slip on it without resistance. This implies that the actual equilibrium position is still the one with minimal potential energy. To simplify, we assume that the floor is flat and possibly tilted. In the (x, y)-plane, the floor is given by the affine function $y = g(x)$, where

$$g(x) := g_0 + g_1 x. \tag{13.25}$$

The constants g_0 and g_1 are supposed given and may vary from one test-problem to another. The chain must hang in the half-plane $D := \{(x, y) : y \geq g(x)\}$. In particular, the hooks must lie in D, which leads to the following compatibility conditions on g_0 and g_1:

$$g_0 \leq 0 \quad \text{and} \quad g_0 + g_1 a \leq b.$$

Then, since the chain is affine between its joints, it is entirely in D if $c_I(x) \leq 0 \in \mathbb{R}^{nj}$, where

$$c_i(x) = g_0 + g_1 x_{i-nb} - y_{i-nb}, \quad \text{for } i = nb + 1, \ldots, nb + nj.$$

Finally, the problem to solve can be written

$$\begin{cases} \min_x e(x) \\ c_i(x) = 0, & i \in E := \{1, \ldots, \text{nb}\} \\ c_i(x) \leq 0, & i \in I := \{\text{nb} + 1, \ldots, \text{nb} + \text{nj}\}. \end{cases}$$

It has the structure of a minimization problem with equality and inequality constraints like (P_{EI}). The Lagrangian of the problem is the function denoted by

$$(x, \lambda) \mapsto \ell(x, \lambda) = e(x) + \lambda^\top c(x),$$

where $\lambda = (\lambda_E, \lambda_I)$ and $\lambda_E \in \mathbb{R}^{\text{nb}}$ and $\lambda_I \in \mathbb{R}^{\text{nj}}$ are the multipliers associated with the equality and inequality constraints. Figure 13.2 represents a typical

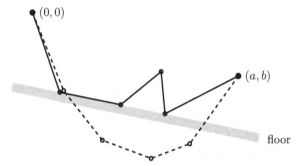

Fig. 13.2. Static equilibrium of the hanging chain

equilibrium position without (the dashed line) or with (the solid line) a floor constraint.

The Simulator

The easiest way of organizing the program is to use *direct communication* (see figure 13.3): a main program (say in the file ch.m) calls the optimization module (say in the file sqp.m), which communicates with a *simulator* (say in the file chs.m) to get information on the problem to solve. The parameter indic monitors the communication with the simulator: the calling procedure uses indic to specify the job to realize by the simulator and this one uses this parameter to inform the calling procedure on the course of the simulation (for example, to indicate failure in the required computation).

In MATLAB, the procedure chs can determine what it has to compute by looking at the number of its input and output arguments when it is called (nargin and nargout), so that there is no need to use the parameter indic on input; however, this variable is still useful on output (its meaning has been specified above). Our implementation of the simulator recognizes the following calls:

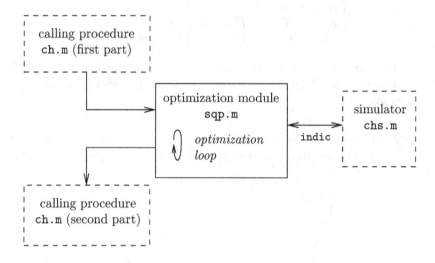

Fig. 13.3. Optimization code organized in direct communication

```
[indic] = chs(xy)
[e,ce,ci,indic] = chs(xy)
[e,ce,ci,g,ae,ai,indic] = chs(xy)
[hl,indic] = chs(xy,lmde,lmdi)
```

In the first case, the simulator just plots the chain specified by the coordinates of the free joints in $xy = x$; in the second case, it computes the function $e = e(x)$ to minimize, the constraint functions $ce = c_E(x)$ and $ci = c_I(x)$; in the third case, it computes $e(x)$, $c_E(x)$, $c_I(x)$, the gradient $g = \nabla e(x)$, and the Jacobian matrices $ae = c_E'(x)$ and $ai = c_I'(x)$; finally in the fourth case, it computes the Hessian of the Lagrangian $hl = \nabla_{xx}^2 \ell(x, \lambda)$, using the coordinates $xy = x$ and the multipliers $lmde = \lambda_E$ and $lmdi = \lambda_I$. Actually, this Hessian does not depend on λ_I since the inequality constraints are linear.

The main module defines the problem's data and calls the optimization procedure after having set its parameters.

It is good practice to write the optimization module independently of the problem to solve, translating in computer language the mathematical objects and operations that are used in this book. The obvious reason is that the module can then be used to solve other optimization problems. Since the simulator needs information on the problem defined in the main module ch (e.g., the length of the bars) and that this information cannot be passed through the optimization module, it must be passed using the **global** statement of MATLAB.

First Session

The goal of this first session is to write the main module ch.m and the simulator chs.m. To do this, just follow the description given in the previous two sections. No knowledge in optimization is required to achieve this goal. Various optimization algorithms will be implemented and tested in other sessions.

A delicate task, which is however crucial for the success of the next sessions, is to check the correctness of the simulator. Both function values (e and c) and derivatives (first $\nabla e(x)$, $c'(x)$, and second $\nabla_{xx}^2 \ell(x, \lambda)$) need to be examined. Function values are best verified by considering simple cases, for which the values of $e(x)$ and $c(x)$ are known (a chain with 2 or 3 bars of equal length, for example).

The consistency between a function and its derivatives is more delicate. In smooth optimization, a standard way of proceeding is to compare the computed derivatives with their approximations by finite differences. For a function $\varphi : \mathbb{R}^n \to \mathbb{R}^m$, one can check that the computed values $\varphi'(x) \cdot e^i = \partial \varphi(x) / \partial x_i \in \mathbb{R}^m$ (e^i denotes the ith basis vector of $\mathbb{R}^n$), for $i = 1, \ldots, n$, are close to one of their approximations

$$\frac{\varphi(x + t_i e^i) - \varphi(x)}{t_i} = \varphi'(x) \cdot e^i + O(t_i),$$

$$\frac{\varphi(x + t_i e^i) - \varphi(x - t_i e^i)}{2t_i} = \varphi'(x) \cdot e^i + O(t_i^2).$$

The second differential quotient is more precise but requires more function evaluations. The choice of t_i is known to be sensitive. It is standard [107] to take

$$t_i = \varepsilon^{1/2} \max(\tau_i, |x_i|),$$

where ε is the machine epsilon, which is the smallest positive floating point number such that $\mathrm{fl}(1 + \varepsilon) \neq 1$ (it is given by the variable eps in MATLAB), and τ_i is a typical size of x_i ($\tau_i = 1$ can be chosen for the data given in this project).

Other ways of assessing the correctness of the derivatives will be available after the implementation of the local algorithms in the sessions of §§ 14.7 and 15.4, and of the line-search in the session of § 17.4 (see the comments after the output printed by the line-search on page 318).

Notes

Section 13.5 has introduced the so-called *quotient* rates of convergence. Sometimes the convergence can only be qualified by the so-called *root* rates. These are defined, discussed and compared with the quotient rates by Ortega and Rheinbold [278].

Theorem 13.6 is not the only result on the convergence of Newton's algorithm. Of particular interest is the Kantorovich theorem, which has the nice feature of not assuming the existence of a zero of F, but proves that such a zero exists, provided some conditions hold at the first iterate z_1 ($F(z_1)$ is small, $F'(z_1)$ is nonsingular) and in some neighborhood (F' does not change too rapidly), see [202, 108].

Exercises

13.1. Find a set X defined by inequality constraints and a point $x \in X$, for which (MF-CQ) holds but not (LI-CQ).

13.2. Find optimization problems in which a solution satisfies: (i) the strong second order sufficient condition of optimality (13.10); (ii) the semi-strong second order sufficient condition of optimality (13.9) but not the strong one (13.10); (iii) the weak second order sufficient condition of optimality (13.8) but not the semi-strong one (13.9).

14 Local Methods for Problems with Equality Constraints

In this chapter, we present and study several local methods for minimizing a nonlinear function subject only to nonlinear equality constraints. This is the problem (P_E) represented in figure 14.1: Ω is an open set of $\mathbb{R}^n$, while

$$(P_E) \quad \begin{cases} \min_x f(x) \\ c(x) = 0 \\ x \in \Omega \end{cases}$$

$$\mathcal{M}_* := \{x \in \Omega : c(x) = 0\}$$

Fig. 14.1. Problem (P_E) and its feasible set

$f : \Omega \to \mathbb{R}$ and $c : \Omega \to \mathbb{R}^m$ are differentiable functions. Since we always assume that c is a submersion, which means that $c'(x)$ is surjective (or onto) for all $x \in \Omega$, the inequality $m < n$ is natural. Indeed, for the Jacobian of the constraints to be surjective, we must have $m \leq n$; but if $m = n$, any feasible point is isolated, which results in a completely different problem, for which the algorithms presented here are hardly appropriate. Therefore, a good geometrical representation of the feasible set of problem (P_E) is that of a submanifold $\mathcal{M}_*$ of $\mathbb{R}^n$, like the one depicted in figure 14.1.

There are several reasons for postponing the study of optimization problems with inequality constraints. First, we tackle difficulties and notation progressively, and prepare the intuition for the general case. Also, the reduced Hessian method (§ 14.5) has no simple equivalent form when inequalities are present. Finally, such problems arise both in their own right and as subproblems in some algorithmic approaches to solve optimization problems with inequality constraints. For instance, nonlinear interior point algorithms sometimes transform an inequality constrained problem into a sequence a equality constrained problems by introducing slack or shift variables and a logarithmic penalization (see [143, 65, 11] for examples). A good mastery of the techniques used to solve problem (P_E) is therefore helpful.

By *local methods*, we mean methods whose convergence is ensured provided the initial iterate is close enough to a solution. In this case, the algorithms presented in chapters 14 and 15 have the nice property to converge quadratically. This feature comes from the linearization of the optimality conditions. Among the quadratically convergent algorithms that have been proposed to solve problem (P_E), we have chosen to describe two of them (and some of their useful variants): Newton's method (§ 14.1) and the reduced Hessian method (§ 14.5). These are probably the most often implemented algorithms. Also, they offer a framework in which different techniques can be used: line-search and trust region globalization techniques, quasi-Newton Hessian approximations, *etc.*

When c is a submersion, the feasible set of (P_E) forms a submanifold of $\mathbb{R}^n$. However, the algorithms studied in this section do not force the iterates to stay in that manifold. For general nonlinear constraints, this would generally require too much computing time. Rather, optimality and feasibility are searched simultaneously, so that optimality is obtained in a time of the same order of magnitude as that needed to obtain feasibility in a code without optimization. This nice feature makes these algorithms very attractive in practice.

According to the first-order optimality conditions (13.1), we know that, when the constraints are qualified at a solution $x_* \in \Omega$ to (P_E), there exists a Lagrange multiplier $\lambda_* \in \mathbb{R}^m$ such that

$$\begin{cases} \nabla f(x_*) + A(x_*)^\top \lambda_* = 0 \\ c(x_*) = 0. \end{cases} \tag{14.1}$$

We have denoted by $A(x) := c'(x)$ the $m \times n$ Jacobian matrix of the constraints: the ith row of $A(x)$ is the transposed gradient $\nabla c_i(x)^\top$ of the ith constraint; hence the (i,j)th element of $A(x)$ is the partial derivative $\partial c_i / \partial x_j(x)$.

14.1 Newton's Method

The Newton Step

We have seen in chapter 13 how Newton's method can be used to solve nonlinear equations (see (13.18)) and to minimize a function (see (13.24)). For optimization problems with equality constraints, it is therefore tempting to compute the step d_k at x_k by means of a quadratic [resp. linear] approximation of the objective function [resp. constraints] at x_k. With such a method, d_k would solve or would compute a stationary point of the quadratic problem

$$\begin{cases} \min_d f'(x_k) \cdot d + \frac{1}{2} f''(x_k) \cdot d^2 \\ c(x_k) + c'(x_k) \cdot d = 0, \end{cases} \tag{14.2}$$

and the next iterate would be $x_{k+1} = x_k + d_k$. Beware of the **nonconvergence of this algorithm!** In some cases, the generated sequence moves away from a solution, no matter how close the initial iterate is to this solution[1].

The right approach consists in dealing simultaneously with the objective minimization and the constraint satisfaction, by working on the optimality conditions (14.1). Actually, these form a system of $n+m$ nonlinear equations in the $n+m$ unknowns (x_*, λ_*), a system that can be solved by Newton's method. This results is a so-called *primal-dual method*, which means that a sequence $\{(x_k, \lambda_k)\}$ is generated, in which x_k approximates a primal solution x_* and λ_k approximates the associated dual solution λ_*.

Let (x_k, λ_k) be the current primal-dual iterate. We use the notation

$$f_k := f(x_k), \quad c_k := c(x_k), \quad A_k := A(x_k) := c'(x_k), \quad \nabla_x \ell_k := \nabla_x \ell(x_k, \lambda_k),$$

and finally denote by

$$L_k := L(x_k, \lambda_k) := \nabla^2_{xx} \ell(x_k, \lambda_k)$$

the Hessian of the Lagrangian ℓ with respect to x at (x_k, λ_k). See (13.2) for a definition of the Lagrangian. Newton's method defines a step in (x, λ) at (x_k, λ_k) by linearizing the system (14.1) at (x_k, λ_k). One finds

$$\begin{pmatrix} L_k & A_k^\top \\ A_k & 0 \end{pmatrix} \begin{pmatrix} d_k \\ \mu_k \end{pmatrix} = - \begin{pmatrix} \nabla_x \ell_k \\ c_k \end{pmatrix}. \tag{14.3}$$

Given a solution (d_k, μ_k) to (14.3), the *Newton method* defines the next iterate (x_{k+1}, λ_{k+1}) by

$$x_{k+1} = x_k + d_k \quad \text{and} \quad \lambda_{k+1} = \lambda_k + \mu_k. \tag{14.4}$$

Since $\nabla_x \ell_k$ is linear with respect to λ_k, (14.3) can be rewritten as follows:

$$\begin{pmatrix} L_k & A_k^\top \\ A_k & 0 \end{pmatrix} \begin{pmatrix} d_k \\ \lambda_k^{\mathrm{QP}} \end{pmatrix} = - \begin{pmatrix} \nabla f_k \\ c_k \end{pmatrix}, \tag{14.5}$$

where we have used the notation

$$\lambda_k^{\mathrm{QP}} := \lambda_k + \mu_k.$$

The superscript 'QP' suggests the fact that, as we shall see below, λ_k^{QP} is the multiplier associated with the constraints of a quadratic problem. The next iterate (x_{k+1}, λ_{k+1}) of Newton's method is in this case

---

[1] See exercise 14.1 for an example, in which f is concave. When f is strongly convex and has a bounded Hessian, one can get convergence with line-search along the direction computed by (14.2). When f is nonconvex, convergence can still be obtained with line-search and the truncated SQP algorithm. This will be clearer with the concepts developed in chapter 17. Nevertheless, as this is shown below, the step computed by (14.2) neglects an important part of the "curvature" of problem (P_E).

$$x_{k+1} = x_k + d_k \quad \text{and} \quad \lambda_{k+1} = \lambda_k^{\text{QP}}. \tag{14.6}$$

This formulation reveals the less important role played by λ_k, compared with that of x_k. Observe indeed in (14.5) that λ_k only appears in the matrix L_k, while x_k is the linearization point of the functions defining the problem.

Osculating Quadratic Problems

Just as in the unconstrained case, the Newton equation (14.3) can be viewed as the optimality system of a quadratic problem (QP), namely

$$\begin{cases} \min_d \nabla_x \ell_k^\top d + \frac{1}{2} d^\top L_k d \\ c_k + A_k d = 0. \end{cases} \tag{14.7}$$

This one is called the *osculating quadratic problem* of (P_E) at (x_k, λ_k). If we consider (14.5) instead of (14.3), we find

$$\begin{cases} \min_d \nabla f_k^\top d + \frac{1}{2} d^\top L_k d \\ c_k + A_k d = 0, \end{cases} \tag{14.8}$$

which is another osculating quadratic problem, whose optimality system is (14.5).

The transformations from (14.3) to (14.7) and from (14.5) to (14.8) call for some comments.

1. Any linear system with a symmetric matrix having the structure of that in (14.5) (the distinguishing feature is the zero $(2,2)$ block of the matrix) can be viewed as the first order optimality conditions of the associated QP in (14.8). This point of view can be fruitful when numerical techniques to solve (14.5) are designed.

2. We know that (14.7) and (14.8) have the same primal solutions. This can also be deduced by observing that their objective functions only differ in the term $\lambda_k^\top A_k d$, which is the constant $-\lambda_k^\top c_k$ anywhere on the feasible set. However, these problems have different dual solutions. With (14.7), we obtain the step μ_k to add to the multiplier λ_k ($\lambda_{k+1} = \lambda_k + \mu_k$), while (14.8) gives directly the new multiplier ($\lambda_{k+1} = \lambda_k^{\text{QP}}$).

3. One can obtain (14.7) directly from (P_E): the constraints are linearized at the current point x_k and the objective function is a quadratic approximation of the *Lagrangian* at (x_k, λ_k) (the constant term $\ell(x_k, \lambda_k)$ of this approximation can be added to the objective function of (14.7), without changing the solution).

4. Note the difference between (14.2) and (14.8). The former takes the Hessian of the objective function; the latter uses the Hessian of the Lagrangian. The difference between these two Hessians comes from the constraint curvature (sum of the terms $(\lambda_k)_i \nabla^2 c_i(x_k)$). In order to have fast

convergence, this curvature must be taken into account. This is all the more important when f is nonconvex.

The validity of (14.7) can be justified a posteriori. Indeed the Lagrangian has a minimum in the subspace tangent to the constraints (if the second-order sufficient conditions of optimality of theorem 13.4 hold); therefore, it makes sense to minimize the quadratic approximation of this Lagrangian, subject to the linearized constraints. Since the same cannot be said of f, (14.2) appears suspect.

We can also make the following remark. To have a chance of being convergent, an algorithm should at least generate a zero displacement when starting at a solution. We see that this property is not enjoyed by (14.2). In fact, if x_k solves (P_E), then $c_k = 0$ and $\nabla f(x_k)^\top d = 0$ for all $d \in N(A_k)$; hence (14.2) amounts to minimizing $\frac{1}{2} d^\top \nabla f(x_k)^2 d$ on $N(A_k)$. If the Hessian of f is not positive semi-definite in the space tangent to the constraints, which may well happen, then $d = 0$ does not solve (14.2) (unbounded problem). In contrast, (14.3) and (14.5) do enjoy this minimal property, insofar as the matrix appearing in these linear systems is nonsingular (see proposition 14.1 below and the comments that follow definition 14.2).

5. No equivalence holds between (14.5) and (14.8): the minimization problem (14.8) may have a stationary point (hence satisfying (14.5)) but no minimum (unbounded problem). Equivalence does hold between (14.5) and (14.8) – or (14.3) and (14.7) – if L_k satisfies

$$d^\top L_k d > 0, \quad \text{for all nonzero } d \text{ in } N(A_k).$$

In fact, in this case, $d \mapsto \nabla f_k^\top d + \frac{1}{2} d^\top L_k d$ is quadratic strictly convex on the affine subspace $\{d : c_k + A_k d = 0\}$. Therefore (14.8) has a unique solution, which solves the optimality equations (14.5). These equations have no other solution (proposition 14.1).

6. From a numerical point of view, the osculating quadratic problem shows that the Newton equations can be solved by minimization algorithms. For large-scale problems, the reduced conjugate gradient algorithm is often used: one computes a restoration step r_k that is feasible for (14.8) (hence satisfying $c_k + A_k r_k = 0$) and then one generates directions in the null space of A_k. We shall come back to this issue in § 14.4 and § 17.2.

Regular Stationary Points

The Newton step can be computed if the linear system that defines it, (14.5) say, is nonsingular. The next proposition gives conditions equivalent to this nonsingularity.

Proposition 14.1 (regular stationary point). *Let A be an $m \times n$ matrix, L be an $n \times n$ symmetric matrix, and*

$$K := \begin{pmatrix} L & A^\top \\ A & 0 \end{pmatrix}. \tag{14.9}$$

Then the following conditions are equivalent:
  (i) K *is nonsingular;*
  (ii) A *is surjective and any* $d \in N(A)$ *satisfying* $Ld \in N(A)^\perp$ *vanishes;*
  (iii) A *is surjective and* $Z^{-\top}LZ^-$ *is nonsingular for some (or any)* $n \times (n{-}m)$ *matrix* Z^- *whose columns form a basis of* $N(A)$.

Proof. $[(i) \Rightarrow (ii)]$ Since K is surjective, so is A. On the other hand, if $d \in N(A)$ satisfies $Ld \in N(A)^\perp = R(A^\top)$, there exists $\mu \in \mathbb{R}^m$ such that $(d, \mu) \in N(K)$, so that $d = 0$.

$[(ii) \Rightarrow (iii)]$ Let Z^- be a matrix like in (iii). If $Z^{-\top}LZ^-u = 0$ for some $u \in \mathbb{R}^{n-m}$, $d := Z^-u \in N(A)$ and $Ld \in N(Z^{-\top}) = R(Z^-)^\perp = N(A)^\perp$, so that $Z^-u = 0$ by (ii). Now $u = 0$ by the injectivity of Z^-.

$[(iii) \Rightarrow (i)]$ It suffices to show that K is injective. Take (d, μ) in its null space. Then $Ad = 0$ and $Ld + A^\top\mu = 0$, which imply $d \in N(A)$ (or $d = Z^-u$ for some u) and $Z^{-\top}Ld = 0$. From (iii), $u = 0$ and $d = 0$. Thus $A^\top\mu = 0$, and $\mu = 0$ by the injectivity of $A^\top$. □

Note that the nonsingularity of L and the surjectivity of A are not sufficient to guarantee the equivalent conditions (i)–(iii). For a counter-example consider

$$L = \begin{pmatrix} 1 & 0 \\ 0 & -1 \end{pmatrix} \quad \text{and} \quad A = \begin{pmatrix} 1 & -1 \end{pmatrix}.$$

The vector $\begin{pmatrix} 1 & 1 & -1 \end{pmatrix}^\top$ is in the null space of K. On the other hand, when A is surjective, condition (iii) is obviously satisfied if $Z^{-\top}LZ^-$ is positive definite, and a fortiori if L is positive definite. Exercise 14.2 gives more information on the spectrum of the matrix K: it is claimed in particular that, when A is surjective, the matrix K always has m negative and m positive eigenvalues (for the intuition, consider the case when $n = m = 1$ and observe that the determinant of K is negative; hence there is always one negative and one positive eigenvalue).

A consequence of exercise 14.2 is that a quadratic function, whose Hessian is the matrix K with a surjective A, is never bounded below. If this function has a stationary point, it is not a minimizer, but a saddle-point. The symmetry of K suggests, however, that a linear system based on this matrix expresses the optimality conditions of a quadratic minimization problem, but this one needs linear equality constraints (using the matrix A) to have a chance of being well-posed: see (14.8) for an example. Actually, a stationary point of this constrained quadratic problem will be a constrained minimizer if and only if the matrix L is positive semi-definite on the null space of A.

The discussion above leads us to introduce the following definition.

Definition 14.2 (regular stationary point). A stationary point (x_*, λ_*) of (P_E) is said to be *regular* if $A_* := c'(x_*)$ is surjective and if $Z_*^{-\top} L_* Z_*^-$ is nonsingular, for some (or any) $n \times (n-m)$ matrix Z_*^- whose columns form a basis of $N(A_*)$.

A regular stationary point is necessarily isolated: it has a neighborhood containing no other stationary point (see exercise 14.3 for a precise statement). Also, a strong primal-dual solution (x_*, λ_*) to (P_E) satisfying (LI-CQ) (i.e., A_* surjective) is a regular stationary point. Indeed, in this case $d^\top L_* d > 0$ for all nonzero $d \in N(A_*)$, so that the so-called *reduced Hessian of the Lagrangian*

$$H_* := Z_*^{-\top} L_* Z_*^-$$

is positive definite. The $(n-m) \times (n-m)$ matrix H_* clearly depends on the choice of the matrix Z_*^-. In some cases, it can be viewed as a Hessian of some function (see exercise 14.4).

The Algorithm

We conclude this section by giving a precise description of Newton's algorithm to solve problem (P_E). As already mentioned, the method generates a primal-dual sequence $\{(x_k, \lambda_k)\} \subset \mathbb{R}^n \times \mathbb{R}^m$.

NEWTON'S ALGORITHM FOR (P_E):

Choose an initial iterate $(x_1, \lambda_1) \in \mathbb{R}^n \times \mathbb{R}^m$.
Compute $c(x_1)$, $\nabla f(x_1)$, and $A(x_1)$.
Set $k = 1$.
1. Stop if $\nabla \ell(x_k, \lambda_k) = 0$ and $c(x_k) = 0$ (optimality is reached).
2. Compute $L(x_k, \lambda_k)$ and find a primal-dual stationary point of the quadratic problem (14.8), i.e., a solution $(d_k, \lambda_k^{\mathrm{QP}})$ to (14.5).
3. Set $x_{k+1} := x_k + d_k$ and $\lambda_{k+1} := \lambda_k^{\mathrm{QP}}$.
4. Compute $c(x_{k+1})$, $\nabla f(x_{k+1})$, and $A(x_{k+1})$.
5. Increase k by 1 and go to 1.

In practice, the stopping criterion in step 1 would test whether $\|\nabla \ell(x_k, \lambda_k)\|$ and $\|c(x_k)\|$ are sufficiently small. This remark holds for all the algorithms of this part of the book.

Before analyzing the convergence properties of this algorithm in § 14.3, we introduce some notation that makes it easier to understand some interesting variants of the method and highlights the structure of the Newton step d_k. How to compute this step is dealt with in § 14.4.

14.2 Adapted Decompositions of $\mathbb{R}^n$

A General Framework

Suppose that c is a submersion on the open set $\Omega \subset \mathbb{R}^n$. Then, the set

$$\mathcal{M}_x := \{y \in \Omega : c(y) = c(x)\}$$

is a submanifold of $\mathbb{R}^n$ with dimension $n-m$ (for the few concepts of differential geometry that we use, we refer the reader to [344, 51, 84, 112] for example). Intuitively, the tangent space to $\mathcal{M}_x$ at x is the set of directions of $\mathbb{R}^n$ along which c does not vary at the first order; it is therefore the null space of the Jacobian matrix

$$A_x := A(x) := c'(x)$$

of c at x. This null space and a complementary subspace decompose $\mathbb{R}^n$ into two subspaces, which make the description and interpretation of the algorithms easier. This decomposition, which we now describe, is shown in figure 14.2.

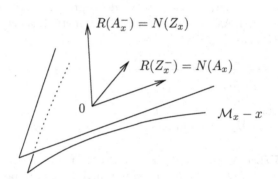

Fig. 14.2. Adapted decomposition of $\mathbb{R}^n$

Consider first the tangent subspace $N(A_x)$. We shall often assume that we have a smooth mapping

$$Z^- : \Omega \to \mathbb{R}^{n \times (n-m)} : x \mapsto Z_x^- := Z^-(x),$$

such that for all $x \in \Omega$, Z_x^- is a *basis* of the tangent subspace. We mean by this that the columns of Z_x^- form a basis of $N(A_x)$ or equivalently:

$$\forall x \in \Omega, \ Z_x^- \text{ is } n \times (n-m) \text{ injective and } A_x Z_x^- = 0. \tag{14.10}$$

Besides, since A_x is surjective, it has a *right inverse*: an $n \times m$ matrix A_x^- satisfying $A_x A_x^- = I_m$. We shall always assume that A_x^- is the value at x of a smooth mapping

$$A^- : \Omega \to \mathbb{R}^{n \times m} : x \mapsto A_x^- := A^-(x).$$

Therefore

$$\forall x \in \Omega, \ A_x^- \text{ is } n \times m \text{ injective and } A_x A_x^- = I_m. \tag{14.11}$$

The range space of A_x^- is a subspace complementary to $N(A_x)$, because $R(A_x^-) \cap N(A_x) = \{0\}$ and $\dim R(A_x^-) + \dim N(A_x) = m + (n-m) = n$.

Thus, $\mathbb{R}^n$ can be written as the direct sum of the subspaces spanned by the columns of Z_x^- and the columns of A_x^-: for all $x \in \Omega$,

$$\mathbb{R}^n = R(Z_x^-) \oplus R(A_x^-).$$

Lemma 14.3 (adapted decomposition of $\mathbb{R}^n$). *Let $Z^- : \Omega \to \mathbb{R}^{n \times (n-m)}$ and $A^- : \Omega \to \mathbb{R}^{n \times m}$ be mappings satisfying respectively (14.10) and (14.11). Then there exists a unique mapping*

$$Z : \Omega \to \mathbb{R}^{(n-m) \times n} : x \mapsto Z_x := Z(x)$$

satisfying for all $x \in \Omega$:

$$Z_x A_x^- = O_{(n-m) \times m}, \tag{14.12}$$

$$Z_x Z_x^- = I_{n-m}. \tag{14.13}$$

This mapping Z is also characterized by the following identity, valid for all $x \in \Omega$:

$$I = A_x^- A_x + Z_x^- Z_x. \tag{14.14}$$

Proof. It can be easily checked that the matrix $X_x = \begin{pmatrix} A_x^- & Z_x^- \end{pmatrix}$ is nonsingular, from which follow the existence and uniqueness of Z_x satisfying (14.12) and (14.13). Next observe from (14.10), (14.11), (14.12) and (14.13) that the matrix $Y_x = \begin{pmatrix} A_x^\top & Z_x^\top \end{pmatrix}^\top$ is the inverse of X_x, since $Y_x X_x = I_n$. Then (14.14) is exactly the identity $X_x Y_x = I_n$. Conversely, this last identity determines Y_x, hence Z_x. □

Figure 14.2 summarizes the properties of the operators A_x, Z_x^-, A_x^-, and Z_x. The manifold $\mathcal{M}_x$ is translated by $-x$, so that the linearization point x is at the origin. To find one's way in this family of operators, a mnemonic trick is welcome: the operators A_x^- and Z_x^-, with a minus exponent, are injective and right inverses; while the operators A_x and Z_x, without a minus exponent, are surjective and left inverses.

Using the identity (14.14), we have for every vector $v \in \mathbb{R}^n$,

$$v = A_x^- A_x v + Z_x^- Z_x v.$$

This identity allows us to decompose a vector v into its *longitudinal* component $Z_x^- Z_x v$, tangent at x to the manifold $\mathcal{M}_x$, and its *transversal* component

$A_x^- A_x v$, which lies in the complementary space $R(A_x^-)$. In view of our preceding development, this decomposition is well-defined, once the matrices Z_x^- and A_x^- have been given. Observe also that $A_x^- A_x$ and $Z_x^- Z_x = I - A_x^- A_x$ are oblique projectors on $R(A_x^-)$ and $R(Z_x^-)$. The orthogonal projectors on these subspaces are

$$A_x^- (A_x^{-\top} A_x^-)^{-1} A_x^{-\top} = I - Z_x^\top (Z_x Z_x^\top)^{-1} Z_x$$

and

$$Z_x^- (Z_x^{-\top} Z_x^-)^{-1} Z_x^{-\top} = I - A_x^\top (A_x A_x^\top)^{-1} A_x.$$

Below, we give some formulae for the computation of the matrices Z_x^- and A_x^- satisfying properties (14.10) and (14.11). These formulae use inverses of matrices, which need not be computed explicitly in algorithms. Likewise, the matrices Z_x^- and A_x^- need not be computed explicitly. What matters is their action (or the action of their transpose) on a vector, which can generally be obtained by solving a linear system. For example, as we shall see, the right inverse A_x^- is usually applied to the vector $c(x)$, whereas $A_x^{-\top}$ is often applied to $\nabla f(x)$.

We now proceed by giving examples of matrices Z_x^-, A_x^-, and Z_x that are frequently used in the algorithms.

Decomposition by Partitioning (or Direct Elimination)

This decomposition has its roots in *optimal control* problems (see § 1.2.2 and § 1.14 for examples of such problems), in which the variables $x = (y, u)$ are partitioned in *state variables* $y \in \mathbb{R}^m$ and *control variables* $u \in \mathbb{R}^{n-m}$. The Jacobian A_x is likewise partitioned in

$$A_x = (B_x \quad N_x),$$

where B_x is an $m \times m$ matrix giving the derivatives of the constraints with respect to the state variables. In the regular case, B_x is nonsingular. Such a decomposition is also used in linear optimization.

The decomposition of $\mathbb{R}^n$ given below is often used for large-scale optimization problems, in which a fixed partitioning of the variables leads to a nonsingular matrix B_x. Note that it is always possible to make a partition of the surjective matrix A_x as above, leading to a nonsingular matrix B_x, provided some permutation of the columns of A_x is performed. There are linear solvers that can select the columns of A_x in order to form a matrix B_x with a reasonably well optimized condition number.

In the framework just described the matrix

$$Z_x^- = \begin{pmatrix} -B_x^{-1} N_x \\ I_{n-m} \end{pmatrix} \tag{14.15}$$

is well defined and satisfies properties (14.10), while the matrix

$$A_x^- = \begin{pmatrix} B_x^{-1} \\ 0 \end{pmatrix} \tag{14.16}$$

is also well defined and satisfies (14.11). The mapping Z given by lemma 14.3 has for its value at x:

$$Z_x = \begin{pmatrix} O & I_{n-m} \end{pmatrix}.$$

Now, let us highlight some other links with the optimal control framework. Assuming that c is of class C^1, the nonsingularity of B_x implies that y, the solution to $c(y, u) = c(x)$ for fixed x, is an implicit function of u: $y = y(u)$ and $c(y(u), u) = c(x)$ for all u in a nonempty open set. Then the basis Z_x^- above is obtained by differentiating the parametrization $u \mapsto (y(u), u)$ of the manifold $\mathcal{M}_x := \{x' \in \Omega : c(x') = c(x)\}$. On the other hand, the displacement

$$-A_x^- c(x) = \begin{pmatrix} -B_x^{-1} c(x) \\ 0 \end{pmatrix}$$

is a Newton step to solve the state equation $c(y, u) = 0$, with fixed control u.

From a computational point of view, we see that, to evaluate $A_x^- c(x)$, it is sufficient to solve the linear system $B_x v = c(x)$, whose solution v gives the first m components of $A_x^- c(x)$. This is less expensive than computing B_x^{-1} explicitly! Likewise, the first m components h of $Z_x^- u$ can be obtained by solving the linear system $B_x h = -N_x u$.

Orthogonal Decomposition

The orthogonal decomposition is obtained by choosing a right inverse A_x^-, whose columns are perpendicular to $N(A_x)$ (they cannot be orthonormal in general), and a tangent basis Z_x^- with orthonormal columns. The condition on A_x^- implies that this matrix has the form $A_x^- = A_x^\top S$, for some matrix S. Since $A_x A_x^- = I$ must hold, A_x^- is necessarily given by

$$A_x^- = A_x^\top (A_x A_x^\top)^{-1}. \tag{14.17}$$

Now, let Z_x^- be an arbitrary orthonormal basis of $N(A_x)$: $A_x Z_x^- = 0$ and $Z_x^{-\top} Z_x^- = I_{n-m}$. To get the matrix Z_x provided by lemma 14.3, let us multiply both sides of the identity (14.14) to the left by $Z_x^{-\top}$, using (14.17). Necessarily

$$Z_x = Z_x^{-\top}.$$

One way of computing the matrices A_x^- and Z_x^- just described, is to use the QR factorization of $A_x^\top$ (see [170] for example):

$$A_x^\top = \begin{pmatrix} Y_x^- & Z_x^- \end{pmatrix} \begin{pmatrix} R_x \\ O \end{pmatrix} = Y_x^- R_x, \tag{14.18}$$

where $\begin{pmatrix} Y_x^- & Z_x^- \end{pmatrix}$ is an orthogonal matrix and R_x is upper triangular. The matrix R_x is nonsingular since A_x is assumed to be surjective. Then, the last $n-m$ columns Z_x^- of the orthogonal factor form an orthonormal basis of $R(Y_x^-)^\perp = R(A_x^\top)^\perp$, which is indeed the null space of A_x. Furthermore, (14.18) and the nonsingularity of R_x show that the columns of $Y_x^- \in \mathbb{R}^{n \times m}$ span $R(A_x^\top) = N(A_x)^\perp$. Since, by multiplying the extreme sides of (14.18) to the left by $Y_x^{-\top}$, it follows that $A_x Y_x^- = R_x^\top$ or $A_x Y_x^- R_x^{-\top} = I_m$, the right inverse of A_x given by (14.17) is necessarily

$$A_x^- = Y_x^- R_x^{-\top}.$$

The orthogonal decomposition just described has the advantage of being numerically stable and of computing a perfectly well-conditioned basis Z_x^-. The QR factorization can be carried out by using Givens rotations or with at most m Householder reflections. Therefore, this is a viable approach when m is not too large.

Oblique Decomposition

Let M be a matrix that is nonsingular on the null space of A_x, meaning that $Z_x^{-\top} M Z_x^-$ is nonsingular for some basis Z_x^- of $N(A_x)$ (this property of M does not depend on the choice of Z_x^-, see proposition 14.1). Then, one can associate with M a right inverse of A_x, defined as follows. Take $v \in \mathbb{R}^m$. Then the quadratic problem in d

$$\begin{cases} \min_d \frac{1}{2} d^\top M d \\ A_x d = v \end{cases} \tag{14.19}$$

has a unique stationary point, which satisfies the optimality conditions

$$\begin{cases} M d + A_x^\top \lambda = 0 \\ A_x d = v, \end{cases} \tag{14.20}$$

for some multiplier $\lambda \in \mathbb{R}^m$. We see that d depends linearly on v. Denoting by $\widehat{A}_x^-$ the matrix representing this linear mapping, i.e., $d = \widehat{A}_x^- v$, the second equation in (14.20) shows that $\widehat{A}_x^-$ is a right inverse of A_x. This matrix $\widehat{A}_x^-$ will be useful to write a simple expression of the Newton displacement to solve (P_E).

An explicit expression of $\widehat{A}_x^-$ can be given by using a basis Z_x^- of the null space of A_x and a right inverse A_x^- of A_x. Then (14.14) and (14.20) show that $d = A_x^- v + Z_x^- u$ for some $u \in \mathbb{R}^{n-m}$. By premultiplying both sides of the first equation of (14.20) by $Z_x^{-\top}$, we obtain $u = -(Z_x^{-\top} M Z_x^-)^{-1} Z_x^{-\top} M A_x^- v$. Finally

$$\widehat{A}_x^- = \left(I - Z_x^- \left(Z_x^{-\top} M Z_x^- \right)^{-1} Z_x^{-\top} M \right) A_x^-. \tag{14.21}$$

Even though A_x^- and Z_x^- appear in this formula, $\widehat{A}_x^-$ does not depend on them (from its definition). From lemma 14.3, there corresponds to the operators Z_x^-

and $\widehat{A}_x^-$ a unique matrix $\widehat{Z}_x$ such that $\widehat{Z}_x\widehat{A}_x^- = 0$ and $\widehat{Z}_x Z_x^- = I$. To give an analytic expression of $\widehat{Z}_x$, observe first that from (14.21), one has

$$Z_x^{-\top}M\widehat{A}_x^- = 0, \tag{14.22}$$

which expresses the fact that the range spaces $R(Z_x^-)$ and $R(\widehat{A}_x^-)$ are "orthogonal" with respect to the matrix M (this would correspond to a proper notion of orthogonality if the matrix M were positive definite). It is then easy to check that

$$\widehat{Z}_x = \left(Z_x^{-\top}MZ_x^-\right)^{-1}Z_x^{-\top}M$$

satisfies the required properties.

To conclude, note that $\widehat{A}_x^-$ may not exist if M is singular on the null space of A_x. Here is a counter-example with $n = 2$ and $m = 1$:

$$M = \begin{pmatrix} 1 & 1 \\ 1 & 0 \end{pmatrix}, \quad A_x = \begin{pmatrix} 1 & 0 \end{pmatrix}, \quad \text{and} \quad Z_x^- = \begin{pmatrix} 0 \\ 1 \end{pmatrix}.$$

Since $Z_x^{-\top}M = A_x$, $\widehat{A}_x^-$ cannot satisfy both $A_x\widehat{A}_x^- = I$ and (14.22). Observe finally that the right inverses (14.16) and (14.17) obtained previously can be recovered from $\widehat{A}_x^-$ by an appropriate choice of M; this is the subject of exercise 14.7.

14.3 Local Analysis of Newton's Method

Local Convergence

In this section, we study the local convergence of the Newton algorithm to solve problem (P_E), introduced in § 14.1. We use the notation

$$A_* = A(x_*) \quad \text{and} \quad L_* = L(x_*, \lambda_*).$$

Quadratic convergence of the primal-dual sequence $\{(x_k, \lambda_k)\}$ will be shown thanks to theorem 13.6. We shall also use proposition 14.1, whose conditions (i)-(iii) imply that the constraints are qualified at the solution x_* in the sense (LI-CQ):

$$A_* \text{ is surjective.} \tag{14.23}$$

A consequence of proposition 14.1 is that, when (x_*, λ_*) is a regular stationary point, the system (14.3) or (14.5) has a unique solution for (x_k, λ_k) close to (x_*, λ_*). Therefore Newton's method is well defined in the neighborhood of regular stationary points.

Theorem 14.4 (convergence of Newton's algorithm). *Suppose that f and c are of class C^2 in a neighborhood of a regular stationary point x_* of (P_E), with associated multiplier λ_*. Then, there exists a neighborhood V of*

(x_*, λ_*) *such that, if the first iterate* $(x_1, \lambda_1) \in V$, *the Newton algorithm defined in* § 14.1 *is well-defined and generates a sequence* $\{(x_k, \lambda_k)\}$ *converging superlinearly to* (x_*, λ_*). *If* f'' *and* c'' *are Lipschitzian in a neighborhood of* x_*, *the convergence of the sequence is quadratic.*

Proof. The result is obtained by applying theorem 13.6 with $z = (x, \lambda)$ and

$$F(z) = \begin{pmatrix} \nabla f(x) + A(x)^\top \lambda \\ c(x) \end{pmatrix}.$$

Clearly, F is of class C^1 in a neighborhood of $z_* = (x_*, \lambda_*)$ and $F'(z_*)$ is non-singular (from proposition 14.1). The superlinear convergence of $\{(x_k, \lambda_k)\}$ to (x_*, λ_*) follows if (x_1, λ_1) is close enough to (x_*, λ_*). If f'' and c'' are Lipschitzian near x_*, so is F' near z_*, and the quadratic convergence of $\{(x_k, \lambda_k)\}$ follows.   □

This theorem tells us that Newton's algorithm makes no distinction between stationary points, provided they are regular in the sense of definition 14.2. The iterates are indeed attracted by such a point, even if it is not a local minimum of (P_E); in particular it can be a maximum. The reason of this property comes from the fact that Newton's algorithm is essentially a method to solve nonlinear equations (here the optimality conditions of (P_E)). When one tries to find a minimizer, this is not a nice property. We shall see, however, that the techniques of chapter 17 tends to overcome this undesirable feature.

Note that the quadratic convergence of the sequence $\{(x_k, \lambda_k)\}$ by no means implies that of $\{x_k\}$ (see exercise 14.8). However, we shall see in chapter 15 (theorem 15.7) that $\{x_k\}$ does converge superlinearly. On the other hand, there are versions of Newton's method that guarantee the quadratic convergence of the primal sequence $\{x_k\}$. Here is an example of such an algorithm.

A Primal Version of the Newton Algorithm

It has already been observed that, in Newton's method, λ_k plays a less crucial role than x_k in the computation of the next iterate (x_{k+1}, λ_{k+1}). If, instead of letting the sequences $\{x_k\}$ and $\{\lambda_k\}$ be generated independently, the dual iterate λ_k is computed from the primal iterate x_k, by means of a function $x \mapsto \lambda(x)$, i.e.,

$$\lambda_k = \lambda(x_k),$$

the algorithm becomes completely primal. Indeed, then the knowledge of x_k entirely determines the next iterate x_{k+1}. We shall show below that the function $\lambda(\cdot)$ can be chosen in such a way that the convergence of $\{x_k\}$ will be quadratic, under natural assumptions. A possible candidate for that function is the *least-squares multiplier*:

$$\lambda^{\text{LS}}(x) := -A^-(x)^\top \nabla f(x), \qquad (14.24)$$

where $A^-(x)$ is a right inverse of $A(x)$. One speaks of least-squares multiplier because $\lambda^{\text{LS}}(x)$ minimizes in λ a weighted ℓ_2 norm of $\nabla_x \ell(x, \lambda)$ (see exercise 14.9).

Let us make precise the algorithm under investigation.

PRIMAL VERSION OF NEWTON'S ALGORITHM FOR (P_E):

Choose an initial iterate $x_1 \in \mathbb{R}^n$.
Compute $c(x_1)$, $\nabla f(x_1)$, and $A(x_1)$.
Set $k = 1$.

1. Compute $\lambda_k = \lambda(x_k)$.
2. Stop if $\nabla \ell(x_k, \lambda_k) = 0$ and $c(x_k) = 0$ (optimality is reached).
3. Compute $L(x_k, \lambda_k)$ and find a solution $(d_k, \lambda_k^{\text{QP}})$ to the linear system

$$\begin{pmatrix} L(x_k, \lambda_k) & A(x_k)^\top \\ A(x_k) & 0 \end{pmatrix} \begin{pmatrix} d_k \\ \lambda_k^{\text{QP}} \end{pmatrix} = - \begin{pmatrix} \nabla f(x_k) \\ c(x_k) \end{pmatrix}. \qquad (14.25)$$

4. Set $x_{k+1} := x_k + d_k$.
5. Compute $c(x_{k+1})$, $\nabla f(x_{k+1})$, and $A(x_{k+1})$.
6. Increase k by 1 and go to 1.

We have used the same notation λ_k^{QP} for the dual solution to (14.25) and (14.5), although their values are different, since here λ_k depends on x_k. Note that although λ_k^{QP} is computed, it has no influence on the value of x_{k+1}.

The next theorem analyses the local convergence of this algorithm.

Theorem 14.5 (convergence of a primal version of Newton's algorithm). *Suppose that f and c are of class C^2 in a neighborhood of a regular stationary point x_* of (P_E), with associated multiplier λ_*. Suppose also that the function $\lambda(\cdot)$ used to set the value of λ_k satisfies $\lambda(x_*) = \lambda_*$ and is continuous at x_*. Then, there exists a neighborhood V of x_* such that, if the first iterate $x_1 \in V$, the above primal version of Newton's algorithm is well-defined, generates a sequence $\{x_k\}$ converging superlinearly to x_*, and $\lambda_k^{\text{QP}} - \lambda_* = o(\|x_k - x_*\|)$. If furthermore f'' and c'' are Lipschitzian in a neighborhood of x_* and if there is a positive constant C such that*

$$\|\lambda(x) - \lambda_*\| \le C\|x - x_*\|, \quad \text{for } x \text{ near } x_*,$$

then the convergence of $\{x_k\}$ is quadratic and $\lambda_k^{\text{QP}} - \lambda_* = O(\|x_k - x_*\|^2)$.

Proof. We mimic the argument used in the proof of theorem 13.6. With the notation

$$F(x, \nu) := \begin{pmatrix} \nabla_x \ell(x, \nu) \\ c(x) \end{pmatrix},$$

and $\mu_k := \lambda_k^{\mathrm{QP}} - \lambda_*$, the linear system (14.25) can be written

$$F'(x_k, \lambda_k) \begin{pmatrix} d_k \\ \mu_k \end{pmatrix} = -F(x_k, \lambda_*).$$

If x_k is in some neighborhood of the regular stationary point x_*, with associated multiplier λ_*, λ_k is near λ_* (continuity of $\lambda(\cdot)$ at x_*). Furthermore, $F'(x_k, \lambda_k) = F'(x_k, \lambda(x_k))$ is nonsingular (see proposition 14.1) and has a bounded inverse on that neighborhood. With the notation $z_{k+1} := (x_{k+1}, \lambda_k^{\mathrm{QP}})$, $z_{k,*} := (x_k, \lambda_*)$, and $z_* := (x_*, \lambda_*)$, and the fact that f and c are of class C^2, one has

$$\begin{aligned} z_{k+1} - z_* &= z_{k,*} - z_* - F'(x_k, \lambda_k)^{-1} F(x_k, \lambda_*) \\ &= F'(x_k, \lambda_k)^{-1} \Big(F'(x_k, \lambda_k)(z_{k,*} - z_*) - F(z_*) \\ &\qquad - \int_0^1 F'(x_* + t(x_k - x_*), \lambda_*) \cdot (z_{k,*} - z_*) \, \mathrm{d}t \Big). \end{aligned}$$

Using $F(z_*) = 0$ and taking norms,

$$\|z_{k+1} - z_*\| \leq C' \Big(\int_0^1 \|F'(x_k, \lambda_k) - F'(x_* + t(x_k - x_*), \lambda_*)\| \, \mathrm{d}t \Big) \|x_k - x_*\|,$$

where C' is a positive constant. Now, since f'', c'', and λ are continuous at x_*, $F'(\cdot, \lambda(\cdot))$ is continuous at x_* and the last estimate gives $z_{k+1} - z_* = o(\|x_k - x_*\|)$, implying the superlinear convergence of x_k to x_* and $\lambda_k^{\mathrm{QP}} - \lambda_* = o(\|x_k - x_*\|)$. If furthermore f'' and c'' are Lipschitzian near x_* and $\lambda(x) - \lambda_* = O(\|x - x_*\|)$, one has $z_{k+1} - z_* = O(\|x_k - x_*\|^2)$, which means that the convergence of $\{x_k\}$ is now quadratic and that $\lambda_k^{\mathrm{QP}} - \lambda_* = O(\|x_k - x_*\|^2)$. □

14.4 Computation of the Newton Step

In this section, we describe three ways of computing the Newton step d_k and the associated multiplier λ_k^{QP}: the direct inversion approach, the dual approach, and the reduced system approach. We are interested both in analytic expressions of $(d_k, \lambda_k^{\mathrm{QP}})$ and computational issues. Each of these methods has its own advantages and drawbacks. It is the last one that most highlights the structure of the Newton step. In each case, one has to find a solution to (14.5), which is recalled here for convenience:

$$\begin{pmatrix} L_k & A_k^\top \\ A_k & 0 \end{pmatrix} \begin{pmatrix} d_k \\ \lambda_k^{\mathrm{QP}} \end{pmatrix} = - \begin{pmatrix} \nabla f_k \\ c_k \end{pmatrix}. \tag{14.26}$$

Below, the matrix of this linear system is supposed nonsingular (see proposition 14.1 for conditions ensuring this property), which implies that A_k is surjective.

The Direct Inversion Approach

The most straightforward approach for computing the Newton step is to consider the linear system (14.26) as a whole, without exploiting its block structure. One should not lose sight of the dimension $n + m$ of this linear system, which can be quite large in practice. Therefore, to make this approach attractive the problem needs to have small dimensions or to have sparse matrices L_k and A_k that can be taken into account. Using this approach could also be a naive but rapid way of computing $(d_k, \lambda_k^{\mathrm{QP}})$ in a personal program, using matrix oriented languages like MATLAB or SCILAB, for instance.

As regards the numerical techniques used to solve the full linear system, observe that, although the matrix in (14.26) is symmetric, it is never positive definite, even at a strong solution to problem (P_E) (see exercise 14.2). Therefore, a Cholesky factorization or conjugate gradient iterations are not adequate algorithms to solve this linear system! Direct linear solvers (i.e., those that factorize the matrix in (14.26)) can be considered, in particular when they can take advantage of the possible sparsity of A_k and L_k. The methods of Bunch and Kaufman [56] for the dense case or the MA27/MA47 solvers of Duff and Reid [114, 115, 116] for the sparse case are often employed. For large-scale problems, iterative solvers with preconditioners have also been developed, see for example [53, 20, 315, 358, 333].

The Dual Approaches

The dual approaches (sometimes called range-space approaches) need to have nonsingular matrices L_k and $A_k L_k^{-1} A_k^\top$. This condition is certainly satisfied if L_k is positive definite (remember that A_k is always assumed surjective in this section).

In the dual approach, the value of d_k is given as a function of λ_k^{QP}, using the first equation of (14.26):

$$d_k = -L_k^{-1}(\nabla f_k + A_k^\top \lambda_k^{\mathrm{QP}}). \tag{14.27}$$

Substituting this expression in the second equation of (14.26) gives the value of the QP multiplier, which is the solution to the linear system

$$(A_k L_k^{-1} A_k^\top)\lambda_k^{\mathrm{QP}} = -A_k L_k^{-1}\nabla f_k + c_k. \tag{14.28}$$

A way of solving (14.26) is then to consider the two linear systems (14.28) and (14.27) one after the other: once λ_k^{QP} has been determined by (14.28), d_k can be evaluated by (14.27). The computational effort depends on how these systems are solved, which should be a consequence of the problem size and structure. If direct solvers are used, one can give a rapid count of the number of linear systems to solve: $m+1$ with the $n \times n$ matrix L_k and one with the $m \times m$ matrix $A_k L_k^{-1} A_k^\top$. Indeed, the calculation can be organized as follows: first, one computes $L_k^{-1} A_k^\top$ and $L_k^{-1}\nabla f_k$; next, λ_k^{QP} is evaluated by solving (14.28); finally, d_k is obtained by (14.27) without having to solve any additional linear system.

When L_k is positive definite, λ_k^{QP} in (14.28) maximizes the dual function associated with the osculating quadratic problem (14.8), which is the function (see also part II)

$$\lambda \mapsto \min_d \left(\frac{1}{2} d^\top L_k d + \nabla f_k^\top d + \lambda^\top (c_k + A_k d) \right). \tag{14.29}$$

On the other hand, d_k given by (14.27) is the solution to this minimization problem in (14.29) with $\lambda = \lambda_k^{\mathrm{QP}}$. This viewpoint gives its name to the approach. It also suggests other ways of solving (14.26), which are often interesting for very large-scale problems such as the Stokes equations in fluid mechanics (see [243] and references therein). We briefly discuss these approaches below.

The *Uzawa algorithm* [13, 134] generates a sequence of multipliers λ converging to λ_k^{QP}. For each λ, the minimization problem in (14.29) is solved, which provides an approximation d of the solution d_k. Next the multiplier is updated by a steepest ascent step on the dual function: $\lambda_+ := \lambda + \alpha(c_k + A_k d)$, where $\alpha > 0$ is an "appropriate" stepsize. This first order method in λ is sometimes too slow. One way of accelerating it in this simple quadratic setting is to use the conjugate gradient (CG) algorithm on the dual function, which is equivalent to solving the linear system (14.28) by CG. Each CG iteration normally requires an accurate solution to a linear system with the matrix L_k, although inexact solution can also be considered (see for example [355]).

Another way of accelerating the Uzawa procedure described above is to substitute in (14.29) the Lagrangian by the augmented Lagrangian (see § 16.3):

$$\lambda \mapsto \min_d \left(\frac{1}{2} d^\top L_k d + \nabla f_k^\top d + \lambda^\top (c_k + A_k d) + \frac{r}{2} \|c_k + A_k d\|_2^2 \right), \tag{14.30}$$

where $r > 0$ is a parameter. The algorithm is similar: $\lambda_+ := \lambda + r(c_k + A_k d)$, where d is now the solution to the minimization problem in (14.30). See [134] for more details.

Time saving is also possible by avoiding an exact minimization of the problem in (14.29) or (14.30) before updating the multiplier (see [303, 118, 54, 92] for instance).

In conclusion, the dual approaches can be appropriate when L_k and $A_k L_k^{-1} A_k^\top$ are nonsingular and a linear system with the matrix L_k is not too difficult to solve. They can also be useful when quasi-Newton techniques are used to approximate L_k^{-1} by positive definite matrices in the nonlinear algorithm (the one that sets problem (14.26)), since then there is no linear system to solve with the matrix L_k, just a matrix-vector product needs to be done.

The Reduced System Approach

In this approach (sometimes called the null-space approach), it is assumed that a decomposition of $\mathbb{R}^n$ has been chosen, similar to those described in § 14.2. The operators $A^-(x)$ and $Z^-(x)$ should take advantage of the features of the problem, in order to avoid expensive operations. We show below that then the optimization aspect contained in (14.26) can be transferred into a single linear system, involving an $(n-m) \times (n-m)$ symmetric matrix: the reduced Hessian of the Lagrangian. This makes the reduced system approach particularly appropriate when $n-m \ll n$. Since the reduced Hessian is positive definite at a strong solution to (P_E), the approach makes it possible to detect convergence to a stationary point that is not a local minimum. Furthermore, the method leads to formulae highlighting the structure of the Newton step d_k.

Let us start by introducing a very useful notion. We have denoted by $Z^-(x)$ an $n \times (n-m)$ matrix, whose columns form a basis of $N(A(x))$, the subspace tangent to the constraint manifold at x. We call *reduced gradient* of f at x for the basis Z^-, the vector of $\mathbb{R}^{n-m}$ defined by

$$g(x) := Z^-(x)^\top \nabla f(x). \tag{14.31}$$

We note $g_k := g(x_k)$. This vector can be interpreted in Riemannian geometry as follows. Equip the manifold $\mathcal{M}_x$ with a Riemannian structure by defining at each point $y \in \mathcal{M}_x$ the scalar product on the tangent space $\gamma_y(Z_y^- u, Z_y^- v) = u^\top v$; then the gradient of $f|_{\mathcal{M}_x}$ at y for this Riemannian metric is just the tangent vector $Z^-(y)g(y)$.

Consider now the computation of d_k. Recalling (14.14), the second equation in (14.26) shows that d_k has the form

$$d_k = -A_k^- c_k + Z_k^- u_k,$$

for some $u_k \in \mathbb{R}^{n-m}$. Then, the first equation in (14.26) gives

$$L_k Z_k^- u_k + A_k^\top \lambda_k^{\mathrm{QP}} = -\nabla f_k + L_k A_k^- c_k.$$

Premultiplying by $Z_k^{-\top}$ to eliminate λ_k^{QP} provides the reduced linear system:

$$H_k u_k = -g_k + Z_k^{-\top} L_k A_k^- c_k, \tag{14.32}$$

where the $(n-m) \times (n-m)$ matrix

$$H_k := Z_k^{-\top} L_k Z_k^-$$

is called the *reduced Hessian of the Lagrangian* at (x_k, λ_k). It depends on the choice of the basis Z_k^-. This matrix is necessarily nonsingular when the matrix in (14.26) is nonsingular (see proposition 14.1). This leads to

$$d_k = -(I - Z_k^- H_k^{-1} Z_k^{-\top} L_k) A_k^- c_k - Z_k^- H_k^{-1} g_k.$$

The operator acting on c_k, namely

$$\widehat{A}_k^- := (I - Z_k^- H_k^{-1} Z_k^{-\top} L_k) A_k^-, \tag{14.33}$$

is the right inverse of A_k defined in (14.21), where M and x have been replaced by L_k and x_k respectively. Finally

$$\boxed{d_k = -\widehat{A}_k^- c_k - Z_k^- H_k^{-1} g_k.} \tag{14.34}$$

This computation reveals the structure of the Newton direction d_k, made up of two terms (see figure 14.3). The first term $\widehat{r}_k := -\widehat{A}_k^- c_k$ is a stationary

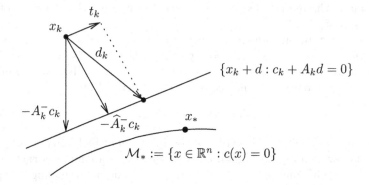

Fig. 14.3. Structure of the Newton step d_k

point of the quadratic problem in $r \in \mathbb{R}^n$:

$$\begin{cases} \min_r \frac{1}{2} r^\top L_k r \\ c_k + A_k r = 0. \end{cases}$$

To see this, just set $\nabla f_k = 0$ in (14.8) and (14.34). This direction aims at reducing $\rho(\cdot) = \|c(\cdot)\|$, an arbitrary norm of the constraints. Indeed, when $c_k \neq 0$, $\widehat{r}_k$ is a descent direction of ρ, since according to lemma 13.1:

$$\rho'(x_k; \widehat{r}_k) = (\| \cdot \|)'(c_k; A_k \widehat{r}_k) = (\| \cdot \|)'(c_k; -c_k) = -\|c_k\| < 0.$$

The second term in the right-hand side of (14.34), $t_k := -Z_k^- H_k^{-1} g_k$, is a stationary point of the quadratic problem in $t \in \mathbb{R}^n$:

$$\begin{cases} \min_t \nabla f_k^\top t + \frac{1}{2} t^\top L_k t \\ A_k t = 0. \end{cases}$$

To see this, just set $c_k = 0$ in (14.8) and (14.34). It is tangent to the manifold $\mathcal{M}_k := \mathcal{M}_{x_k}$ and aims at decreasing the function f. Indeed, when H_k is positive definite and $g_k \neq 0$, t_k is a descent direction of f at x_k, since

$$f'(x_k) \cdot t_k = \nabla f(x_k)^\top (-Z_k^- H_k^{-1} g_k) = -g_k^\top H_k^{-1} g_k < 0.$$

We shall come back to this issue in § 14.6, when comparing the direction d_k with directions generated by other algorithms.

On the influence of L_k on the direction d_k, we can observe the following.

1. The second-order information used in the Newton direction is entirely contained in the part $Z_k^{-\top} L_k$ of L_k. This can be seen in formula (14.34): only this part enters the matrices $\widehat{A}_k^-$ and H_k. In particular, the direction d_k is not changed if we add to L_k a matrix of the form $A_k^\top S_k A_k$, where S_k is an arbitrary symmetric $m \times m$ matrix.

2. If we multiply L_k by a number $\alpha \neq 0$, the transversal part $-\widehat{A}_k^- c_k$ of the direction is not affected, while the longitudinal part $-Z_k^- H_k^{-1} g_k$ is divided by α. In other words, the "size" of L_k only acts on the tangential part of d_k.

Consider now the computation of λ_k^{QP}. Premultiply the first equation of (14.26) by $A_k^{-\top}$ and use formula (14.34) of d_k to find

$$\boxed{\lambda_k^{\mathrm{QP}} = -\widehat{A}_k^{-\top} \nabla f_k + A_k^{-\top} L_k \widehat{A}_k^- c_k.} \tag{14.35}$$

This multiplier, as well as the first term in (14.35),

$$\widehat{\lambda}_k := -\widehat{A}_k^{-\top} \nabla f_k, \tag{14.36}$$

are sometimes called *second-order multipliers*, since they involve second-order derivatives of the functions f and c, via the Hessian of the Lagrangian L_k. These are estimates of the optimal multiplier, since $\lambda_k^{\mathrm{QP}} = \widehat{\lambda}_k = \lambda_*$ when $x_k = x_*$, a stationary point. Such is also the case of

$$\lambda_k^{\mathrm{LS}} := -A_k^{-\top} \nabla f_k,$$

called the *first-order multiplier* or *least-squares multiplier* (see (14.24)). It is said to be of first-order because it only involves the first derivatives of the data.

With this section, we have concluded the description of Newton's algorithm to solve equality constrained optimization problems. Next comes the description of an algorithm, also proceeding by linearizations, but different from Newton's method. It can be seen as a kind of nonlinear block Gauss-Seidel approach.

14.5 Reduced Hessian Algorithm

There is an algorithm to solve problem (P_E), different from Newton's method, that also enjoys local quadratic convergence. In optimization, its existence can be suggested by the following considerations.

When c is a submersion on Ω, the feasible set

$$\mathcal{M}_* = \{x \in \Omega : c(x) = 0\}$$

is a manifold of dimension $n-m$. Then (P_E) has only $n-m$ degrees of freedom and a natural question is whether there exists a method where the matrix containing the second-order information (second derivatives of f and c or their quasi-Newton approximation) is only $(n-m) \times (n-m)$. This is certainly the case if the iterates x_k are forced to stay in $\mathcal{M}_*$. Indeed, such an algorithm can be obtained by taking a parametrization of $\mathcal{M}_*$ around x_* and applying Newton's method in the parameter space, which has dimension $n-m$. However, requiring $x_k \in \mathcal{M}_*$ is not realistic: it is often computationally expensive and, anyway, it cannot be realized exactly when c is an arbitrary nonlinear function. What is desired is a method with the following properties:

- the only matrix containing second-order information is $(n-m) \times (n-m)$,
- the iterates x_k are not forced to satisfy the constraints at each iteration,
- the speed of convergence is quadratic.

In this section, we show how to introduce such an algorithm. We shall see that this approach is particularly attractive when $n-m \ll n$ and quasi-Newton techniques are employed. Throughout the section, we assume that the stationary point x_* we are seeking is regular (see definition 14.2).

The Reduced Optimality System

The first stage leading to the definition of the algorithm is to provide an optimality system of reduced size, with fewer equations than in (14.1). This stage is optional but, by eliminating the multiplier from (14.1), it leads to a more concise presentation.

Premultiply the first equation of (14.1) by $Z^-(x_*)^\top$ to find, with (14.10), the *reduced optimality system*:

$$\begin{cases} g(x_*) = 0 \\ c(x_*) = 0, \end{cases} \tag{14.37}$$

where g is the reduced gradient of f, defined by (14.31). The multiplier λ_* no longer appears in this system, which counts $(n-m)+m = n$ equations for the n unknowns x_*.

Note that the two systems (14.1) and (14.37) have the same solutions x_*. Indeed, we have just shown that (14.37) can be obtained from (14.1). On the other hand, we deduce from the first equation of (14.37) that

$$\nabla f(x_*) \in N(Z^-(x_*)^\top) = R(Z^-(x_*))^\perp = N(A(x_*))^\perp = R(A(x_*)^\top).$$

Therefore there exists $\lambda_* \in \mathbb{R}^m$ such that $\nabla f(x_*) + A(x_*)^\top \lambda_* = 0$. This is the first equation of (14.1). Thus, there is no loss of solutions by considering (14.37) instead of (14.1).

Solving the Reduced Optimality System by a Decoupling Technique

The reduced Hessian method essentially consists in performing one Newton-like step to solve the second equation of (14.37), followed by one Newton-like step to solve the first equation. This resembles a nonlinear block Gauss-Seidel method. There is an important difference however. We shall show that, to yield local quadratic convergence, the first step can be an arbitrary Newton-like step, but the second one must have a very specific form. In particular, this second step must be tangent to the manifold defined by the second equation in (14.37).

The algorithm generates two sequences of iterates, $\{x_k\}$ and $\{y_k\}$, both converging to the same solution x_*. Local convergence is studied more easily if the method is thought of generating the sequence $\{y_k\}$. It is this sequence that converges (almost) quadratically. Curiously the sequence $\{x_k\}$ converges slightly less rapidly, but the algorithm is easier to implement in terms of the sequence $\{x_k\}$. We now introduce the method by considering the sequence $\{y_k\}$, while $\{x_k\}$ appears as an intermediate sequence.

Starting with an iterate $y_k \in \Omega$, we first perform a Newton-like step that aims at solving the second equation of (14.37). For this, we use a right inverse of the Jacobian of c. This gives an intermediate point x_k, defined by

$$x_k = y_k - A^-(y_k)c(y_k).$$

Note that, if $m = n$, then $A^-(y_k)$ is the inverse of $A(y_k)$ and the step $-A^-(y_k)c(y_k)$ is exactly the Newton step at y_k to solve $c(x) = 0$ (compare with (13.17) and (13.19)). When $m < n$, which is our situation, every right inverse $A^-(y_k)$ produces a particular solution $x_k - y_k$ to the constraint equation, linearized at y_k.

We are now interested in making a Newton-like step from x_k to solve the first equation of (14.37). The point x_k is supposed to be in Ω. Observe first that the reduced gradient can be written $g(x) = Z_x^{-\top}\nabla_x\ell(x,\lambda_*)$, where λ_* is the multiplier associated with the solution x_*. By optimality, $\nabla_x\ell(x_*,\lambda_*) = 0$. Hence, assuming that Z_x^- is continuous at x_* and using lemma 13.2, one has

$$g'(x_*) = Z_*^{-\top}L_*, \tag{14.38}$$

where we have set $Z_*^- = Z^-(x_*)$ and $L_* = L(x_*,\lambda_*)$, as usual. If (x_*,λ_*) is a regular stationary point, the *reduced Hessian of the Lagrangian* at (x_*,λ_*),

$$H_* := Z_*^{-\top}L_*Z_*^-,$$

is nonsingular (see proposition 14.1), so that $g'(x_*)$ is surjective. Therefore g is a submersion in a neighborhood of x_*, which is supposed to contain Ω. As above, we can therefore take a right inverse $B^-(x_k)$ of $g'(x_k)$ and define the next iterate by

$$y_{k+1} = x_k - B^-(x_k)g(x_k).$$

We have just described a procedure for computing y_{k+1} from y_k, with x_k as an intermediate iterate.

We now raise the following question. Is it possible to find a matrix mapping $x \mapsto B^-(x)$ so as to obtain fast convergence of the sequence $\{y_k\}$ to x_*? To answer this question, we introduce the functions φ and $\psi : \mathbb{R}^n \to \mathbb{R}^n$ defined by:

$$\varphi(y) = y - A^-(y)c(y)$$
$$\psi(x) = x - B^-(x)g(x).$$

Then, the procedure we are analyzing can be viewed as fixed point iterations: $y_{k+1} = (\psi \circ \varphi)(y_k)$. As a result, if $B^-(\cdot)$ can be determined in such a way that $(\psi \circ \varphi)'(x_*) = 0$, the algorithm is likely to converge quadratically (see exercise 14.10). The next lemma specifies the value of $B_*^- := B^-(x_*)$ to get this property.

Lemma 14.6 (condition of quadratic convergence of a decoupling method). *Suppose that g and c are differentiable at x_*, that $A^-(\cdot)$ and $B^-(\cdot)$ are continuous at x_*, and that (x_*, λ_*) is a regular stationary point of (P_E). Then*

$$(\psi \circ \varphi)'(x_*) = 0 \quad \Longleftrightarrow \quad B_*^- = Z_*^- H_*^{-1},$$

where $H_* := Z_*^{-\top} L_* Z_*^-$, for some basis Z_*^- of $N(A_*)$.

Proof. Set $B_* = Z_*^{-\top} L_*$ and $C_* = (\psi \circ \varphi)'(x_*)$. Then, with the assumptions and lemma 13.2:
$$C_* = (I - B_*^- B_*)(I - A_*^- A_*).$$
If $(\psi \circ \varphi)'(x_*) = 0$, then $C_* Z_*^- = 0$, which gives

$$B_*^- B_* Z_*^- = Z_*^-.$$

We deduce $B_*^- = Z_*^- H_*^{-1}$. Conversely, if $B_*^- = Z_*^- H_*^{-1}$, we have $A_* B_*^- = 0$. Then

$$\begin{pmatrix} A_* \\ B_* \end{pmatrix} C_* = 0.$$

Since the operator applied to C_* is nonsingular, we have $C_* = (\psi \circ \varphi)'(x_*) = 0$. $\qquad\square$

The Algorithm

Lemma 14.6 suggests designing the algorithm that generates the sequences $\{x_k\}$ and $\{y_k\}$ as follows:

$$x_k = y_k - A^-(y_k)c(y_k)$$
$$y_{k+1} = x_k - Z^-(x_k)H_k^{-1}g(x_k).$$

Here H_k is an $(n-m) \times (n-m)$ matrix approximating the reduced Hessian H_* of the Lagrangian, or $Z^-(x_k)^\top L(x_k, \lambda_k) Z^-(x_k)$, for a certain multiplier λ_k.

As such, this algorithm can be very time-consuming because the constraints must be linearized at the two points x_k and y_k, and also the two right inverses $A^-(y_k)$ and $Z^-(x_k)$ must be computed. Even though it is crucial to compute g at x_k and c at y_k, theorem 13.6 states that good convergence can be preserved if the operators involving first derivatives are evaluated at other points; the important thing is that these points converge to the solution. Since the reduced gradient must be evaluated at x_k, and since it involves a basis $Z^-(x_k)$ of the tangent space, the constraints must be linearized at x_k anyway. However, A^- can be evaluated at x_k instead of y_k. This avoids linearizing the constraints at y_k. Stating the algorithm in terms of the sequence $\{x_k\}$, we then obtain

$$y_{k+1} = x_k - Z^-(x_k)H_k^{-1}g(x_k)$$
$$x_{k+1} = y_{k+1} - A^-(x_k)c(y_{k+1}).$$

Finally, setting $g_k = g(x_k)$, $A_k^- = A^-(x_k)$, $Z_k^- = Z^-(x_k)$ and

$$t_k = -Z_k^- H_k^{-1} g_k, \tag{14.39}$$

the algorithm can be stated in a very concise manner:

$$\boxed{x_{k+1} = x_k + t_k - A_k^- c(x_k + t_k).} \tag{14.40}$$

As with the Newton method (14.34), the first phase of Algorithm (14.40) consists in performing a displacement tangent to the manifold $\mathcal{M}_k$ at x_k. In the second phase, the algorithm aims at getting the next iterate x_{k+1} closer to the manifold $\mathcal{M}_*$ by taking the displacement $-A_k^- c(x_k + t_k)$, in which the constraints are evaluated at $x_k + t_k$, after the tangent step.

Although the reduced Hessian algorithm, which is summarized in the recurrence (14.40), should be quite clear, we formally state it below.

REDUCED HESSIAN ALGORITHM FOR (P_E):

Choose an initial iterate $x_1 = y_1 \in \mathbb{R}^n$.
Compute $c(x_1)$, $\nabla f(x_1)$, and $A(x_1)$.
Set $k = 1$.

1. Compute the reduced gradient $g(x_k)$ by (14.31).
2. Stop if $g(x_k) = 0$ and $c(y_k) = 0$ (optimality is reached).
3. Compute the reduced Hessian of the Lagrangian H_k, or an approximation to it, and the tangent step t_k by (14.39).

4. Evaluate the constraint at $y_{k+1} := x_k + t_k$.
5. Compute the new iterate x_{k+1} by (14.40), $\nabla f(x_{k+1})$ and $A(x_{k+1})$.
6. Increase k by 1 and go to 1.

Note that this algorithm is essentially primal, since it can be expressed only in terms of the primal sequence $\{x_k\}$. A multiplier estimate λ_k is however often necessary, either to evaluate the reduced Hessian of the Lagrangian at (x_k, λ_k) in step 3 or to update a quasi-Newton approximation to it (see chapter 18). The cheapest one is the least-squares multiplier defined by (14.24).

Simplified Newton Method

Algorithm (14.40) would be simpler if, in the second phase, the constraints were evaluated at x_k. It would then be written

$$\boxed{x_{k+1} = x_k + t_k - A_k^- c_k.} \tag{14.41}$$

This algorithm is sometimes called the *simplified Newton method* because it only uses the reduced Hessian of the Lagrangian H_k, not the full Hessian L_k (compare with (14.34) or see §14.6). It has a slower convergence speed than (14.40): under natural assumptions, $\{x_k\}$ converges quadratically in two steps (see exercise 14.11). On the other hand, there are examples showing that the sequence $\{x_k\}$ may not converge quadratically in one step (see [63, 373]). To get good convergence, it is therefore important to evaluate the constraints at $x_k + t_k$, after the tangent displacement.

Local Convergence

The next theorem states that the sequence $\{y_k\} \equiv \{x_k + t_k\}$ of Algorithm (14.40) converges superlinearly if the matrix H_k appearing in the tangent step t_k satisfies the estimate

$$H_k - H_* = O(\|x_k - x_*\|).$$

If H_k is set to $Z^-(x_k)^\top L(x_k, \lambda_k) Z^-(x_k)$, it depends on x_k and λ_k and this condition is satisfied if $\lambda_k - \lambda_* = O(\|x_k - x_*\|)$ and if the functions f'', c'', and Z^- are Lipschitzian near x_*. This leaves a certain freedom for the choice of the multiplier λ_k. For example, one can take $\lambda_k = \lambda^{\text{LS}}(x_k)$, the least-squares multiplier defined by (14.24). It is easy to check that $\lambda^{\text{LS}}(x_*) = \lambda_*$, and thus $\lambda^{\text{LS}}(x_k) - \lambda_* = O(\|x_k - x_*\|)$ if A^- and f' are Lipschitzian near x_*. With this value of the multiplier, Algorithm (14.40) becomes entirely primal, in the sense that the algorithm only constructs the sequence $\{x_k\}$, the multiplier being reduced to an auxiliary vector, itself depending on x_k.

The result given below is slightly weaker than theorem 14.5 stating the convergence of $\{x_k\}$ in the primal variant of Newton's algorithm. For that algorithm, the sequence $\{x_k\}$ converges quadratically if $\lambda_k - \lambda_* = O(\|x_k - x_*\|)$.

The proof of theorem 14.7 uses the notation $O(\cdot)$ as explained at the end of § 13.5. At first, it may disconcert the reader. For example, the first estimate obtained in the proof, namely (14.43), means that there exists a positive constant C such that, if x_k is in some neighborhood of x_*:

$$\|y_{k+1} - x_* - (x_k - x_*) + Z_k^- H_k^{-1} Z_*^{-\top} L_*(x_k - x_*)\| \leq C\|x_k - x_*\|^2.$$

The point x_k is considered as an arbitrary point in that neighborhood and, despite the presence of the iteration index k, there is no reference to a particular sequence. The estimate obtained at the end of the proof, namely $x_{k+2} - x_* = O(\|x_k - x_*\|^2)$, implies that if x_1 and x_2 are in a sufficiently small neighborhood of x_*, then x_3 and x_4 are in that neighborhood (because for example $\|x_3 - x_*\| \leq (C\|x_1 - x_*\|)\|x_1 - x_*\| \leq \|x_1 - x_*\|$ if $\|x_1 - x_*\|$ is sufficiently small). Therefore, by induction, all the estimates can now be applied to all the generated sequences. The interest of this notation is to provide very concise proofs (for another example, see exercise 14.11).

Theorem 14.7 (convergence of the reduced Hessian algorithm). *Suppose that f and c are twice differentiable at a regular stationary point x_* of problem (P_E) (this allows the use of the operators $Z^-(x)$ and $A^-(x)$ introduced in § 14.2, for x near x_*) and that the reduced gradient g is differentiable near x_*. Suppose also that c', g', Z^- and A^- are Lipschitzian near x_*, and that the matrix H_k used in (14.40) satisfies $H_k - H_* = O(\|x_k - x_*\|)$. Then, there exists a neighborhood V of x_* such that, when the first iterate $x_1 \in V$, Algorithm (14.40) is well defined and generates a sequence $\{x_k\}$ converging quadratically in two steps to x_*. Furthermore, the sequence $\{y_k\}$ converges superlinearly to x_* with the estimate*

$$y_{k+1} - x_* = O(\|x_{k-1} - x_*\| \, \|y_k - x_*\|). \tag{14.42}$$

Proof. Remark first that, when x_k is close to x_*, by assumption, H_k is close to H_*, which is nonsingular (x_* is regular). Thus, H_k is nonsingular and the iteration is well defined. Also $\{H_k^{-1}\}$ is bounded when x_k remains in some neighborhood of x_*.

Remembering that $y_{k+1} = x_k + t_k$ and using $g(x_*) = 0$, (14.38), and the Lipschitz continuity of g', we have

$$\begin{aligned} y_{k+1} - x_* &= x_k - x_* - Z_k^- H_k^{-1} g_k \\ &= x_k - x_* - Z_k^- H_k^{-1} Z_*^{-\top} L_*(x_k - x_*) \\ &\quad + O(\|x_k - x_*\|^2). \end{aligned} \tag{14.43}$$

But $H_k^{-1} - H_*^{-1} = -H_k^{-1}(H_k - H_*)H_*^{-1} = O(\|x_k - x_*\|)$, so that, with the Lipschitz continuity of Z^-, the following holds

$$y_{k+1} - x_* = (I - Z_*^- H_*^{-1} Z_*^{-\top} L_*)(x_k - x_*) + O(\|x_k - x_*\|^2). \tag{14.44}$$

This implies in particular that $y_{k+1} - x_* = O(\|x_k - x_*\|)$. We also have $x_{k+1} = y_{k+1} - A_k^- c(y_{k+1})$. Therefore, using successively $c(x_*) = 0$, the Lipschitz continuity of c' and A^-, (14.14), (14.44), and (14.13), we obtain

$$
\begin{aligned}
x_{k+1} - x_* &= y_{k+1} - x_* - A_k^- A_*(y_{k+1} - x_*) + O(\|y_{k+1} - x_*\|^2) \\
&= y_{k+1} - x_* - A_*^- A_*(y_{k+1} - x_*) + O(\|x_k - x_*\| \, \|y_{k+1} - x_*\|) \\
&= Z_*^- Z_*(y_{k+1} - x_*) + O(\|x_k - x_*\| \, \|y_{k+1} - x_*\|) \qquad (14.45) \\
&= Z_*^-(Z_* - H_*^{-1} Z_*^{-\top} L_*)(x_k - x_*) + O(\|x_k - x_*\|^2). \qquad (14.46)
\end{aligned}
$$

The operator acting on $(x_k - x_*)$ in (14.46) is nonzero in general but its square vanishes, because $(Z_* - H_*^{-1} Z_*^{-\top} L_*) Z_*^- = 0$. From this observation, we deduce the estimate

$$
x_{k+2} - x_* = O(\|x_k - x_*\|^2),
$$

which shows the two-step quadratic convergence of the sequence $\{x_k\}$.

Using (14.44), (14.45) (at the previous iteration), and observing that

$$
(I - Z_*^- H_*^{-1} Z_*^{-\top} L_*) Z_*^- = 0,
$$

we obtain (14.42). The superlinear convergence of $\{y_k\}$ follows.   □

At this point it is reasonable to wonder why the convergence of the sequence $\{y_k\}$ is not quadratic. Since Algorithm (14.40) uses the second derivatives of f and c, it is legitimate to expect quadratic convergence. The above proof clarifies this, indeed: the constraints are not linearized at y_k, but at the neighboring points x_{k-1} and x_k. Then, passing from y_k to y_{k+1} involves the right inverse $A^-(x_{k-1})$ instead of $A^-(y_k)$, which perturbs the speed of convergence. If the right inverse $A^-(y_{k+1})$ were used in place of $A^-(x_k)$, an $O(\|y_{k+1} - x_*\|^2)$ would appear in (14.45) instead of an $O(\|x_k - x_*\| \, \|y_{k+1} - x_*\|)$ and quadratic convergence would ensue. Numerically, it is not clear that the computing time of $A(y_k)$ and $A^-(y_k)$ would be balanced by the quadratic convergence thus recovered, which is why the algorithm is often stated in the form (14.40).

Beware of the different behavior of the sequences $\{x_k\}$ and $\{y_k\}$. Even though they are generated by the same algorithm and both converge to the same point x_*, the first one is slower than the second one. This may look surprising, but examples do exist, in which the sequence $\{x_k\}$ does not converge quadratically (see [63]).

Newton and Quasi-Newton Versions

We have already mentioned that the reduced Hessian method is a very attractive approach when $n-m$ is much smaller than n. This is particularly true for their quasi-Newton versions. In these algorithms the $(n-m) \times (n-m)$ reduced Hessian $H_k = Z^-(x_k)^\top L(x_k, \lambda_k) Z^-(x_k)$ is approximated by a matrix updated by a quasi-Newton formula (see chapters 4.4 and 18). Only this "small" matrix needs to be updated to collect all the necessary second-order information on the problem that provides superlinear convergence. Furthermore, the small order of these updated matrices makes it possible to rapidly obtain a good approximation of the reduced Hessian.

In the Newton version, H_k must be computed. The interest of the reduced Hessian method is then less clear. One way of computing H_k is to evaluate first $L(x_k, \lambda_k) Z^-(x_k)$, by computing $n-m$ directional derivatives of the gradient of the Lagrangian along the columns of $Z^-(x_k)$, and then premultiplying the matrix thus obtained by $Z^-(x_k)^\top$. This computation is conceivable, but the knowledge of $L(x_k, \lambda_k) Z^-(x_k)$ would allow the use of Newton's method, which does not require any other information on the Hessian of the Lagrangian (see remark 1 on page 235); furthermore, Newton's method does not require a re-evaluation of the constraints after the tangent step.

Another way of getting second-order information in the reduced Hessian algorithm is to approximate H_k by computing the directional derivatives of the reduced gradient g along the $n-m$ columns of $Z^-(x_k)$. Note that $\tilde{H}_k := g'(x_k) Z^-(x_k)$ is usually different from H_k, although, in view of formula (14.38), $g'(x_*) Z_*^-$ does equal $Z_*^{-\top} L_* Z_*^-$. Now $\tilde{H}_k$ satisfies the estimate $\tilde{H}_k - H_* = O(\|x_k - x_*\|)$ (with sufficiently smooth data), so that theorem 14.7 can be applied. Note also that $\tilde{H}_k$ is not necessarily a symmetric matrix. This property depends in particular on the choice of the bases Z^-: if $Z^-(x)$ is computed by partitioning $A(x)$ (i.e., using formula (14.15)), then $\tilde{H}_k$ is symmetric; but in general it is not so when orthonormal bases are used (see [149]).

14.6 A Comparison of the Algorithms

Table 14.1 and figure 14.4 compare the form and speed of convergence of the three algorithms described in this chapter: Newton (14.6) with (14.5) or (14.34)–(14.35), simplified Newton (14.41), and reduced Hessian (14.40).

In all algorithms, the longitudinal step (tangent to the manifold $\mathcal{M}_k$) is identical and is written

$$t_k = -Z_k^- H_k^{-1} g_k.$$

When H_k is positive definite, this step is opposite to the gradient of f, seen as a function defined on the manifold $\mathcal{M}_k$ equipped at x_k with the scalar product (Riemannian structure on $\mathcal{M}_k$):

$$\gamma_{x_k}(Z_{x_k}^- u, Z_{x_k}^- v) = u^\top H_k v.$$

Algorithms	Longitudinal displacement	Transversal displacement	Speed of convergence
Newton	t_k	$-\widehat{A}_k^- c_k$	quadratic
Simplified Newton	t_k	$-A_k^- c_k$	2-step quadratic
Reduced Hessian	t_k	$-A_k^- c(x_k+t_k)$	"almost" quadratic

Table 14.1. Comparison of local methods

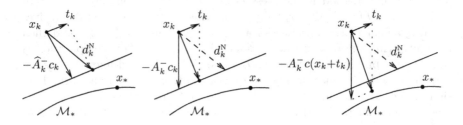

Fig. 14.4. Comparison of the Newton (d_k^N), simplified Newton, and reduced Hessian steps

When H_k is set to $Z_k^{-\top} L_k Z_k^-$ and H_k is positive definite, t_k can also be viewed as the unique solution to the quadratic problem in t:

$$\begin{cases} \min_t \nabla f_k^\top t + \frac{1}{2} t^\top L_k t \\ A_k t = 0, \end{cases}$$

This interpretation shows that t_k does not depend on the choice of the basis Z_k^-, despite the use of this matrix in the formula above. The algorithms presented in table 14.1 therefore only differ in the choice of the restoration operator, A_k^- or $\widehat{A}_k^-$, and in the points where the constraints are evaluated, x_k or x_k+t_k.

First let us compare the two forms of Newton's method: standard (step given by (14.34)), and simplified (step given by (14.41)). We see that the two displacements have the same form, but the operator acting on $c_k = c(x_k)$ is $\widehat{A}_k^-$ in the first case, and A_k^- in the second (both are right inverses of A_k). It has been observed (§ 14.2) that $\widehat{A}_k^-$ only depends on the problem's data (see problem 14.19), while A_k^- is the concern of the user of the algorithm. Theorems 14.4 and 14.5 have shown that the choice $\widehat{A}_k^-$ leads to quadratically convergent methods. On the other hand, it is easy to check that the convergence of $\{x_k\}$ with (14.41) is only two-step quadratic when the right inverse A_k^- is arbitrary: one-step quadratic convergence is never guaranteed (see exercise 14.11). Therefore Newton's method is the most effective. Note finally that one can view the simplified Newton method as an algorithm neglecting the part $Z_k^{-\top} L_k A_k^-$ of L_k in the standard Newton method (see formula (14.33)). Newton's algorithm gains in efficiency from getting more information on the Hessian of the Lagrangian.

As for the reduced Hessian algorithm (14.40), it is very close to the simplified Newton method (14.41). The algorithms differ in the point at which the constraints are evaluated: $x_k + t_k$ in (14.40) and x_k in (14.41). The reduced Hessian method can thus be viewed as a technique to compensate a possible bad choice of right inverse A_k^- by a re-evaluation of the constraints after the tangent step. As shown by theorem 14.7, this yields a good speed of convergence for the sequence $\{x_k + t_k\}$, a property that is not shared with the simplified Newton algorithm.

14.7 The Hanging Chain Project II

The goal of the second session is to implement one of the local algorithms introduced in this chapter and to understand its behavior on the hanging chain test problem presented in § 13.8 (we assume here that the main program and the simulator have been written in MATLAB). Various algorithms can be implemented. Below, we concentrate our comments on the standard Newton method described on page 221 in § 14.1, because it is this algorithm that is the easiest to extend to inequality constrained problems. We shall gain experience on its features, its efficiency, and shall reveal its weak points (some of them will be fixed in the next chapters).

We refer the reader to figure 13.3 for the general flowchart of the program. In this session, we start to write the optimization function sqp, which is assumed to be in the file sqp.m. We want to have an implementation that can be used to solve other optimization problems than the hanging chain test problem. This is a good reason for using the mathematical notation of this chapter inside sqp.m, not the language linked to the test problem. In our implementation, the function sqp has the following form

```
function [x,lme,lmi,info] = ...
    sqp (simul,x,lme,lmi,f,ce,ci,g,ae,ai,hl,options)
```

Some of the input or output arguments can be empty, depending on the presence of equality and/or inequality constraints; in particular, the variables in connection with the inequality constraints can be ignored for the while. The *input arguments* are the following: simul is a string giving the name of the simulator (here 'chs'); x is the initial value of the primal variable x (position of the joints); lme and lmi are the initial values of the multiplier λ_E and λ_I associated with the equality and inequality constraints; f, ce, and ci are the values of the objective function f to minimize (the energy) and of the equality and inequality constraint functions c_E and c_I (lengths of the bars and floor constraint) at the initial point x; g, ae, and ai are the values of the gradient of f and the Jacobian matrices A_E and A_I of c_E and c_I at the initial point x; hl is the Hessian of the Lagrangian at the initial (x, λ) or an approximation to it; and the structure options is aimed at tuning the behavior of the solver. Standard options include upper bounds on the number of iterations and simulations (options.iter and options.simul), the required tolerances on the KKT conditions (options.tol(1:4), see below), the output channel for printing (options.fout), *etc.* Other options will be discussed in other sessions. The *output arguments* are as follows: x, lme, and lmi are the final values of the primal and dual (multipliers) variables found by sqp; and info is a structure providing various information on the course of the optimization realized by the solver, telling in particular whether optimality has been reached, up to the required precision specified by the options.tol input argument, and in any case the reason why the solver has stopped.

We have already said on page 228 that the Newton algorithm aims at finding a stationary point, i.e., a pair (x_*, λ_*) satisfying the optimality conditions (13.1), not necessarily a local minimum. Therefore, it makes sense to have a stopping criterion based on these conditions. In our code, we stop the iterations as soon as, for some norms, the current iterate (x, λ) satisfies

$$\|\nabla f(x) + A(x)^\top \lambda\| \leq \texttt{options.tol(1)}$$
$$\|c_E(x)\| \leq \texttt{options.tol(2)}$$
$$\|c_I(x)^+\| \leq \texttt{options.tol(3)}$$
$$\max(\|\lambda_I^-\|, \|A_I^\top c_I(x)\|) \leq \texttt{options.tol(4)}.$$

where $t^+ = \max(0, t)$, $t^- = \max(0, -t)$, and $A_I = \mathrm{Diag}(\lambda_I)$.

Writing the MATLAB function sqp implementing the Newton algorithm of page 221 is actually extremely simple. The core of the function is only a few lines long. The time consuming operation is the one to solve the linear system in step 2, but for a small problem this is straightforward. The easiest way of doing this operation is to form the matrix K in (14.9) and to use the standard linear solver of MATLAB (see § 14.4 for other possibilities). Since hl and ae are the variables containing respectively the Hessian of the Lagrangian and the Jacobian of the equality constraints, steps 2 and 3 of the algorithm are simply made up of

```
K = [hl ae'; ae zeros(me)];
d = -K\[g;ce];
x = x + d(1:n);
lme = d(n+1:n+me);
```

where me $= m_E$ is the number of equality constraints, n $= n$ is the number of variables, and the final values of x and lme are the updated iterates x_+ and λ_+.

Algorithmic Details, Errors to Avoid, Difficulties to Overcome

The solver sqp offers the user the possibility to set the initial value of x and λ. This is interesting when it is desirable to restart the solver from a known approximate solution (recall that the method is primal-dual so that both x and λ must be specified). More generally, requiring to initialize x is sensible, since the user often knows an approximate solution to the problem. This is less clear for λ, since the multipliers have sometimes a less direct "physical" meaning or, perhaps, this meaning is known but the value of λ is still difficult to determine. Therefore, it is sometimes wise to let the solver choose the initial multiplier. For an equality constrained problem, one often computes the initial λ as the solution to the linear least-squares problem

$$\min_{\lambda \in \mathbb{R}^m} \frac{1}{2} \|\nabla_x \ell(x, \lambda)\|_2^2. \tag{14.47}$$

This is motivated by the fact that the gradient of the Lagrangian vanishes at a solution. The convex quadratic problem above always has a solution (theorem 19.1), which is the least-squares multiplier (14.24) when $c'_E(x)$ is surjective.

The Newton algorithm is structured as an iteration loop, which contains the piece of code given above. Of course the simulator simul must be called at each iteration after having computed x_+ and λ_+, in order to update the values of hl, ae, g, and ce and to check optimality.

Writing an optimization software is a special computer science activity in the sense that the realized code has to control the convergence of a sequence. In some cases, the sequence may diverge simply because the conditions of convergence are not satisfied, not because of an error in the code. Since convergence requires an unpredictable number of iterations, it is sometimes difficult to tell on a particular case whether the behavior of the solver is correct. To certify the correctness of the function sqp, a good idea is to try it on problems with an increasing difficulty and to check the quadratic convergence of the generated sequences, as explained below.

- Try first to start sqp at the solution to a trivial problem: for example, the chain with 2 bars of length 5, with $(a, b) = (6, 0)$, whose single joint should be at position $(3, -4)$. The solver should stop without making any iteration, so that this test case checks only the validity of the stopping criterion and the simulator.

- Try next to start sqp near the solution to an easy problem: for example, the chain with 3 bars of length 5, with $(a, b) = (11, 0)$, whose joints should be at position $(3, -4)$ and $(8, -4)$. Convergence should be obtained in very few iterations, if the initial nodes are at positions $(2, -5)$ and $(9, -3)$. Our code converges in 5 iterations with options.tol(1:4) set to 1.e-10.

The Newton algorithm of page 221 is known to converge quadratically if the initial primal-dual iterate (x_1, λ_1) is sufficiently close to a regular stationary point (theorem 14.4). Checking that quadratic convergence actually occurs is a good way of verifying that the implementation of both the algorithm and the simulator has been done properly. The very definition of quadratic convergence of a sequence $\{z_k\}$ makes use of the limit point z_* to which it converges (see § 13.5). Since, in the course of the optimization, the limit point $z_* := (x_*, \lambda_*)$ of the generated sequence $\{z_k\} := \{(x_k, \lambda_k)\}$ is not known, the definition cannot be directly applied. The idea is then to observe the behavior of another sequence, whose limit point is zero (hence known!) and that also converges quadratically. Below, we consider the following two possibilities.

- For Newton's method, a natural object to look at is the function of which the algorithm tries to find a zero. For an equality constrained optimization problem, it is the function $z := (x, \lambda) \in \mathbb{R}^{n+m} \mapsto F(z) = (\nabla_x \ell(x, \lambda), c(x)) \in \mathbb{R}^{n+m}$. When $z_* := (x_*, \lambda_*)$ is a regular stationary point (definition 14.2), $F'(z_*)$ is nonsingular and it is not difficult to show that $(z_k - z_*) \sim F(z_k)$ in the sense of (13.11). Therefore $F(z_k) \to 0$ quadratically in Newton's algorithm.

- Another vector that tends to zero is the step $s_k := z_{k+1} - z_k$. By lemma 13.5, $\{s_k\}$ also converges quadratically to zero in Newton's method.

Let us check quadratic convergence of our implementation on the following test case.

Test case 1a: second hook at $(a, b) = (1, -0.3)$, lengths of the bars: $L = (0.4, 0.3, 0.25, 0.2, 0.4)$, and initial positions of the chain joints: $(0.2, -0.5)$, $(0.4, -0.6)$, $(0.6, -0.8)$, and $(0.8, -0.6)$.

The results obtained with test case 1a are shown in figure 14.5. Convergence

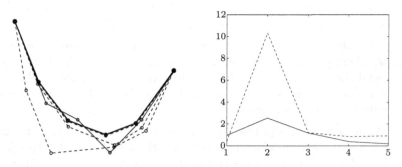

Fig. 14.5. Test case 1a

with options.tol$(1:4) = 10^{-10}$ is obtained in 6 iterations. The picture on the left shows the initial position of the chain (thin solid bars), the 5 intermediate positions (dashed bars) and the final position (bold solid bars). The picture on the right gives a plot of the ratios $\|F(z_{k+1})\|_2 / \|F(z_k)\|_2^2$ and $\|s_{k+1}\|_2 / \|s_k\|_2^2$, for $k = 1, \ldots, 5$. The boundedness of these ratios leaves no doubt on the quadratic convergence of the sequence $\{z_k\}$ to its limit.

Experimenting with the Newton Method

The test case 1a reveals the ideal behavior of Newton's method: quadratic convergence is obtained when the initial position of the chain is close to a regular solution. This solution is a strict local minimum (the smallest eigenvalue of the reduced Hessian of the Lagrangian $Z_*^{-\top} L_* Z_*^-$, for some orthonormal basis Z_*^-, is positive) and probably the global one.

Other solutions can be found by Newton's method with the same data, and those are not local minima. This is the case with the following two starting points.

Test case 1b: identical to test case 1a, except that the initial positions of the chain joints are $(0.2, 0.5)$, $(0.4, 0.6)$, $(0.6, 0.8)$, and $(0.8, 0.6)$.

Test case 1c: identical to test case 1a, except that the second hook at $(a, b) = (0.8, -0.3)$ and that the initial positions of the chain joints are $(0.3, 0.3)$, $(0.5, 0.4)$, $(0.3, 0.4)$, and $(0.6, 0.3)$.

The resulting equilibria are shown in figure 14.6. The picture on the left

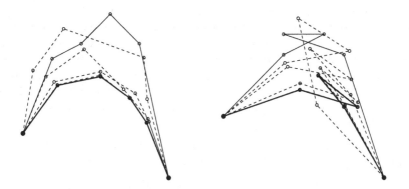

Fig. 14.6. Test cases 1b and 1c: a maximum (left) and a stationary point (right)

shows a local maximum (the largest eigenvalue of the reduced Hessian of the Lagrangian is negative). The right hand side picture shows a stationary point that is neither a minimum nor a maximum (the 3×3 reduced Hessian of the Lagrangian has two negative eigenvalues and a positive one).

The next two examples have been built to show cases without convergence.

Test case 1d: identical to test case 1a, except that the initial positions of the chain joints are $(0.2, -0.5)$, $(0.4, 1.0)$, $(0.6, -0.8)$, and $(0.8, -0.6)$ (hence, only the y-coordinate of the second joint has been modified).

Test case 2a: second hook at $(a, b) = (2, 0)$, lengths of the bars: $L = (1, 1)$, and initial position of the chain joint: $(1.5, -0.5)$.

The results are shown in figure 14.7. In the left picture, we have only plotted

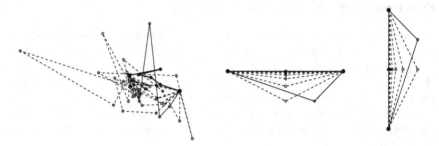

Fig. 14.7. Test cases 1d, 2a, and 2b: non convergence in (x, λ) (left), non convergence in λ (middle), and convergence in (x, λ) (right)

the position of the chain at the first 10 iterations, since apparently Newton's method does not converge. The generated sequence has a typical erratic behavior. By chance, one of these iterates may fall into the neighborhood of convergence of a stationary point, but this does not occur during the first 50 iterations. The middle picture is more puzzling, since it looks as if the algorithm converges. This is actually the case for the primal variables x (giving the position of the chain), which converge to the single feasible joint $(1, 0)$, but the dual variables diverge (their norm blows up). This reflects the fact that the optimal solution does not satisfy the KKT conditions (the Jacobian of the equality constraint in not surjective at the solution and there is no optimal multipliers); in fact, a weighty chain formed of two horizontal bars is not physically possible. The situation is quite different for the similar test case 2b below.

Test case 2b: second hook at $(a, b) = (0, -2)$, lengths of the bars: $L = (1, 1)$, and initial position of the chain joint: $(0.5, -0.5)$.

The result is shown in the right hand side picture in figure 14.7: convergence in both (x, λ) is obtained in 17 iterations.

We conclude with the following test case and let the reader guess whether the position of the chain given in figure 14.8 is a local minimum.

Test case 3: second hook at $(a, b) = (0, -1)$ and lengths of the bars: $L = (0.5, 0.5, 2.0, 0.4, 0.4)$.

Notes

The operators A^-, Z^-, and Z defined in §14.2 were introduced by Gabay [137]. They have allowed us to use the same formalism for the optimal control and orthogonal settings. We have seen that convergence results need to have a smooth map $x \mapsto (A_x^-, Z_x^-, Z_x)$. It is usually difficult to guarantee this smoothness in a large region (for example there is no continuous basis

Fig. 14.8. Test case 3: is this a stable static equilibrium position?

mapping $x \mapsto Z_x^-$ on a sphere of even dimension). Even locally, standard procedures such as the QR factorization presented in § 14.2 may compute a noncontinuous basis mapping [83]. This issue has been examined by several authors, who have proposed procedures for computing a smoothly varying sequence of matrices Z_k^- when approaching a solution: see [83, 157, 24, 68]. The connection between the symmetry of $g'(x)Z^-(x)$ and the choice of basis of the tangent space is discussed in [149; § 3].

The accuracy of the computation of the Newton step by the reduced system approach (see § 14.4) crucially depends on the choice of operators A^- and Z^-. When these are obtained from the partitioning of A into $(B\ N)$, with a nonsingular B, and from a Gaussian factorization of B, Fletcher and Jonhson [129] recommend to use Gaussian elimination on the whole matrix $A^\top$ to get

$$A^\top = \begin{pmatrix} L_1 \\ L_2 \end{pmatrix} U,$$

where L_1 is unit lower triangular and U is upper triangular. The elements of L_1 and L_2 can be guaranteed to be not bigger than 1 in absolute value (e.g., because the elements of $N^\top$ are taken into account in the choice of the pivots). This approach provides well conditioned basis Z^- and a solution to the Newton system that is less sensitive to the ill-conditioning of A and that of the reduced Hessian of the Lagrangian.

The presentation of the reduced Hessian method given in § 14.5 follows [145]. This algorithm, condensed in formula (14.40), was introduced by Coleman and Conn [81], who proved convergence of the sequence $\{x_k\}$. Superlinear (or quadratic) convergence of the sequence $\{y_k\}$ was observed independently by Hoyer [197], Gilbert [145], and Byrd [64]. The simplified Newton method (14.41) has been studied by many authors: Murray and Wright [271], Powell [292], Gabay [138], Nocedal and Overton [276], Byrd and Nocedal [67], to mention a few. Newton's method on the reduced system (14.37) is considered by Goodman [177], who analyses its links with Newton's algorithm (14.5)–(14.6).

Exercises

14.1. *Nonconvergence with a step computed by (14.2).* Consider the problem in $x = (x_1, x_2) \in \mathbb{R}^2$:
$$\begin{cases} \min_x -ax_1^2 + 2x_2 \\ x_1^2 + x_2^2 = 1, \end{cases}$$
where $a \in {]0, 1[}$. Show that the unique solution $x_* = (0, -1)$ to this problem can be repulsive for an algorithm based on (14.2): for x on the constraint manifold, arbitrary close to (but different from) the solution, and for a stationary point d of (14.2), $x + d$ is further from the solution than x.

14.2. *Inertia of the matrix K in (14.9).* The *inertia* i of a matrix is the triple (n_-, n_0, n_+) formed by the numbers of its negative, null, and positive eigenvalues respectively. Let K be the matrix defined in (14.9), where L is an $n \times n$ symmetric matrix and A is an $m \times n$ surjective matrix (hence $m \leq n$). Show that
$$i(K) = i(Z^{-\top} L Z^-) + (m, 0, m),$$
where the columns of Z^- form a basis of $N(A)$ (see [90, 72, 179, 244] for related results).

[*Hint*: Prove the following claims and conclude: (i) $n_0(K) = n_0(Z^{-\top} L Z^-)$; ($ii$) there is no restriction in assuming that $Z^{-\top} L Z^-$ is nonsingular (use a perturbation argument, for instance), which is supposed from now on; (iii) $i(K) = i(Z^{-\top} L Z^-) + i(\Sigma)$, where
$$\Sigma := \begin{pmatrix} S & I_m \\ I_m & 0 \end{pmatrix}$$
for some $m \times m$ symmetric matrix S (use the matrix $\widehat{A}^-$ defined by (14.21) and Sylvester's law of inertia: $i(PKP^\top) = i(K)$ if P is nonsingular); (iv) $i(\Sigma) = (m, 0, m)$.]

14.3. *Regular stationary points are isolated.* Let (x_*, λ_*) be a regular stationary point of problem (P_E). Show that there is a neighborhood of (x_*, λ_*) in $\mathbb{R}^n \times \mathbb{R}^m$ containing no other stationary point than (x_*, λ_*).

14.4. *A view of the reduced Hessian of the Lagrangian.* Let $f : \Omega \to \mathbb{R}$ and $c : \Omega \to \mathbb{R}^m$ be twice differentiable functions defined in a neighborhood Ω of a point $x_* \in \mathbb{R}^n$ and denote $\ell(x, \lambda) := f(x) + \lambda^\top c(x)$, for $(x, \lambda) \in \Omega \times \mathbb{R}^m$, and $L_* := \nabla^2_{xx} \ell(x_*, \lambda_*)$. Suppose that $\nabla_x \ell(x_*, \lambda_*) = 0$ for some $\lambda_* \in \mathbb{R}^m$ (it is not assumed that $c(x_*) = 0$) and that $A_* := c'(x_*)$ is surjective. Let Z_*^- be an $n \times (n-m)$ matrix whose columns form a basis of $N(A_*)$. Show that one can find a twice differentiable parametric representation $\varphi : U \subset \mathbb{R}^{n-m} \to \mathcal{M}_{x_*} \subset \mathbb{R}^n$ of the manifold $\mathcal{M}_{x_*} := \{x \in \Omega : c(x) = c(x_*)\}$ around x_* defined in a neighborhood U of 0, such that $\varphi(0) = x_*$, $\nabla(f \circ \varphi)(0) = 0$, and $\nabla^2(f \circ \varphi)(0) = Z_*^{-\top} L_* Z_*^-$ is the reduced Hessian of the Lagrangian.

14.5. *Right inverse and complementary subspace.* Let A be an $m \times n$ surjective matrix and $\mathcal{S}$ be a subspace of $\mathbb{R}^n$, complementary to $N(A)$ (i.e., $N(A) \cap \mathcal{S} = \{0\}$ and $\dim \mathcal{S} = m$). Show that there exists a unique right inverse A^- of A such that $R(A^-) = \mathcal{S}$.

14.6. *On the orthogonal decomposition.* Let A be an $m \times n$ surjective matrix, A^- be a right inverse of A and Z^- be a matrix whose columns form a basis of $N(A)$. Show that $A^-A + Z^-Z^{-\top} = I_n$ if and only if $A^- = A^\top(AA^\top)^{-1}$ (i.e., A^- is the unique right inverse of A whose range space is perpendicular to $N(A)$) and $Z^{-\top}Z^- = I_{n-m}$ (i.e., the columns of Z^- are orthonormal).

14.7. *On the oblique right inverse.* Let A be an $m \times n$ surjective matrix. Find an $n \times n$ symmetric matrix M, that is positive definite in the null space of A, such that the right inverse $\widehat{A}^-$ of A defined by (14.20) is the one given by formula (14.16). The same question to recover the right inverse given by formula (14.17).

14.8. *Quadratic convergence of $\{(x_k, y_k)\}$ without linear convergence of $\{x_k\}$.* Let $y_1 \in \,]0, 1[$ and consider the sequence $\{(x_k, y_k)\}_{k \geq 1} \in \mathbb{R}^2$ generated by $y_{k+1} = y_k^2$, $x_{k+1} = x_k$ if k is odd and $x_{k+1} = y_{k+1}^2$ if k is even. Show that $\{(x_k, y_k)\}$ converges quadratically to $(0, 0)$, while $\{x_k\}$ does not even converge linearly to 0.

14.9. *Least-squares multiplier.* Suppose that $A(x) = c'(x)$ is surjective and let $A^-(x)$ be a right inverse of $A(x)$. Find a least-squares problem, to which the least-squares multiplier $\lambda^{\mathrm{LS}}(x) = -A^-(x)^\top \nabla f(x)$ is the solution.

[*Hint:* The *least-squares problem* has the form $\min_{\lambda \in \mathbb{R}^m} \|M \nabla_x \ell(x, \lambda)\|_2$, for some nonsingular matrix M to be found.]

14.10. *Quadratically convergent fixed point iterations.* Let $\Psi : \mathbb{R}^n \to \mathbb{R}^n$ be a $C^{1,1}$ map in the neighborhood of one of its fixed points x_* (i.e., $\Psi(x_*) = x_*$). Suppose that $\Psi'(x_*) = 0$. Show that if x_1 is sufficiently close to x_*, then the sequence generated by $x_{k+1} = \Psi(x_k)$, for $k \geq 1$, converges quadratically to x_*.

14.11. *Convergence of the simplified Newton method.* Suppose that f and c are twice differentiable at a regular stationary point x_* of problem (P_E) (this allows the use of the operators $Z^-(x)$ and $A^-(x)$ introduced in § 14.2, for x near x_*) and that the reduced gradient g is differentiable near x_*. Suppose also that c', g', Z^- and A^- are Lipschitzian near x_*, and that the matrix H_k used in the simplified Newton method (14.41) satisfies $H_k - H_* = O(\|x_k - x_*\|)$. Then, there exists a neighborhood V of x_* such that, when the first iterate $x_1 \in V$, Algorithm (14.41) is well defined and generates a sequence $\{x_k\}$ converging quadratically in two steps to x_*.

[*Hint:* Show that $x_{k+1} - x_* = Z_*^-(Z_* - H_*^{-1}Z_*^{-\top}L_*)(x_k - x_*) + O(\|x_k - x_*\|^2)$, applying a technique similar to the one used in the proof of theorem 14.7, and conclude.]

15 Local Methods for Problems with Equality and Inequality Constraints

In this chapter, we consider the general minimization problem (P_{EI}), with equality and inequality nonlinear constraints, which we recall in figure 15.1. The notation used to describe this problem was given in the introduction,

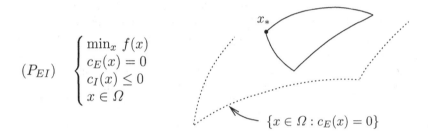

$$(P_{EI}) \quad \begin{cases} \min_x f(x) \\ c_E(x) = 0 \\ c_I(x) \leq 0 \\ x \in \Omega \end{cases}$$

$$\{x \in \Omega : c_E(x) = 0\}$$

Fig. 15.1. Problem (P_{EI}) and its feasible set

on page 193. As in chapter 14, we always suppose that c_E is a submersion (i.e., $c'_E(x)$ is surjective or onto for all x in the *open* set Ω); hence the set $c_E^{-1}(0) := \{x \in \Omega : c_E(x) = 0\}$ is a submanifold of $\mathbb{R}^n$. The feasible set of (P_{EI}), denoted by

$$X := \{x \in \Omega : c_E(x) = 0, \ c_I(x) \leq 0\},$$

is then the part of this manifold formed of the points also satisfying the inequality constraints $c_i(x) \leq 0$ for all $i \in I$. The set delimited by the curves of $c_E^{-1}(0)$ in figure 15.1 is a typical example of feasible set for problem (P_{EI}). We have put the solution x_* on the boundary of this set, but nothing imposes that this actually occurs. The solution could just as well be inside the curved triangle without touching the solid lines. Finding a solution like the one in figure 15.1 is usually more difficult than when there is no active inequality constraints (and when this fact is known). An additional fearsome difficulty, not present in problem (P_E), is indeed linked to the determination of the active constraints at the solution.

Let us recall the first-order optimality conditions of problem (P_{EI}): when the constraints are qualified at a solution $x_* \in X$, there exists a Lagrange multiplier vector $\lambda_* \in \mathbb{R}^m$ such that

$$
\text{(KKT)} \quad
\begin{cases}
(a) \; \nabla f(x_*) + A(x_*)^\top \lambda_* = 0 \\
(b) \; c_E(x_*) = 0, \quad c_I(x_*) \leq 0 \\
(c) \; (\lambda_*)_I \geq 0 \\
(d) \; (\lambda_*)_I^\top c_I(x_*) = 0.
\end{cases}
\tag{15.1}
$$

This chapter is organized as follows. In § 15.1, the SQP algorithm is introduced as a Newton-like approach to solve the KKT system (15.1). We shall stress the fact that, in the presence of nonconvexity, the solution to the osculating quadratic problem has to be selected with care. In § 15.2, we give conditions ensuring primal-dual quadratic convergence. First, the case when strict complementarity holds is examined. The active constraints at the solution are shown to be identified by the osculating quadratic problem as soon as the primal-dual iterate is in some neighborhood of a regular stationary point. The algorithm then reduces to Newton's method for the problem where the active constraints are considered as equality constraints, so that the local convergence result of theorem 14.4 can be applied. Next, we focus on the case without strict complementarity and show that quadratic convergence still holds, although the active constraint are no longer necessarily correctly identified by the osculating quadratic program. Necessary and sufficient conditions for primal superlinear convergence are given in § 15.3.

15.1 The SQP Algorithm

Introduction of the Algorithm

The Sequential Quadratic Programming (SQP) algorithm is a form of Newton's method to solve problem (P_{EI}) that is well adapted to computation. We have seen in chapter 14 that, to introduce such an algorithm, it is a good idea to start with the linearization of the optimality conditions and we follow the same approach here. Let us linearize (15.1) at the current point (x_k, λ_k), denoting by (d_k, μ_k) the change in the variables. This one solves the following system of equalities *and* inequalities in the unknown (d, μ):

$$
\begin{cases}
L_k d + A_k^\top \mu = -\nabla_x \ell_k \\
(c_k + A_k d)^\# = 0 \\
(\lambda_k + \mu)_I \geq 0 \\
(\lambda_k + \mu)_I^\top (c_k)_I + (\lambda_k)_I^\top (A_k d)_I = 0.
\end{cases}
\tag{15.2}
$$

As before, we use the notation $c_k := c(x_k)$, $A_k := A(x_k) := c'(x_k)$, $\nabla_x \ell_k = \nabla_x \ell(x_k, \lambda_k)$ and $L_k := \nabla_{xx}^2 \ell(x_k, \lambda_k)$. The notation $(\cdot)^\#$ was defined on page 194.

Because of its inequalities, (15.2) is not simple to solve. The key observation is that a good interpretation can be obtained if we add to the last equation the term $(\mu)_I^\top (A_k d)_I$. Compared with the others, this term is negligible when the steps μ_k and d_k are small, which should be the case when the iterates are close to a solution to (P_{EI}). Introducing the unknown $\lambda^{\mathrm{QP}} := \lambda_k + \mu$, the modified system (15.2) can then be written

$$\begin{cases} L_k d + A_k^\top \lambda^{\mathrm{QP}} = -\nabla f_k \\ (c_k + A_k d)^\# = 0 \\ (\lambda^{\mathrm{QP}})_I \geq 0 \\ (\lambda^{\mathrm{QP}})_I^\top (c_k + A_k d)_I = 0. \end{cases} \tag{15.3}$$

A remarkable fact, easy to check, is that (15.3) is the optimality system of the following *osculating quadratic problem* (QP)

$$\begin{cases} \min_d \ \nabla f(x_k)^\top d + \tfrac{1}{2} d^\top L_k d \\ c_E(x_k) + A_E(x_k) d = 0 \\ c_I(x_k) + A_I(x_k) d \leq 0. \end{cases} \tag{15.4}$$

This QP is easily obtained from (P_{EI}). Its constraints are those of (P_{EI}), linearized at x_k. Its objective function is hybrid, with $\nabla f(x_k)$ in the linear part and the Hessian of the Lagrangian in its quadratic part. The osculating quadratic problem (14.8), associated with the equality constrained problem (P_E), has made us familiar with the structure of (15.4).

We call *Sequential Quadratic Programming* (SQP) the algorithm generating a sequence $\{(x_k, \lambda_k)\}$ of approximations of (x_*, λ_*) by computing at each iteration a primal-dual stationary point $(d_k, \lambda_k^{\mathrm{QP}})$ of the quadratic problem (15.4), and by setting $x_{k+1} = x_k + d_k$ and $\lambda_{k+1} := \lambda_k^{\mathrm{QP}}$.

SEQUENTIAL QUADRATIC PROGRAMMING (SQP):

An initial iterate (x_1, λ_1) is given.
Compute $c(x_1)$, $\nabla f(x_1)$, and $A(x_1)$.
Set $k = 1$.
1. Stop if the KKT conditions (15.1) holds at $(x_*, \lambda_*) \equiv (x_k, \lambda_k)$ (optimality is reached).
2. Compute $L(x_k, \lambda_k)$ and find a primal-dual stationary point of (15.4), i.e., a solution $(d_k, \lambda_k^{\mathrm{QP}})$ to (15.3).
3. Set $x_{k+1} := x_k + d_k$ and $\lambda_{k+1} := \lambda_k^{\mathrm{QP}}$.
4. Compute $c(x_{k+1})$, $\nabla f(x_{k+1})$, and $A(x_{k+1})$.
5. Increase k by 1 and go to 1.

This algorithm assumes that the QP (15.4) always has a solution or, equivalently, that it is feasible and bounded (theorem 19.1). Adapted remedies must be implemented when this does not happen, such as the elastic mode of [156], which deals with infeasible linearized constraints.

What is gained with this formulation of Newton's method is that (15.4) is simpler to solve than (15.2). In fact, various quadratic programming techniques can be used to solve (15.4): active-set strategies, interior-point methods, dual approaches, *etc.* We also see that the combinatorial aspect of the original problem, which lies in the determination of the active inequality constraints, is transferred to the QP (15.4), where it is simpler to deal with than in the original nonlinear problem. However, the SQP algorithm has its own cost, which should not be overlooked. Indeed, all constraints must be linearized, including the inactive inequalities, which should play no role when the iterates are close to a solution. If these are many, the algorithm may loose some efficiency. Careful implementations use techniques to deal more efficiently with this situation (see for example [324, 301]).

Discarding Parasitic Displacements

The implementation of the SQP algorithm and the analysis of its local convergence are more complex than when only equality constraints are present. In fact, the quadratic problem (15.4) may be infeasible (its feasible set may be empty) or unbounded (the optimal value is $-\infty$), or it may have multiple local solutions (a nonconvexity effect), even in the neighborhood of a solution (x_*, λ_*) to (P_{EI}). This may happen even when (x_*, λ_*) enjoys nice properties such as the second-order sufficient conditions of optimality, strict complementarity, and constraint qualification. Here is an example.

Example 15.1. We want to minimize the logarithm of $(1+x)$ for x restricted to the interval $[0, 3]$. In canonical form, the problem is

$$\begin{cases} \min_x \log(1 + x) \\ -x \leq 0 \\ x - 3 \leq 0. \end{cases}$$

The logarithm has been used to introduce nonconvexity in the problem, since by the monotonicity of the logarithmic function, it is equivalent to minimize $(1+x)$ or $\log(1+x)$. It is easily checked that this problem has a unique primal-dual solution $(x_*, \lambda_*) = (0, (1, 0))$, which satisfies the second-order sufficient conditions of optimality, strict complementarity, and the constraint qualification (LI-CQ). It is therefore a "good" solution. However, the osculating QP (15.4) at this solution can be written

$$\begin{cases} \min_d d - \frac{1}{2}d^2 \\ -d \leq 0 \\ -3 + d \leq 0. \end{cases}$$

This problem has three primal-dual stationary points (d, λ): a local minimum $(0, (1, 0))$, a maximum $(1, (0, 0))$ and a global minimum $(3, (0, 2))$. It would be unbounded without the constraint $x \leq 3$ in the original problem, which is inactive at the solution. Among these stationary points, only the first one is suitable: it gives a zero displacement (which is to be expected from an algorithm started at a solution!), and optimal multipliers. The other two stationary points are parasitic.   □

The situation of this example can only occur if L_k is not positive definite. Otherwise, problem (15.4) is strictly convex and therefore has a unique solution as soon as the feasible set is nonempty. The convergence results given in § 15.2 assume that the parasitic solutions to the QP, like those revealed in the example, are discarded. Specifically, this is done by assuming that d_k is the minimum norm solution to the QP.

15.2 Primal-Dual Quadratic Convergence

We first analyze the well-posedness of the SQP algorithm and the convergence of the generated primal-dual sequences, when the first iterate is chosen in some neighborhood of a "regular" stationary point (a notion that is made precise in the statement of theorem 15.2 below) that satisfies strict complementarity. At such a stationary point, (LI-CQ) holds.

Theorem 15.2 highlights an interesting property of the SQP algorithm: in some neighborhood of a stationary point satisfying the assumptions above, the active constraints of the osculating quadratic problem (15.4) are the same as those of (P_{EI}). We have said that the identification of the active constraints is a major difficulty when solving inequality constrained problems and that, in the SQP algorithm, this difficulty is transferred to the osculating quadratic problem (QP), where it is easier to deal with. The result below tells us more: the active constraints of an osculating QP at one iteration are likely to be the same at the next iteration, at least close to a regular stationary point. Numerically, this means that, at least asymptotically, it is advantageous to solve the osculating QP's by algorithms that can take advantage of a good guess of the active constraints. Then, the combinatorial problem of determining which are the active constraints at the solution no longer occurs during the last iterations of the SQP algorithm.

Observe that, as this was already the case for equality constrained problems, the SQP algorithm may well generate a sequence that converges to a stationary point of (P_{EI}) that is not a minimum point of the problem. Observe indeed that, at any stationary point (x_*, λ_*) of (P_{EI}), $(0, \lambda_*)$ is a primal-dual solution to the quadratic problem, so that the SQP algorithm suggests not leaving x_*. This is due to the fact that SQP has been designed by linearizing the optimality conditions and therefore the algorithm makes no distinction between minima, maxima, or other stationary points.

Theorem 15.2 (primal-dual quadratic convergence of the SQP algorithm). *Suppose that f and c are of class C^2 in a neighborhood of a* *stationary point x_* of (P_{EI}), with associated multiplier λ_*. Suppose also that strict complementarity holds and that $(x_*, (\lambda_*)_{E \cup I_*^0})$ is a regular stationary point of the equality constrained problem*

$$\begin{cases} \min_x f(x) \\ c_i(x) = 0, \quad for\ i \in E \cup I_*^0, \end{cases} \tag{15.5}$$

in the sense of definition 14.2. Consider the SQP algorithm, in which d_k is a minimum norm stationary point of the osculating quadratic problem (15.4). Then there is a neighborhood V of (x_*, λ_*) such that, if the first iterate $(x_1, \lambda_1) \in V$:

(i) the SQP algorithm is well defined and generates a sequence $\{(x_k, \lambda_k)\}$ that converges superlinearly to (x_*, λ_*);

(ii) the active constraints of the osculating quadratic problem (15.4) are those of problem (P_{EI});

(iii) if, in addition, f and c are of class $C^{2,1}$ in a neighborhood of x_*, the convergence of $\{(x_k, \lambda_k)\}$ is quadratic.

Proof. The idea of the proof is to show that, close to (x_*, λ_*), the selected minimum norm stationary point of the osculating quadratic problem (15.4) and the primal-dual Newton step for (15.5) are identical. The result then follows from theorem 14.4.

Suppose that (x, λ) is close to (x_*, λ_*). Since $(x_*, (\lambda_*)_{E \cup I_*^0})$ is a regular stationary point of (15.5), $c'_{E \cup I_*^0}(x_*)$ is surjective and the quadratic program in $\tilde{d}$

$$\begin{cases} \min_{\tilde{d}} \nabla f(x)^{\top} \tilde{d} + \frac{1}{2} \tilde{d}^{\top} L(x, \lambda) \tilde{d} \\ c_i(x) + c'_i(x) \cdot \tilde{d} = 0, \quad for\ i \in E \cup I_*^0 \end{cases} \tag{15.6}$$

has a unique primal-dual stationary point. We denoted it by $(\tilde{d}, \tilde{\lambda}_{E \cup I_*^0})$ and form with $\tilde{\lambda}_{E \cup I_*^0}$ a vector $\tilde{\lambda} \in \mathbb{R}^m$, by setting $\tilde{\lambda}_i = 0$ for $i \in I \backslash I_*^0$.

Let us show that $(\tilde{d}, \tilde{\lambda})$ is a stationary point of the osculating quadratic problem (15.4), if $(x, \lambda) := (x_k, \lambda_k)$ is in some neighborhood of (x_*, λ_*). We only need to show that $c_i(x) + c'_i(x) \cdot \tilde{d} \leq 0$ for $i \in I \backslash I_*^0$ and $\lambda_i \geq 0$ for $i \in I_*^0$. From theorem 14.4, $(x + \tilde{d}, \tilde{\lambda})$ is close to (x_*, λ_*), when (x, λ) is close to (x_*, λ_*). Therefore, for $i \in I_*^0$, $\tilde{\lambda}_i \geq 0$, since $(\lambda_*)_i > 0$ by strict complementarity. On the other hand, $\tilde{d}$ is small, so that $c_i(x) + c'_i(x) \cdot \tilde{d} \leq 0$ for $i \in I \backslash I_*^0$. Hence $(\tilde{d}, \tilde{\lambda})$ is a stationary point of (15.4). We deduce from this that, for (x, λ) close to (x_*, λ_*), the SQP algorithm is well defined and d is small (it is a minimum norm stationary point and $\tilde{d}$ is small by theorem 14.4).

Let us now show that the pair $(d, \lambda^{\text{QP}}) := (d_k, \lambda_k^{\text{QP}})$ formed of the minimum norm solution to the QP and its associated multiplier is in fact $(\tilde{d}, \tilde{\lambda})$, if (x, λ) is in some neighborhood of (x_*, λ_*). From theorem 14.4, this will conclude the proof. For (x, λ) close to (x_*, λ_*) and $i \in I \backslash I_*^0$, $c_i(x) + c'_i(x) \cdot d < 0$, so that $\lambda_i^{\text{QP}} = 0 = \tilde{\lambda}_i$. Because of the uniqueness of the stationary point of (15.6), it remains to show that $c_i(x) + c'_i(x) \cdot d = 0$ for all $i \in I_*^0$ and (x, λ)

close to (x_*, λ_*). If this is not the case, there would exist an index $j \in I_*^0$ and a sequence $(x, \lambda) \to (x_*, \lambda_*)$, such that $c_j(x) + c_j'(x) \cdot d < 0$. Then $\lambda_j^{\mathrm{QP}} = 0$ and

$$\nabla f(x) + L(x, \lambda)d + \sum_{i \in (E \cup I_*^0) \setminus \{j\}} \lambda_i^{\mathrm{QP}} \nabla c_i(x) = 0.$$

Since $\nabla f(x) + L(x, \lambda)d \to \nabla f(x_*)$ (d is smaller than $\tilde{d}$, which converges to 0) and $c_{E \cup I_*^0}'(x_*)$ is surjective, λ_i^{QP} for $i \in (E \cup I_*^0) \setminus \{j\}$ would converge to some limit, $\bar{\lambda}_i$ say. Taking the limit in the equation above would give

$$\nabla f(x_*) + \sum_{i \in (E \cup I_*^0) \setminus \{j\}} \bar{\lambda}_i \nabla c_i(x_*) = 0.$$

Therefore, we would have found two different multipliers: $\bar{\lambda}$ (we set $\bar{\lambda}_i = 0$ for $i \notin (E \cup I_*^0) \setminus \{j\}$) and λ_* ($\bar{\lambda} \neq \lambda_*$ since $\bar{\lambda}_j = 0$ and $(\lambda_*)_j > 0$ by strict complementarity). This would be in contradiction with the uniqueness of the multiplier, which follows from the surjectivity of $c_{E \cup I_*^0}'(x_*)$. $\qquad \square$

It is clear from the proof of theorem 15.2 that it is not really necessary to take for d_k, a minimum norm stationary point of the osculating quadratic problem (15.4), some $d_k^{\min}$ say. The result is still true if the SQP algorithm ensures that $d_k \to 0$ when $d_k^{\min} \to 0$. For example, it would suffice to compute a stationary point d_k satisfying an estimate of the form $\|d_k\| \leq C \|d_k^{\min}\|$, for some positive constant C.

Theorem 15.4 below considers the case when strict complementarity does not hold, but assumes that (x_*, λ_*) satisfies the second order sufficient conditions of optimality and linear independence of the active constraint gradients (LI-CQ). The result is also local, in the sense that the first iterate (x_1, λ_1) is supposed to be close enough to (x_*, λ_*). The proof of this result is more difficult. This is because one can no longer use theorem 14.4 as in the preceding proof: the SQP step may be different from the Newton step on (15.5), however close to (x_*, λ_*) the current iterate (x, λ) can be. In other words, the property of local identification of the active constraints by the osculating quadratic problem no longer holds when complementarity is not strict. Here is an example.

Example 15.3. Consider the problem in $x \in \mathbb{R}$:

$$\begin{cases} \min_x x^2 + x^4 \\ x \leq 0. \end{cases}$$

The solution is $x_* = 0$ and $\lambda_* = 0$, so that strict complementarity does not hold. On the other hand, the constraint is qualified at x_* in the sense of (LI-CQ) and the second order sufficient conditions of optimality hold. The osculating quadratic problem at x (it does not depend on λ since the constraint is linear) is the problem in $d \in \mathbb{R}$:

$$\begin{cases} \min_d (2x + 4x^3)d + (1 + 6x^2)d^2 \\ x + d \leq 0. \end{cases}$$

If $x > 0$, $x + d = 0$ and the solution is obtained in one step. But if $x < 0$, $x + d = 4x^3/(1 + 6x^2) \in \ ]2x/3, 0[$, so that the linearized constraint is inactive and the SQP step is different from the Newton step on (15.5). In this case, however, the convergence is cubic in x (also in (x, λ)): $|x + d|/|x|^3 \leq 4$.

$\square$

The preceding example suggests that fast convergence can still be obtained even without strict complementarity. This is confirmed by the following theorem.

Theorem 15.4 (primal-dual quadratic convergence of the SQP algorithm). *Suppose that f and c are of class $C^{2,1}$ in a neighborhood of a local solution x_* to (P_{EI}). Suppose also that the constraint qualification (LI-CQ) is satisfied at x_* and denote by λ_* the associated multiplier. Finally, suppose that the second-order sufficient condition of optimality (13.8) is satisfied. Consider the SQP algorithm, in which d_k is a minimum norm stationary point of the osculating quadratic problem (15.4). Then there exists a neighborhood V of (x_*, λ_*) such that, if the first iterate $(x_1, \lambda_1) \in V$, the SQP algorithm is well defined and the sequence $\{(x_k, \lambda_k)\}$ converges quadratically to (x_*, λ_*).*

Proof. The following lemma is assumed (see [308]).

Lemma 15.5. *Under the conditions of theorem 15.4, there exists a neighborhood of (x_*, λ_*) such that (15.3) has a local solution and the local solution (d_k, λ_k^{QP}) with d_k of minimum norm satisfies:*

$$\|d_k\| + \|\lambda_k^{QP} - \lambda_*\| \leq C(\|x_k - x_*\| + \|\lambda_k - \lambda_*\|).$$

From this lemma, the algorithm is well defined if (x_k, λ_k) remains close to (x_*, λ_*). This will result from the estimates obtained below.

Let us set

$$\delta_k = \|x_k - x_*\| + \|\lambda_k - \lambda_*\|.$$

From lemma 15.5, we have

$$d_k = O(\delta_k) \quad \text{and} \quad \lambda_{k+1} - \lambda_* = O(\delta_k), \tag{15.7}$$

where d_k is a minimum-norm solution to (15.4) and $\lambda_{k+1} = \lambda_k^{QP}$ is the associated multiplier. We deduce that, for $i \in I \backslash I_*^0$ and δ_k small enough, we have

$$c_i(x_k) + c_i'(x_k) \cdot d_k < 0.$$

Hence $(\lambda_{k+1})_i = 0$, and with the set of indices

$$J = E \cup I_*^0,$$

the optimality of d_k is expressed by

$$L_k d_k + A_J(x_k)^\top (\lambda_{k+1})_J + \nabla f_k = 0.$$

A Taylor expansion of the left-hand side, using $\nabla_x \ell(x_*, \lambda_*) = 0$, $x_{k+1} = x_k + d_k$ and (15.7), leads to

$$
\begin{aligned}
0 &= \nabla_x \ell(x_k, \lambda_*) + L(x_k, \lambda_k)d_k + A_J(x_k)^\top(\lambda_{k+1} - \lambda_*)_J \\
&= L_*(x_{k+1} - x_*) + A_J(x_*)^\top(\lambda_{k+1} - \lambda_*)_J + O(\delta_k^2).
\end{aligned}
\tag{15.8}
$$

Expand likewise the constraints of the osculating quadratic problem: we have for $i \in J$

$$c_i(x_k) + c_i'(x_k) \cdot d_k = c_i'(x_*) \cdot (x_{k+1} - x_*) + (\gamma_k)_i, \tag{15.9}$$

where $(\gamma_k)_i = O(\delta_k^2)$.

From the assumption, $A_J(x_*)$ is surjective, so we can find a vector $v_k \in \mathbb{R}^m$ such that

$$A_J(x_*) v_k = (\gamma_k)_J \quad \text{and} \quad v_k = O(\delta_k^2).$$

The last estimate can be obtained by taking a minimum-norm v_k satisfying the first equation. With the notation

$$w_k = x_{k+1} - x_* + v_k,$$

(15.9) becomes for $i \in J$:

$$c_i(x_k) + c_i'(x_k) \cdot d_k = c_i'(x_*) \cdot w_k. \tag{15.10}$$

The complementarity conditions of the osculating quadratic problem can be written

$$(\lambda_{k+1})_i (c_i(x_k) + c_i'(x_k) \cdot d_k) = 0, \quad \text{for all } i \in I. \tag{15.11}$$

Hence, if $(\lambda_*)_i > 0$ and δ_k small enough, we have $c_i(x_k) + c_i'(x_k) \cdot d_k = 0$. Then we obtain from (15.10)

$$
\begin{cases}
c_i'(x_*) \cdot w_k = 0 \text{ if } i \in E \cup I_*^{0+} \\
c_i'(x_*) \cdot w_k \le 0 \text{ if } i \in I_*^{00}.
\end{cases}
\tag{15.12}
$$

This shows that w_k lies in the critical cone C_*, defined by (13.6). From the second-order sufficiency condition, we then have for a constant $C_1 > 0$:

$$C_1 \|w_k\|^2 \le w_k^\top L_* w_k. \tag{15.13}$$

Now compute $w_k^\top L_* w_k$. From (15.8) and $v_k = O(\delta_k^2)$,

$$w_k^\top L_* w_k = -(\lambda_{k+1} - \lambda_*)_J^\top A_J(x_*) w_k + O(\|w_k\| \delta_k^2) \le C_2 \|w_k\| \delta_k^2,$$

since $(\lambda_{k+1} - \lambda_*)_J^\top A_J(x_*) w_k = 0$ thanks to (15.11) and (15.12). With (15.13), we then obtain

$$C_1 \|w_k\| \le C_2 \delta_k^2.$$

Since $v_k = O(\delta_k^2)$, we deduce

$$x_{k+1} - x_* = O(\delta_k^2).$$

On the other hand, this estimate, (15.8) and the injectivity of $A_J(x_*)^\top$ show that

$$(\lambda_{k+1} - \lambda_*)_J = O(\delta_k^2).$$

Since $(\lambda_{k+1})_i = (\lambda_*)_i = 0$ for $i \in I \backslash I_*^0$, these last two estimates show the quadratic convergence of the sequence $\{(x_k, \lambda_k)\}$. $\square$

15.3 Primal Superlinear Convergence

Theorem 15.4 gives conditions for the quadratic convergence of $\{(x_k, \lambda_k)\}$. Actually, this implies neither quadratic nor superlinear convergence for $\{x_k\}$ (see exercise 14.8). Nevertheless, the following result (theorem 15.7) shows that, for the SQP algorithm using the Hessian of the Lagrangian in the quadratic programs (15.4), the sequence $\{x_k\}$ converges superlinearly. This result is interesting because it is often desirable to have fast convergence of this sequence.

We consider for this an algorithm slightly more general than the one described in § 15.1, which encompasses the quasi-Newton versions of the method. We suppose that $\{x_k\}$ is generated by

$$x_{k+1} = x_k + d_k,$$

where d_k is a stationary point of the quadratic problem

$$\begin{cases} \min_d \nabla f(x_k)^\top d + \frac{1}{2} d^\top M_k d \\ (c(x_k) + A(x_k)d)^\# = 0. \end{cases} \tag{15.14}$$

This is the same problem as (15.4), but the Hessian of the Lagrangian L_k is replaced by a symmetric matrix M_k. Incidentally, note that the multiplier λ_k is no longer explicitly used in the algorithm. Theorem 15.7 gives a necessary and sufficient condition on M_k to guarantee superlinear convergence of $\{x_k\}$.

The optimality conditions of (15.14) are (λ_k^{QP} is the multiplier associated with the constraints):

$$\begin{cases} (a)\ \nabla f_k + M_k d_k + A_k^\top \lambda_k^{\mathrm{QP}} = 0 \\ (b)\ (c_k + A_k d_k)^\# = 0 \\ (c)\ (\lambda_k^{\mathrm{QP}})_I \ge 0 \\ (d)\ (\lambda_k^{\mathrm{QP}})_I (c_k + A_k d_k)_I = 0 \end{cases} \tag{15.15}$$

We shall need the orthogonal projector onto the critical cone C_* at a solution x_* to (P_{EI}) (see (13.6)). We denote this (nonlinear) projector by P_*. It is well defined since C_* is a nonempty closed convex set.

Lemma 15.6. *If $\lambda \in \mathbb{R}^m$ is such that $\lambda_{I_*^{00}} \geq 0$ and $\lambda_{I\setminus I_*^0} = 0$, then $P_* A_*^\top \lambda = 0$.*

Proof. Take $\lambda \in \mathbb{R}^m$ as in the terms of the lemma and $h \in C_*$. Then $(A_* h)_{E \cup I_*^{0+}} = 0$, $(A_* h)_{I_*^{00}} \leq 0$, and we have

$$(0 - A_*^\top \lambda)^\top (h - 0) = -\lambda^\top A_* h = -\lambda_{I_*^{00}}^\top (A_* h)_{I_*^{00}} \geq 0.$$

The characterization (13.12) of the projection yields the result.   □

Theorem 15.7 (primal superlinear convergence of the SQP algorithm). *Suppose that f and c are twice differentiable at $x_* \in \Omega$. Suppose also that (x_*, λ_*) is a primal-dual solution to (P_{EI}) satisfying (LI-CQ) and the second-order sufficient condition of optimality (13.8). Consider the sequence $\{(x_*, \lambda_*)\}$ generated by the recurrence $x_{k+1} = x_k + d_k$ and $\lambda_{k+1} = \lambda_k^{QP}$, where (d_k, λ_k^{QP}) is a primal-dual solution to (15.14). Suppose that $\{(x_k, \lambda_k)\}$ converges to (x_*, λ_*). Then $\{x_k\}$ converges superlinearly if and only if*

$$P_*(L_* - M_k)d_k = o(\|d_k\|), \tag{15.16}$$

where P_* is the orthogonal projector onto the critical cone C_*.

Proof. Using $(15.15)_a$, $\nabla_x \ell(x_*, \lambda_*) = 0$ and $\lambda_{k+1} \to \lambda_*$, we have

$$\begin{aligned}
-M_k d_k &= \nabla_x \ell(x_k, \lambda_{k+1}) \\
&= \nabla_x \ell(x_*, \lambda_{k+1}) + L(x_*, \lambda_{k+1})(x_k - x_*) + o(\|x_k - x_*\|) \\
&= A_*^\top(\lambda_{k+1} - \lambda_*) + L_*(x_k - x_*) + o(\|x_k - x_*\|).
\end{aligned}$$

Hence

$$(L_* - M_k)d_k = A_*^\top(\lambda_{k+1} - \lambda_*) + L_*(x_{k+1} - x_*) + o(\|x_k - x_*\|). \tag{15.17}$$

To show that condition (15.16) is necessary, assume that $x_{k+1} - x_* = o(\|x_k - x_*\|)$. Then (15.17) gives

$$(L_* - M_k)d_k = A_k^\top(\lambda_{k+1} - \lambda_*) + o(\|x_k - x_*\|).$$

Project with P_*, which is Lipschitzian (see (13.15)), and observe that, from $(15.15)_c$ and $(15.15)_d$, $(\lambda_{k+1} - \lambda_*)$ satisfies for large k the conditions on λ of lemma 15.6:

$$P_*(L_* - M_k)d_k = P_* A_*^\top(\lambda_{k+1} - \lambda_*) + o(\|x_k - x_*\|) = o(\|x_k - x_*\|).$$

Condition (15.16) follows, because $(x_k - x_*) \sim d_k$ by lemma 13.5.

Conversely, let us show that condition (15.16) is sufficient. For $i \in J := E \cup I_*^0$, we have

$$c_i(x_k) + c_i'(x_k) \cdot d_k = c_i'(x_*) \cdot (x_{k+1} - x_*) + (\gamma_k)_i,$$

where $(\gamma_k)_i = o(\|x_k - x_*\|) + o(\|d_k\|)$. Since $A_J(x_*)$ is surjective, $(\gamma_k)_J = A_J(x_*)v_k$, for some $v_k = o(\|x_k - x_*\|) + o(\|d_k\|)$. With the notation

$$w_k := x_{k+1} - x_* + v_k,$$

there holds

$$c_i(x_k) + c_i'(x_k) \cdot d_k = c_i'(x_*) \cdot w_k, \quad \text{for } i \in J.$$

Now $c_i(x_k) + c_i'(x_k) \cdot d_k = 0$ for $i \in E \cup I_*^{0+}$ and k large enough, so that

$$\begin{cases} c_i'(x_*) \cdot w_k = 0 \text{ if } i \in E \cup I_*^{0+} \\ c_i'(x_*) \cdot w_k \leq 0 \text{ if } i \in I_*^{00}. \end{cases}$$

This implies that $w_k \in C_*$ for large k (see (13.6)) and that, for some constant $C_1 > 0$,

$$C_1 \|w_k\|^2 \leq w_k^\top L_* w_k, \quad \text{for large } k. \tag{15.18}$$

On the other hand, for $i \in I_*^{00}$, from $(15.15)_d$, we have $0 = (\lambda_{k+1})_i (c_i(x_k) + c_i'(x_k) \cdot d_k) = (\lambda_{k+1})_i (c_i'(x_*) \cdot w_k)$ and $(\lambda_*)_i = 0$. While for $i \in I \backslash I_*^0$, $(\lambda_{k+1} - \lambda_*)_i = 0$. Therefore

$$(\lambda_{k+1} - \lambda_*)^\top A_* w_k = 0, \quad \text{for large } k.$$

Now, with this equation, (15.17), $v_k = o(\|x_k - x_*\|) + o(\|d_k\|)$, the fact that $u^\top v \leq u^\top P_* v$, for all $v \in \mathbb{R}^n$ and all $u \in C_*$ (see (13.14)), and (15.16), we find that

$$\begin{aligned} w_k^\top L_* w_k &= w_k^\top L_*(x_{k+1} - x_*) + O(\|w_k\| \, \|v_k\|) \\ &= w_k^\top (L_* - M_k)d_k + o(\|w_k\| \, \|x_k - x_*\|) + o(\|w_k\| \, \|d_k\|) \\ &\leq w_k^\top P_*(L_* - M_k)d_k + o(\|w_k\| \, \|x_k - x_*\|) + o(\|w_k\| \, \|d_k\|) \\ &= o(\|w_k\| \, \|x_k - x_*\|) + o(\|w_k\| \, \|d_k\|). \end{aligned}$$

With (15.18), $w_k = o(\|x_k - x_*\|) + o(\|d_k\|)$; hence

$$x_{k+1} - x_* = o(\|x_k - x_*\|) + o(\|d_k\|).$$

The property $x_{k+1} - x_* = o(\|x_k - x_*\|)$ follows easily.   □

When there are no inequality constraints, P_* is the orthogonal projector onto the null space $N(A_*)$. It is then linear. Given a basis Z_*^- of $N(A_*)$, it can be written

$$P_* = Z_*^- (Z_*^{-\top} Z_*^-)^{-1} Z_*^{-\top}.$$

Since Z_*^- is injective and $Z_*^{-\top} Z_*^-$ is nonsingular, condition (15.16) can be written

$$Z_*^{-\top}(L_* - M_k)d_k = o(\|d_k\|) \quad \text{or} \quad (Z_*^{-\top} L_* - Z_k^{-\top} M_k)d_k = o(\|d_k\|).$$

To write the last condition, we have supposed that $Z^-(\cdot)$ is continuous at x_* and that $\{M_k\}$ is bounded. This shows that the important part of M_k is $Z_k^{-\top} M_k$, which reminds us that only the part $Z_k^{-\top} L_k$ of L_k plays a role in the definition of the Newton direction for equality constrained problems (see observation 1 on page 235).

15.4 The Hanging Chain Project III

In this third session, we resume the project on the determination of the static equilibrium position of a hanging chain, started in § 13.8 and developed in § 14.7. Our present objective is to implement the local SQP algorithm, presented on page 257, to be able to take into account the floor constraint. The algorithm is quite similar to the Newton method implemented in the second session. The main difference is that the solver of linear equations has to be replaced by a solver of quadratic optimization problems. This simple change will have several consequences that are discussed in this section.

It is a good idea to keep the work done in the second session and to use $\mathtt{mi} = m_I$ as a flag that makes the sqp function select the type of solver (linear or quadratic), depending on the presence of inequality constraints. Solving a linear system is indeed much simpler than solving a quadratic optimization problem, so that the sqp function must be allowed to take advantage of the absence of inequality constraints.

Modifications to Bring to the sqp Function

Most of the work has been done in the previous session. There are only two modifications to bring to the function sqp.

The main change consists in substituting a quadratic optimization solver (to solve (15.4)) for the linear solver previously used in sqp (see chapter 14). Writing a solver of quadratic optimization problems is a difficult task. Fortunately, in our case, the MATLAB solver quadprog can be used, so that we can concentrate on other aspects of the SQP algorithm. Quadprog first finds an initial feasible point by solving a linear optimization problem and then uses an active set method to find a solution to the quadratic problem. It can detect infeasibility and unboundedness.

A second change deals with the determination of the initial dual solution $\lambda = (\lambda_E, \lambda_I)$. Since it is known that λ_I must be nonnegative, it is better now to determine λ as a solution to the bound constrained least-squares problem

$$\min_{\substack{\lambda=(\lambda_E,\lambda_I)\in\mathbb{R}^m \\ \lambda_I\geq 0}} \frac{1}{2}\|\nabla_x\ell(x,\lambda)\|_2^2,$$

instead of using (14.47). This convex quadratic optimization problem always has a solution (theorem 19.1). It can be solved by quadprog.

Checking the Correctness of the SQP Solver

There is little change to make an error on the part of the simulator dealing with the inequality constraints, since these are very simple. Nevertheless, it is better to check it and to verify the implementation of the quadratic solver.

The same strategy as in the case with equality constrained problems can be followed: trying to solve more and more difficult problems and check the quadratic convergence of the generated sequence.

Let us check the quadratic convergence on the following variant of test case 1a, in which we add a floor constraint.

Test case 1e: same data as for the test case 1a (namely second hook at $(a, b) = (1, -0.3)$ and bars of lengths $L = (0.4, 0.3, 0.25, 0.2, 0.4)$) with an additional floor with parameters $(g_0, g_1) = (-0.35, -0.2)$ (see the definition of the floor in (13.25)). The initial positions of the chain joints are $(0.1, -0.3)$, $(0.4, -0.5)$, $(0.6, -0.4)$, and $(0.7, -0.5)$.

The results obtained with test case 1e are shown in figure 15.2. Convergence

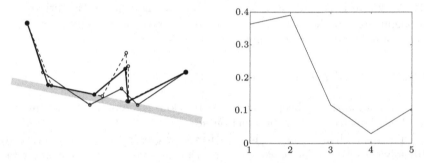

Fig. 15.2. Test case 1e

with `options.tol(1 : 4)` $= 10^{-10}$ is again obtained in 6 iterations. The picture on the left uses the same conventions as before: the thin solid bars represent the initial position of the chain, the dashed bars correspond to the 5 intermediate positions (hardly distinguishable), and the bold solid bars are those of the final optimal position. This one is a local minimum (the multipliers associated with the inequality constraints are positive and the critical cone is reduced to $\{0\}$). The picture on the right gives a plot of the ratios $\|s_{k+1}\|_2 / \|s_k\|_2^2$, where $s_k = z_{k+1} - z_k$, for $k = 1, \ldots, 5$. The boundedness of these ratios shows without any doubt that the sequence $\{z_k\} = \{(x_k, \lambda_k)\}$ converges quadratically to its limit, as predicted by the theory (theorems 15.2 and 15.4).

Experimenting with the SQP Algorithm

A first observation, with unpleasant consequences, is that `quadprog` is aimed at computing a local minimum of a quadratic problem, not an arbitrary stationary point (note that finding a solution or a stationary point of a non-convex quadratic problem is NP-hard, see [354] for example). Therefore, it is quite frequent to find situations where `quadprog` fails to find a stationary point, as required by the SQP algorithm. For example, with test case 1f below, which is test case 1e with the initial position of the chain given in test case 1a, the first osculating quadratic problem is unbounded.

Test case 1f: same data as for the test case 1e; but the initial positions of the chain joints are $(0.2, -0.5)$, $(0.4, -0.6)$, $(0.6, -0.8)$, and $(0.8, -0.6)$.

The unboundedness of an osculating quadratic problem can occur only when its feasible set is unbounded and the Hessian of the Lagrangian L is not positive definite at the current iterate. Hence, taking a positive definite approximation of L cures the difficulty. This can be obtained by using a quasi-Newton approximation of L; this technique is considered in chapter 18. Another possibility is add to L a (not too large) positive diagonal matrix E, such that $L+E$ is positive definite (for example by using a modified Cholesky factorization of L [154, 201]). Figure 15.3 shows the results obtained with this technique.

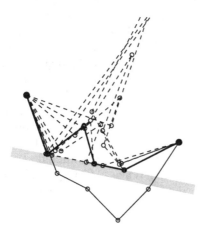

Fig. 15.3. Test case 1f

The optimal chain is actually a local minimum (the critical cone is reduced to $\{0\}$ and the energy is $e = -0.489$), different from the one obtained in figure 15.2 (in which $e = -0.518$). Observe that, although the initial position of the chain is not feasible for the floor constraint, the subsequent positions are all feasible. This is due to the affinity of the floor constraint (see (13.25) and exercise 15.1).

Another difficulty arises when the linearized constraints are incompatible, leading to an infeasible osculating quadratic problem. This difficulty is encountered at the first iteration with the initial chain given in test case 1g below. Remedies for this kind of situations exist, see [351, 341, 156] and the references thereof.

Test case 1g: same data as for the test case 1e; but the initial positions of the chain joints are $(0.1, -0.3)$, $(0.3, -0.4)$, $(0.6, -0.4)$, and $(0.7, -0.4)$.

Notes

The SQP algorithm, in a form equivalent to the one introduced in § 15.1 on page 257, was first proposed by Wilson [359; 1963]. This author was mainly concerned with the extension of the simplex method, first to quadratic programming, and then to nonlinear convex optimization problems. The algorithm was obtained by searching for a saddle point of a quadratic approximation of the Lagrangian in the primal and dual variables. No convergence proof was given. See also the introduction of this part of the book, on page 191, for other references on the origin of the SQP algorithm.

The local quadratic convergence of theorem 15.2 is due to several authors; see for example [307], in which various classes of algorithms are considered. Theorem 15.4 is taken from [38]; further refinements can be found in [40].

The criterion (15.16) for superlinear convergence dates back to Dennis and Moré [104], who introduced a similar condition to characterize the superlinear convergence of sequences generated by quasi-Newton methods in unconstrained optimization (see theorem 4.11). It was extended to problems with equality constraints by Boggs, Tolle, and Wang [36], under a somewhat strong assumption (linear convergence of the sequence $\{x_k\}$). The possibility of getting rid of this assumption has been observed by many authors. The generalization to inequality constrained problems given in theorem 15.7 is due to Bonnans [40], who uses a projector varying along the iterations; in contrast, we use the projector P_* onto the critical cone.

The local convergence of the SQP algorithm has been extended to different contexts, such as semi-infinite programming [180], infinite dimension programming [3, 4, 5, 6, 219]. When (MF-CQ) holds, but not (LI-CQ), the optimal multiplier may not be unique, so that the limit behavior of the multiplier sequence $\{\lambda_k^{QP}\}$ is difficult to predict; this situation is analyzed in [367, 183, 7, 8].

Exercise

15.1. Consider the SQP algorithm applied to problem (P_{EI}) in which the ith constraint, for some $i \in E \cup I$ (equality or inequality constraint), is affine (i.e., $c_i(x+d) = c_i(x) + c_i'(x){\cdot}d$ for all x and $d \in \mathbb{R}^n$). Let (x, λ) be the current iterate and define x_+ by $x_+ := x + \alpha d$, where d is a solution to the osculating quadratic problem (15.4) (we drop the index k) and $\alpha \in {]}0, 1]$. Show that x_+ is feasible for the ith constraint (i.e., $c_i(x_+) = 0$ if $i \in E$, or $c_i(x_+) \le 0$ if $i \in I$) if either x is feasible for the ith constraint or if $\alpha = 1$.

16 Exact Penalization

16.1 Overview

The algorithms studied in chapters 14 and 15 generate converging sequences if the first iterate is close enough to a regular stationary point (see theorems 14.4, 14.5, 14.7, 15.2, and 15.4). Such an iterate is not necessarily at hand, so it is important to have techniques that allow the algorithms to force convergence, even when the starting point is far from a solution. This is known as the *globalization* of a local algorithm. The term is a little ambiguous, since it may suggest that it has a link with the search of global minimizers of (P_{EI}). This is not at all the case (for an entry point on *global optimization*, see [200]).

There are (at least) two classes of techniques to globalize a local algorithm: *line-search* and *trust-region*; we shall only consider the line-search approach in this survey. Both techniques use the same idea: the progress made from one iterate x_k to the next one x_{k+1} towards the solution is measured by means of an auxiliary function, called the *merit function* (the novel notion of *filter*, not discussed in this part, looks like a promising alternative; see [130] for the original paper). In unconstrained optimization, "the" appropriate merit function is of course the objective f itself. Here, the measure has to take into account the two, usually contradictory, goals in (P_{EI}): minimizing f and satisfying the constraints. Accordingly, the merit function has often the following form

$$f(x) + p(x),$$

where p is a function penalizing the constraint violation: p is zero on the feasible set and positive outside. Instead of merit functions, one also speaks of *penalty functions*, although the latter term is usually employed when the penalty function is minimized by algorithms for unconstrained optimization. As we shall see, the approach presented here is more subtle: truly constrained optimization algorithms are used (like those in chapters 14 and 15); the merit function only intervenes as a tool for measuring the adequacy of the step computed by the local methods. It is not used for computing the direction itself. The main advantage is that the ill-conditioning encountered with penalty methods is avoided, and the fast speed of convergence of the local methods is ensured close to a solution.

As many merit functions exist, a selection must be made. We shall only study those that do not use the derivatives of f and c. These are the most widely encountered in optimization codes and their numerical effectiveness has been demonstrated. To start with, let us examine some common examples of merit/penalty functions. We denote by $\|\cdot\|_2$ the ℓ_2 norm and by $\|\cdot\|_P$ an arbitrary norm.

(a) Quadratic penalization:

$$f(x) + \frac{\sigma}{2}\|c(x)^{\#}\|_2^2. \tag{16.1}$$

(b) Lagrangian:

$$f(x) + \mu^{\top}c(x).$$

(c) Augmented Lagrangian (case $I = \emptyset$):

$$f(x) + \mu^{\top}c(x) + \frac{\sigma}{2}\|c(x)\|_2^2. \tag{16.2}$$

Augmented Lagrangian (general case):

$$f(x) + \mu_E^{\top}c_E(x) + \frac{\sigma}{2}\|c_E(x)\|_2^2$$
$$+ \sum_{i \in I}\left(\mu_i \max\left(\frac{-\mu_i}{\sigma}, c_i(x)\right) + \frac{\sigma}{2}\left[\max\left(\frac{-\mu_i}{\sigma}, c_i(x)\right)\right]^2\right). \tag{16.3}$$

(d) Nondifferentiable augmented function:

$$f(x) + \sigma\|c(x)^{\#}\|_P.$$

These functions have quite different features. One important property that distinguishes them is the *exactness* of the penalization, which is the subject of the present chapter. The concept of exact penalization is sometimes ambiguous – or at least varies from author to author. We adopt the following definition.

A function $\Theta : \Omega \to \mathbb{R}$ is called an *exact penalty function* at a local minimum x_* of (P_{EI}) if x_* is a local minimum of Θ. The converse implication (x_* is a local minimum of (P_{EI}) whenever it minimizes Θ locally) is not generally possible unless feasibility of x_* is assumed. The example in figure 16.1 is an illustration: x'_* is a local minimum of the functions (a) or (d) with $\sigma \geq 0$ but, being infeasible, it is not a solution to the minimization problem. The reason why the concept of exactness is so important for globalizing the SQP algorithm will be discussed in chapter 17.

Table 16.1 gives some properties of the merit functions (a)–(d). This deserves some comments.

- As far as the differentiability of Θ_{σ} is concerned, we assume that f and c are of class C^{∞}. We see that, in general, the presence of inequality constraints decreases the degree of differentiability of the merit functions. In this respect, the Lagrangian (b) is an exception.

$$\begin{cases} \min_x 0 \\ c(x) = 0 \end{cases}$$

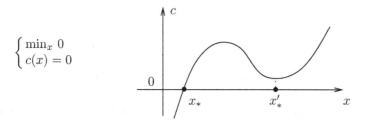

Fig. 16.1. Exactness and feasibility

Function	Differentiability	Exactness	Conditions for exactness	Threshold of σ depends on
(a)	C^1	no		
(b)	C^∞	yes	(P_{EI}) convex $\mu = \lambda_*$	
(c)	C^1	yes	$\mu = \lambda_*$ σ large	2nd derivatives
(d)	C^0	yes	σ large	1st derivatives

Table 16.1. Comparison of some merit functions

- We also see that only functions (b)–(d) can be exact. The quadratic penalty function is hardly ever exact: if $I = \emptyset$, it is differentiable and its gradient at a solution x_* is $\nabla f(x_*)$, which is usually nonzero. As we shall see in the following sections, the Lagrangian (b) is exact for convex problems and the augmented Lagrangian (c) is exact for nonconvex problems provided the penalty parameter σ is large enough.

- To be exact, both functions (b) and (c) need to have $\mu = \lambda_*$. From an algorithmic point of view, this means that the value of μ must be continually modified in order to approximate the unknown optimal multiplier λ_*. Algorithms using the Lagrangians do not minimize the same function at each iteration, which can raise convergence difficulties.

- Another shortcoming of (c) is that the threshold of σ, beyond which the penalization becomes exact, involves the eigenvalues of the Hessian of the Lagrangian. It is therefore not easily accessible to computation, and certainly out of reach if the Hessian of the Lagrangian is not explicitly computed, as in the quasi-Newton versions of the algorithms. Nevertheless, many algorithms use this function (for example, those described in [85]).

- Finally, the conditions for the exactness of function (d) are less restrictive and this is the main reason why this merit function is often used for globalizing the SQP algorithm, as in chapter 17. We shall see in particular that the threshold of σ can easily be estimated during the iterations, with the help of an estimate of the optimal multiplier. Function (d) is nonsmooth, however.

In the rest of this chapter, we study some properties of the merit functions (b)–(d), focusing on their exactness.

16.2 The Lagrangian

In this section, problem (P_{EI}) is assumed to be convex: f and the c_i's, $i \in I$, are convex, and c_E is affine. In this case, the Lagrangian of the problem is exact at a solution x_*, providing the multiplier is set to a dual solution λ_*. Actually, proposition 16.1 below shows a little more than that: ℓ has a saddle-point at (x_*, λ_*), a concept made precise in the next definition.

Let X and Y be two sets and let $\varphi : X \times Y \to \mathbb{R}$ be a function. We say that $(x_*, y_*) \in X \times Y$ is a *saddle-point* of φ on $X \times Y$ when

$$\varphi(x_*, y) \le \varphi(x_*, y_*) \le \varphi(x, y_*), \quad \text{for all } x \in X \text{ and } y \in Y.$$

Thus, $x \mapsto \varphi(x, y_*)$ is minimal at x_* and $y \mapsto \varphi(x_*, y)$ is maximal at y_*.

Recall that the *Lagrangian* of problem (P_{EI}) is the function

$$(x, \mu) \in \Omega \times \mathbb{R}^m \mapsto \ell(x, \mu) = f(x) + \mu^\top c(x). \tag{16.4}$$

If a feasible point x_* minimizes $\ell(\cdot, \mu)$, then $0 = \nabla_x \ell(x_*, \mu)$, which indicates that x_* will be a solution to (P_{EI}) provided μ is a dual solution. The following result shows that, for convex problems, the primal-dual solutions to (P_{EI}) are saddle-points of ℓ on $\Omega \times \{\mu \in \mathbb{R}^m : \mu_I \ge 0\}$. The way is then open to computing primal-dual solutions to (P_{EI}) with algorithms computing saddle-points. We shall not proceed in that way but it is useful to bear this point of view in mind. In addition, this result shows that $x \mapsto \ell(x, \lambda_*)$ is an exact penalty function for convex problems.

Proposition 16.1 (saddle-point of the Lagrangian). *Suppose that problem (P_{EI}) is convex, that x_* is a solution, and that f and c are differentiable at x_*. Suppose also that there exists a multiplier λ_* such that the optimality conditions (KKT) are satisfied. Then (x_*, λ_*) is a saddle-point of the Lagrangian defined in (16.4) on $\Omega \times \{\mu \in \mathbb{R}^m : \mu_I \ge 0\}$.*

Proof. Take $\mu \in \{\mu \in \mathbb{R}^m : \mu_I \ge 0\}$. We have

$$
\begin{aligned}
\ell(x_*, \mu) &= f(x_*) + \mu_I^\top c_I(x_*) && \text{[because } c_E(x_*) = 0\text{]} \\
&\le f(x_*) && \text{[because } \mu_i c_i(x_*) \le 0 \text{ for } i \in I\text{]} \\
&= f(x_*) + \lambda_*^\top c(x_*) && \text{[because } c_E(x_*) = 0 \text{ and } (\lambda_*)_I^\top c_I(x_*) = 0\text{]} \\
&= \ell(x_*, \lambda_*).
\end{aligned}
$$

On the other hand, since $(\lambda_*)_I \geq 0$ and (P_{EI}) is convex, the function $x \in \Omega \mapsto \ell(x, \lambda_*)$ is convex. According to the assumptions, this function is differentiable at x_* and, in view of the optimality conditions (KKT), we have $\nabla_x \ell(x_*, \lambda_*) = 0$. We deduce that this function is minimal at x_*: $\ell(x_*, \lambda_*) \leq \ell(x, \lambda_*)$, for all $x \in \Omega$. □

16.3 The Augmented Lagrangian

The Lagrangian (16.4) is not an exact penalty function if the problem is nonconvex. For example, the nonconvex problem

$$\begin{cases} \min_x \ \log(x) \\ x \geq 1 \end{cases}$$

has the unique primal-dual solution $(x_*, \lambda_*) = (1, 1)$ and its Lagrangian $\ell(x, \lambda_*) = \log(x) + 1 - x$ is concave with a *maximum* at $x = 1$.

The augmented Lagrangian ℓ_r obviates this shortcoming. In fact we shall prove a local version of proposition 16.1: if $\mu = \lambda_*$ and r is large enough, $\ell_r(\cdot, \mu)$ has a strict local minimum at a strong solution to the optimization problem (P_{EI}).

The augmented Lagrangian is best introduced by using a perturbation technique as in duality theory, but this is beyond the scope of this review. Here we follow a more intuitive approach, starting with the case where only equality constraints are present. In this case, one takes

$$\ell_r(x, \mu) = f(x) + \mu_E^\top c_E(x) + \frac{r}{2}\|c_E(x)\|_2^2. \tag{16.5}$$

This is the standard Lagrangian ℓ, augmented by the term $(r/2)\|c_E(x)\|_2^2$. This term penalizes the constraint violation and makes $\ell_r(\cdot, \mu)$ convex around the point x_* in a subspace complementary to the tangent space $N(A_E(x_*))$. This creates a basin around a strong solution to (P_E), making the penalization exact (this point of view is developed in exercise 16.2).

To deal with inequality constraints, we first transform (P_{EI}) by introducing slack variables $s \in \mathbb{R}^{m_I}$:

$$\begin{cases} \min_{(x,s)} \ f(x) \\ c_E(x) = 0 \\ c_I(x) + s = 0 \\ s \geq 0. \end{cases}$$

Next, this problem is *approached* by using the augmented Lagrangian associated with its equality constraints:

$$\min_x \ \min_{s \geq 0} \ \left(f(x) + \mu_E^\top c_E(x) + \frac{r}{2}\|c_E(x)\|_2^2 + \mu_I^\top(c_I(x)+s) + \frac{r}{2}\|c_I(x)+s\|_2^2 \right).$$

The augmented Lagrangian associated with (P_{EI}) is the function of x and μ defined by the minimal value of the optimization problem in $s \geq 0$ above:

$$\ell_r(x,\mu) := \min_{s \geq 0}\left(f(x) + \mu_E^\top c_E(x) + \frac{r}{2}\|c_E(x)\|_2^2\right.$$
$$\left. + \mu_I^\top(c_I(x)+s) + \frac{r}{2}\|c_I(x)+s\|_2^2\right).$$

Actually, the minimization in s can be carried out explicitly since the minimized function of s is quadratic with a positive diagonal Hessian. More precisely, discarding terms independent of s, the objective can be written $\frac{r}{2}\|s+c_I(x)+\mu_I/r\|_2^2$, so that the minimizer is the projection of $-c_I(x)-\mu_I/r$ on the positive orthant, namely $s = \max(-c_I(x) - \mu_I/r, 0)$. Adding $c_I(x)$, one finds

$$c_I(x) + s = \max\left(\frac{-\mu_I}{r}, c_I(x)\right).$$

Substituting $c_I(x) + s$ by this value in the objective of the problem above yields an explicit formula for the *augmented Lagrangian*. This is the function $\ell_r : \Omega \times \mathbb{R}^m \to \mathbb{R}$, defined for $(x,\mu) \in \Omega \times \mathbb{R}^m$ and $r \in \mathbb{R}_+^* := \{t \in \mathbb{R} : t > 0\}$ by

$$\ell_r(x,\mu) = f(x) + \mu^\top \tilde{c}_r(x,\mu) + \frac{r}{2}\|\tilde{c}_r(x,\mu)\|_2^2, \qquad (16.6)$$

where $\tilde{c}_r : \Omega \times \mathbb{R}^m \to \mathbb{R}^m$ is defined by

$$(\tilde{c}_r(x,\mu))_i = \begin{cases} c_i(x) & \text{if } i \in E \\ \max\left(\frac{-\mu_i}{r}, c_i(x)\right) & \text{if } i \in I. \end{cases} \qquad (16.7)$$

The coefficient r is called the *augmentation parameter*. This augmented Lagrangian (16.6) has therefore a structure very similar to the one associated with the equality constraint problem (P_E), see (16.5), with c_E substituted by the non-differentiable function $\tilde{c}_r$ introduced above.

Despite the nonsmoothness of the max operator in (16.7), the augmented Lagrangian is differentiable in x, provided that f and c have that property. The easiest way of verifying this claim is to write the terms associated with the inequalities in (16.6) as follows

$$\mu_I^\top(\tilde{c}_r(x,\mu))_I + \frac{r}{2}\|(\tilde{c}_r(x,\mu))_I\|_2^2 = \frac{1}{2r}\sum_{i \in I}\left(\max(0, \mu_i + rc_i(x))^2 - \mu_i^2\right).$$

This is a differentiable function of x, since $\max(0, \cdot)$ is squared. A straightforward computation then leads to

$$\nabla_x \ell_r(x,\mu) = \nabla f(x) + c'(x)^\top(\mu + r\tilde{c}_r(x,\mu)). \qquad (16.8)$$

Second differentiability in x is also ensured around a primal solution satisfying some strong conditions. Let x_* be a solution to (P_{EI}) and let λ_*

be a multiplier associated with x_*. Using the complementarity conditions $(\lambda_*)_I^\top c_I(x_*) = 0$ and the nonnegativity of $(\lambda_*)_I$, it is not difficult to see that, *for x close to x_*,* there holds

$$\ell_r(x, \lambda_*) = \ell(x, \lambda_*) + \frac{r}{2} \sum_{i \in E \cup I_*^{0+}} c_i(x)^2 + \frac{r}{2} \sum_{i \in I_*^{00}} (c_i(x)^+)^2. \qquad (16.9)$$

Because of the operator $(\cdot)^+$ in (16.9), $\ell_r(\cdot, \lambda_*)$ may not be twice differentiable at x_*. In the case of *strict complementarity*, however, $I_*^{00} = \emptyset$ and the last sum disappears, so that the augmented Lagrangian can be written (for x close to x_*)

$$\ell_r(x, \lambda_*) = \ell(x, \lambda_*) + \frac{r}{2} \sum_{i \in E \cup I_*^0} c_i(x)^2.$$

Locally, equality and active inequality constraints are then treated in the same way and $\ell_r(\cdot, \lambda_*)$ is smooth around x_* (provided f and c are smooth). The next proposition gathers these differentiability properties.

Proposition 16.2 (differentiability of the augmented Lagrangian). *If f and c are differentiable at x, then the augmented Lagrangian ℓ_r, defined by (16.6), is differentiable at x and its gradient is given by (16.8). If (x_*, λ_*) is a KKT point for (P_{EI}) satisfying strict complementarity and if $(f, c_{E \cup I_*^0})$ is p times differentiable (with $p \geq 0$ integer) in some neighborhood of x_*, then the augmented Lagrangian is p times differentiable is some (possibly smaller) neighborhood of x_*.*

The next result gives conditions for (x_*, λ_*) to be a saddle-point of ℓ_r on $V \times \mathbb{R}^m$, where V is a neighborhood of x_* in Ω. Compared with proposition 16.1, the result is local in x, but global in μ, and the minimum in x is strict. As before, this result implies that, if r is large enough (but finite!), $\ell_r(\cdot, \lambda_*)$ is an exact penalty function for (P_{EI}).

Proposition 16.3 (saddle-point of the augmented Lagrangian). *Suppose that f and $c_{E \cup I^0}$ are twice differentiable at a local minimum x_* of (P_{EI}) at which the KKT conditions hold, and that the semi-strong second-order sufficient condition of optimality (13.9) is satisfied for some multiplier λ_*. Then there exist a neighborhood V of x_* in Ω and a number $\underline{r} > 0$ such that, for all $r \geq \underline{r}$, (x_*, λ_*) is a saddle-point of ℓ_r on $V \times \mathbb{R}^m$. More precisely, we have for all $(x, \mu) \in (V \setminus \{x_*\}) \times \mathbb{R}^m$:*

$$\ell_r(x_*, \mu) \leq \ell_r(x_*, \lambda_*) < \ell_r(x, \lambda_*). \qquad (16.10)$$

Proof. Let us first show that λ_* maximizes $\ell_r(x_*, \cdot)$ for any $r > 0$. We have for $\mu \in \mathbb{R}^m$:

$$\ell_r(x_*, \mu) = f(x_*) + \sum_{\substack{i \in I \\ c_i(x_*) \geq -\mu_i/r}} \left(\mu_i c_i(x_*) + \frac{r}{2} c_i(x_*)^2 \right) - \sum_{\substack{i \in I \\ c_i(x_*) < -\mu_i/r}} \frac{\mu_i^2}{2r}.$$

The maximum in μ can be obtained term by term. If $c_i(x_*) = 0$, the maximum in the right-hand side is $f(x_*)$, obtained for all $\mu_i \geq 0$. If $c_i(x_*) < 0$, this maximum is again $f(x_*)$, obtained for $\mu_i = 0$. Since $(\lambda_*)_I$ satisfies these conditions, we have

$$\ell_r(x_*, \mu) \leq f(x_*) = \ell_r(x_*, \lambda_*), \quad \text{for all } \mu \in \mathbb{R}^m.$$

Let us now show the second statement, dealing with the *strict* local minimality of x_*. Note that we need to prove the inequality on the right in (16.10) for only a single value of r, $\underline{r} > 0$ say, because then, this inequality will hold for any $r \geq \underline{r}$ and any $x \in V$ (independent of r). Indeed, $\ell_r(x_*, \lambda_*) = f(x_*)$ does not depend on r and, for fixed x, $r \mapsto \ell_r(x, \lambda_*)$ is nondecreasing (this is a clear consequence of the way the augmented Lagrangian was introduced, just before the proposition).

We prove this by contradiction, assuming that there is a sequence of positive numbers $r_k \to \infty$ and a sequence of points $x_k \to x_*$, with $x_k \neq x_*$ such that, for $k \geq 1$:

$$\ell_{r_k}(x_k, \lambda_*) \leq \ell_{r_k}(x_*, \lambda_*). \tag{16.11}$$

Taking a subsequence if necessary, we have for $k \to \infty$:

$$\frac{x_k - x_*}{\|x_k - x_*\|} \to d, \quad \text{with } \|d\| = 1.$$

Hence, setting $\alpha_k := \|x_k - x_*\|$, we have

$$x_k = x_* + \alpha_k d + o(\alpha_k).$$

Our aim now is to show that d is a critical direction. We do this by appropriate expansions in the left-hand side of (16.11): second order expansion of the Lagrangian and first order expansion of the constraints in both sums of (16.9). To simplify the notation, we introduce $L_* = \nabla^2_{xx}\ell(x_*, \lambda_*)$. From the smoothness of f and c and the optimality of (x_*, λ_*), we have

$$\ell(x_k, \lambda_*) = \ell(x_*\lambda_*) + \frac{\alpha_k^2}{2} d^\top L_* d + o(\alpha_k^2),$$
$$c_i(x_k) = \alpha_k c_i'(x_*) \cdot d + o(\alpha_k), \quad \text{for } i \in E \cup I_*^0.$$

Injecting these estimates in (16.11), using (16.9) and $\ell_{r_k}(x_*, \lambda_*) = \ell(x_*, \lambda_*)$, provides

$$\frac{\alpha_k^2}{2} d^\top L_* d + o(\alpha_k^2) + \frac{r_k}{2} \sum_{i \in E \cup I_*^{0+}} (\alpha_k c_i'(x_*) \cdot d + o(\alpha_k))^2$$
$$+ \frac{r_k}{2} \sum_{i \in I_*^{00}} \left([\alpha_k c_i'(x_*) \cdot d + o(\alpha_k)]^+\right)^2 \leq 0. \tag{16.12}$$

Dividing by $\alpha_k^2 r_k$ and taking the limit yield

$$c_i'(x_*) \cdot d = 0, \quad \text{if } i \in E \cup I_*^{0+}$$
$$c_i'(x_*) \cdot d \leq 0, \quad \text{if } i \in I_*^{00}.$$

Therefore d is a nonzero critical direction.

On the other hand, (16.12) also implies that

$$\frac{\alpha_k^2}{2} d^\top L_* d + o(\alpha_k^2) \leq 0.$$

Dividing by α_k^2 and taking the limit show that $d^\top L_* d \leq 0$, which contradicts assumption (13.9) since $d \in C_* \backslash \{0\}$. $\qquad \square$

In the previous result, the semi-strong second-order sufficient condition of optimality (13.9) is assumed. If only the weak condition (13.8) holds, $\ell_r(\cdot, \lambda_*)$ may not have a local minimum at x_*, whatever the choice of $\lambda_* \in \Lambda_*$ and the value of r. An example is given in exercise 16.4.

16.4 Nondifferentiable Augmented Function

We now consider the following merit function for problem (P_{EI}):

$$\Theta_\sigma(x) = f(x) + \sigma \| c(x)^{\#} \|_P, \qquad (16.13)$$

which we call the *nondifferentiable augmented function*. In (16.13), $\sigma > 0$ is called the *penalty parameter*, the operator $(\cdot)^{\#}$ was defined on page 194, and $\| \cdot \|_P$ is a norm, and is arbitrary for the moment. We denote by $\| \cdot \|_D$ the *dual norm* of $\| \cdot \|_P$, with respect to the Euclidean scalar product. It is defined by

$$\| v \|_D = \sup_{\| u \|_P = 1} v^\top u.$$

We therefore have the *generalized Cauchy-Schwarz inequality*:

$$|u^\top v| \leq \| u \|_P \| v \|_D, \quad \text{for all } u \text{ and } v. \qquad (16.14)$$

See exercise 16.5 for some examples of dual norms.

Because of the norm $\| \cdot \|_P$ and of the operator $(\cdot)^{\#}$, Θ_σ is usually non-differentiable; but when f and c are smooth, Θ_σ has directional derivatives; this is a consequence of lemma 13.1. It so happens that this differentiability concept will be sufficient for our development.

Let $v \in \mathbb{R}^m$ be such that $v_I \leq 0$ and denote by $P_v : \mathbb{R}^m \to \mathbb{R}^m$ the operator defined by $P_v u = (\cdot^{\#})'(v; u)$, that is

$$(P_v u)_i = \begin{cases} u_i & \text{si } i \in E \\ u_i^+ & \text{if } i \in I \text{ and } v_i = 0 \\ 0 & \text{if } i \in I \text{ and } v_i < 0. \end{cases}$$

This notation allows us to write concisely the directional derivative of Θ_σ at a feasible point.

Lemma 16.4. *If f and c have a directional derivative at x in the direction $h \in \mathbb{R}^n$, then Θ_σ has also a directional derivative at x in the direction h. If, in addition, x is feasible for (P_{EI}), we have*

$$\Theta'_\sigma(x; h) = f'(x; h) + \sigma \|P_{c(x)} c'(x; h)\|_P.$$

Proof. The directional differentiability of $\Theta_\sigma = f + \sigma(\| \cdot \|_P \circ (\cdot)^\# \circ c)$ comes from lemma 13.1, the assumptions on f and c, and the fact that $(\cdot)^\#$ and $\| \cdot \|_P$ are Lipschitzian and have directional derivatives.

If x is feasible, $c(x)^\# = 0$ and we have from lemma 13.1,

$$\Theta'_\sigma(x; h) = f'(x; h) + \sigma(\| \cdot \|_P)'(0; (c^\#)'(x; h)).$$

On the other hand,

$$(c^\#)'(x; h) = (\cdot^\#)'(c(x); c'(x; h)) = P_{c(x)} c'(x; h)$$

and

$$(\| \cdot \|_P)'(0; v) = \lim_{t \to 0+} \frac{1}{t}(\|tv\|_P - 0) = \|v\|_P.$$

The result follows. $\square$

Necessary Conditions of Exactness

In this subsection, we examine which conditions are implied by the fact that a *feasible* point x_* is a minimum point of Θ_σ. We quote three such properties in proposition 16.5: x_* is also a minimum point of (P_{EI}), there exists a multiplier λ_* associated with x_*, and σ must be sufficiently large. The second property shows that the exactness of Θ_σ plays a similar role as a constraint qualification assumption, since it implies the existence of a dual solution.

For the third property mentioned above, we need an assumption on the norm $\| \cdot \|_P$ used in Θ_σ. The norm $\|v\|_P$ must decrease if one sets to zero some of the I-components of $v \in \mathbb{R}^m$:

$$u_i = \begin{cases} v_i & \text{if } i \in E \\ 0 \text{ or } v_i & \text{if } i \in I \end{cases} \implies \|u\|_P \le \|v\|_P. \tag{16.15}$$

Clearly, ℓ_p norms, $1 \le p \le \infty$, satisfy this property; but it is not necessarily satisfied by an arbitrary norm (see exercise 17.1). Also, the claim on σ in proposition 16.5 may not be correct if $\| \cdot \|_P$ does not satisfy (16.15) (see exercise 16.6).

Proposition 16.5 (necessary conditions of exactness). *If x_* is feasible for (P_{EI}) and Θ_σ has a local minimum (resp. strict local minimum) at x_*, then x_* is a local minimum (resp. strict local minimum) of (P_{EI}). If, in addition, f and c are Gâteaux differentiable at x_*, then there exists a multiplier λ_* such that the necessary optimality conditions (KKT) hold. If, in addition, the norm $\| \cdot \|_P$ satisfies (16.15) and (LI-CQ) holds at x_*, then $\sigma \ge \|\lambda_*\|_D$.*

Proof. If x_* is a local minimum of Θ_σ, there exists a neighborhood V of x_* such that
$$\Theta_\sigma(x_*) \le \Theta_\sigma(x), \quad \text{for all } x \in V.$$

Since $x_* \in X$ and $\Theta_\sigma|_X = f|_X$, we have

$$f(x_*) \le f(x), \quad \text{for all } x \in V \cap X,$$

which shows that x_* is a local minimum of (P_{EI}). The above inequality is strict for $x \ne x_*$, if x_* is a strict local minimum of Θ_σ.

Now suppose f and g are Gâteaux differentiable at x_*. Then Θ_σ has directional derivatives at x_* (lemma 16.4). Since x_* is a local minimum of Θ_σ, we have $\Theta'_\sigma(x_*; h) \ge 0$ for all $h \in \mathbb{R}^m$. But x_* is feasible; hence, by lemma 16.4:

$$\nabla f(x_*)^\top h + \sigma \|P_{c(x_*)}(A(x_*)h)\|_P \ge 0, \quad \text{for all } h \in \mathbb{R}^m. \tag{16.16}$$

We deduce
$$P_{c(x_*)}(A(x_*)h) = 0 \quad \Longrightarrow \quad \nabla f(x_*)^\top h \ge 0.$$

Thus, $h = 0$ solves the linear program

$$\begin{cases} \min_h \nabla f(x_*)^\top h \\ A_E(x_*)h = 0, \\ A_{I_*^0}(x_*)h \le 0. \end{cases}$$

The constraints of this problem being qualified (by (A-CQ)), we deduce the existence of a multiplier $\lambda_* \in \mathbb{R}^m$ such that

$$\begin{cases} \nabla f(x_*) + A(x_*)^\top \lambda_* = 0 \\ (\lambda_*)_{I_*^0} \ge 0 \\ (\lambda_*)_{I \setminus I_*^0} = 0. \end{cases}$$

Since x_* is feasible, (KKT) holds with (x_*, λ_*).

Finally, suppose that the norm $\|\cdot\|_P$ satisfies (16.15) and that (LI-CQ) holds. Take again (16.16) and use the first-order optimality condition to obtain
$$\lambda_*^\top A(x_*)h \le \sigma \|P_{c(x_*)}A(x_*)h\|_P, \quad \text{for all } h \in \mathbb{R}^n.$$

Set $J = E \cup I_*^0$, and remember that $(\lambda_*)_i = 0$ if $i \notin J$. For an arbitrary v in $\mathbb{R}^m$, we have $\lambda_*^\top v = (\lambda_*)_J^\top v_J$ and, from (LI-CQ), we can find $h \in \mathbb{R}^n$ such that $A_J(x_*)h = v_J$. We deduce that

$$\lambda_*^\top v = (\lambda_*)_J^\top A_J(x_*)h = \lambda_*^\top A(x_*)h \le \sigma \|P_{c(x_*)}A(x_*)h\|_P \le \sigma \|v\|_P,$$

where the last inequality uses property (16.15) of the norm. Then $\lambda_*^\top v \le \sigma \|v\|_P$, and since v is arbitrary, we have $\|\lambda_*\|_D \le \sigma$. $\qquad\square$

Sufficient Conditions of Exactness

In practice, we are more interested in having conditions that ensure the exactness of Θ_σ and this is what we focus on now. We shall show that the necessary condition obtained on σ in proposition 16.5 is sharp: if x_* is a strong solution to problem (P_{EI}) with associated multiplier λ_*, x_* also minimizes Θ_σ provided $\sigma > \|\lambda_*\|_D$ (the strict inequality is not needed for convex problems). This result holds without any particular assumption on the norm $\|\cdot\|_P$.

The necessary conditions of exactness of Θ_σ were obtained by expressing the fact that, if x_* minimizes Θ_σ, the directional derivative $\Theta'_\sigma(x_*; h)$ must be nonnegative for all $h \in \mathbb{R}^n$ (see the proof of proposition 16.5). Now we want to exhibit values of σ such that Θ_σ has a minimum at x_*. Function Θ_σ is nondifferentiable and nonconvex. Therefore, it is not sufficient to show that $\Theta'_\sigma(x_*; h) \geq 0$ for all $h \in \mathbb{R}^n$ in order to ensure its exactness. One cannot impose $\Theta'_\sigma(x_*; h) > 0$ for all $h \in \mathbb{R}^n$ either, since this may never occur for any value of σ (for example, when $E \neq \emptyset$ and $I = \emptyset$, $\Theta'_\sigma(x_*; h) = 0$ for any h in the space tangent to the constraint manifold). Therefore, we shall use either a technical detour (for convex problems) or a direct proof like the one of proposition 16.3 (for nonconvex problems).

In proposition 16.7 below, we consider the case of convex problems and in proposition 16.8 the case of nonconvex problems. To prove the exactness of the nondifferentiable function Θ_σ for convex problems, we simply use the fact that, if σ is large enough, Θ_σ is above the differentiable Lagrangian (16.4) (lemma 16.6), which is known to be exact at x_* (proposition 16.1). Observe that lemma 16.6 does not assume convexity.

Lemma 16.6. *If $\sigma \geq \|\lambda\|_D$ and $\lambda_I \geq 0$, then $\ell(\cdot, \lambda) \leq \Theta_\sigma(\cdot)$ on $\mathbb{R}^n$.*

Proof. First observe that $\lambda_I \geq 0$ implies $\lambda_I^\top c_I(x) \leq \lambda_I^\top c_I(x)^+$. Then, for all $x \in \mathbb{R}^n$,

$$\ell(x, \lambda) \leq f(x) + \lambda^\top c(x)^\# \leq f(x) + \|\lambda\|_D \|c(x)^\#\|_P \leq \Theta_\sigma(x).$$

$\square$

Proposition 16.7 (sufficient conditions of exactness, convex problems). *Suppose that problem (P_{EI}) is convex and that f and c are differentiable at a solution x_* to (P_{EI}) with an associated multiplier λ_*. Then Θ_σ has a global minimum at x_* as soon as $\sigma \geq \|\lambda_*\|_D$.*

Proof. According to proposition 16.1, $\ell(\cdot, \lambda_*)$ is minimized by x_* and, by lemma 16.6, it is dominated by Θ_σ ($\sigma \geq \|\lambda_*\|_D$ and $(\lambda_*)_I \geq 0$). Therefore

$$\begin{aligned}
\Theta_\sigma(x_*) &= f(x_*) \\
&= \ell(x_*, \lambda_*) \\
&\leq \ell(x, \lambda_*), \quad \text{for all } x \in \mathbb{R}^n \\
&\leq \Theta_\sigma(x), \quad \text{for all } x \in \mathbb{R}^n.
\end{aligned}$$

$\square$

The same technical detour could be used for highlighting sufficient conditions of exactness of Θ_σ for nonconvex problems: if $\sigma > \|\lambda_*\|_D$, Θ_σ is above the augmented Lagrangian (16.6) in some neighborhood of x_*, so that the exactness of Θ_σ follows that of the augmented Lagrangian (proposition 16.3). This strategy is proposed in exercise 16.7. The direct proof given below has the advantage of being valid even when only the weak second order sufficient condition of optimality (13.8) holds at x_* (in contrast, the semi-strong condition (13.9) is assumed in proposition 16.3 and exercise 16.7).

Proposition 16.8 (sufficient conditions of exactness). *Suppose that f and $c_{E\cup I_*^0}$ are twice differentiable at a local minimum x_* of (P_{EI}) at which the KKT conditions hold, that the weak second-order sufficient condition of optimality (13.8) is satisfied, and that*

$$\sigma > \sup_{\lambda_* \in \Lambda_*} \|\lambda_*\|_D,$$

where Λ_* is the nonempty set of multipliers associated with x_*. Then Θ_σ has a strict local minimum at x_*.

Proof. We prove the result by contradiction, assuming that x_* is not a strict minimum of Θ_σ. Then, there exists a sequence $\{x_k\}$ such that $x_k \neq x_*$, $x_k \to x_*$ and

$$\Theta_\sigma(x_k) \leq \Theta_\sigma(x_*), \quad \forall k \geq 1. \tag{16.17}$$

Since the sequence $\{(x_k - x_*)/\|x_k - x_*\|\}$ is bounded (here $\|\cdot\|$ denotes an arbitrary norm), it has a subsequence such that $(x_k - x_*)/\|x_k - x_*\| \to d$, where $\|d\| = 1$. Denoting $\alpha_k = \|x_k - x_*\|$, one has

$$x_k = x_* + \alpha_k d + o(\alpha_k).$$

Because Θ_σ is Lipschitzian in a neighborhood of x_*:

$$\Theta_\sigma(x_k) = \Theta_\sigma(x_* + \alpha_k d) + o(\alpha_k).$$

Now (16.17) shows that $\Theta_\sigma'(x_*; d) \leq 0$. Then, from lemma 16.4, one can write

$$f'(x_*) \cdot d + \sigma \|P_{c(x_*)}(c'(x_*) \cdot d)\|_P \leq 0. \tag{16.18}$$

This certainly implies that

$$f'(x_*) \cdot d \leq 0. \tag{16.19}$$

On the other hand, from the assumptions, there is an optimal multiplier λ_* such that $\sigma > \|\lambda_*\|_D$. Using the first order optimality conditions, including the nonnegativity of $(\lambda_*)_I$ and the complementarity conditions $(\lambda_*)_I^\top c_I(x_*) = 0$, one has

$$-f'(x_*) \cdot d = \lambda_*^\top (c'(x_*) \cdot d)$$
$$\leq \lambda_*^\top P_{c(x_*)}(c'(x_*) \cdot d)$$
$$\leq \|\lambda_*\|_D \|P_{c(x_*)}(c'(x_*) \cdot d)\|_P.$$

Then (16.18) and $\sigma > \|\lambda_*\|_D$ imply that $P_{c(x_*)}(c'(x_*) \cdot d) = 0$, i.e.,

$$\begin{cases} c_i'(x_*) \cdot d = 0 \text{ for } i \in E \\ c_i'(x_*) \cdot d \leq 0 \text{ for } i \in I_*^0. \end{cases}$$

These and (16.19) show that d is a nonzero critical direction.

Now, let λ_* be the multiplier depending on d, determined by the weak second-order sufficient condition of optimality (13.8). According to theorem 13.4, one has

$$d^\top \nabla_{xx}^2 \ell(x_*, \lambda_*) d > 0.$$

The following Taylor expansion (use $\nabla_x \ell(x_*, \lambda_*) = 0$)

$$\ell(x_k, \lambda_*) = \ell(x_*, \lambda_*) + \frac{\alpha_k^2}{2} d^\top \nabla_{xx}^2 \ell(x_*, \lambda_*) d + o(\alpha_k^2)$$

allows us to see that, for k large enough,

$$\ell(x_k, \lambda_*) > \ell(x_*, \lambda_*). \tag{16.20}$$

Then, for large indices k, there holds

$$\begin{aligned}
\Theta_\sigma(x_k) &\leq \Theta_\sigma(x_*) &&\text{[by (16.17)]} \\
&= f(x_*) \\
&= \ell(x_*, \lambda_*) \\
&< \ell(x_k, \lambda_*) &&\text{[by (16.20)]} \\
&\leq \Theta_\sigma(x_k) &&\text{[by lemma 16.6 and } \sigma \geq \|\lambda_*\|_D\text{],}
\end{aligned}$$

which is the expected contradiction.   □

Notes

The augmented Lagrangian (16.2) for equality constrained problems was first proposed by Arrow and Solow [14; 1958]. Hestenes [191; 1969] and Powell [288; 1969] both used this function to introduce the so-called *method of multipliers*, which has popularized this type of penalization. The augmented Lagrangian (16.3) or (16.6), adapted to inequality constrained problems, was proposed by Rockafellar [310, 311; 1971-74] and Buys [62; 1972]. It was further extended to constraints of the form $c(x) \in K$, where c is a vector-valued function and

K is a closed convex cone, by Shapiro and Sun [330; 2004]. This penalty function is usually no more than continuously differentiable, even if the problem data are infinitely differentiable. Many developments have been carried out to overcome this drawback, proposing augmentation terms with a different structure (for entry points see [17, 16; 1999-2000], which deal with primal penalty functions, and [109, 110, 111; 1999-2001], which consider primal-dual penalty functions). Surveys on the augmented Lagrangian can be found in [26, 169].

The exact penalty function (16.13) goes back at least to Eremin [119; 1966] and Zangwill [374; 1967]. Its connection with problem (P_{EI}) has been studied by many authors, see Pietrzykowski [284], Charalambous [74], Ioffe [198], Han and Mangasarian [186], Bertsekas [26], Fletcher [126], Bonnans [39, 41], Facchinei [120], Burke [60], Pshenichnyj [301], Bonnans and Shapiro [50], and the references therein.

Exercises

16.1. *Finsler's lemma* [123] *and its limit case* [9]. Let M be an $n \times n$ symmetric matrix that is positive definite on the null space of a matrix A (i.e., $u^\top M u > 0$ for all nonzero $u \in N(A)$). Show that there exists an $r_0 \in \mathbb{R}$ such that, for all $r \geq r_0$, $M + r A^\top A$ is positive definite.

[*Hint*: Use an argument by contradiction.]

Suppose now that the symmetric matrix M is only positive semidefinite on the null space of A (i.e., $u^\top M u \geq 0$ for all $u \in N(A)$). Show that the following claims are equivalent: (*i*) $v \in N(A)$ and $v^\top M v = 0$ imply that $Mv = 0$, and (*ii*) there exists an $r_0 \in \mathbb{R}$ such that, for all $r \geq r_0$, $M + r A^\top A$ is positive semidefinite. Find a matrix M that is positive semidefinite on the null space of A, for which these properties (*i*) and (*ii*) are not satisfied.

[*Hint*: For (*i*) $\Rightarrow$ (*ii*), use with care an argument by contradiction.]

Consequence: If M is nonsingular and positive semidefinite (but not positive definite) on the null space of A, it cannot enjoy property (*ii*) (since (*i*) does not hold).

16.2. *Augmented Lagrangian for equality constrained problems.* Consider problem (P_E) with functions f and c of class C^2 and the associated augmented Lagrangian $\ell_r(x, \lambda) = f(x) + \lambda^\top c(x) + \frac{r}{2}\|c(x)\|_2^2$. By a direct computation of $\nabla_x \ell_r(x_*, \lambda_*)$ and $\nabla^2_{xx} \ell_r(x_*, \lambda_*)$, show that, if r is large enough, $\ell_r(\cdot, \lambda_*)$ has a strict local minimum at a point x_* satisfying ($SC2$).

[*Hint*: Use Finsler's lemma (exercise 16.1).]

16.3. *Fletcher's exact penalty function* [124]. Consider problem (P_E), in which f and c are smooth, and c is a submersion. Denote by $A^-(x)$ a right inverse of the constraint Jacobian $A(x) := c'(x)$ and assume that A^- is a smooth function of x. Let $\lambda^{\mathrm{LS}}(x) := -A^-(x)^\top \nabla f(x)$ be the associated least-squares multiplier. For $r \in \mathbb{R}$, consider the function $\varphi_r : \mathbb{R}^n \to \mathbb{R}$ defined by

$$\varphi_r(x) = f(x) + \lambda^{\mathrm{LS}}(x)^\top c(x) + \frac{r}{2}\|c(x)\|_2^2. \tag{16.21}$$

Let (x_*, λ_*) be a pair satisfying the second-order sufficient conditions of optimality $(SC2)$ of problem (P_E). Show that there exists an $r_0 \in \mathbb{R}$, such that for $r \geq r_0$, φ_r has a strict local minimum at x_*.

[*Hint*: Prove the following claims, in which $A_* := A(x_*)$, $A_*^- := A^-(x_*)$, and $L_* := \nabla^2_{xx}\ell(x_*, \lambda_*)$, and conclude: (*i*) $\lambda^{\mathrm{LS}}(x_*) = \lambda_*$; (*ii*) $\nabla\varphi_r(x_*) = 0$; (*iii*) $(\lambda^{\mathrm{LS}})'(x_*) = -A_*^{-\top}L_*$ and $\nabla^2\varphi_r(x_*) = L_* - (A_*^\top A_*^{-\top}L_* + L_*A_*^- A_*) + rA_*^\top A_*$; (*iv*) $\nabla^2\varphi_r(x_*)$ is positive definite if r is large enough.]

16.4. *Counter-example for proposition* 16.3. Consider the problem in $\mathbb{R}^3$:

$$\begin{cases} \min_x x_3 \\ x_3 \geq (x_1 + x_2)(x_1 - x_2) \\ x_3 \geq (x_2 + 3x_1)(2x_2 - x_1) \\ x_3 \geq (2x_2 + x_1)(x_2 - 3x_1). \end{cases}$$

Show that: (*i*) $x_* = 0$ is the unique solution to the problem and that the associated multiplier set is $\Lambda_* = \{\lambda \in \mathbb{R}^3_+ : \lambda_1 + \lambda_2 + \lambda_3 = 1\}$; (*ii*) the weak second order sufficient condition of optimality (13.8) is satisfied, but not the semi-strong ones (13.9); (*iii*) for any $\lambda_* \in \Lambda_*$ and $r \geq 0$, the augmented Lagrangian (16.6) has not a minimum at x_*.

Consequence: When the semi-strong second order sufficient conditions of optimality (13.9) do not hold at x_*, the augmented Lagrangian $\ell_r(\cdot, \lambda_*)$ function may not have a local minimum at x_*, for any $\lambda_* \in \Lambda_*$ and $r \geq 0$.

16.5. *Dual norms.* (*i*) The ℓ_p norm on $\mathbb{R}^n$ is defined by

$$\|u\|_p := \begin{cases} \left(\sum_{i=1}^n |u_i|^p\right)^{\frac{1}{p}} & \text{if } 1 \leq p < \infty \\ \max_{1 \leq i \leq n} |u_i| & \text{if } p = \infty. \end{cases}$$

Show that the dual norm of $\|\cdot\|_p$ is the norm $\|\cdot\|_{p'}$, where p' is uniquely defined by

$$\frac{1}{p} + \frac{1}{p'} = 1.$$

(*ii*) Let Q be a symmetric positive definite matrix and define the norm $\|u\|_P = (u^\top Q u)^{\frac{1}{2}}$. Show that its dual norm is given by $\|v\|_D = (v^\top Q^{-1} v)^{\frac{1}{2}}$.

16.6. *Counter-example for proposition* 16.5. Consider the problem

$$\min\left\{\frac{1}{2}\|x\|_2^2 : x \in \mathbb{R}^2, \ x_1 \leq 0, \ x_2 + 1 \leq 0\right\}.$$

Show that the unique primal-dual solution to this problem is $x_* = (0, -1)$ and $\lambda_* = (0, 1)$. Show that $x \mapsto \|x\|_P = (x_1^2 + x_2^2 + \sqrt{3}x_1x_2)^{1/2}$ is a norm that does not satisfy (16.15), and that $\|\lambda_*\|_D = 2$. Show that $\Theta_\sigma(x) = \frac{1}{2}\|x\|_2^2 + \sigma\|(x_1, x_2 + 1)^+\|_P$ has a minimum at x_* when $\sigma \geq 1$.

Consequence: The exactness of Θ_σ does not imply $\sigma \geq \|\lambda_*\|_D$ if the norm $\|\cdot\|_P$ does not satisfy (16.15).

16.7. *A variant of proposition* 16.8. (*i*) Let x_* be feasible for (P_{EI}) and $\lambda \in \mathbb{R}^m$ be such that $\lambda_I \geq 0$ and $\lambda_I^\top c_I(x_*) = 0$; let $r > 0$ and $\sigma > \|\lambda\|_D$. Show that there exists a neighborhood V of x_* in Ω such that for all $x \in V$, there holds $\ell_r(x, \lambda) \leq \Theta_\sigma(x)$.

(*ii*) Suppose that f and $c_{E \cup I_*^0}$ are twice differentiable at a local minimum x_* of (P_{EI}) at which the KKT conditions hold, that the semi-strong second-order sufficient condition of optimality (13.9) holds for some optimal multiplier λ_*, and that $\sigma > \|\lambda_*\|_D$. Show, using (*i*), that Θ_σ has a strict local minimum at x_*.

16.8. ℓ_1 *penalty function.* Suppose that f and $c_{E \cup I_*^0}$ are twice differentiable at a local minimum x_* of (P_{EI}) at which the KKT conditions hold and that the weak second-order sufficient condition of optimality (13.8) is satisfied. Positive scalars σ_i ($i \in E \cup I$) are given and the following penalty function is considered:

$$\Theta_\sigma^1(x) = f(x) + \sum_{i \in E} \sigma_i |c_i(x)| + \sum_{i \in I} \sigma_i c_i(x)^+.$$

Show that, if $\sigma_i > |(\lambda_*)_i|$, for $i \in E \cup I$ and all optimal multiplier λ_*, then x_* is a strict local minimum of Θ_σ^1.

[*Hint*: Use the norm $v \mapsto \|v\|_P := \sum_i \sigma_i |v_i|$ and proposition 16.8.]

Remark: The ℓ_1-penalty function offers a natural way of controlling the magnitude of penalty parameters σ_i, when one such parameter is associated with each constraint.

16.9. *Nondifferentiable augmented Lagrangian* ([37] for equality constrained problems; [41] for an alternative to (16.22)). Suppose that f and $c_{E \cup I_*^0}$ are twice differentiable at a local minimum x_* of (P_{EI}) at which the KKT conditions hold. Let be given $\mu \in \mathbb{R}^m$ and $\sigma \in \mathbb{R}_+$. Suppose one of the following:

(*i*) either the weak second-order sufficient condition of optimality (13.8) is satisfied and $\sigma > \sup\{\|\lambda_* - \mu\|_D : \lambda_* \in \Lambda_*\}$,

(*ii*) or the semi-strong second-order sufficient condition of optimality (13.9) holds for some optimal multiplier λ_* and $\sigma > \|\mu - \lambda_*\|_D$.

Then $\Theta_{\mu,\sigma} : \mathbb{R}^n \to \mathbb{R}$ defined by

$$\Theta_{\mu,\sigma}(x) := f(x) + \mu^\top c(x)^\# + \sigma\|c(x)^\#\|_P \tag{16.22}$$

has a strict local minimum at x_*.

[*Hint*: Under assumptions (*i*) use a technique similar to the one in the proof of proposition 16.8; under assumptions (*ii*) follow the same strategy as in exercise 16.7.]

17 Globalization by Line-Search

There is no guarantee that the local algorithms in chapters 14 and 15 will converge when they are started at a point x_1 far from a solution x_* to problem (P_E) or (P_{EI}). They can generate erratic sequences, which may by chance enter the neighborhood of a solution and then converge to it; but most often, the sequences will not converge. There exist several ways of damping this uncoordinated behavior and modifying the computation of the iterates so as to force their convergence. Two classes of techniques can be distinguished among them: line-search and trust-region. The former is presented in this chapter.

In methods with line-search, the iterates are generated by the recurrence

$$x_{k+1} = x_k + \alpha_k d_k,$$

where d_k is a direction in $\mathbb{R}^n$ and $\alpha_k > 0$ is a stepsize, computed by a *line-search* technique (see chapter 3), whose aim is to decrease a merit function. In this chapter, we consider algorithms in which d_k solves or approximately solves the osculating quadratic program (14.8)/(15.4) of the Newton/SQP algorithm in chapters 14/15 and the merit function is the function Θ_σ in chapter 16. For convenience, we recall the definition of Θ_σ:

$$\Theta_\sigma(x) = f(x) + \sigma \| c(x)^\# \|_P, \tag{17.1}$$

where $\| \cdot \|_P$ denotes an arbitrary norm and the notation $(\cdot)^\#$ was introduced on page 194. Properties of function Θ_σ are studied in chapter 16; remember that this function is usually nondifferentiable.

Let us stress the originality of this approach, which uses the solution to the osculating quadratic program to minimize Θ_σ. If d_k were an arbitrary descent direction of the nondifferentiable merit function Θ_σ, for example the steepest-descent direction, the resulting algorithm would not necessarily converge (see § 9.2.1). We shall show, however, that the difficulty coming from nonsmoothness does not occur if the search direction d_k solves the osculating quadratic problem (15.4). As for the stepsize, the value $\alpha_k = 1$ is preferred, in order to preserve the quadratic convergence of the local method. We shall see that the unit stepsize is actually accepted when x_k is close to a strong solution to (P_{EI}), provided some modifications of the search direction or the merit function are made. Therefore, the final algorithm can also be viewed as a quadratically convergent method for minimizing the structured nonsmooth function Θ_σ, a speed of convergence that cannot be obtained with general purpose nondifferentiable algorithms like those presented in part II of this book.

The concept of exactness plays an important part in the success of the approach we have just outlined. Without this property, it might indeed have been necessary to adapt σ continually to make the solution d_k to the quadratic problem a descent direction of the merit function Θ_σ. This is illustrated for an equality constraint problem in figure 17.1 (a single constraint and two

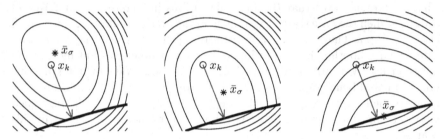

Fig. 17.1. Importance of exactness: σ too small (l), giving descent (m), giving exactness (r)

variables). The figure provides three pictures showing the level curves of Θ_σ for three increasing values of σ ($\bar{x}_\sigma$ is the minimizer of Θ_σ). They also show the constraint manifold (the bold curve at the bottom) and the Newton direction at x_k (the arrow). We assume that the current iterate x_k is close to x_* (hence the figure gives greatly enlarged views) and that the multiplier λ_k is also close to λ_*, so that the Newton direction d_k points towards x_* (this is a consequence of the quadratic convergence result in chapter 14). We can see that d_k is an ascent direction of Θ_σ if σ is not large enough (left-hand picture). In this case, there is no hope in finding a positive stepsize α_k along d_k that provides a decrease in Θ_σ. In the middle picture, σ is large enough to make d_k a descent direction of Θ_σ, although not large enough to make Θ_σ exact at x_*. In the right-hand picture, the penalty parameter σ is large enough to have $\bar{x}_\sigma = x_*$ (exactness of Θ_σ) and this gives d_k a greater chance of being a descent direction of Θ_σ. As we shall see, other conditions must also be satisfied. Observe finally that the nondifferentiability of Θ_σ manifests itself in the pictures by the lack of smoothness of its level curves when they cross the constraint manifold.

To get descent property of d_k, it will be necessary to increase σ at some iterations, but the exactness property of Θ_σ for a finite value of σ will allow the algorithm to do this finitely often. This is a very desirable property, which makes the proof of convergence possible. As soon as σ is fixed, Θ_σ

plays the role of an immutable reference, which is able to appreciate the progress towards the solution, whatever may happen to the iterates.

This chapter describes and analyzes two classes of algorithms. Line-search SQP algorithms (§ 17.1) are based on the SQP direction of chapter 15 and use line-search on Θ_σ to enforce its convergence. We derive conditions that ensure the descent property of the SQP direction on Θ_σ and study the global convergence of the algorithm. This analysis assumes the strict convexity of the osculating quadratic program defining the SQP direction (as well as its feasibility), which may require not using the Hessian of the Lagrangian, but a positive definite approximation thereof (chapter 18 explains how to generate quasi-Newton approximations). The truncated SQP algorithm of § 17.2 is presented as a line-search method that can use the exact Hessian of the Lagrangian (although we restrict the analysis to equality constrained problems). In this case, it is the way to solve the quadratic program approximately (discarding tangent negative curvature information) that allows the algorithm to generate descent directions of the merit function Θ_σ. The so-called Maratos effect (nonadmissibility of the unit stepsize asymptotically) is discussed in § 17.3, and the most common remedies for this phenomenon are described.

17.1 Line-Search SQP Algorithms

The quadratic program (QP) considered in this section is slightly more general than (15.4): the Hessian of the Lagrangian $L(x_k, \lambda_k)$ is replaced by some $n \times n$ symmetric matrix M_k. This allows us to include the Newton and the quasi-Newton versions of SQP in the same framework. On the other hand, the descent property of the QP solution and convergence of the line-search SQP algorithm often require the positive definiteness of M_k. The osculating quadratic problem in d becomes:

$$
\begin{cases}
\min_d \nabla f(x_k)^\top d + \frac{1}{2} d^\top M_k d \\
c_E(x_k) + A_E(x_k)d = 0 \\
c_I(x_k) + A_I(x_k)d \leq 0.
\end{cases}
\tag{17.2}
$$

A stationary point d_k of this QP satisfies, for some multiplier $\lambda_k^{\mathrm{QP}} \in \mathbb{R}^m$, the optimality conditions:

$$
\begin{cases}
(a) \ \nabla f_k + M_k d_k + A_k^\top \lambda_k^{\mathrm{QP}} = 0 \\
(b) \ (c_k + A_k d_k)^\# = 0 \\
(c) \ (\lambda_k^{\mathrm{QP}})_I \geq 0 \\
(d) \ (\lambda_k^{\mathrm{QP}})_I^\top (c_k + A_k d_k)_I = 0.
\end{cases}
\tag{17.3}
$$

For short, we have set $\nabla f_k = \nabla f(x_k)$, $c_k = c(x_k)$, and $A_k = A(x_k) = c'(x_k)$.

Let us now outline the line-search SQP algorithm that uses Θ_σ as a merit function. The description includes references to numerical techniques, whose sense will be clarified further in the section. The analysis of this algorithm is the subject of this section.

LINE-SEARCH SQP:

Choose an initial iterate $(x_1, \lambda_1) \in \mathbb{R}^n \times \mathbb{R}^m$.
Compute $f(x_1)$, $c(x_1)$, $\nabla f(x_1)$, and $A_1 := c'(x_1)$.
Set $k = 1$.

1. Stop if the KKT conditions (13.1) holds at $(x_*, \lambda_*) \equiv (x_k, \lambda_k)$ (optimality is reached).
2. Compute a symmetric matrix M_k, approximating the Hessian of the Lagrangian, and find a primal-dual stationary point $(d_k, \lambda_k^{\mathrm{QP}})$ of the quadratic problem (17.2) (i.e., a solution to the optimality conditions (17.3)), which is assumed to be feasible.
3. Adapt σ_k if necessary (the update rule must satisfy (17.9) to ensure convergence, but a rule similar to the one on page 295 is often used).
4. Choose $\alpha_k > 0$ along d_k so as to obtain a "sufficient" decrease in Θ_{σ_k} (for example, use the line-search technique given on page 296).
5. Set $x_{k+1} := x_k + \alpha_k d_k$ and update $\lambda_k \to \lambda_{k+1}$.
6. Compute $\nabla f(x_{k+1})$ and $A_{k+1} := c'(x_{k+1})$.
7. Increase k by 1 and go to 1.

This algorithm does not specify how to update the dual variables λ_k. Some authors do a line-search on λ with the help of a primal-dual merit function, which therefore involves λ-values. Others compute λ_{k+1} from x_{k+1} as in the primal algorithm of § 14.3. Another possibility is also to take

$$\lambda_{k+1} := \lambda_k + \alpha_k(\lambda_k^{\mathrm{QP}} - \lambda_k), \tag{17.4}$$

where α_k is the stepsize used for the primal variables. It has already been said that the role of λ_k is less important than that of x_k, because it intervenes in the algorithm only through the matrix M_k (for example the Hessian of the Lagrangian) in (17.2). The few requirements on the way the new multiplier is determined reflects in some way this fact.

General assumptions for this section. We assume throughout this section that f and c are differentiable in an open set containing the segments $[x_k, x_{k+1}]$ that link the successive iterates. We also assume that the quadratic problem (17.2) is always feasible (i.e., its constraints are compatible).

In practice, the last assumption on the feasibility of (17.2) is far from always being satisfied at each iteration. Therefore, carefully written   codes use techniques and heuristics for dealing with infeasible quadratic programs. For more computational efficiency, it is also often better to have a different penalty factor associated with each constraint, as in exercise 16.8. For simplicity, we keep a merit function with a single penalty parameter σ, knowing that an extension is possible without difficulty.

Decrease in Θ_σ Along d_k

The merit function Θ_σ decreases from x_k along d_k if d_k is a descent direction of Θ_σ at x_k (we saw in lemma 16.4 that Θ_σ has directional derivatives), meaning that

$$\Theta'_\sigma(x_k; d_k) < 0.$$

We focus on this issue in this subsection.

The next proposition identifies three conditions that make d_k a descent direction of Θ_σ: σ is large enough, M_k is positive definite, and x_k is not a stationary point of (P_{EI}). Such a result is useful for the quasi-Newton versions of SQP, where the positive definiteness of M_k is preserved. To hold, the result needs the following assumption on the norm $\|\cdot\|_P$ used in Θ_σ:

$$v \mapsto \|v^\#\|_P \text{ is convex.} \tag{17.5}$$

This hypothesis is weaker than (16.15) (see exercise 17.1).

Proposition 17.1 (descent property). *If $(d_k, \lambda_k^{\mathrm{QP}})$ satisfies the optimality conditions (17.3) and if $\|\cdot\|_P$ satisfies (17.5), then*

$$\Theta'_\sigma(x_k; d_k) \le \nabla f_k^\top d_k - \sigma\|c_k^\#\|_P = -d_k^\top M_k d_k + (\lambda_k^{\mathrm{QP}})^\top c_k - \sigma\|c_k^\#\|_P. \tag{17.6}$$

If, in addition, $\sigma \ge \|\lambda_k^{\mathrm{QP}}\|_D$, we have

$$\Theta'_\sigma(x_k; d_k) \le -d_k^\top M_k d_k.$$

Hence $\Theta'_\sigma(x_k; d_k) < 0$, if $\sigma \ge \|\lambda_k^{\mathrm{QP}}\|_D$, if M_k is positive definite, and if x_k is not a stationary point of problem (P_{EI}).

Proof. Since a norm has directional derivatives and is Lipschitzian (like any convex function), the function $v \to \|v^\#\|_P$ has directional derivatives. From (17.5) and (17.3)$_b$, we have for $t \in \,]0, 1[$:

$$\begin{aligned}
\|(c_k + tA_k d_k)^\#\|_P &= \|[(1-t)c_k + t(c_k + A_k d_k)]^\#\|_P \\
&\le (1-t)\|c_k^\#\|_P + t\|(c_k + A_k d_k)^\#\|_P \\
&= (1-t)\|c_k^\#\|_P.
\end{aligned}$$

Therefore

$$(\|\cdot^\#\|_P)'(c_k; A_k d_k) = \lim_{t \to 0+} \frac{1}{t}(\|(c_k + tA_k d_k)^\#\|_P - \|c_k^\#\|_P) \le -\|c_k^\#\|_P.$$

Then, with $(17.3)_a$, $(17.3)_b$ and $(17.3)_d$, we prove (17.6):

$$
\begin{aligned}
\Theta'_\sigma(x_k; d_k) &\leq \nabla f_k^\top d_k - \sigma \|c_k^{\#}\|_P \\
&= -d_k^\top M_k d_k - (\lambda_k^{\text{QP}})^\top A_k d_k - \sigma \|c_k^{\#}\|_P \\
&= -d_k^\top M_k d_k + (\lambda_k^{\text{QP}})^\top c_k - \sigma \|c_k^{\#}\|_P.
\end{aligned}
$$

If $\sigma \geq \|\lambda_k^{\text{QP}}\|_D$, using $(17.3)_c$ and the generalized Cauchy-Schwarz inequality (16.14), we have

$$
(\lambda_k^{\text{QP}})^\top c_k - \sigma \|c_k^{\#}\|_P \leq (\lambda_k^{\text{QP}})^\top c_k^{\#} - \sigma \|c_k^{\#}\|_P \leq (\|\lambda_k^{\text{QP}}\|_D - \sigma)\|c_k^{\#}\|_P \leq 0.
$$

Now, the second inequality of the proposition is obtained from (17.6). If $\Theta'_\sigma(x_k; d_k) = 0$ and M_k is positive definite, then $d_k = 0$. From (17.3), it follows that x_k is stationary, with λ_k^{QP} as its associated multiplier.   □

Note that equality holds in (17.6) if there are only equality constraints (see the proof of lemma 17.4 below), but this is not necessarily the case when $I \neq \emptyset$ (this is the subject of exercise 17.2). Therefore, algorithms requiring the computation of $\Theta'_{\sigma_k}(x_k; d_k)$ often use the negative upper bound given by the right-hand side of (17.6):

$$
\Delta_k := \nabla f_k^\top d_k - \sigma_k \|c_k^{\#}\|_P = -d_k^\top M_k d_k + (\lambda_k^{\text{QP}})^\top c_k - \sigma_k \|c_k^{\#}\|_P. \tag{17.7}
$$

We have indexed σ by k, since its value will have to be modified at some iterations.

Update of the Penalty Parameter σ_k

A consequence of proposition 17.1 is that when x_k is nonstationary, when M_k is positive definite, and when σ_k satisfies

$$
\sigma_k > \|\lambda_k^{\text{QP}}\|_D, \tag{17.8}
$$

then $\Delta_k < 0$ and the solution d_k to the osculating quadratic problem is a descent direction of Θ_{σ_k} at x_k, meaning that $\Theta'_{\sigma_k}(x_k; d_k) < 0$. Inequality (17.8) reminds us of the exactness condition $\sigma > \|\lambda_*\|_D$ found for Θ_σ in chapter 16 and is therefore natural: by maintaining (17.8) at each iteration, the algorithm ensures the exactness of Θ_σ at convergence ($\sigma_k = \sigma$ for large k and $\lambda_k^{\text{QP}} \to \lambda_*$).

To maintain (17.8) at each iteration, it is necessary to modify σ_k sometimes (the evolution of λ_k^{QP} cannot be known when the algorithm is started). Global convergence will show that this inequality has to be imposed with some safeguard, given by the positive constant $\bar{\sigma}$ below. To keep some generality, we shall just specify the properties that an adequate adaptation rule for σ_k must enjoy:

$$\begin{cases} (a) \ \sigma_k \geq \|\lambda_k^{\mathrm{QP}}\|_D + \bar\sigma, \quad \text{for all } k \geq 1, \\ (b) \ \text{there exists an index } k_1 \text{ such that:} \\ \quad \text{if } k \geq k_1 \text{ and } \sigma_{k-1} \geq \|\lambda_k^{\mathrm{QP}}\|_D + \bar\sigma, \text{ then } \sigma_k = \sigma_{k-1}, \\ (c) \ \text{if } \{\sigma_k\} \text{ is bounded, } \sigma_k \text{ is modified finitely often.} \end{cases} \quad (17.9)$$

Property (a) means that a little more than (17.8) must hold at each iteration. With (b), we assume that, after finitely many steps, σ_{k-1} is modified only when necessary, to obtain (a). Finally, (c) requires that each modification of σ_k is significant, so as to stabilize the sequence $\{\sigma_k\}$: asymptotically, the merit function should no longer depend on the iteration index.

It can be checked that the following rule, proposed by Mayne and Polak [250], satisfies these properties (the constant 1.5 is given to be specific; actually, any constant > 1 is appropriate):

> **if** $\sigma_{k-1} \geq \|\lambda_k^{\mathrm{QP}}\|_D + \bar\sigma$
> **then** $\sigma_k = \sigma_{k-1}$
> **else** $\sigma_k = \max(1.5\,\sigma_{k-1}, \|\lambda_k^{\mathrm{QP}}\|_D + \bar\sigma)$.

Having a large parameter σ_k is harmless for the theoretical convergence, but can be disastrous in practice; so it must sometimes be decreased. In this case, the properties in (17.9) may no longer be satisfied and convergence may no longer be guaranteed. Nevertheless, an update rule like the one below is often used (the constants 1.1 and 1.5 can be replaced by any constant > 1):

UPDATE RULE FOR σ_k:
if $\sigma_{k-1} \geq 1.1\,(\|\lambda_k^{\mathrm{QP}}\|_D + \bar\sigma)$,
then $\sigma_k = (\sigma_{k-1} + \|\lambda_k^{\mathrm{QP}}\|_D + \bar\sigma)/2$;
else if $\sigma_{k-1} \geq \|\lambda_k^{\mathrm{QP}}\|_D + \bar\sigma$,
$\quad$ **then** $\sigma_k = \sigma_{k-1}$.
$\quad$ **else** $\sigma_k = \max(1.5\,\sigma_{k-1}, \|\lambda_k^{\mathrm{QP}}\|_D + \bar\sigma)$;

In this rule, when the previous penalty factor σ_{k-1} exceeds 1.1 times the minimal threshold $\|\lambda_k^{\mathrm{QP}}\|_D + \bar\sigma$, the new factor σ_k is set to the arithmetic mean of this threshold and of σ_{k-1}.

It is often better to use a different penalty factor for each constraint (in particular, when the constraints have very different orders of magnitude). This is done by taking as a penalty function $\Theta_\sigma(x) = f(x) + \|Sc(x)^\#\|_P$, where $S = \mathrm{Diag}(\sigma_1, \dots, \sigma_n)$. The case of the ℓ_1 norm is considered in exercise 16.8.

Line-Search

The determination of the stepsize $\alpha_k > 0$ along d_k, forcing the decrease in Θ_{σ_k}, must be done in a precise manner (see § 3 for unconstrained problems). We shall enforce satisfaction of the following *Armijo condition* [12]: ω being a fixed constant in $]0, \frac{1}{2}[$, one determines $\alpha > 0$ such that

$$x_k + \alpha d_k \in \Omega \quad \text{and} \quad \Theta_{\sigma_k}(x_k + \alpha d_k) \leq \Theta_{\sigma_k}(x_k) + \omega \alpha \Delta_k. \qquad (17.10)$$

The requirement $\omega < \frac{1}{2}$ comes from the necessity of having asymptotic admissibility of the unit stepsize (see § 17.3); it is essential neither for consistency of (17.10) nor for global convergence ($\omega \in \,]0,1[$ would be sufficient). The value of Δ_k in (17.10) should ideally be $\Theta'_{\sigma_k}(x_k, d_k)$, but since this directional derivative is not easy to compute, we take the negative upper bound given by (17.7).

Since $\Theta'_{\sigma_k}(x_k; d_k) \leq \Delta_k < 0$ and $\omega < 1$, one can easily verify that it is possible to find $\alpha_k > 0$ satisfying (17.10). However, this Armijo condition does not eliminate unduly small α_k's, which might impair convergence of the iterates to a stationary point. This explains the following line-search algorithm. A constant $\beta \in \,]0, \frac{1}{2}]$ is chosen.

BACKTRACKING LINE-SEARCH:

Set $i = 0$, $\alpha_{k,0} = 1$.
1. If (17.10) is satisfied with $\alpha = \alpha_{k,0}$, set $\alpha_k = \alpha$ and exit.
2. Choose $\alpha_{k,i+1} \in [\beta \alpha_{k,i}, (1 - \beta)\alpha_{k,i}]$.
3. Increase i by 1 and go to 1.

Taking for example $\beta = \frac{1}{2}$, the stepsize selected by this algorithm is the first element encountered in the list $\{1, \frac{1}{2}, \frac{1}{4}, \frac{1}{8}, \cdots\}$ satisfying (17.10). Taking the first of these stepsizes does prevent α from being too small. The determination of $\alpha_{k,i+1}$ in the interval $[\beta \alpha_{k,i}, (1-\beta)\alpha_{k,i}]$ should be done using *interpolation* formulas.

Global Convergence with Positive Definite Hessian Approximations

In this subsection, we analyze the global convergence of the line-search SQP algorithm given on page 292, when σ_k is adapted by a rule satisfying properties (17.9), the stepsize α_k is determined by the line-search algorithm on page 296, and the matrices M_k used in the osculating quadratic program (17.2) are maintained positive definite, in such a way that

$$\{M_k\} \text{ and } \{M_k^{-1}\} \text{ are bounded.} \qquad (17.11)$$

This is a strong assumption. For example, it is not known whether it is satisfied in the quasi-Newton versions of SQP. Besides, if $M_k = L(x_k, \lambda_k)$, positive definiteness is not guaranteed. We shall, however, accept this assumption, which allows a simple convergence proof.

Theorem 17.2 (global convergence of the line-search SQP algorithm). *Suppose that f and c are of class $C^{1,1}$ in Ω and that $\| \cdot^\# \|_P$ is* *convex. Consider the line-search SQP algorithm on page 292, using symmetric positive definite matrices M_k satisfying (17.11) and an update rule of σ_k satisfying (17.9). Then, starting the algorithm at a point $x_1 \in \Omega$, one of the following situations occurs:*

(i) *the sequence $\{\sigma_k\}$ is unbounded, in which case $\{\lambda_k^{\mathrm{QP}}\}$ is also unbounded;*

(ii) *there exists an index k_2 such that $\sigma_k = \sigma$ for $k \geq k_2$, and at least one of the following situations occurs:*

(a) $\Theta_\sigma(x_k) \to -\infty$,

(b) $\mathrm{dist}(x_k, \Omega^c) \to 0$,

(c) $\nabla_x \ell(x_k, \lambda_k^{\mathrm{QP}}) \to 0$, $c_k^{\#} \to 0$, $(\lambda_k^{\mathrm{QP}})_I \geq 0$ *and* $(\lambda_k^{\mathrm{QP}})_I^{\top}(c_k)_I \to 0$.

Proof. If $\{\sigma_k\}$ is unbounded, we see from rule $(17.9)_b$ that $\{\lambda_k^{\mathrm{QP}} : \sigma_k \neq \sigma_{k-1}\}$ is unbounded. If $\{\sigma_k\}$ is bounded, rule $(17.9)_c$ shows that there exists an index k_2 such that $\sigma_k = \sigma$ for all $k \geq k_2$. It remains to show that one of the situations $(ii$-$a)$, $(ii$-$b)$, or $(ii$-$c)$ occurs. For this, we suppose that $(ii$-$a)$ and $(ii$-$b)$ do not hold and show $(ii$-$c)$.

Each iteration after k_2 forces the decrease in the same function Θ_σ. Since $\Theta_\sigma(x_k) \geq C > -\infty$, Armijo's condition (17.10) shows that

$$\alpha_k \Delta_k \to 0.$$

Then, if we show $\alpha_k \geq \underline{\alpha} > 0$, the result $(ii$-$c)$ will follow. Indeed, from $\Delta_k \to 0$, (17.6) and $(17.9)_a$, we deduce

$$d_k^{\top} M_k d_k \to 0 \quad \text{and} \quad c_k^{\#} \to 0.$$

Because M_k is positive definite and has a bounded inverse, $d_k \to 0$. Then, from $(17.3)_a$ and the boundedness of M_k, we see that $\nabla_x \ell(x_k, \lambda_k^{\mathrm{QP}}) \to 0$. On the other hand, $(17.3)_c$ shows that $(\lambda_k^{\mathrm{QP}})_I \geq 0$. Finally, $\Delta_k = \nabla f_k^{\top} d_k - \sigma \|c_k^{\#}\|_P \to 0$ and $c_k^{\#} \to 0$ imply that $\nabla f_k^{\top} d_k \to 0$ and, using $(17.3)_a$, $(\lambda_k^{\mathrm{QP}})^{\top} A_k d_k \to 0$. Hence, from $(17.3)_d$ and $(17.3)_b$,

$$\begin{aligned}
(\lambda_k^{\mathrm{QP}})_I^{\top}(c_k)_I &= -(\lambda_k^{\mathrm{QP}})_I^{\top}(A_k d_k)_I \\
&= (\lambda_k^{\mathrm{QP}})_E^{\top}(A_k d_k)_E + o(1) \\
&= -(\lambda_k^{\mathrm{QP}})_E^{\top}(c_k)_E + o(1) \\
&= o(1),
\end{aligned}$$

because $\{\lambda_k^{\mathrm{QP}}\}$ is bounded and $(c_k)_E \to 0$.

Therefore, it remains to prove that $\alpha_k \geq \underline{\alpha} > 0$, for all k and some constant $\underline{\alpha}$. We can consider the indices k of $\mathcal{K} := \{k \geq k_2 : \alpha_k < 1\}$. Then from the rule determining the stepsize, $\alpha_k \in [\beta \bar{\alpha}_k, (1-\beta)\bar{\alpha}_k]$ for some $\bar{\alpha}_k \in \,]0, 1]$ satisfying

$$\alpha_k + \bar{\alpha}_k d_k \notin \Omega \quad \text{or} \quad \Theta_\sigma(x_k + \bar{\alpha}_k d_k) > \Theta_\sigma(x_k) + \omega \bar{\alpha}_k \Delta_k.$$

For large k, the first condition is impossible because $d_k \to 0$ would then imply that $\mathrm{dist}(x_k, \Omega^c) \to 0$. Hence, for large $k \in \mathcal{K}$, we have

$$\Theta_\sigma(x_k + \bar\alpha_k d_k) > \Theta_\sigma(x_k) + \omega\bar\alpha_k \Delta_k. \tag{17.12}$$

Let us expand the left-hand side of (17.12). Using the smoothness of f and c, $\bar\alpha_k \le 1$, the convexity of $\| \cdot^\# \|_P$ (hence its Lipschitz continuity), $(17.3)_b$, and finally (17.6)–(17.7), we have successively

$$f(x_k + \bar\alpha_k d_k) = f_k + \bar\alpha_k \nabla f_k^\top d_k + O(\bar\alpha_k^2 \|d_k\|^2)$$
$$c(x_k + \bar\alpha_k d_k) = c_k + \bar\alpha_k A_k d_k + O(\bar\alpha_k^2 \|d_k\|^2)$$
$$= (1 - \bar\alpha_k)c_k + \bar\alpha_k(c_k + A_k d_k) + O(\bar\alpha_k^2 \|d_k\|^2)$$
$$\|c(x_k + \bar\alpha_k d_k)^\#\|_P \le (1 - \bar\alpha_k)\|c_k^\#\|_P + \bar\alpha_k\|(c_k + A_k d_k)^\#\|_P + O(\bar\alpha_k^2 \|d_k\|^2)$$
$$= (1 - \bar\alpha_k)\|c_k^\#\|_P + O(\bar\alpha_k^2 \|d_k\|^2)$$
$$\Theta_\sigma(x_k + \bar\alpha_k d_k) \le \Theta_\sigma(x_k) + \bar\alpha_k \Delta_k + C_1 \bar\alpha_k^2 \|d_k\|^2.$$

Then (17.12) yields

$$-(1 - \omega)\bar\alpha_k \Delta_k \le C_1 \bar\alpha_k^2 \|d_k\|^2.$$

But $\Delta_k = -d_k^\top M_k d_k + (\lambda_k^{\mathrm{QP}})^\top c_k - \sigma\|c_k^\#\|_P \le -d_k^\top M_k d_k \le -C_2\|d_k\|^2$ (boundedness of $\{M_k^{-1}\}$), so that we deduce from the above inequality:

$$\bar\alpha_k \ge (C_2/C_1)(1 - \omega) > 0,$$

because $\omega < 1$. The positive lower bound on α_k can therefore be taken as $\underline{\alpha} := \beta(C_2/C_1)(1 - \omega)$. This concludes the proof. $\square$

Among the situations described in theorem 17.2, only situation (ii-c) is satisfactory. In this case, every cluster point of $\{(x_k, \lambda_k^{\mathrm{QP}})\}$ satisfies the optimality conditions (KKT). Unfortunately, any of the other situations may occur. For example, (i) may occur in the example in figure 16.1 when $\{x_k\}$ converges to x'_*, a point where λ_* is not defined. Situation (ii-a) will occur if, outside of the feasible set, f decreases more rapidly than $\|c(\cdot)^\#\|_P$ increases, and if x_1 is taken far enough from the feasible set; the example

$$\min\{-x^2 : x = 0\},$$

with $\| \cdot \|_P = | \cdot |$, is such. Finally, situation (ii-b) occurs if Ω contains no stationary point.

17.2 Truncated SQP

In this section, we consider another globalization technique of the Newton algorithm to solve the problem with only equality constraints:

$$(P_E) \quad \begin{cases} \min_x f(x) \\ c(x) = 0. \end{cases}$$

The local algorithm was introduced in § 14.1 and we refer the reader to § 14.4 (in the subsection entitled "The reduced system approach") for the notation. In contrast to the approach used in the previous section, we do not replace here the Hessian of the Lagrangian by a positive definite approximation. This was useful to ensure the well-posedness of the osculating quadratic program and the decrease in Θ_σ along the computed direction. Instead, we describe an algorithm that directly exploits the curvature of the problem (i.e., the second derivatives of f and c) gathered in the Hessian of the Lagrangian, even in the presence of nonconvexity.

Here also, the computed direction will be a descent direction of the merit function Θ_σ, which allows global convergence. Therefore, it must differ from Newton's direction, but the modification only needs to be done at points where the *reduced* Hessian of the Lagrangian is not positive definite. This form of weak nonconvexity can therefore be detected by the algorithm, which is a nice feature. The idea is similar to the truncated Newton algorithm in unconstrained optimization (see § 6.4): the truncated conjugate gradient (CG) algorithm is used to solve, sometimes approximately, the reduced linear system (see (14.32))

$$H_k u_k = v_k, \tag{17.13}$$

where

$$H_k := Z_k^{-\top} L_k Z_k^{-} \quad \text{and} \quad v_k := -g_k + Z_k^{-\top} L_k A_k^{-} c_k. \tag{17.14}$$

Note that the reduced Hessian of the Lagrangian H_k is symmetric but may be indefinite. By the truncated CG, the algorithm aims at collecting only the "positive definite part" of H_k. This is obtained by stopping the CG iterations certainly before a conjugate direction w is a negative curvature direction for H_k (more precisely, before $w^\top H_k w$ becomes less than an appropriate positive threshold). Let us denote by $\tilde{u}_k$ the approximate solution to (17.13) computed by the truncated CG algorithm. We shall show that the search direction

$$d_k = -A_k^{-} c_k + Z_k^{-} \tilde{u}_k \tag{17.15}$$

is then a descent direction of Θ_σ provided σ is larger than an easily computable threshold. Another interesting property of this approach is that, since H_k is positive definite around a strong solution to (P_E), the CG iterations can be pursued up to completion close to such a solution, so that local quadratic convergence is not prevented.

Let us look at this in more detail.

Truncated CG Iterations

The *truncated conjugate gradient* (TCG) *algorithm* to solve (17.13) is presented below. For clarity, we drop the index k of the Newton algorithm and denote by i the CG iteration index (in superscript). For $i = 0, \ldots, j$, Algorithm TCG generates iterates u^i, approximating the solution to (17.13), residuals $r^i := Hu^i - v$, and conjugate directions w^i. The algorithm can be stopped at any iteration (global convergence of the truncated SQP method will not be affected by this), but it must certainly be interrupted at u^j if the next conjugate direction w^j is a *quasi-negative curvature* direction for H. This means that the following inequality does not hold with $i = j$:

$$(w^i)^\top Hw^i \geq \nu \|w^i\|_2^2. \tag{17.16}$$

The threshold $\nu > 0$ is assumed to be independent of the index k, although an actual implementation would use a more sophisticated rule for setting this parameter, allowing small values when approaching a solution. Hence, Algorithm TCG simply discards quasi-negative directions. It is in this way that nonconvexity is dealt with.

ALGORITHM TCG FOR (17.13):

1. Choose $\nu > 0$. Set $u^0 = 0$ and $r^0 = -v$, where v is defined by (17.14).
2. For $i = 0, 1, \ldots$ do the following:
    2.1. If desired or if $r^i = 0$, stop to iterate and go to step 3 with $j = i$.
    2.2. Compute a new conjugate direction:
    $$w^i = \begin{cases} -r^i & \text{if } i = 0 \\ -r^i + \frac{\|r^i\|^2}{\|r^{i-1}\|^2} w^{i-1} & \text{if } i \geq 1. \end{cases}$$
    2.3. Compute $p^i = Hw^i$.
    2.4. If (17.16) does not hold, go to step 3 with $j = i$.
    2.5. Compute the new iterate $u^{i+1} = u^i + t^i w^i$ and the new residual $r^{i+1} = r^i + t^i p^i$, with the stepsize
    $$t^i = \frac{\|r^i\|^2}{(w^i)^\top p^i}.$$
3. Take as the approximate solution to (17.13):
    $$\tilde{u} = \begin{cases} v & \text{if } j = 0 \\ u^j & \text{if } j \geq 1. \end{cases}$$

Observe that, since the first iterate of Algorithm TCG is $u^0 = 0$, the first CG direction is $w^0 = -r^0 = v$, the right-hand side of (17.13). This is important for the analysis that follows. Another key point is that the directions w^i are conjugate: $w^{i_1} Hw^{i_2} = 0$ for $i_1 \neq i_2$. Note finally that Algorithm TCG chooses to output the approximate solution u^j currently obtained when $j \geq 1$ (it is different from zero), but $\tilde{u} = w^0 = v$ when $j = 0$ ($u^0 = 0$ in this case).

Lemma 17.3. *The vector $\tilde{u}$ computed by Algorithm TCG has the form*

$$\tilde{u} = Jv, \tag{17.17}$$

where J is the identity matrix when $j = 0$ and

$$J = \sum_{i=0}^{j-1} \frac{w^i (w^i)^\top}{(w^i)^\top H w^i} \tag{17.18}$$

when $j \geq 1$. Furthermore $\|J\|_2 \leq \max\left(1, \frac{i}{\nu}\right)$.

Proof. If $i = 0$, $u = v$ and the result follows. Otherwise Algorithm TCG generates conjugate directions w^0, ..., w^{j-1}. By orthogonality of r^i and w^{i-1}, by the fact that the algorithm starts with $u^0 = 0$, and by conjugacy of the directions w^i, one has for $1 \leq i \leq j$:

$$\begin{aligned}
\|r^i\|^2 &= -(w^i)^\top r^i \\
&= -(w^i)^\top (H u^i - v) \\
&= -(w^i)^\top H \left(\sum_{l=0}^{i-1} t^l w^l\right) + (w^i)^\top v \\
&= (w^i)^\top v.
\end{aligned}$$

Also, $\|r^0\|^2 = (w^0)^\top v$. Therefore

$$\tilde{u} = \sum_{i=0}^{j-1} t^i w^i = \sum_{i=0}^{j-1} \frac{(w^i)^\top v}{(w^i)^\top H w^i} \, w^i = \left(\sum_{i=0}^{j-1} \frac{w^i (w^i)^\top}{(w^i)^\top H w^i}\right) v.$$

This proves (17.18).

The upper bound on $\|J\|_2$ comes from the fact that $\|vv^\top\|_2 = \|v\|_2^2$ and (17.16). □

Note that, when $j \geq 1$, the matrix J is positive semi-definite with rank j. In view of (17.13) and (17.17), this matrix appears as a kind of "pseudo-inverse of the positive definite part" of H.

Descent Property

In the next lemma, we give conditions ensuring that the direction d_k given by (17.15) is a descent direction of Θ_{σ_k}. For this, it is convenient to give another expression of d_k by introducing the following right inverse of A_k:

$$\bar{A}_k^- := (I - Z_k^- J_k Z_k^{-\top} L_k) A_k^-. \tag{17.19}$$

This is the right inverse $\widehat{A}_k^-$ in (14.33), in which H_k^{-1} has been substituted by its approximation J_k. Then

$$d_k = \tilde{r}_k + \tilde{t}_k, \tag{17.20}$$

where

$$\tilde{r}_k = -\widetilde{A}_k^- c_k \quad \text{and} \quad \tilde{t}_k = -Z_k^- J_k g_k.$$

We also use the multiplier associated with $\widetilde{A}_k^-$:

$$\tilde{\lambda}_k = -\widetilde{A}_k^{-\top} \nabla f_k. \tag{17.21}$$

How to compute this multiplier efficiently is dealt with in the next subsection.

Lemma 17.4 (descent property). *Suppose that f and c are differentiable at x_k. Let d_k be given by (17.15), where $\tilde{u}_k$ is the approximate solution to (17.13) computed by Algorithm TCG. Then Θ_{σ_k} has a directional derivative in the direction d_k, whose value is given by*

$$\Theta'_{\sigma_k}(x_k; d_k) = -g_k^\top J g_k + \tilde{\lambda}_k^\top c_k - \sigma_k \|c_k\|_P. \tag{17.22}$$

It is negative if x_k is nonstationary and $\sigma_k > \|\tilde{\lambda}_k\|_D$.

Proof. Since a norm is Lipschitz continuous and has directional derivatives, $\|\cdot\|_P \circ c$ has directional derivatives (see lemma 13.1). Using the fact that d_k satisfies the linearized constraints (i.e., $A_k d_k = -c_k$), one has $(\|\cdot\|_P \circ c)'(x_k; d_k) = (\|\cdot\|_P)'(c_k; -c_k) = -\|c_k\|_P$. Therefore

$$\Theta'_{\sigma_k}(x_k; d_k) = \nabla f_k^\top d_k - \sigma_k \|c_k\|_P.$$

Using (17.20) and (17.21), we get (17.22).

Suppose now that $\sigma_k > \|\tilde{\lambda}_k\|_D$. Since $\tilde{\lambda}_k^\top c_k \leq \|\tilde{\lambda}_k\|_D \|c_k\|_P$, we obtain

$$\Theta'_{\sigma_k}(x_k; d_k) \leq -g_k^\top J_k g_k + (\|\tilde{\lambda}_k\|_D - \sigma_k)\|c_k\|_P \leq 0.$$

If $\Theta'_{\sigma_k}(x_k; d_k) = 0$, it follows that $c_k = 0$ and $g_k^\top J_k g_k = 0$. If the number of CG iterations $j_k = 0$, then $J_k = I$, hence $g_k = 0$ and x_k is stationary. It remains to show that j_k cannot be ≥ 1 when $\Theta'_{\sigma_k}(x_k; d_k) = 0$. If $j_k \geq 1$, one would have $v_k \neq 0$ (see step 2.1 of Algorithm TCG) and therefore $g_k \neq 0$ (since $c_k = 0$). But with the structure of J_k and the fact that $w_k^0 = v_k = -g_k$ when $c_k = 0$, one would have $g_k^\top J_k g_k \geq (g_k^\top w_k^0)^2/((w_k^0)^\top H_k w_k^0) = \|g_k\|^4/(g_k^\top H g_k) > 0$, which would contradict the fact that $g_k^\top J_k g_k = 0$. □

Computation of $\tilde{\lambda}_k$

Let us drop the index k. From (17.21) and (17.19), the definition of $\tilde{\lambda}$ involves the matrix J:

$$\tilde{\lambda} = -A^{-\top}(\nabla f - LZ^-Jg).$$

We do not want to store this matrix, however. In fact, to compute $\tilde{\lambda}$, one has to evaluate $\bar{u} = Jg$, which is the approximate solution to

$$H\bar{u} = g, \tag{17.23}$$

obtained by using the same conjugate directions w^i and the same products $p^i = Hw^i$, $i = 0, \ldots, j-1$, as those used to compute the approximate solution $\tilde{u}$ to (17.13) by Algorithm TCG. The computation of $\tilde{u}$ and $\bar{u}$ can be made in parallel, hence avoiding the need to store the conjugate directions w^i (or J) or the need to compute twice the Hessian-vector products $p^i = Hw^i$. This is what Algorithm TCG2 below does. Its outputs are $\tilde{u}$ and $\bar{u}$.

ALGORITHM TCG2 FOR (17.13) AND (17.23):

1. Choose $\nu > 0$. Set $u^0 = 0$, $r^0 = -v$, $\bar{u}^0 = 0$, and $\bar{r}^0 = -g$, where v is defined by (17.14).
2. For $i = 0, 1, \ldots$ do the following:
   2.1. If desired or if $r^i = 0$, stop to iterate and go to step 3 with $j = i$.
   2.2. Compute a new conjugate direction:

$$w^i = \begin{cases} -r^i & \text{if } i = 0 \\ -r^i + \dfrac{\|r^i\|^2}{\|r^{i-1}\|^2}\, w^{i-1} & \text{if } i \geq 1. \end{cases}$$

   2.3. Compute $p^i = Hw^i$.
   2.4. If (17.16) does not hold, go to step 3 with $j = i$.
   2.5. Compute the new iterates $u^{i+1} = u^i + t^i w^i$ and $\bar{u}^{i+1} = \bar{u}^i + \bar{t}^i w^i$ and the new residuals $r^{i+1} = r^i + t^i p^i$ and $\bar{r}^{i+1} = \bar{r}^i + \bar{t}^i p^i$, with the stepsizes

$$t^i = \frac{\|r^i\|^2}{(w^i)^\top p^i} \quad \text{and} \quad \bar{t}^i = -\frac{(\bar{r}^i)^\top w^i}{(w^i)^\top p^i}.$$

3. Take as the approximate solution to (17.13) and (17.23):

$$\tilde{u} = \begin{cases} v & \text{if } j = 0 \\ u^j & \text{if } j \geq 1 \end{cases} \quad \text{and} \quad \bar{u} = \begin{cases} g & \text{if } j = 0 \\ \bar{u}^j & \text{if } j \geq 1. \end{cases}$$

It may occur that the linear system (17.23) is solved before (17.13). In this case, the stepsizes $\bar{t}^i$ vanish and $\bar{u}^i$ is no longer modified. It is easy to verify that $\tilde{\lambda}$ is obtained from $\bar{u}$ by:

$$\tilde{\lambda} = -A^{-\top}(\nabla f - LZ^-\bar{u}).$$ (17.24)

Indeed, since $\bar{u}^0 = 0$, one has for $1 \leq i \leq j$:

$$(w^i)^\top \bar{r}^i = (w^i)^\top (H\bar{u}^i - g) = (w^i)^\top H \left(\sum_{l=0}^{i-1} \bar{t}^l w^l \right) - (w^i)^\top g = -(w^i)^\top g.$$

Hence

$$\bar{u} = \sum_{i=0}^{j-1} \bar{t}^i w^i = \sum_{i=0}^{j-1} \frac{(w^i)^\top g}{(w^i)^\top H w^i} w^i = Jg.$$

The Truncated SQP Algorithm and its Global Convergence

The truncated SQP algorithm to solve problem (P_E) generates a sequence $\{x_k\}_{k \geq 1}$ by the recurrence

$$x_{k+1} = x_k + \alpha_k d_k,$$

where the direction $d_k \in \mathbb{R}^n$ is determined by (17.15), with $\tilde{u}_k$ computed by Algorithm TCG2, and the stepsize $\alpha_k > 0$ is determined by a line-search on the merit function Θ_{σ_k}.

According to lemma 17.4, d_k is a descent direction of Θ_{σ_k} provided x_k is nonstationary and $\sigma_k > \|\tilde{\lambda}_k\|_D$. This requires a modification of σ_k at some iterations and we assume that a rule respecting conditions similar to (17.9) is adopted: for some fixed constant $\bar{\sigma} > 0$, the following holds

$$\left\{ \begin{array}{l} (a) \ \sigma_k \geq \|\tilde{\lambda}_k\|_D + \bar{\sigma}, \quad \text{for all } k \geq 1, \\ (b) \ \text{there exists an index } k_1 \text{ such that:} \\ \qquad \text{if } k \geq k_1 \text{ and } \sigma_{k-1} \geq \|\tilde{\lambda}_k\|_D + \bar{\sigma}, \text{ then } \sigma_k = \sigma_{k-1}, \\ (c) \ \text{if } \{\sigma_k\} \text{ is bounded, } \sigma_k \text{ is modified finitely often.} \end{array} \right.$$ (17.25)

Since at a nonstationary iterate x_k, d_k is a descent direction of Θ_{σ_k}, one can determine a stepsize $\alpha_k > 0$ such that the following Armijo inequality holds

$$\Theta_{\sigma_k}(x_k + \alpha_k d_k) \leq \Theta_{\sigma_k}(x_k) + \omega \alpha_k \Theta'_{\sigma_k}(x_k; d_k),$$ (17.26)

where ω is a constant chosen in $]0, \frac{1}{2}[$. As in the line-search SQP algorithm on page 292, the stepsize is determined in step 4 below by *backtracking*.

We can now summarize the overall TSQP algorithm to solve the equality constrained problem (P_E).

ALGORITHM TSQP:

Choose an initial iterate $(x_1, \lambda_1) \in \mathbb{R}^n \times \mathbb{R}^m$.
Compute $f(x_1)$, $c(x_1)$, $\nabla f(x_1)$, and $A(x_1)$.
Set the constants $\nu > 0$ (quasi-negative curvature threshold), $\omega \in \left]0, \frac{1}{2}\right[$ (slope modifier in the Armijo condition), $\bar{\sigma} > 0$ (penalty parameter threshold), and $\beta \in \left]0, \frac{1}{2}\right]$ (backtracking safeguard parameter).
Set $k = 1$.

1. *Stopping test*: Stop if $c_k = 0$ and $g_k = 0$.
2. *Step computation*:
   - Compute the restoration step $r_k = -A_k^- c_k$.
   - Compute the reduced gradient $g_k = Z_k^{-\top} \nabla f_k$ and the right-hand side of (17.13) $v_k = -g_k - Z_k^{-\top} L_k r_k$.
   - Run Algorithm TCG2 to compute $\tilde{u}_k$ and $\bar{u}_k$.
   - Compute the full step $d_k = r_k + Z_k^- \tilde{u}_k$ and the multiplier $\tilde{\lambda}_k$ by (17.24).
3. *Penalty parameter setting*: Update σ_k such that (17.25) holds.
4. *Backtracking line-search*:
   - Set $\alpha = 1$.
   - While α does not satisfy Armijo's inequality (17.26), pick a new stepsize α in $[\beta\alpha, (1-\beta)\alpha]$.
   - Set $\alpha_k = \alpha$.
5. *New iterates*: Set $x_{k+1} = x_k + \alpha_k d_k$ and $\lambda_{k+1} = \lambda_{k+1}^{\mathrm{LS}}$.
6. Increase k by 1 and go to 1.

Before proving the global convergence of this algorithm, let us make some observations. In a practical algorithm, the stopping test in step 1 would be replaced by a condition checking that c_k and g_k are sufficiently small. In practice, in step 4, the new stepsize chosen in the interval $[\beta\alpha, (1-\beta)\alpha]$ during the line-search should be obtained by *interpolation*. In step 5, we have set the new multiplier λ_{k+1} to the least-squares multiplier

$$\lambda_k^{\mathrm{LS}} := -A_k^{-\top} \nabla f_k.$$

This makes Algorithm TSQP close to the primal version of Newton's algorithm analyzed in theorem 14.5. Another possibility would have been to choose $\lambda_{k+1} = \tilde{\lambda}_k$. Observe however that, even if the CG iterations of Algorithm TCG2 solve (17.13) and (17.23) exactly, $\tilde{\lambda}_k \neq \lambda_k^{\mathrm{QP}}$ (in this case $\tilde{\lambda}_k = \widehat{\lambda}_k$ given by (14.36), compare with (14.35)), so that with that choice of λ_{k+1}, Algorithm TSQP does not reduce to Newton's algorithm in a neighborhood of a strong solution.

Theorem 17.5 (global convergence of the line-search truncated SQP algorithm). *Suppose that the functions f and c are twice continuously differentiable with Lipschitz continuous first derivatives. Suppose also that the sequences $\{\nabla f_k\}$, $\{L_k\}$, $\{A_k^-\}$, and $\{Z_k^-\}$ generated by Algorithm TSQP are bounded. Then the sequence of penalty parameters $\{\sigma_k\}$ is stationary for sufficiently large k: $\sigma_k = \sigma$. If furthermore $\{\Theta_\sigma(x_k)\}$ is bounded below, the sequences $\{c_k\}$ and $\{g_k\}$ converge to 0.*

Proof. We denote by $C_1, C_2, \dots$ positive constants, independent of k. We can assume that $\|c_k\| + \|g_k\| > 0$ for all $k \geq 1$, because otherwise the conclusion is clear.

Note first, that the assumptions imply the boundedness of $\{\tilde{\lambda}_k\}$ (use (17.24), the boundedness of $\{A_k^-\}$, $\{\nabla f_k\}$, $\{L_k\}$, $\{Z_k^-\}$, and that of $\{J_k\}$ given by lemma 17.3). Then by $(17.25)_b$, $\{\sigma_k\}$ is also bounded, hence stationary for large enough k (use $(17.25)_c$). From Armijo's inequality (17.26), $\Theta_\sigma(x_k)$ is decreasing. It is also bounded below (by assumption), hence it converges. This implies that $\alpha_k \Theta_\sigma'(x_k; d_k)$ tends to 0, or equivalently (use lemma 17.4 and $(17.25)_a$)

$$\alpha_k g_k^\top J_k g_k \to 0 \quad \text{and} \quad \alpha_k c_k \to 0. \tag{17.27}$$

Let us now show that $\{\alpha_k\}$ is bounded away from 0. From the line-search (step 4), when $\alpha_k < 1$, there is a stepsize $\overline{\alpha}_k \in\ ]0, 1]$ such that $\alpha_k \in [\beta\overline{\alpha}_k, (1-\beta)\overline{\alpha}_k]$ and

$$\Theta_\sigma(x_k + \overline{\alpha}_k d_k) > \Theta_\sigma(x_k) + \omega\overline{\alpha}_k \Theta_\sigma'(x_k; d_k).$$

Using the smoothness of f and c and the fact that d_k satisfies the linearized constraints, one has successively

$$f(x_k + \overline{\alpha}_k d_k) = f(x_k) + \overline{\alpha}_k f'(x_k) \cdot d_k + O(\overline{\alpha}_k^2 \|d_k\|^2),$$
$$c(x_k + \overline{\alpha}_k d_k) = (1 - \overline{\alpha}_k)c(x_k) + O(\overline{\alpha}_k^2 \|d_k\|^2),$$
$$\Theta_\sigma(x_k + \overline{\alpha}_k d_k) \leq \Theta_\sigma(x_k) + \overline{\alpha}_k \Theta_\sigma'(x_k; d_k) + C_1\overline{\alpha}_k^2 \|d_k\|^2.$$

Therefore $(\omega - 1)\Theta_\sigma'(x_k; d_k) < C_1\overline{\alpha}_k\|d_k\|^2$ or

$$g_k^\top J_k g_k + \|c_k\|_P < C_2\overline{\alpha}_k\|d_k\|^2, \tag{17.28}$$

where $C_2 = C_1/((1-\omega)\min(1, \overline{\sigma}))$. With the boundedness of $\{A_k^-\}$, $\{Z_k^-\}$, $\{L_k\}$, and $\{J_k\}$, we have $d_k = O(\|J_k^{1/2} v_k\| + \|c_k\|_P)$ and, due to the form of v_k, $d_k = O(\|J_k^{1/2} g_k\| + \|c_k\|_P)$. Then, inequality (17.28) becomes

$$g_k^\top J_k g_k + \|c_k\|_P < C_3\overline{\alpha}_k(g_k^\top J_k g_k + \|c_k\|_P^2).$$

From (17.27), $\alpha_k c_k \to 0$ and therefore for large k

$$g_k^\top J_k g_k < C_3\overline{\alpha}_k g_k^\top J_k g_k.$$

This inequality shows that $g_k^\top J_k g_k \neq 0$ when $\alpha_k < 1$ and k is large enough and that $\{\overline{\alpha}_k\}$ is bounded away from zero. Since $\alpha_k \geq \beta \overline{\alpha}_k$, $\{\alpha_k\}$ is also bounded away from zero.

From (17.27)

$$g_k^\top J_k g_k \to 0 \quad \text{and} \quad c_k \to 0. \tag{17.29}$$

It remains to show that $g_k \to 0$. Assume the opposite: there is a constant $\gamma > 0$ and subsequence $\mathcal{K}$ such that $\|g_k\| \geq \gamma$ for $k \in \mathcal{K}$. Using the first term of the expression (17.18) of J_k when $j_k \geq 1$, $w_k^0 = v_k$, and the boundedness of $\{H_k\}$, one can write

$$g_k^\top J_k g_k \geq \min\left(\|g_k\|_2^2, \frac{(g_k^\top v_k)^2}{v_k^\top H_k v_k}\right) \geq \min\left(\gamma^2, C_4 \frac{(g_k^\top v_k)^2}{\|v_k\|^2}\right).$$

The numerator can be bounded below as follows:

$$\begin{aligned}
(g_k^\top v_k)^2 &= [-\|g_k\|^2 + O(\|g_k\|\,\|c_k\|)]^2 \\
&= \|g_k\|^4 + O(\|g_k\|^3\,\|c_k\|) + O(\|g_k\|^2\,\|c_k\|^2) \\
&\geq \frac{1}{2}\|g_k\|^4 - C_5\|g_k\|^2\,\|c_k\|^2 \\
&\geq \|g_k\|^2(\tfrac{1}{2}\gamma^2 - C_5\|c_k\|^2),
\end{aligned}$$

which is positive for large k in $\mathcal{K}$. For the denominator, we use the upper bound:

$$\|v_k\|^2 \leq 2\|g_k\|^2 + C_6\|c_k\|^2 \leq \|g_k\|^2(2 + C_6\|c_k\|^2/\gamma^2).$$

Therefore for large k in $\mathcal{K}$:

$$g_k^\top J_k g_k \geq \min\left(\gamma^2, \frac{\frac{1}{2}\gamma^2 - C_5\|c_k\|^2}{2 + C_6\|c_k\|^2/\gamma^2}\right).$$

This is in contradiction with (17.29).   □

17.3 From Global to Local

In this section, we analyze conditions under which the line-search algorithms of the present chapter can transform themselves into the "local" algorithms of chapter 14. In view of the quadratic convergence of the local methods, this "mutation" is highly desirable. Because the direction generated by the local algorithm is used as a descent direction of some merit function, this transformation will occur if the line-search accepts the unit stepsize during the last iterations. This property is referred to as the *asymptotic admissibility of the unit stepsize*. We shall see that it is not guaranteed without certain modifications of the algorithms, which are therefore crucial for their efficiency.

For simplicity, we assume in this section that the problem has only equality constraints:

$$(P_E) \quad \begin{cases} \min_x f(x) \\ c(x) = 0. \end{cases}$$

Since our study is asymptotic, assuming convergence of the sequence $\{(x_k, \lambda_k)\}$ to a primal-dual solution (x_*, λ_*), this simplification amounts to assuming that the active constraints are identified after finitely many iterations, in which case problem (P_{EI}) reduces locally to a problem with only equality constraints (theorem 15.2 tells us something about this).

The Maratos Effect

The merit function Θ_σ introduced in § 16.4 and defined by

$$\Theta_\sigma(x) = f(x) + \sigma \|c(x)\|_P$$

does not necessarily accept unit stepsizes asymptotically. This is known as the *Maratos effect*. We mean by this that when d_k solves the quadratic problem

$$\begin{cases} \min_d \nabla f(x_k)^\top d + \frac{1}{2} d^\top M_k d \\ c(x_k) + A(x_k) d = 0, \end{cases} \tag{17.30}$$

we may have

$$\Theta_\sigma(x_k + d_k) > \Theta_\sigma(x_k), \tag{17.31}$$

however close to (x_*, L_*) the current pair (x_k, M_k) may be.

The following counter-example demonstrates this fact. There, the considered iterate x_k is on the constraint manifold: $c(x_k) = 0$. We have seen in proposition 17.1 that, if $\sigma_k \geq \|\lambda_k^{QP}\|_D$ and M_k is positive definite, Θ_{σ_k} decreases along the Newton direction d_k, which means that, for small stepsizes, the decrease in f along d_k compensates the increase in $\|c\|_P$. In the counter-example, this compensation not longer holds for stepsizes close to 1.·

Counter-example 17.6. Consider the problem on $\mathbb{R}^2$

$$\begin{cases} \min_x -x_1 + \tau(x_1^2 + x_2^2 - 1) \\ x_1^2 + x_2^2 - 1 = 0, \end{cases}$$

where $\tau \in \mathbb{R}$. Its unique solution is $x_* = (1, 0)$ and the associated multiplier is $\lambda_* = \frac{1}{2} - \tau$. The Hessian of the Lagrangian at the solution is $L_* = I$.

Suppose now that the step d at x is given by the osculating quadratic problem, defined at a feasible point x with the matrix $M = L_* = I$:

$$\begin{cases} \min_d -d_1 + \frac{1}{2}\|d\|_2^2 \\ x^\top d = 0. \end{cases}$$

Its solution for $x = (\cos\theta, \sin\theta)$ lying on the constraint is

$$d = \begin{pmatrix} \sin^2 \theta \\ -\sin\theta\cos\theta \end{pmatrix}$$

and $c(x + \alpha d) = \alpha^2 \sin^2 \theta$. Hence, if $\| \cdot \|_P = | \cdot |$,

$$\Theta_\sigma(x) = -\cos\theta$$
$$\Theta_\sigma(x + \alpha d) = -\cos\theta - \alpha\sin^2\theta + (\tau + \sigma)\alpha^2 \sin^2\theta.$$

Then $\Theta_\sigma(x + d) > \Theta_\sigma(x)$ whenever $\tau + \sigma > 1$ (and $\theta \neq 0$). Because $\sigma \geq |\lambda_*| \equiv |\frac{1}{2} - \tau|$ is needed to have an exact penalty, Θ_σ increases for a unit stepsize if $\tau > \frac{3}{4}$.

Figure 17.2 shows the level curves of Θ_σ around the solution for $\tau = 1$ and $\sigma = 0.6$, as well as the Newton step d from an x on the constraint manifold (the bold curve), rather close to the solution $(1, 0)$. One clearly observes that $\Theta_\sigma(x + d) > \Theta_\sigma(x)$. $\qquad\square$

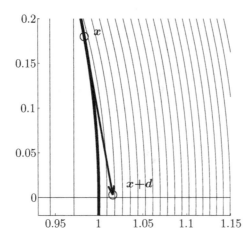

Fig. 17.2. Example with a Maratos effect

This phenomenon somehow reveals a discrepancy between Θ_σ and the osculating quadratic problem used to compute d_k. Since this model is good (it yields local quadratic convergence), the blame must be put on the merit function, or on the way in which it is used. In the rest of this section, we analyze different remedies for the Maratos effect and prove that they are effective close to a solution. The Maratos effect can also occur far from a solution and it is then more difficult to deal with. The first remedy consists in modifying the step d_k by adding to it a small displacement, called a *second order correction*, that does not prevent quadratic convergence. Another possibility is to modify the merit function, which is considered next.

Modification of the Step: Second Order Correction

Example 17.6 has shown that there are situations in which, even close to the solution, the increase in $\|c(\cdot)\|_P$ from x_k to $x_k + d_k$ is not compensated by a decrease in f, resulting finally in an increase in Θ_σ. The remedy for the Maratos effect presented in this subsection consists in adding to d_k a small correcting step $e_k \in \mathbb{R}^n$, whose aim is to decrease $\|c(\cdot)\|_P$. This additional step is defined by

$$e_k = -A_k^- c(x_k + d_k), \tag{17.32}$$

where A_k^- is some right inverse of the Jacobian matrix $A_k = c'(x_k)$, which is assumed to be surjective. Hence, e_k is a constraint-restoration step at $x_k + d_k$. Figure 17.3 shows the second order correction for counter-example 17.6: the small step e from $x + d$ to $x + d + e$.

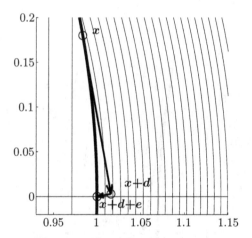

Fig. 17.3. Second order correction

One speaks of *second-order correction* because $c(x_k + d_k) = O(\|d_k\|^2)$ and therefore $e_k = O(\|d_k\|^2)$ is of order 2 in d_k. This modification of d_k preserves a possible quadratic convergence since, assuming $x_k + d_k - x_* = O(\|x_k - x_*\|^2)$, we have

$$x_k + d_k + e_k - x_* = (x_k + d_k - x_*) + e_k = O(\|x_k - x_*\|^2),$$

owing to the preceding estimate of e_k and to the fact that $d_k \sim (x_k - x_*)$ (lemma 13.5).

Because e_k is computed by evaluating c at a point different from x_k, it cannot be guaranteed that $d_k + e_k$ is a descent direction of Θ_{σ_k} at x_k. Therefore, a line-search along this direction may be impossible. The least expensive approach is then to determine a stepsize $\alpha_k > 0$ along the arc

$$\alpha \mapsto p_k(\alpha) = x_k + \alpha d_k + \alpha^2 e_k.$$

It has the descent direction d_k as a tangent at $\alpha = 0$ and visits $x_k + d_k + e_k$ for $\alpha = 1$. The stepsize α_k can be computed in the same way as along d_k, forcing at each iteration the inequality

$$\Theta_{\sigma_k}(x_k + \alpha_k d_k + \alpha_k^2 e_k) \le \Theta_{\sigma_k}(x_k) + \omega \alpha_k \Theta'_{\sigma_k}(x_k; d_k), \tag{17.33}$$

for some $\alpha_k \in \,]0,1]$. It is easy to verify that this inequality can always be satisfied, provided d_k is a descent direction of Θ_{σ_k} at x_k.

In the next proposition, we give conditions under which the unit stepsize $\alpha_k = 1$ is accepted in (17.33) when x_k is near a strong solution to (P_E). Part of these conditions is related to the matrix M_k, which must satisfy (17.34). This condition is of the form $t_k \ge o(\|d_k\|^2)$, for some real numbers t_k, which means that there must exist a sequence of real numbers $\{s_k\}$, such that $t_k \ge s_k$ and $s_k = o(\|d_k\|^2)$ when $k \to \infty$. Observe that this condition is satisfied when M_k is "large enough". This is not surprising, since then the tangent step is small (see remark 2 on page 235) and the total step d_k is close to the restoration step, along which the unit stepsize is known to be accepted by the norm of the constraints (see exercise 17.4). Observe also that condition (17.34) is satisfied when M_k is the Hessian of the Lagrangian (with convergent multipliers), which corresponds to Newton's method.

Proposition 17.7 (admissibility of the unit step-size with a second order correction). *Suppose that f and c are of class C^2 in a neighborhood of a solution x_* to (P_E) satisfying the second-order sufficient conditions of optimality and at which $A_* = c'(x_*)$ is surjective. Let $\{x_k\}$ be a sequence converging to x_*, let d_k satisfy the first-order optimality conditions of the osculating quadratic problem (17.30), and let e_k be defined by (17.32). Suppose also that*

- *$\{A_k^-\}$ is bounded and $d_k \to 0$,*

- *the matrix M_k used in the osculating quadratic problem (17.30) over-estimates the Hessian of the augmented Lagrangian $L_*^r := L_* + rA_*^\top A_*$, in the sense that*

$$d_k^\top (M_k - L_*^r) d_k \ge o(\|d_k\|^2), \tag{17.34}$$

where $r \ge 0$ is such that L_*^r is positive definite (such an r always exists under the assumptions already stated, see exercise 16.1),

- *the penalty parameter σ_k used in Θ_{σ_k} satisfies*

$$\|\lambda_k^{\mathrm{QP}}\|_D \le \sigma_k \le \hat\sigma, \tag{17.35}$$

where λ_k^{QP} is a multiplier associated with the constraints of (17.30) and $\hat\sigma$ is a constant.

Then, for $\omega < \frac{1}{2}$ and large enough k, there holds

$$\Theta_{\sigma_k}(x_k + d_k + e_k) \le \Theta_{\sigma_k}(x_k) + \omega \Theta'_{\sigma_k}(x_k; d_k).$$

Proof. Despite the nondifferentiability of Θ_{σ_k}, one can obtain an expansion of $\Theta_{\sigma_k}(x_k+d_k+e_k)$ with a precision of order $o(\|d_k\|^2)$. This one follows from an expansion of $f(x_k+d_k+e_k)$ and $c(x_k+d_k+e_k)$ about x_k. Using the smoothness assumptions on f and c, the constraint in (17.30), the definition of e_k in (17.32), the boundedness of $\{A_k^-\}$, and the optimality of (x_*, λ_*), we have successively

$$c(x_k+d_k) = c_k + A_k d_k + \frac{1}{2}c''(x_*) \cdot d_k^2 + o(\|d_k\|^2),$$

$$= \frac{1}{2}c''(x_*) \cdot d_k^2 + o(\|d_k\|^2),$$

$$e_k = O(\|c(x_k+d_k)\|)$$

$$= O(\|d_k\|^2),$$

$$c(x_k+d_k+e_k) = c(x_k+d_k) + A_k e_k + o(\|e_k\|)$$

$$= o(\|d_k\|^2),$$

$$-A_k^{-\top}\nabla f_k = \lambda_* - A_k^{-\top}(\nabla f_k + A_k^\top \lambda_*)$$

$$= \lambda_* + o(1),$$

$$\nabla f_k^\top e_k = -(A_k^{-\top}\nabla f_k)^\top c(x_k+d_k)$$

$$= \lambda_*^\top c(x_k+d_k) + o(\|d_k\|^2)$$

$$= \frac{1}{2}\lambda_*^\top \left(c''(x_*) \cdot d_k^2\right) + o(\|d_k\|^2),$$

$$f(x_k+d_k+e_k) = f_k + \nabla f_k^\top(d_k + e_k) + \frac{1}{2}d_k^\top \nabla^2 f(x_*)d_k + o(\|d_k\|^2)$$

$$= f_k + \nabla f_k^\top d_k + \frac{1}{2}d_k^\top L_* d_k + o(\|d_k\|^2).$$

With these estimates, the boundedness of $\{\sigma_k\}$, and the fact that, when there are only equality constraints, the directional derivative of Θ_{σ_k} in the direction d_k can be written $\Theta'_{\sigma_k}(x_k; d_k) = \nabla f_k^\top d_k - \sigma_k\|c_k\|_P$ (see the proof of lemma 17.4), one gets

$$\Theta_{\sigma_k}(x_k+d_k+e_k) - \Theta_{\sigma_k}(x_k) - \omega\Theta'_{\sigma_k}(x_k; d_k)$$

$$= \nabla f_k^\top d_k + \frac{1}{2}d_k^\top L_* d_k - \sigma_k\|c_k\|_P - \omega\Theta'_{\sigma_k}(x_k; d_k) + o(\|d_k\|^2)$$

$$= (1 - \omega)\Theta'_{\sigma_k}(x_k; d_k) + \frac{1}{2}d_k^\top L_* d_k + o(\|d_k\|^2). \tag{17.36}$$

We have to show that the right-hand side of (17.36) is nonpositive asymptotically.

Using the optimality conditions of (17.30), the Cauchy-Schwarz inequality (16.14), and the bounds in (17.35), the directional derivative $\Theta'_{\sigma_k}(x_k; d_k) = \nabla f_k^\top d_k - \sigma_k\|c_k\|_P$ can also be written

$$\Theta'_{\sigma_k}(x_k; d_k) = -d_k^\top M_k d_k + (\lambda_k^{\mathrm{QP}})^\top c_k - \sigma_k\|c_k\|_P \le -d_k^\top M_k d_k. \tag{17.37}$$

Since $d_k^\top L_* d_k \leq d_k^\top L_*^r d_k$ for a nonnegative r, (17.36) becomes with (17.37) and (17.34):

$$\Theta_{\sigma_k}(x_k+d_k+e_k) - \Theta_{\sigma_k}(x_k) - \omega\Theta'_{\sigma_k}(x_k; d_k)$$
$$\leq \left(\frac{1}{2} - \omega\right)(-d_k^\top M_k d_k) - \frac{1}{2}d_k^\top(M_k - L_*^r)d_k + o(\|d_k\|^2)$$
$$\leq \left(\frac{1}{2} - \omega\right)(-d_k^\top M_k d_k) + o(\|d_k\|^2).$$

For large k, the right-hand side is nonpositive since, by (17.34) and the positive definiteness of L_*^r, $d_k^\top M_k d_k \geq d_k^\top L_*^r d_k + o(\|d_k\|^2) \geq C\|d_k\|^2$, for some positive constant C and large k.   □

The result of proposition 17.7 has many variants. It is usually easy to prove them by adapting the arguments used in the proof above (basically by cleverly combining Taylor expansions of an appropriate order). For example, one can avoid using the Hessian of the augmented Lagrangian by replacing condition (17.34) by

$$d_k^\top P_*^\top(M_k - L_*)P_* d_k \geq o(\|d_k\|^2) + o(\|c_k\|),$$

where P_* denotes a projection operator on $N(A_*)$. The proof of this claim has been left as an exercise.

Computing the correction step e_k can be time-consuming for some applications, since this requires a new evaluation of the constraints at $x_k + d_k$. When x_k is far from a solution, this step can also be very large, perturbing uselessly the SQP step d_k. Therefore meticulous implementations of the line-search SQP algorithm usually have a test for deciding whether e_k must be computed and the arc-search detailed above must be substituted for the less expensive line-search. Counter-example 17.6 has shown that the Maratos effect occurs when x_k is on the constraint manifold. On the other hand, truncation of the unit stepsize is unlikely to occur in the neighborhood of a solution when the transversal part of the step prevails. To see this, observe that when c has its values in $\mathbb{R}^n$, the unit stepsize is accepted along Newton's direction to solve $c(x) = 0$ when one uses $x \mapsto \|c(x)\|_P$ as a merit function (see exercise 17.4). These observations suggest that there may be a danger of small stepsize only when the restoration step is small with respect to the tangent step. The next proposition confirms this viewpoint. It shows that the unit stepsize is accepted asymptotically for the iterations satisfying the inequality

$$\|r_k\| \geq C_{\mathrm{ME}}\|t_k\|, \tag{17.38}$$

where C_{ME} is a positive constant and $\|\cdot\|$ is an arbitrary norm. To write this inequality, we have decomposed the full step d_k into $d_k = r_k + t_k$, where the restoration step is written $r_k = -A_k^- c_k$, for some right inverse A_k^- of A_k, and the tangent step $t_k \in R(Z_k^-)$ satisfies $\nabla f_k^\top t_k \leq 0$.

Proposition 17.8 (admissibility of the unit step-size at restoration prevailing iterations). *Suppose that f and c are of class C^1 in a neighborhood of a stationary point x_* of (P_E). Let $\{x_k\}$ be a sequence converging to x_* and $d_k = r_k + t_k$, where $r_k = -A_k^- c(x_k)$ and $t_k \in R(Z_k^-)$ satisfies $\nabla f(x_k)^\top t_k \leq 0$. Suppose that $\{A_k^-\}$ and $\{\sigma_k\}$ are bounded, that $\sigma_k \geq \|A_k^{-\top} \nabla f(x_k)\|_D + \bar{\sigma}$ for some constant $\bar{\sigma} > 0$, and that $\omega < 1$. Then, for large indices k for which (17.38) holds with a positive constant C_{ME}, one has*

$$\Theta_{\sigma_k}(x_k + d_k) \leq \Theta_{\sigma_k}(x_k) + \omega \Theta'_{\sigma_k}(x_k; d_k).$$

Proof. Here, as we shall see, first-order expansions are sufficient. Using the fact that $d_k = O(\|r_k\|)$ for the considered indices, one has

$$f(x_k + d_k) = f_k + \nabla f_k^\top d_k + o(\|r_k\|)$$
$$c(x_k + d_k) = c_k + A_k d_k + o(\|r_k\|)$$
$$= o(\|r_k\|).$$

Therefore, using $\Theta'_{\sigma_k}(x_k; d_k) = \nabla f_k^\top d_k - \sigma_k \|c_k\|_P$ (see the proof of lemma 17.4), $\nabla f_k^\top t_k \leq 0$, $\omega < 1$, $\nabla f_k^\top r_k \leq \|A_k^{-\top} \nabla f_k\|_D \|c_k\|_P$, and $r_k = O(\|c_k\|_P)$:

$$\Theta_{\sigma_k}(x_k + d_k) - \Theta_{\sigma_k}(x_k) - \omega \Theta'_{\sigma_k}(x_k; d_k)$$
$$= (1 - \omega)\nabla f_k^\top d_k - (1 - \omega)\sigma_k \|c_k\|_P + o(\|r_k\|)$$
$$\leq (1 - \omega)\left(\|A_k^{-\top} \nabla f_k\|_D - \sigma_k\right) \|c_k\|_P + o(\|r_k\|)$$
$$\leq -(1 - \omega)\bar{\sigma}\|c_k\|_P + o(\|c_k\|_P),$$

which is negative for large k.   □

A consequence of this result is that, optimization codes implementing the second order correction often decide to compute e_k and to do an arc-search, only at iterations where (17.38) does not hold. The constant C_{ME} is determined by heuristics.

Modification of the Merit Function: Nondifferentiable Augmented Lagrangian

Another way of getting the asymptotic admissibility of the unit stepsize is to change the merit function. Remember that d_k is obtained by minimizing a quadratic model of the Lagrangian subject to linearized constraints. Hence, taking

$$\ell_{\mu,\sigma}(x) = f(x) + \mu^\top c(x) + \sigma \|c(x)\|_P$$

as a merit function should be convenient, insofar as μ is close enough to λ_* and σ is small enough. The validity of this intuition is confirmed by proposition 17.9 below.

Beforehand, observe that the problem

$$\begin{cases} \min_x f(x) + \mu^\top c(x) \\ c(x) = 0, \quad x \in \Omega \end{cases}$$

is clearly equivalent to (P_E). Now, let x_* be a solution to (P_E), with associated multiplier λ_*. Then x_* is still a solution to the problem above, with associated multiplier $\lambda_* - \mu$. Therefore, the results of §16.4 imply that $\ell_{\mu,\sigma}$ is exact if

$$\sigma > \|\lambda_* - \mu\|_D.$$

On the other hand, one easily computes

$$\ell'_{\mu,\sigma}(x_k; d_k) = -d_k^\top M_k d_k + (\lambda_k^{\text{QP}} - \mu)^\top c_k - \sigma\|c_k\|_P,$$

which is therefore negative if M_k is positive definite and

$$\sigma \geq \|\lambda_k^{\text{QP}} - \mu\|_D.$$

Figure 17.4 shows the level curves of $\ell_{\mu,\sigma}$ for counter-example 17.6, with

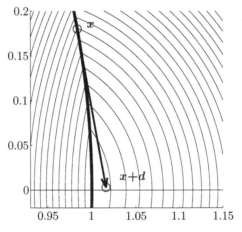

Fig. 17.4. Nondifferentiable augmented Lagrangian

$\tau = 1$, $\mu = -0.55$, and $\sigma = 0.1$.

Proposition 17.9 (admissibility of the unit step-size with a nondifferentiable augmented Lagrangian). *Suppose that f and c are of class C^2 in a neighborhood of a solution x_* to (P_E), satisfying the second-order sufficient conditions of optimality. Let $\{x_k\}$ be a sequence converging to x_*, and d_k be a stationary point of the osculating quadratic problem (17.30). In this last problem, suppose that the matrix M_k over-estimates $L_*^r = L_* + rA_*^\top A_*$ in the sense that*

$$d_k^\top (M_k - L_*^r) d_k \geq o(\|d_k\|^2), \tag{17.39}$$

where $r \geq 0$ is such that L_*^r is positive definite (such an r always exists under the assumptions already stated, see exercise 16.1). Assume also that $d_k \to 0$, that $\omega < \frac{1}{2}$, and that $\sigma_k \geq \|\lambda_k^{QP} - \mu_k\|_D$. Then there exists $\varepsilon > 0$ such that, if $\|\mu_k - \lambda_*\| \leq \varepsilon$ and $0 \leq \sigma_k \leq \varepsilon$, we have for large enough k

$$\ell_{\mu_k,\sigma_k}(x_k+d_k) \leq \ell_{\mu_k,\sigma_k}(x_k) + \omega\ell'_{\mu_k,\sigma_k}(x_k; d_k).$$

Proof. The following expansions are easily obtained:

$$f(x_k+d_k) = f_k + \nabla f_k^\top d_k + \frac{1}{2}d_k^\top \nabla^2 f(x_*)d_k + o(\|d_k\|^2).$$

$$c(x_k+d_k) = \frac{1}{2}c''(x_*) \cdot d_k^2 + o(\|d_k\|^2).$$

We can then write

$$\ell_{\mu_k,\sigma_k}(x_k+d_k) - \ell_{\mu_k,\sigma_k}(x_k) - \omega\ell'_{\mu_k,\sigma_k}(x_k; d_k)$$

$$= \nabla f_k^\top d_k + \frac{1}{2}d_k^\top \nabla^2 f(x_*)d_k + \frac{1}{2}\mu_k^\top c''(x_*) \cdot d_k^2 - \mu_k^\top c_k - \sigma_k\|c_k\|_P$$

$$\quad - \omega\ell'_{\mu_k,\sigma_k}(x_k; d_k) + O(\sigma_k\|d_k\|^2) + o(\|d_k\|^2)$$

$$= (1-\omega)\ell'_{\mu_k,\sigma_k}(x_k; d_k) + \frac{1}{2}d_k^\top L_* d_k$$

$$\quad + O((\|\mu_k - \lambda_*\|_D + \sigma_k)\|d_k\|^2) + o(\|d_k\|^2)$$

$$\leq (1-\omega)\ell'_{\mu_k,\sigma_k}(x_k; d_k) + \frac{1}{2}d_k^\top L_*^r d_k + C_1\varepsilon\|d_k\|^2 + o(\|d_k\|^2)$$

$$\leq \left(\frac{1}{2} - \omega\right)\ell'_{\mu_k,\sigma_k}(x_k; d_k) - \frac{1}{2}d_k^\top(M_k - L_*^r)d_k + C_1\varepsilon\|d_k\|^2 + o(\|d_k\|^2)$$

$$\leq -C_2\left(\frac{1}{2} - \omega\right)\|d_k\|^2 + C_1\varepsilon\|d_k\|^2 + o(\|d_k\|^2)$$

$$\leq 0,$$

if k is large enough and $\varepsilon > 0$ is small enough. We have used the uniform positive definiteness of M_k, which comes from the positive definiteness of L_*^r and from (17.39). $\qquad\square$

We refer the reader to the original paper [37] and to [146, 10] for examples of use of the nondifferentiable augmented Lagrangian in implementable algorithms.

17.4 The Hanging Chain Project IV

This is the fourth session dealing with the problem of finding the static equilibrium of chain made of rigid bars that stays above a given tilted floor. The problem was introduced in §13.8 and developed in §§14.7 and 15.4. We now consider the implementation of the globalization technique presented in this chapter. This will provide more robustness to the SQP solver and will give it a tendency to avoid the stationary points that are not local minima.

We propose to use the merit function (17.1) in which $\|\cdot\|_P$ is the ℓ_1 norm $\|v\|_1 := \sum_{i=1}^m |v_i|$:

$$\Theta_\sigma(x) = f(x) + \sigma\|c(x)^\#\|_1. \tag{17.40}$$

This norm satisfies the assumption (17.5) required by proposition 17.1 (see exercise 17.1). The dual norm of the ℓ_1 norm is the ℓ_∞ norm $\|w\|_\infty := \max_{1 \leq i \leq m} |w_i|$ (see exercise 16.5).

We assume that the osculating quadratic program has the form (17.2), with a matrix M_k that is symmetric positive definite. This property of M_k is important in order to get a primal solution d_k to (17.2) that is a descent direction of the exact merit function Θ_σ defined by (17.40) (see proposition 17.1). Since the Hessian of the Lagrangian $L_k := \nabla^2_{xx}\ell(x_k, \lambda_k)$ is not necessarily positive definite, we propose to take for M_k a modification of L_k obtained by adding to it a small positive diagonal matrix (using, for example, a modified Cholesky factorization [154, 201]). Using a positive definite quasi-Newton approximation to L_k is another possibility that will be examined in chapter 18.

Modifications to Bring to the sqp Function

It is interesting to keep the possibility of using the algorithms defined in the previous sessions by introducing flags. In our code, we use `options.imode` `(1:2)`, which has the following meanings:

- `imode(1)`: 0 (M_k is a quasi-Newton approximation to L_k), 1 ($M_k = L_k$), 2 ($M_k = L_k + E_k$, where E_k is a small positive diagonal matrix that makes M_k positive definite),

- `imode(2)`: 0 (with line-search), 1 (with unit stepsize).

If we compare the local SQP algorithm on page 257, implemented in the previous sessions, and the version with line-search on page 292, we see that we essentially have to add the steps 3, 4, and 5 of the latter algorithm to the sqp function.

- The determination of the penalty parameter σ_k in step 3 can be done by the update rule of page 295. At the first iteration, we take $\sigma_1 = \|\lambda_1^{\mathrm{QP}}\|_D + \bar\sigma$ and set the constant $\bar\sigma$ to $\max(\sqrt{\mathtt{eps}}, \|\lambda_1^{\mathrm{QP}}\|_D/100)$.

- The determination of a stepsize α_k along d_k in step 4 can be done like in the backtracking line-search of page 296, with $\beta = 0.1$ and $\alpha_{k,i+1}$ determined by *interpolation*, i.e., as the minimizer of the quadratic function $\alpha \mapsto \xi(\alpha)$ satisfying $\xi(0) = \Theta_{\sigma_k}(x_k)$, $\xi'(0) = \Delta_k$, and $\xi(\alpha_{k,i}) = \Theta_{\sigma_k}(x_k + \alpha_{k,i}d_k)$.

- We set the new multiplier λ_{k+1} by (17.4).

It is better not to limit the number of stepsize trials in the line-search, since this number, which is most often 1, can be large at some difficult iteration. However, the line-search algorithm may cycle when there is an error in the simulator or when rounding errors occur at the end of a minimization. Therefore, some arrangements have to be implemented to prevent this cycling. In our code, the line-search is interrupted when the norm of the step $\alpha_{k,i}\|d_k\|_\infty$ to get improvement in the merit function becomes smaller than a prescribed value options.dxmin given on entry in the solver.

It is important to take care over the output printed by the code, since it provides meaningful information on the course of the optimization. Here is the text, in connection with the line-search, that our code prints at each iteration.

```
iter 11,   simul 14,   merit -1.47914e+00,   slope -7.59338e-02
   Armijo's line-search
   1.0000e+00    8.47489e-01    8.47489e-01
   1.0000e-01    1.49986e-03    1.49986e-02
   4.1753e-02   -1.60114e-03   -3.83479e-02
```

The value of Δ_k defined by (17.7), which approximates $\Theta'_{\sigma_k}(x_k; d_k)$, is given after the keyword slope, and should always be negative. Each line of the table below the phrase "Armijo's line-search" corresponds to a stepsize trial: $\alpha_{k,i}$ is in the first column, $\Theta_{\sigma_k}(x_k + \alpha_{k,i}d_k) - \Theta_{\sigma_k}(x_k)$ in the second, and $(\Theta_{\sigma_k}(x_k + \alpha_{k,i}d_k) - \Theta_{\sigma_k}(x_k))/\alpha_{k,i}$ in the last one. We see in the first column that the unit stepsize $\alpha_{k,1} = 1$ is tried first and that it is determined next by interpolation with the safeguard $\beta = 0.1$. The last column is useful to detect a possible inconsistency in the simulator (or in the sqp function). If d_k is not a descent direction of the merit function Θ_{σ_k} (it should be a descent direction if M_k is positive definite and if nothing is wrong in the simulator and in the sqp function, see proposition 17.1), there is a large number of stepsize trials $\alpha_{k,i}$ tending to zero. Then, the value in the last column should tend to Δ_k (this is actually certainly correct if there is no inequality constraint, since then $\Delta_k = \Theta'_{\sigma_k}(x_k; d_k)$, see the comment after proposition 17.1).

Question: Tell why the last value in the third column of the table after the phrase "Armijo's line-search" above is often approximately half that of Δ_k (like here: $3.83479/7.59338 \simeq 0.505$).

Experimenting with the SQP Algorithm

The first observation is good news: line-search really helps to force convergence. For example, test case 1d (page 249), which diverges without line-search, now converges to the global minimum. Figure 17.5 shows the result with the usual convention: the thin solid bars represent the initial position of the chain, the dashed bars correspond to the intermediate positions, and the

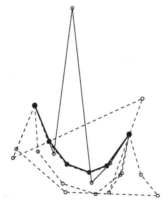

Fig. 17.5. Test case 1d with line-search

bold solid bars are those of the final optimal position. For clarity, we have not represented all the intermediate positions of the 10 iterations required to get convergence, but 1 out of 2.

The second observation is that line-search helps the SQP algorithm to avoid stationary points that are not local minima. For example if we apply the present algorithm with line-search to test case 1b (page 249), the generated sequence now converges to the global minimum of the problem, not to the global maximum as before. The left picture in figure 17.6 shows the result (1

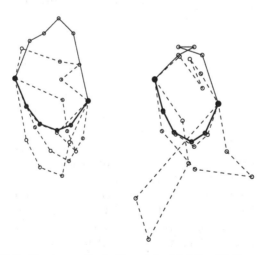

Fig. 17.6. Test cases 1b (left) and 1c (right) with line-search

iteration out of 3). The same phenomenon occurs with test case 1c (page 249), whose convergence to the global minimum is shown in the right hand side picture of figure 17.6.

A third observation: the convergence is smoother with line-search. This is not a very precise concept, but we mean by this that the behavior of the generated sequence is less erratic. Consider for example test case 1f (page 269). The result is shown in figure 17.7. If we compare with figure 15.3, we see that

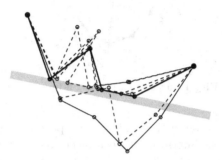

Fig. 17.7. Test case 1f with line-search

the second iterate is now closer the the initial one: the stepsize is actually less than 1 ($\alpha_1 = 0.1$) only at the first iteration. This additional function evaluation is beneficial since the total number of function evaluations is less than the one without line-search (10 instead of 11, not a major improvement, admittedly).

Notes

The use of the exact penalty function (17.1) to globalize the SQP algorithm was proposed by Pshenichnyj (see for example [302]), Han [185; 1977] (with the ℓ_1 norm), and others. The TSQP algorithm described in § 17.2 is taken from [75; 2003]. Another way of dealing with nonconvex problems is to modify the Hessian of the Lagrangian, using a modified Cholesky factorization (see for example [133] and the references therein).

The "effect" described in § 17.3 was discovered by Maratos [247; 1978] and counter-example 17.6 is adapted from [73]. Second-order correction strategies were proposed by Boggs, Tolle, and Wang [36], Coleman and Conn [82], Fletcher [127], Gabay [138], Mayne and Polak [250]. The use of the non-differentiable augmented Lagrangian was proposed by Bonnans [37]. Note that Fletcher's exact penalty function (16.21) also accepts the unit stepsize asymptotically, but it involves first derivatives, so that its use may lead to expensive algorithms if a number of different stepsizes are required during the line-search or to algorithmic remedies for avoiding expensive operations;

see [299, 33, 34]. Other approaches include the "watchdog" technique [73] and the nonmonotone line-search [281, 46].

To conclude this chapter let us briefly mention and/or review other contributions dealing with the use of second derivatives within SQP, techniques for solving the QP, and algorithmic modifications for tackling large-scale problems: Betts and Frank [29] add a positive multiple of the identity matrix to the full Hessian of the Lagrangian when the factorization of the KKT matrix reveals nonpositive definiteness of the reduced Hessian of the Lagrangian; Bonnans and Launay [45]; Murray and Prieto [270]; Gill, Murray, and Saunders [155]; Leibfritz and Sachs [225]; Facchinei and Lucidi [121]; Boggs, Kearsley, and Tolle [32, 31] propose solving the QP by an interior point method that can be prematurely halted by a pseudo-trust-region constraint, although their method uses line-search for its globalization; Sargent and Ding [321] also use an interior point method to solve the QP inexactly within a line-search approach, but discard the Hessian of the Lagrangian if it fails to yield a descent direction of the merit function; Byrd, Gilbert, and Nocedal [65] combine SQP with an interior point approach on the nonlinear problem and use trust regions for the globalization.

Exercises

17.1. *Norm assumptions.* Let $\| \cdot \|$ be an arbitrary norm on $\mathbb{R}^m$ and consider the following properties (the operators $| \cdot |$ and $(\cdot)^+$ act componentwise; the statements are valid for all u and $v \in \mathbb{R}^m$ when this makes sense):

(i) $\| |u| \| = \|u\|$;
(ii) $|u| \leq |v| \implies \|u\| \leq \|v\|$;
(iii) $u_i = v_i$ or $0 \implies \|u\| \leq \|v\|$;
(iv) $0 \leq u \leq v \implies \|u\| \leq \|v\|$;
(v) $u \leq v \implies \|u^+\| \leq \|v^+\|$;
(vi) $v \mapsto \|v^+\|$ is convex.

Show that $(i) \Leftrightarrow (ii) \Rightarrow (iii) \Rightarrow (iv) \Leftrightarrow (v) \Leftrightarrow (vi)$, but that none of the other implications holds in general. Show that (vi) may not hold for an arbitrary norm.

Remark: These implications show that assumptions (16.15) and (17.5) on the norm $\| \cdot \|_P$ are satisfied with the ℓ_p norms, $1 \leq p \leq \infty$, since ℓ_p norms satisfy (i). They also show that (16.15) is more restrictive than (17.5).

17.2. *On the directional derivative of Θ_σ.* Find a one-dimensional example, in which $\Theta'_\sigma(x; d) < \nabla f(x)^\top d - \sigma \|c(x)^\#\|_P$, where d is the solution to the osculating quadratic problem (17.2) (hence the inequality in (17.6) may be strict).

[*Hint*: Equality holds if $I = \emptyset$.]

17.3. *Descent direction for the exact penalization of the Lagrangian.* Consider the exact penalty function $\Theta_{\mu,\sigma} : \mathbb{R}^n \to \mathbb{R}$ defined for $\mu \in \mathbb{R}^m$ and $\sigma > 0$ by

$$\Theta_{\mu,\sigma}(x) := f(x) + \mu^{\mathsf{T}} c(x)^{\#} + \sigma \|c(x)^{\#}\|_P,$$

where the norm $\| \cdot \|_P$ satisfies (17.5) (see also exercise 16.9). Let (d_k, λ_k^{QP}) satisfy the optimality conditions (17.3). Show that d_k is a descent direction of $\Theta_{\mu,\sigma}$ at x_k, provided x_k is not a stationary point of (P_{EI}), M_k is positive definite, $\sigma \geq \|\lambda_k^{QP} - \mu\|_D$, and $\mu_I \geq 0$.

17.4. *Admissibility of the unit stepsize for Newton's method.* Consider the problem of finding a root x_* of the equation $F(x) = 0$, where $F : \mathbb{R}^n \to \mathbb{R}^n$ is a smooth function. Newton's method consists in updating x by $x_+ = x + d$, where d solves $F'(x)d = -F(x)$ (see § 13.7). Let $\| \cdot \|$ be an arbitrary norm and consider $\varphi(x) = \|F(x)\|$ as a merit function for this problem. Suppose that $F'(x_*)$ is nonsingular. Show that, for any constant $\omega \in]0, 1[$, there is a neighborhood V of x_*, such that if $x \in V$, $\varphi(x + d) \leq \varphi(x) + \omega \varphi'(x; d)$.

18 Quasi-Newton Versions

In this chapter we discuss the quasi-Newton versions of the algorithms presented in chapters 14, 15 and 17. Just as in the case of unconstrained problems (see § 4.4), the quasi-Newton approach is useful when one does not want to compute second order derivatives of the functions defining the optimization problem to solve. This may be motivated by various reasons, which are actually the same as in unconstrained optimization: computing second derivatives may demand too much human investment, or their computing time may be too important, or the problem dimensions may not allow the storage of Hessian matrices (in the latter case, limited-memory quasi-Newton methods will be appropriate). Generally speaking, quasi-Newton methods require more iterations to converge, but each iteration is faster (at least for equality constrained problems).

An exhaustive study of quasi-Newton versions is impossible here. Actually, the abundance of proposed algorithms reflects the difficulty of the problem; a fully satisfactory solution to which, if any, has not been found yet. Thus, after outlining the way quasi-Newton methods are used for problems with constraints (§ 18.1), we shall limit ourselves to two approaches. In § 18.2, we describe the one most often implemented so far. It is due to M.J.D. Powell [293] and can be used for problems with inequality constraints. It has the advantage of simplicity, but it is not completely satisfactory conceptually. After having justified this claim, we give in § 18.3 another approach, reflecting the works and concerns of the author. It has several variants and is presently only applicable to equality constraints. The one presented here is safe, in that it extends a well-established technique in unconstrained optimization: Wolfe's line-search (see chapter 3 and its § 3.4). However, its two constraint-linearizations per iteration can reveal expensive for some problems. We then outline some extensions less greedy in computing time.

18.1 Principles

Let us recall some principles underlying quasi-Newton techniques (see also chapter 4) and see how these can be applied to problems with constraints. As mentioned above, one seeks rapidly convergent algorithms, without computing second derivatives of the objective and constraint functions f and c.

Typically, if superlinear convergence is impossible, the algorithm is considered as badly designed.

To derive these algorithms, one starts from Newton's version and one observes that the second-order information is entirely contained in one single matrix. In our case it will be the Hessian of the Lagrangian or the reduced Hessian, depending on the considered algorithm. One then defines the quasi-Newton algorithm, using the same quadratic problems as in Newton's method, but replacing the second-order derivatives by a matrix appropriately updated at each iteration. The role of this update is to build up the second-order information, upon observation of the variation of certain quantities computed from the first derivatives of f and c. The change in these first derivatives give indeed information on the second-order derivatives.

To be more concrete, let us consider equality constrained problems:

$$(P_E) \quad \begin{cases} \min_x f(x) \\ c(x) = 0, \quad x \in \Omega. \end{cases}$$

In Newton's method (§ 14.1), the direction $d_k \in \mathbb{R}^n$, giving the change in the current primal iterate x_k, is computed by solving the problem for d (see (14.8))

$$\begin{cases} \min_d \nabla f_k^\top d + \frac{1}{2} d^\top L_k d \\ c_k + A_k d = 0. \end{cases}$$

In this problem, the Hessian of the Lagrangian $L_k := \nabla^2_{xx}\ell(x_k, \lambda_k)$ comes into play: it is the $n \times n$ matrix gathering all the second-order derivatives. Since the computation of L_k is undesired, it is approximated by an $n \times n$ matrix M_k. We then have to solve at each iteration a problem formally identical to the above:

$$\begin{cases} \min_d \nabla f_k^\top d + \frac{1}{2} d^\top M_k d \\ c_k + A_k d = 0. \end{cases} \tag{18.1}$$

A new iterate (x_{k+1}, λ_{k+1}) is computed as explained in § 14.1 and M_k is updated to obtain M_{k+1}. Two sequences are thus generated: $\{(x_k, \lambda_k)\}$ converging to the solution (x_*, λ_*) to the problem, and $\{M_k\}$ approximating the Hessian of the Lagrangian. In general, this latter sequence does not converge to L_* but is a sufficiently good approximation of it in certain directions, so as to ensure the algorithm's superlinear convergence.

The process is analogous for deriving the quasi-Newton version of the reduced Hessian algorithm (§ 14.5). We know that, locally (see (14.40)), the algorithm generates a sequence $\{x_k\}$ by the recurrence

$$x_{k+1} = x_k + t_k - A_k^- c(x_k + t_k),$$

where $t_k = -Z_k^- H_k^{-1} g_k$ is the tangential component of the displacement, and $g_k = Z_k^{-\top} \nabla f_k$ is the reduced gradient. Here, the only matrix involving second derivatives is the reduced Hessian of the Lagrangian $H_k = Z_k^{-\top} L_k Z_k^-$.

As before, this matrix is approximated by an $(n-m) \times (n-m)$ matrix M_k, which results in an algorithm generating $\{x_k\}$ by

$$x_{k+1} = x_k + t_k - A_k^- c(x_k + t_k), \quad \text{with } t_k = -Z_k^- M_k^{-1} g_k,$$

the sequence $\{M_k\}$ being generated by an appropriate formula.

The question is now how to update the matrices M_k. Two properties are important in the choice of the update formula.

1. First, symmetry is a natural requirement. In fact, M_k must approximate a Hessian, or a reduced Hessian; both are symmetric matrices. Imposing symmetry to the matrices M_k therefore increases their chance of getting correct values.

2. Second, it is recommended to have matrices M_k positive definite.

   This property is better justified for the reduced Hessian method, where M_k approximates the reduced Hessian of the Lagrangian. Indeed, from the second-order optimality conditions, this matrix is positive semi-definite at a solution; it is normal to require the same property to M_k. Besides, if the approximation M_k of H_k is positive definite, the tangent direction $t_k = -Z_k^- M_k^{-1} g_k$ is a descent direction of f or of Θ_σ (the exact penalty function of § 16.4) at x_k. Globalization of the algorithm will be facilitated (see chapter 17).

   In the quasi-Newton method (18.1), imposing positive definiteness of the approximation M_k of L_k is more questionable (see § 18.2). In fact, the Hessian of the Lagrangian may not be positive definite at a solution. However, just as for the reduced Hessian method, positive definiteness of M_k produces a d_k that is a descent direction of Θ_σ (see proposition 17.1) and thus facilitates globalization of the algorithm.

Because of these two arguments, the update of M_k is often done using the *BFGS formula* (see chapter 4), which defines M_{k+1} by

$$M_{k+1} = M_k - \frac{M_k \delta_k \delta_k^\top M_k}{\delta_k^\top M_k \delta_k} + \frac{\gamma_k \gamma_k^\top}{\gamma_k^\top \delta_k}. \tag{18.2}$$

In this formula, γ_k and δ_k are two vectors of appropriate dimension, supposed to gather information on the Hessian approximated by the matrices M_k.

Observe that this formula does yield the desired properties. First, M_{k+1} is symmetric if M_k is such. Second, it is easy to check that, if M_k is symmetric positive definite, so is M_{k+1} if and only if the following *curvature condition* is satisfied

$$\gamma_k^\top \delta_k > 0. \tag{18.3}$$

Condition (18.3) is clearly necessary because M_{k+1} satisfies the so-called *quasi-Newton equation*:

$$\gamma_k = M_{k+1} \delta_k. \tag{18.4}$$

The sufficiency of (18.3) is shown in theorem 4.5. The BFGS formula, therefore, makes it easy to generate symmetric positive definite matrices. If M_1 has these properties (most often M_1 is chosen as a multiple of the identity matrix), it is sufficient to have (18.3) at each iteration. As we shall see, the necessity of satisfying (18.3) is actually a source of difficulties; but the qualities of BFGS have been a strong incentive for researchers to overcome them.

In unconstrained optimization, the choice of the vectors γ_k and δ_k comes naturally: $\delta_k = x_{k+1} - x_k$ is the change in x and $\gamma_k = \nabla f_{k+1} - \nabla f_k$ is the corresponding change in the gradient of f. In constrained optimization, the situation is not so simple. These two vectors will be chosen according to two criteria.

Let M_* be the matrix that M_k should approximate (M_* is the Hessian of the Lagrangian at a solution L_* for SQP and the reduced Hessian of the Lagrangian at a solution $H_* = Z_*^{-\top} L_* Z_*^-$ for the reduced Hessian algorithm). Since M_{k+1} satisfies the quasi-Newton equation (18.4), M_* should satisfy this equation as well, at least to first order. This requires from γ_k and δ_k that they collect information on the matrix M_* we want to approximate. Actually, an asymptotic analysis shows that an estimate of the type

$$\frac{\gamma_k - M_* \delta_k}{\|\delta_k\|} \to 0, \quad \text{when } k \to \infty \tag{18.5}$$

is necessary. Anyway, it is easy to show that this estimate holds in unconstrained optimization when $\gamma_k = \nabla f_{k+1} - \nabla f_k$, $\delta_k = x_{k+1} - x_k$, $M_* = \nabla^2 f(x_*)$ and $x_k \to x_*$. The *asymptotic criterion* (18.5) is the first condition guiding the search for good vectors γ_k and δ_k.

The second criterion is to give the possibility to realize the curvature condition (18.3). Remember that, in unconstrained optimization, this condition is a consequence of the *Wolfe line-search*: we seek a stepsize $\alpha_k > 0$ along a descent direction d_k of the objective f, so as to satisfy

$$f(x_k + \alpha_k d_k) \le f(x_k) + \omega_1 \alpha_k \nabla f(x_k)^\top d_k$$

and

$$\nabla f(x_k + \alpha_k d_k)^\top d_k \ge \omega_2 \nabla f(x_k)^\top d_k.$$

The constants ω_1 and ω_2 must satisfy $0 < \omega_1 < \omega_2 < 1$. The first inequality forces a decrease in the objective function and the second, besides preventing too small stepsizes α_k, implies the curvature condition (18.3) for the vectors $\gamma_k = \nabla f_{k+1} - \nabla f_k$ and $\delta_k = x_{k+1} - x_k$ (just subtract $\nabla f(x_k)^\top d_k$ from both sides and use $-(1-\omega_2) \nabla f(x_k)^\top d_k > 0$). It is this approach that will be adopted in the reduced Hessian method (§ 18.3). It is more clumsy for Newton's method (see [10]) and will be only sketched here (and also in § 18.3).

18.2 Quasi-Newton SQP

Recall that, locally, SQP computes a displacement d_k at x_k by solving the quadratic program for d

$$\begin{cases} \min_d \nabla f(x_k)^\top d + \frac{1}{2} d^\top M_k d \\ c_E(x_k) + A_E(x_k)d = 0 \\ c_I(x_k) + A_I(x_k)d \leq 0. \end{cases} \quad (18.6)$$

In the quasi-Newton version of the algorithm [184, 293], M_k becomes a symmetric positive definite matrix, updated at each iteration by the BFGS formula (18.2) using two vectors γ_k and δ_k of $\mathbb{R}^n$. Let us specify these vectors.

As shown in the local analysis of chapter 15, M_k should approximate the Hessian of the Lagrangian. It therefore appears to be reasonable to take $\gamma_k = \gamma_k^\ell$, the variation of the gradient of the Lagrangian when x varies by δ_k:

$$\gamma_k^\ell = \nabla_x \ell(x_{k+1}, \lambda_{k+1}) - \nabla_x \ell(x_k, \lambda_{k+1}) \quad \text{and} \quad \delta_k = x_{k+1} - x_k. \quad (18.7)$$

In γ_k^ℓ, we have fixed the multiplier to the value λ_{k+1}, supposed to be closer to λ_* than its current estimation λ_k. With the above values of γ_k and δ_k, we have the estimate (18.5) where $M_* = L_*$, as soon as f and c are of class C^2 and $(x_k, \lambda_{k+1}) \rightarrow (x_*, \lambda_*)$. Indeed

$$\gamma_k^\ell = L_*\delta_k + \left(\int_0^1 \left(\nabla_{xx}^2 \ell(x_k + t\delta_k, \lambda_{k+1}) - L_* \right) dt \right) \delta_k = L_*\delta_k + o(\|\delta_k\|).$$

This shows that the first selection criterion of the pair (γ_k, δ_k) is satisfied with $\gamma_k = \gamma_k^\ell$.

According to the second selection criterion, the curvature condition $(\gamma_k^\ell)^\top \delta_k > 0$ should be feasible with an appropriate choice of x_{k+1}. It is natural to seek x_{k+1} along d_k by a line-search decreasing a merit function (see chapter 17). As a matter of fact, this is what Wolfe's line-search does when there are no constraints. Here, however, the Lagrangian may have a negative curvature at x_k along the line $\{x_k + \alpha d_k : \alpha \in \mathbb{R}\}$ and be unbounded from below on this affine manifold. As a result, the curvature condition might be impossible to satisfy with the above strategy: it may well happen that $(\gamma_k^\ell)^\top \delta_k \leq 0$ with $x_{k+1} = x_k + \alpha_k d_k$, for any stepsize $\alpha_k > 0$. The situation is therefore different from that in unconstrained optimization. Wolfe's line-search cannot be extended to constrained problems in a straightforward manner.

To overcome this difficulty, M.J.D. Powell proposed in [293] computing the vector γ_k by modifying γ_k^ℓ when the scalar product $(\gamma_k^\ell)^\top \delta_k$ is not positive enough. In a first phase, a stepsize α_k is computed along d_k to decrease a merit function (for example the exact penalty function of § 16.4), which gives the next iterate $x_{k+1} = x_k + \alpha_k d_k$. Then γ_k^ℓ and δ_k are defined by (18.7) and one takes $\gamma_k^P \in \mathbb{R}^n$ as a convex combination of γ_k^ℓ and $M_k\delta_k$:

$$\gamma_k^{\mathrm{P}} := \theta \gamma_k^{\ell} + (1-\theta) M_k \delta_k. \tag{18.8}$$

The choice of $M_k \delta_k$ as "emergency" vector obtained with $\theta = 0$ in (18.8), comes from the facts that $\delta_k^{\top} M_k \delta_k > 0$ and $M_{k+1} = M_k$ if $\gamma_k = M_k \delta_k$ (see formula (18.2)). In order to modify γ_k^{ℓ} the least possible, to preserve the most possible information from the problem data, the parameter θ is taken as large as possible in $[0, 1]$ while satisfying

$$(\gamma_k^{\mathrm{P}})^{\top} \delta_k \geq \kappa \, \delta_k^{\top} M_k \delta_k,$$

where the constant $\kappa \in \,]0, 1[$ is suggested to be set to 0.2 in [293, 292]. Since M_k is assumed to be positive definite, this inequality is satisfied for $\theta = 0$. A simple computation gives

$$\theta = \begin{cases} 1 & \text{if } (\gamma_k^{\ell})^{\top} \delta_k \geq \kappa \, \delta_k^{\top} M_k \delta_k \\ (1-\kappa) \, \dfrac{\delta_k^{\top} M_k \delta_k}{\delta_k^{\top} M_k \delta_k - (\gamma_k^{\ell})^{\top} \delta_k} & \text{otherwise.} \end{cases} \tag{18.9}$$

Then M_k is updated by the BFGS formula using $\gamma_k = \gamma_k^{\mathrm{P}}$. This technique is known as *Powell's correction*.

An algorithm combining this technique with those of the preceding chapters is given below.

Quasi-Newton SQP:

Choose an initial iterate $(x_1, \lambda_1) \in \mathbb{R}^n \times \mathbb{R}^m$.

Compute $f(x_1)$, $c(x_1)$, $\nabla f(x_1)$, and $A(x_1) = c'(x_1)$.

Set the constants $\omega \in \,]0, \frac{1}{2}[$ (slope modifier in the Armijo condition), $\bar{\sigma} > 0$ (penalty parameter threshold), and $\beta \in \,]0, \frac{1}{2}]$ (backtracking safeguard parameter).

Set $k = 1$.

1. *Stopping test*: Stop if $\nabla \ell(x_k, \lambda_k) = 0$ and $c(x_k) = 0$ (optimality is reached).
2. *Matrix update*:
   - If $k = 1$, initialize M_1 to an $n \times n$ symmetric positive definite matrix.
   - If $k > 1$, compute $\gamma_{k-1} = \gamma_{k-1}^{\mathrm{P}}$ and δ_{k-1} by formulas (18.7), (18.8) and (18.9); update M_k from M_{k-1} by the BFGS formula (18.2).
3. *Step computation*: Find the unique primal-dual solution $(d_k, \lambda_k^{\mathrm{QP}})$ to (18.6), which is supposed feasible.
4. *Penalty parameter setting*: Update the penalty parameter σ_k so as to satisfy (17.9).

5. *Backtracking line-search on the exact penalty function* Θ_{σ_k}:
   - Set $\alpha = 1$.
   - While α does not satisfy Armijo's inequality (17.26), pick a new stepsize α in $[\beta\alpha, (1-\beta)\alpha]$.
   - Set $\alpha_k = \alpha$.
6. *New iterates*: $x_{k+1} = x_k + \alpha_k d_k$ and update $\lambda_k \to \lambda_{k+1}$.
7. Increase k by 1 and go to 1.

In step 3, the constraints of the quadratic problem are assumed consistent (nonempty feasible set). If such is not the case, techniques can be used to modify this problem so as to increase its chances of being consistent (see for example [293, 322, 351, 61]). To be complete, this algorithm should also include some technique described in § 17.3 to avoid the Maratos effect (undue shortening of the stepsize). This point is fairly important for practical efficiency of the algorithm; we omit these aspects to alleviate the presentation. For the update of λ_k in step 6, see the comments after the line-search SQP algorithm on page 292.

No strong convergence result can be given for the above algorithm. In fact, even for unconstrained problems, global convergence cannot be proven without assuming convexity; see § 4.5 and also [290, 66]. Nevertheless, the speed of convergence can be analyzed, global convergence being assumed to hold. Some assumptions on the problem's data are necessary for this analysis: the functions f and c must be of class $C^{2,1}$, strict complementarity must hold, the Jacobian of the active constraints at the solution must be surjective, and the second-order sufficient conditions of optimality must be satisfied. This list of assumptions is long but acceptable. In contrast, less attractive hypotheses must be made on the behavior of the algorithm: if, in addition to the above, one assumes

- the unit stepsize $\alpha_k = 1$ is asymptotically accepted by the line-search (no Maratos effect),
- the sequence of matrices $\{M_k\}$ is bounded,
- these matrices M_k are uniformly positive definite in the subspace tangent to the active constraints at the solution,

then the sequence $\{x_k\}$ converges (locally) R-superlinearly to x_* [292].

Even though this result implies a fairly elaborate analysis, it is certainly not satisfactory; among other things, R-superlinear convergence is not so strong as Q-superlinear convergence, which can be reasonably expected from a quasi-Newton method (see § 4.7). Generally speaking, the algorithm works well, though; it is implemented in many software libraries. Its robustness is not completely above suspicion, since some convergence difficulties may be encountered (see [297], [298; p. 125], and the example given at the end of § 18.4). These observations led various authors to tackle the problem again; many other approaches have been, and are still, proposed.

An apparently promising approach is the following. It was explored by many authors [184, 346, 347, 69, 10]. To simplify the analysis, we assume only equality constraints. Problem (P_E) can then be rewritten

$$\begin{cases} \min_x f(x) + \frac{r}{2}\|c(x)\|_2^2 \\ c(x) = 0, \quad x \in \Omega. \end{cases}$$

The Lagrangian of this problem is the *augmented Lagrangian* ℓ_r of § 16.3. We know that, if the augmentation factor r is taken large enough, the resulting Hessian is positive definite at a point satisfying the second-order sufficient conditions of optimality (see exercise 16.2). Applying the quasi-Newton version of SQP to this problem involves matrices M_k that are now supposed to approximate the Hessian of the augmented Lagrangian; positive definiteness of M_k becomes a natural requirement. In the first studies of this approach [184, 346, 161], M_k was updated with the pair $(\gamma_k, \delta_k) = (\gamma_k^r, \delta_k)$ defined by

$$\gamma_k^r = \nabla_x \ell_r(x_{k+1}, \lambda_k^{\mathrm{QP}}) - \nabla_x \ell_r(x_k, \lambda_k^{\mathrm{QP}}) \quad \text{and} \quad \delta_k = x_{k+1} - x_k.$$

This vector choice suffers from serious shortcomings, though: (i) one does not know the threshold $\bar{r}$ from which the Hessian of ℓ_r at the solution becomes positive definite; (ii) yet, a too big r-value raises important numerical difficulties [346, 276], and (iii) far from the solution, there may exist no r-value for which $(\gamma_k^r)^\top \delta_k > 0$.

Some difficulties of this approach can be remedied, using the structure of the Hessian of ℓ_r at the solution:

$$\nabla_{xx}^2 \ell_r(x_*, \lambda_*) = L_* + r A_*^\top A_*.$$

This formula suggests that we take a pair (γ_k, δ_k), where $\gamma_k = \gamma_k^s$ is the variation of the gradient of the Lagrangian, to which is added a term taking the augmentation into account [347]:

$$\gamma_k^s = \gamma_k^\ell + r A_k^\top A_k \delta_k \quad \text{and} \quad \delta_k = x_{k+1} - x_k. \tag{18.10}$$

The scalar product of γ_k^s and δ_k is

$$(\gamma_k^s)^\top \delta_k = (\gamma_k^\ell)^\top \delta_k + r\|A_k \delta_k\|_2^2,$$

so that we obtain $(\gamma_k^s)^\top \delta_k > 0$ for r large enough, providing that $A_k \delta_k \neq 0$. Clearly, this strategy will fail if $A_k \delta_k$ is zero within roundoff errors, with $(\gamma_k^\ell)^\top \delta_k \leq 0$. Byrd, Tapia and Zhang [69] introduced a safeguard in this strategy, replacing $A_k^\top A_k \delta_k$ by δ_k in γ_k^s when $A_k \delta_k$ is small and $(\gamma_k^\ell)^\top \delta_k$ is not positive enough. Then positivity of $(\gamma_k^s)^\top \delta_k$ can be recovered as above with r large enough. The numerical experiments of [69] have shown that the approach is numerically competitive with Powell's correction. Besides, it enjoys a nice local convergence property: if the sequence $\{x_k, \lambda_k^{\mathrm{QP}}\}$ converges to a primal-dual solution (x_*, λ_*) satisfying $(SC2)$ and such that A_* is surjective, and if the unit stepsize is asymptotically accepted, then the convergence of $\{x_k\}$ is R-superlinear; it is even Q-superlinear if the augmentation parameter r stabilizes to a large enough value. However, this result is still not quite satisfactory because, once again, the threshold on r is not known and the rules updating r given in [69] do not guarantee that the assumptions of this theorem are satisfied.

An interesting aspect in the approach by Byrd, Tapia and Zhang [69] is to give an update rule of the augmentation parameter r. This rule allows a convenient management of the transversal component of the matrix M_k (i.e., its action on the range space of $A_k^\top$). However, the safeguard needed when the displacement δ_k is tangent to the constraint manifold ($A_k \delta_k = 0$) reveals that the algorithm does not completely master the longitudinal component of M_k (its action on the null space of A_k). This observation motivated the study in [10]. It is suggested there using the line-search to adapt the longitudinal displacement, so as to ensure the curvature condition (18.3), without any need for a safeguard. The resulting algorithm can be viewed as an extension of Wolfe's line-search to equality constrained problems. Actually, the search is no longer done along a half-line but along a piecewise-linear path, as in the algorithm described in § 18.3 below. The numerical experiments reported in [10] show that this technique is more robust than the one described above.

18.3 Reduced Quasi-Newton Algorithm

Consider the problem with only equality constraints

$$(P_E) \quad \begin{cases} \min_x f(x) \\ c(x) = 0, \quad x \in \Omega \end{cases}$$

and apply to it the reduced Hessian algorithm introduced in § 14.5. We shall give a quasi-Newton version of the algorithm, which locally (i.e., close to a solution) is purely primal: it requires no multiplier. When the line-search comes into play, a multiplier is necessary to adapt the penalty parameter of the merit function.

Recall that the reduced Hessian algorithm generates the sequence of iterates $\{x_k\}$ by the recurrence

$$x_{k+1} = x_k + t_k - A_k^- c(x_k + t_k), \quad \text{with } t_k = -Z_k^- M_k^{-1} g_k,$$

where g_k is the reduced gradient at x_k and M_k approximates the reduced Hessian of the Lagrangian. We choose to generate the sequence of matrices $\{M_k\}$ by the BFGS formula, in accordance with the motivations given in § 18.1. Note that these matrices are here $(n-m) \times (n-m)$. The question now is to determine the vectors γ_k and δ_k of $\mathbb{R}^{n-m}$ that are used in the BFGS formula, such that M_k approximates correctly the reduced Hessian of the Lagrangian $H_* := Z_*^{-\top} L_* Z_*^-$.

Choosing the Pair (γ_k, δ_k)

We already observed in (14.38) that the derivative of the reduced gradient g at a stationary point x_* is

$$g'(x_*) = Z_*^{-\top} L_*.$$

The reduced Hessian of the Lagrangian is therefore a part of $g'(x_*)$, namely its restriction to the directions $h \in R(Z_*^-)$, tangent to the constraint manifold at x_*. One then understands that variations of g along directions tangent to the constraint manifold

$$\mathcal{M}_k := \{y \in \Omega : c(y) = c(x_k)\}$$

are convenient to collect information on H_k.

Let us show that, locally, the following pair (γ_k, δ_k) can be used:

$$\gamma_k = g(x_k + t_k) - g(x_k) \quad \text{and} \quad \delta_k = Z_k t_k. \tag{18.11}$$

In this formula, $Z_k \in \mathbb{R}^{(n-m) \times n}$ is the unique operator associated with Z_k^- and A_k^- by lemma 14.3. Observe first that δ_k gives the components of $t_k \in R(Z_k^-)$ in the tangent basis formed by the columns of Z_k^-: $t_k = Z_k^- \delta_k$. Then, if Z^- is continuous and g is C^1, we find when $x_k \to x_*$ and $t_k \to 0$:

$$
\begin{aligned}
\gamma_k &= \left(\int_0^1 g'(x_k + \alpha t_k) \, d\alpha \right) t_k \\
&= Z_*^{-\top} L_* t_k + \left(\int_0^1 (g'(x_k + \alpha t_k) - g'(x_*)) \, d\alpha \right) t_k \\
&= H_* \delta_k + o(\|\delta_k\|).
\end{aligned}
\tag{18.12}
$$

The asymptotic criterion (18.5) is therefore satisfied with (γ_k, δ_k) given by (18.11) and $M_* = H_*$.

The second selection criterion of the pair (γ_k, δ_k), the curvature condition $\gamma_k^\top \delta_k > 0$, comes much less easily when x_k is far from a solution, even if this one satisfies $(SC2)$ (i.e., H_* is positive definite). Of course, Powell's correction introduced in § 18.2 could also be used; but our aim in this section is to show that, just as in unconstrained optimization, line-searches are able to provide the curvature condition, and thus to extract from the problem data more accurate information on the reduced Hessian of the Lagrangian. The eventual hope is to obtain an algorithm converging more rapidly.

Before proceeding, we mention that other pairs (γ_k, δ_k) than (18.11) can also be chosen (a list of such pairs is given by Nocedal and Overton in [276]). Here is another possibility, often used:

$$\gamma_k = Z_k^{-\top} \left(\nabla_x \ell(x_k + t_k, \lambda_k) - \nabla_x \ell(x_k, \lambda_k) \right) \quad \text{and} \quad \delta_k = Z_k t_k. \tag{18.13}$$

In the expression of γ_k above, λ_k is a certain multiplier (for example, the least-squares multiplier λ_k^{LS}, $\widehat{\lambda}_k$, or λ_k^{QP}; see § 14.1). If f and c are C^2 and if Z^- is continuous, the asymptotic criterion (18.5) is satisfied with $M_* = H_*$, when (x_k, λ_k) converges to (x_*, λ_*) and $t_k \to 0$.

As far as local properties are concerned, all these pairs are asymptotically equivalent. In contrast, it really seems that the global behavior of algorithms, far from the solution, is strongly influenced by the choice of these pairs [148]. In [149], a geometric argument is developed, revealing a relationship to be respected between the structure of (γ_k, δ_k) and the choice of basis Z_k^-. With the basis (14.15) obtained by variable partitioning (see § 14.2), for example, the pair (18.11) can be used. On the other hand, if one insists on using orthonormal bases (for example because the bases (14.15) are ill conditioned), the pair (18.13) is better. In this section, we only consider the pair (18.11).

Curvilinear Search

Now, the question at stake is whether the curvature condition $\gamma_k^\top \delta_k > 0$, with γ_k and δ_k given by (18.11), can be obtained with a line-search from x_k along t_k. The answer is positive, if x_k is close to a solution satisfying $(SC2)$; this is an easy consequence of (18.12), because H_* is positive definite. However, the answer is negative in general, as shown by the following counter-example [146].

Counter-example 18.1. Consider the minimization of a linear function on the unit circle of $\mathbb{R}^2$ ($x_{(i)}$ denotes the ith component of $x \in \mathbb{R}^2$):

$$\begin{cases} \min_x x_{(2)} \\ c(x) \equiv \frac{1}{2}(\|x\|_2^2 - 1) = 0, \quad x \in \mathbb{R}^2. \end{cases}$$

With $x = (x_{(1)}, x_{(2)})$, associate the vector $\tilde{x} := (x_{(2)}, -x_{(1)})$. Suppose that x is in the positive orthant ($x_{(1)} > 0$ and $x_{(2)} > 0$) and that the matrix M (a scalar in this example because $n-m = 1$) is 1. Then the Jacobian of the constraint $A_x = x^\top$ is surjective and we can take as a basis of $\mathbb{R}^2$: $(Z_x^-, A_x^-) = (\tilde{x}, x/\|x\|_2^2)$. We then have $Z_x = \tilde{x}^\top/\|x\|_2^2$. The reduced gradient can be written $g(x) = -x_{(1)}$, the tangent displacement is $t = x_{(1)}\tilde{x}$ and thus $g(x + \alpha t) = -(x_{(1)} + \alpha x_{(1)}x_{(2)})$. Since $\delta = x_{(1)} > 0$, the curvature condition $\gamma^\top \delta > 0$ is equivalent to $g(x+\alpha t) > g(x)$, i.e., $-\alpha x_{(1)}x_{(2)} > 0$; this holds for no $\alpha > 0$. □

Although the curvature condition may not be obtained from a line-search along t_k, it can be realized by moving along the path $\alpha \mapsto p_k(\alpha)$ defined by the differential equation in figure 18.1. Note that the derivative of $\alpha \mapsto c(p_k(\alpha))$ vanishes since, by (14.10), there holds

$$(c \circ p_k)'(\alpha) = A(p_k(\alpha))Z^-(p_k(\alpha))\delta_k = 0.$$

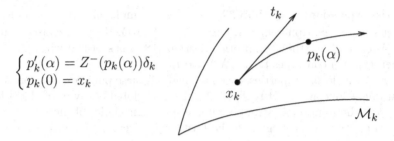

$$\begin{cases} p'_k(\alpha) = Z^-(p_k(\alpha))\delta_k \\ p_k(0) = x_k \end{cases}$$

Fig. 18.1. Curvilinear search

Therefore, c is constant along the path: for all α

$$c(p_k(\alpha)) = c(x_k),$$

which means that this path lies in the manifold $\mathcal{M}_k$, which is "parallel" to the constraint manifold ($c(x_k)$ is usually nonzero). The next result shows that, not only the curvature condition can be obtained along p_k, but f can be decreased significantly.

Proposition 18.2 (Wolfe conditions along a curvilinear path). *Suppose that the path $\alpha \mapsto p_k(\alpha)$ exists for $\alpha \geq 0$ large enough. Suppose also that f and p_k are continuously differentiable, that f is bounded from below along the path p_k, and that $g_k^\top \delta_k < 0$. Take $0 < \omega_1 < \omega_2 < 1$. Then the following inequalities*

$$f(p_k(\alpha_k)) \leq f(x_k) + \omega_1 \alpha_k g_k^\top \delta_k \quad and \quad g(p_k(\alpha_k))^\top \delta_k \geq \omega_2 g_k^\top \delta_k \quad (18.14)$$

hold for some $\alpha_k > 0$.

Proof. Set $\xi_k = f \circ p_k$. Then, inequalities (18.14) can be written

$$\xi_k(\alpha_k) \leq \xi_k(0) + \omega_1 \alpha_k \xi'_k(0) \quad and \quad \xi'_k(\alpha_k) \geq \omega_2 \xi'_k(0).$$

These are precisely the Wolfe conditions on the function $\alpha \in \mathbb{R}_+ \mapsto \xi_k(\alpha)$. Because this function is C^1, bounded from below on $\mathbb{R}_+$, and $\xi'_k(0) = g_k^\top \delta_k < 0$, the result follows from theorem 3.7. □

In the *generalized Wolfe conditions* (18.14), f can be replaced by the nondifferentiable augmented Lagrangian (see exercise 16.9 and the end of § 17.3)

$$\ell_{\mu,\sigma}(x) = f(x) + \mu^\top c(x) + \sigma\|c(x)\|_P. \quad (18.15)$$

Indeed we have $f(p_k(\alpha)) = \ell_{\mu,\sigma}(p_k(\alpha))$ for all α, since c stays constant along the path p_k. Note, however, that the displacement along p_k only takes care of the decrease in f and neglects the second objective of (P_E): to satisfy the constraints. It is therefore necessary to add to this displacement a step restoring the constraints. We shall see below how to do this in a more realistic algorithm. Using the merit function (18.15), a step satisfying the generalized Cauchy-Schwarz conditions along the curvilinear path $p_k(\cdot)$, and an additional constraint restoration step, it is then possible to show a convergence result.

Except in very special cases, a curvilinear search along the path p_k is out of question. With really nonlinear constraints, this would imply prohibitive computation costs. One may then ask whether similar conditions could be satisfied along a piecewise-linear path approximating p_k and simpler to compute. This is the subject of the next subsection.

Piecewise Line-Search

A simple approximation of p_k is obtained by taking an integration scheme of the differential equation defining p_k, like the explicit Euler scheme for example (see figure 18.2). Denote by

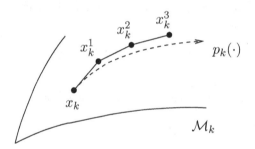

Fig. 18.2. Piecewise linear search

$$\alpha_k^0 := 0 < \alpha_k^1 < \cdots < \alpha_k^{i_k} =: \alpha_k$$

the discretization stepsizes. As we shall see below, to obtain the curvature condition $\gamma_k^\top \delta_k > 0$, these stepsizes cannot be given a priori, but must satisfy a simple rule. Denote by x_k^i the point approximating $p_k(\alpha_k^i)$ (with $x_k^0 = x_k$ and $x_k^{i_k} = y_{k+1}$). The explicit Euler scheme can be written

$$x_k^{i+1} = x_k^i + (\alpha_k^{i+1} - \alpha_k^i)t_k^i, \quad i = 0, \ldots, i_k - 1, \qquad \text{where } t_k^i = Z^-(x_k^i)\delta_k.$$

Since the points x_k^i may no longer lie on the manifold $\mathcal{M}_k$, the variation of c must be taken into account in the search for the stepsize. We shall do so by forcing the decrease in a rather general penalty function, of the form

$$\Theta(x) = f(x) + \phi(c(x)), \tag{18.16}$$

where $\phi : \mathbb{R}^m \to \mathbb{R}_+$ is a continuous convex function, satisfying $\phi(0) = 0$. For example, the exact penalty function Θ_σ of § 16.4 or the function $\ell_{\mu,\sigma}$ defined by (18.15) can be taken. With these assumptions, ϕ is Lipschitzian and has directional derivatives. Then Θ enjoys the same properties when c is smooth (lemma 13.1). In particular,

$$\Theta'(x_k^i; t_k^i) = g(x_k^i)^\top \delta_k + \phi'(c(x_k^i); A(x_k^i)t_k^i) = g(x_k^i)^\top \delta_k.$$

Setting $i = 0$ in this relation and assuming $g_k \neq 0$, we see that t_k is a descent direction of Θ at x_k.

The rule determining the stepsizes α_k^i can now be specified. We refer to it as a *piecewise line-search* (PLS) technique.

PIECEWISE LINE-SEARCH (PLS):

We have at hand the current iterate $x_k^0 \equiv x_k$.
Set $i = 0$ and $\alpha_k^0 = 0$.

1. A stepsize $\alpha_k^{i+1} > \alpha_k^i$ is computed so as to decrease Θ well enough, by satisfying

$$\Theta(x_k^i + (\alpha_k^{i+1} - \alpha_k^i)t_k^i) \leq \Theta(x_k^i) + \omega_1(\alpha_k^{i+1} - \alpha_k^i)\Theta'(x_k^i; t_k^i).$$

2. Set $x_k^{i+1} = x_k^i + (\alpha_k^{i+1} - \alpha_k^i)t_k^i$.
3. Test the curvature condition at the new point x_k^{i+1}: if

$$g(x_k^{i+1})^\top \delta_k \geq \omega_2 g_k^\top \delta_k$$

holds, stop the PLS.
4. Increase i by 1 and go to 1.

Let us just check that it is always possible to find a stepsize $\alpha_k^{i+1} > \alpha_k^i$ such that the inequality in step 1 holds. For this, it is sufficient to show that t_k^i is a descent direction for Θ at x_k^i. Such is indeed the case: $\Theta'(x_k^i; t_k^i) = g(x_k^i)^\top \delta_k < \omega_2 g_k^\top \delta_k$ (by construction, the curvature condition is not satisfied), and $g_k^\top \delta_k < 0$.

Under natural assumptions [146, 10] (for example, α_k^{i+1} should not be arbitrarily close to α_k^i), this PLS can be shown to stop after a finite number i_k of cycles. A point $y_{k+1} = x_k^{i_k}$ is then produced, which satisfies

$$\Theta(y_{k+1}) \leq \Theta(x_k) + \omega_1 \nu_k \quad \text{and} \quad g(y_{k+1})^\top \delta_k \geq \omega_2 g_k^\top \delta_k. \tag{18.17}$$

The quantity $\nu_k < 0$ collects the contributions of all previous steps 1 to the decrease in Θ:

$$\nu_k = \sum_{i=0}^{i_k-1} (\alpha_k^{i+1} - \alpha_k^i)\Theta'(x_k^i; t_k^i).$$

As explained in § 17.3, it is desirable that this search accepts asymptotically the unit stepsize, i.e., when x_k is close to x_*, one would like $y_{k+1} = x_k + t_k$ to satisfy inequalities (18.17). Such a property would spare time in the computation of the tangential displacement α_k, since trying $\alpha_k = 1$ first would have some chance of success. Since M_k approximates the reduced Hessian of the Lagrangian, $x_k + t_k$ will be an approximation of the minimum point of the Lagrangian on $x_k + R(Z_k^-)$. In order to accept the unit stepsize, Θ must therefore be a correct approximation of the Lagrangian and, for the same reasons as in unconstrained optimization and § 17.3, ω_1 should be taken in $]0, \frac{1}{2}[$. A merit function enjoying the above properties is the nondifferentiable augmented Lagrangian (18.15), providing that $\mu = \mu_k \in \mathbb{R}^m$ is chosen so as to tend to the optimal multiplier. Note that this penalty function is exact if

$$\sigma > \|\lambda_* - \mu\|_D.$$

A Reduced Quasi-Newton Algorithm

The PLS described above is essentially in charge of managing the longitudinal displacement of the iterate; it disregards the transversal displacement, the one decreasing the norm of the constraints. After the PLS has produced y_{k+1}, the algorithm needs therefore a restoration step of the constraints, with a linesearch to find the new iterate x_{k+1}. If this restoration is performed along the direction

$$\bar{r}_k = -A^-(y_{k+1})c(y_{k+1}),$$

the merit function must decrease along that direction. But we have

$$\ell'_{\mu_k,\sigma_k}(y_{k+1}; \bar{r}_k) = (\lambda^{\mathrm{LS}}(y_{k+1}) - \mu_k)^\top c(y_{k+1}) - \sigma_k\|c(y_{k+1})\|_P,$$

where λ^{LS} is the least-squares multiplier defined by (14.24). This computation suggests taking an update rule for σ_k such that, for some constant $\bar{\sigma} > 0$, there holds:

$$\sigma_k \geq \|\lambda^{\mathrm{LS}}(y_{k+1}) - \mu_k\|_D + \bar{\sigma}.$$

In this case, we have $\ell'_{\mu_k,\sigma_k}(y_{k+1}; \bar{r}_k) \leq -\bar{\sigma}\|c(y_{k+1})\|_D$, which is negative if y_{k+1} is not feasible.

When checking the above inequality, a transversal stepsize $\beta_k > 0$ along $\bar{r}_k$ can be determined by a backtracking technique: β_k has the form β^{b_k}, where $\beta \in]0, 1[$ and b_k is the smallest nonnegative integer such that

$$\ell_{\mu_k,\sigma_k}(y_{k+1} + \beta_k\bar{r}_k) \leq \ell_{\mu_k,\sigma_k}(y_{k+1}) + \omega_1\beta_k\ell'_{\mu_k,\sigma_k}(y_{k+1}; \bar{r}_k).$$

Altogether, one obtains the following algorithm, which summarizes the techniques described in this section.

REDUCED QUASI-NEWTON ALGORITHM:

Choose an initial iterate $x_1 \in \mathbb{R}^n$.
Compute $f(x_1)$, $c(x_1)$, $\nabla f(x_1)$, and $A(x_1) = c'(x_1)$.
Set the constants $\omega_1 \in]0, \frac{1}{2}[$ and $\omega_2 \in]\omega_1, 1[$ (constants for the PLS),
$\bar{\sigma} > 0$ (penalty parameter threshold), and $\beta \in]0, \frac{1}{2}]$ (backtracking safeguard parameter).
Set $k = 1$, $\sigma_1 = 0$, and $\mu_1 = 0$.

0. Initialize M_1 to an $(n-m) \times (n-m)$ symmetric positive definite matrix, approximating the reduced Hessian of the Lagrangian.
1. *Longitudinal (or tangent) displacement*:
   1.1. Compute the tangent direction $t_k = -Z_k^- M_k^{-1} g_k$.
   1.2. Compute the intermediate iterate y_{k+1} by PLS, started with $\alpha_k^1 = 1$ and using the penalty function $\Theta = \ell_{\mu_k, \sigma_k}$.
2. *Transversal (or restoration) displacement*:
   2.1. If necessary, adapt the multiplier μ_k and the penalty parameter σ_k.
   2.2. Find a stepsize $\beta_k > 0$ along $\bar{r}_k$ as described above; this gives the next iterate x_{k+1}.
3. *Stopping test*: Stop if $g(x_{k+1}) = 0$ and $c(y_{k+1}) = 0$ (optimality is reached).
4. *Matrix update*: Update $M_k \to M_{k+1}$ by the BFGS formula (18.2), using $\gamma_k = g(y_{k+1}) - g(x_k)$ and $\delta_k = \alpha_k Z_k t_k$.
5. Increase k by 1 and go to 1.

A more detailed version of this algorithm can be found in [146], as well as a study of some of its properties.

Update Criterion

Even if $\alpha_k = 1$ is accepted by the PLS, step 1.2 of the above reduced quasi-Newton algorithm needs to linearize the constraints at the intermediate point y_{k+1}. This makes a total of at least two constraint linearizations per iteration: one at x_k and one at y_{k+1}. A linearization means the computation of the constraint Jacobian A and sometimes the computation of the operator Z^- driving the reduced gradient. This can be very costly in certain problems. Therefore reduced quasi-Newton algorithms have been sought that do not require these two linearizations.

In these algorithms, one strives to update the reduced matrix from the pair (γ_k, δ_k) defined by

$$\gamma_k = g(x_{k+1}) - g(x_k) \quad \text{and} \quad \delta_k = Z_k t_k. \tag{18.18}$$

A variant can be used for γ_k, corresponding to the choice (18.13):

$$\gamma_k = Z_k^{-\top}(\nabla_x \ell(x_{k+1}, \lambda_k) - \nabla_x \ell(x_k, \lambda_k)).$$

Because the reduced gradient at $x_k + t_k$ is no longer necessary, the constraints no longer need to be linearized at this point, as in the reduced Hessian method (§ 14.5).

Using without precautions the pair (18.18) to update the matrix M_k may result in an algorithm with poor efficiency. In fact, the asymptotic criterion (18.5) selecting good pairs (γ_k, δ_k) may no longer hold. Assume that the new iterate is computed as in the reduced Hessian method of § 14.5: $x_{k+1} = x_k + t_k + r_k$, where $t_k = -Z_k^- M_k^{-1} g_k$ and $r_k := -A_k^- c(x_k + t_k)$. An expansion similar to (18.12) gives for γ_k defined by (18.18) and appropriate smoothness hypotheses:

$$\begin{aligned}
\gamma_k &= Z_*^{-\top} L_*(x_{k+1} - x_k) \\
&\quad + \left(\int_0^1 (g'(x_k + \alpha(x_{k+1} - x_k)) - g'(x_*)) \, d\alpha \right)(x_{k+1} - x_k) \\
&= Z_*^{-\top} L_* t_k + Z_*^{-\top} L_* r_k + o(\|t_k + r_k\|) \\
&= H_* \delta_k + Z_*^{-\top} L_* r_k + o(\|\delta_k\|) + o(\|r_k\|).
\end{aligned}$$

To obtain the desired estimate, one should have

$$r_k = o(\|t_k\|).$$

One calls *update criterion* a condition having the form of the above estimate, which is therefore used to measure the appropriateness of an update. In this spirit, Nocedal and Overton [276] and Gilbert [144] have independently proposed comparing the length of the tangential displacement t_k and the transversal one r_k. The matrix M_k is then updated if the following *update criterion* is satisfied:

$$\|r_k\| \leq \mu_k \|t_k\|. \tag{18.19}$$

According to the previous estimate of γ_k, it is desirable to have $\mu_k \to 0$ in (18.19). In principle, the sequence $\{\mu_k\}$ can be given a priori if the algorithm is started close to a solution [276]. But with an arbitrary initial iterate, it is a good idea to let the algorithm itself manage the parameter μ_k [144, 147]. The update criterion (18.19) can also be seen as a means of selecting those iterations where the cheap γ_k of (18.18), and the safe γ_k of (18.11), are similar because $x_k + t_k$ and x_{k+1} are closer and closer together, in terms of the distance $\|t_k\|$ separating $x_k + t_k$ and x_k.

The update criterion (18.19) works well in theory [276, 144, 147, 182], but gives sometimes disappointing numerical results. This seems due to the fact that M_k is not updated often enough.

Research on this subject is not closed and other approaches are still proposed [30, 368, 357].

18.4 The Hanging Chain Project V

This is the fifth and last session on the implementation of the SQP algorithm and its application to the problem of determining the static equilibrium of a hanging chain staying above a tilted flat floor. This MATLAB project has been developed in §§ 13.8, 14.7, 15.4, and 17.4. This session is dedicated to the implementation of a quasi-Newton version of the algorithm along the lines of § 18.2. We shall also introduce the concept of performance profile, which will help us to compare the numerical efficiency of the Newton and quasi-Newton SQP approaches.

Modifications to Bring to the sqp Function

The method we focus on is summarized on page 328. This algorithm is basically the same as the one that has been implemented so far, except that, instead of computing the Hessian of the Lagrangian $L_k := \nabla^2_{xx}\ell(x_k, \lambda_k)$ in the simulator, it uses a BFGS approximation M_k to it. This is the mechanism that has to be added to the sqp function (we make it active when options.imode(1) is set to 0, see § 17.4). Here are some more details.

In our implementation, we set the initial matrix M_1 used at the first iteration to the identity matrix. At the second iteration, instead of updating $M_1 = I$ by the BFGS formula to obtain M_2, we update $M'_1 := \eta_1 I$, where the positive number η_1 aims at giving to M'_1 a good scaling. It is standard to choose η_1 by forcing M'_1 to verify a scalar version of the quasi-Newton equation $\gamma_1 = M'_1\delta_1$, where $\delta_1 := x_2 - x_1$ and γ_1 reflects the change in the gradient of the Lagrangian from (x_1, λ_2) to (x_2, λ_2) (see below). Taking the scalar product of this equation with δ_1 and γ_1 yields the following possible values for η_1:

$$\eta'_1 := \frac{\gamma_1^\top \delta_1}{\|\delta_1\|_2^2} \quad \text{and} \quad \eta''_1 := \frac{\|\gamma_1\|_2^2}{\gamma_1^\top \delta_1}.$$

It is assumed here that $\gamma_1^\top \delta_1 > 0$ (see below). It is difficult to give good reasons to favor one of these formula. In our code, we choose $\eta_1 = \eta''_1$, since this value is larger that η'_1 (by the Cauchy-Schwarz inequality) and that a larger matrix M_1 gives more chance to the unit stepsize to be accepted (since the step is usually smaller; this argument is taken from a discussion in [152]).

For $k \geq 1$, the pair of vectors (γ_k, δ_k), used to update M_k (or M'_1 if $k = 1$) into M_{k+1} by the BFGS formula (18.2), is formed of $\gamma_k := \gamma_k^P$, the *Powell correction* of γ_k^ℓ (see (18.7), (18.8), and (18.9), in which κ is set to 0.2), and $\delta_k := x_{k+1} - x_k$. Note that γ_k^P is defined by (18.8) with $M_1 = I$, not with $M_1 = \eta''_1 I$, which depends on γ_1.

Performance Profiles

Even though the number of test cases we have proposed is very small, we can draw some conclusions from the performance profiles à la Dolan and

Moré [113] of two SQP algorithms with line-search: the one with the modified Hessian of the Lagrangian tested in § 17.4 and the one with a quasi-Newton approximation to this Hessian considered in this section.

Performance profiles are used to compare the efficiency of a collection $\mathcal{S}$ of solvers on a set $\mathcal{P}$ of test problems. The comparison is summarized by one curve per solver, which is definitely easier to read than a table of values. The idea is the following. Let

$$\tau_{p,s} := \text{performance of the solver } s \text{ on the problem } p.$$

Here, a *performance* refers to a positive value that reflects an aspect of the efficiency of a solver, such as the number of function evaluations or the computing time that it requires to solve a particular problem to a given precision. This value has to be smaller when the solver is more efficient. The *relative performance* of a solver s (with respect to the other solvers) on a problem p is the ratio

$$\rho_{p,s} = \frac{\tau_{p,s}}{\min\{\tau_{p,s'} : s' \in \mathcal{S}\}}.$$

Of course $\rho_{p,s} \geq 1$. On the other hand, it is assumed that $\rho_{p,s} \leq \bar{\rho}$ for all problems p and solvers s, which can be ensured only by setting $\rho_{p,s}$ to the large number $\bar{\rho}$ if the solver s cannot solve the problem p. Actually, we shall consider that s fails to solve p if and only if $\rho_{p,s} = \bar{\rho}$. The *performance profile* of the solver s (relative to the other solvers) is then the function

$$t \in [1, \bar{\rho}] \mapsto \wp_s(t) := \frac{|\{p \in \mathcal{P} : \rho_{p,s} \leq t\}|}{|\mathcal{P}|} \in [0, 1],$$

where $|\cdot|$ is used to denote the number of elements of a set (its cardinality).

Only three facts need to be kept in mind to have a good interpretation of these upper-semi-continuous piecewise-constant nondecreasing functions:

- $\wp_s(1)$ gives the fraction of problems on which the solver s is the best; note that two solvers may have an even score and that all the solvers may fail to solve a given problem, so that it is not guaranteed to have $\sum_{s \in \mathcal{S}} \wp_s(1) = 1$;

- by definition of $\bar{\rho}$, $\wp_s(\bar{\rho}) = 1$; on the other hand, for small $\varepsilon > 0$, $\wp_s(\bar{\rho} - \varepsilon)$ gives the fraction of problems that the solver s can solve; this value is independent of the performance under consideration;

- the value $\wp_s(t)$ may be given an interpretation by inverting the function $t \mapsto \wp_s(t)$: for the fraction $\wp_s(t)$ of problems in $\mathcal{P}$, the performance of the solver s is never worse than t times that of the best solver (this one usually depends on the considered problem); in this respect the argument at which $\wp_s$ reaches its "almost maximal" value $\wp_s(\bar{\rho} - \varepsilon)$ is meaningful.

With performance profiles, the relative efficiency of each solver appears at a glance: the higher is the graph of $\wp_s$ the better is the solver s.

Experimenting with the SQP Algorithm

We have discussed in the introduction of this chapter the reasons why a quasi-Newton method can be advantageous, and they are numerous. However, this technique is often less precise than an algorithm using the Hessian of the Lagrangian. For example, among the 10 test cases we have defined in the previous sessions (labeled 1a-1g, 2a, 2b, and 3), the quasi-Newton-SQP algorithm is only able to solve one of them (test case 2b) with `options.tol(1:4)` set to 10^{-10}, while the modified-Newton-SQP algorithm solves 8 of them. For this reason, we set `options.tol(1:4)` to 10^{-6} in the numerical experiments of this section.

We have plotted in figure 18.3 performance profiles of two solvers on

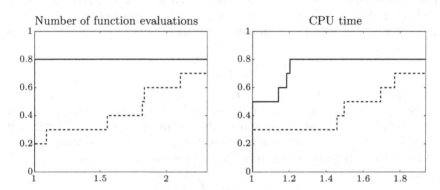

Fig. 18.3. Performance profiles for two versions of the SQP algorithm with linesearch: `SQP-mn` (with modified Hessian, solid line) and `SQP-qn` (with a quasi-Newton approximation, dashed line)

the collection of 10 test cases presented in this hanging chain project. Both solvers globalize the SQP algorithm by line-search. The first one (solid line), say `SQP-mn`, takes a positive definite modification of the Hessian of the Lagrangian in its osculating QP's (see § 17.4) and the second one (dashed line), say `SQP-qn`, is based on the BFGS approximation to this Hessian, as discussed above.

The first considered performance (left picture in figure 18.3) is the *number of function evaluations*, which is identical to the number of stepsize trials in the line-search (gradients for both solvers and Hessians for `SQP-mn` are evaluated at each iteration, that is to say each time a stepsize is accepted by the line-search). The value $\bar{\rho}$ at which both curves take the value 1 is set slightly above 2.3, beyond the rightmost abscissa shown in the picture. We see that `SQP-mn` fails on 20 % of the problems (hence on 2 problems: 1g because of an infeasible QP and 3a) and that `SQP-qn` fails on three problems (1c because the updated matrices and their inverses blow up – this is further discussed below, 1g, and 3a). When it does not fail, `SQP-mn` is always the winner (because its performance profile is constant). This is not surprising since the solver uses the Hessian of the Lagrangian, which contains much more information than its quasi-Newton approximation used in SQP-qn. However, looking at abscissa 1, we see that the two solvers have an even score on 2 problems (20 % of them), which is a good result for SQP-qn since this solver does not take advantage of the second derivatives.

If we consider the *CPU time* as the performance criterion (right hand side picture in figure 18.3), we see that the solver SQP-qn improves with respect to SQP-mn. It becomes the best solver on 30 % of the problems (there is no even CPU time scores, since the time is measured with many digits of precision). It does not beat SQP-mn, however (we see in the picture that the CPU time spent by SQP-mn is never worse than approximately 1.2 times the one spent by SQP-qn). This is very likely due to the small dimension of the test cases and to the fact that the Hessian of the Lagrangian of the hanging chain problem is very sparse and therefore not time consuming to compute in SQP-mn.

Let us now consider the test case 1c (page 249), on which the quasi-Newton SQP solver SQP-qn fails. The final position of the chain is shown in the left picture of figure 18.4 (we have selected one out of 3 intermedi-

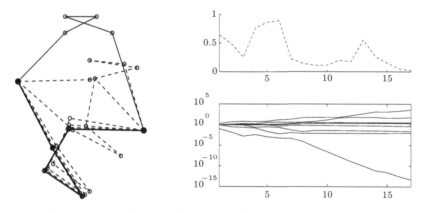

Fig. 18.4. Test case 1c with quasi-Newton Hessian approximations: the final position of the chain (left), Powell's θ_k (right, above), eigenvalues of M_k (right, below)

ate positions). This is not at all a stationary point of the problem: neither optimality nor feasibility are reached. Actually, as shown by the right hand side plot (the one below) in figure 18.4, the generated matrices M_k have their smallest eigenvalue that tends to zero and their largest that blows up. We have also plotted (the picture above in the right hand side) the value of the parameter θ_k in (18.8) along the iterations: it is never equal to 1, which means that $(\gamma_k^\ell)^\top \delta_k$ is never considered to be sufficiently large (it is most often negative). Close to the final position, the algorithm enters a vicious circle: the bad quality of M_k induces an unsatisfactory search direction d_k;

then the line-search determines a very small step-size α_k along d_k, so that the next iterate is close to the previous one; repetitive application of formula (18.8) then produces a sequence of γ_k^P with $(\gamma_k^P)^\top \delta_k$ tending to zero; this deteriorates again the matrix M_k. It is amusing to note that by taking the more conservative value $\kappa = 0.23$ (or a larger one) instead of the classical $\kappa = 0.2$ in (18.9), the generated sequence converges to the global minimum. It is a matter of chance. We hope that this example may serve as a motivation for the algorithms developed in § 18.3.

Part IV

Interior-Point Algorithms
for Linear and Quadratic Optimization

J. Frédéric BONNANS

Khachiyan [207] proved in 1979 that a *linear program* (optimization problem with linear objective and constraints) could be solved in *polynomial time*, thus resolving a dozens year old conjecture. It is only after Karmarkar's works [204] on *interior-point methods*, in 1984, that polynomial algorithms became competitive with the method used until then, the *simplex* algorithm of Dantzig [94], which was shown in 1970 to be non-polynomial, by Klee and Minty [216].

More recently, the attention focused on a family of *primal-dual* interior-point algorithms, called *central-path algorithms*. They enjoy at the same time the best complexity estimate known so far, namely $O(\sqrt{n}\bar{L})$ (see below for the meaning of this quantity) and quadratic convergence. Most efficient codes these days are based on this type of algorithm. In spite of recent progress in the theoretical analysis of algorithms, there is a substantial gap between theoretical estimates of speeds of convergence and practical performances. In fact, interior-point algorithms have the remarkable property of converging in a few iterations on most practical examples.

The aim of these notes is to give an introduction to the simplex method, still frequently used, and to primal-dual path-following methods. The latter will be stated in the framework of monotone linear complementarity problems; this allows the treatment of convex linear-quadratic problems by interior-point methods. We will also give some complements on the largest-step algorithm, the complexity theory for problems with integer data, and on the Karmakar algorithm.

Linear and Linear-Quadratic Optimization

Consider an optimization problem with quadratic objective and linear constraints, of the type

$$\operatorname*{Min}_{x \in \mathbb{R}^n} c^\top x + \tfrac{1}{2} x^\top H x; \quad Ax = b, \quad x \geq 0, \qquad (QP)$$

where H is an $n \times n$ symmetric positive semidefinite matrix, A is a $p \times n$ matrix, and $b \in \mathbb{R}^p$. The above format for the constraints is called *standard form*. The problem is said to be *linear* if the objective is linear, i.e. if $H = 0$. Define the *value* of the problem to be the infimum value of the objective on the feasible set. With the preceding problem, which we will call *primal*, is associated the *dual problem*,

$$\operatorname*{Max}_{x,\lambda,s} -b^\top \lambda - \tfrac{1}{2} x^\top H x; \quad c + H x + A^\top \lambda = s, \quad s \geq 0. \qquad (QD)$$

The primal and dual values are equal, except in the singular case where neither the primal nor the dual have feasible points (Corollary 19.13). In case the primal and dual values are finite (hence equal), the primal and dual problems do have optimal solutions (Theorems 19.1 and 19.12) and a point x

is a primal solution if and only if there exists (s, λ) such that (x, s, λ) solves the following *optimality system*

$$\begin{cases} xs = 0, \\ Ax = b, \quad c + Hx + A^\top\lambda = s, \\ x \geq 0, \quad s \geq 0. \end{cases} \qquad (OS)$$

We use in the above display the componentwise multiplication of vectors: xs is defined as the vector with ith components $x_i s_i$.

For linear problems, a key concept is that of *basic point*, defined as being a feasible point having at most p nonzero components. Indeed, if a linear problem has optimal solutions at all, one of them is basic (Proposition 19.2). The simplex algorithm, presented in Chap. 19, generates a sequence of basic points and, after finitely many iterations, obtains an optimal one.

A solution (x, λ, s) of the optimality system is *strictly complementary* if $x + s > 0$, or equivalently $\max(x_i, s_i) > 0$, for all $i \in \{1, \cdots, n\}$. If a linear problem has a finite value, a strictly complementary optimal solution (non-basic in general) can be shown to exist (Proposition 19.16).

Central-Path Algorithms

Let $\mathbf{1}$ be the vector with all components equal to 1, its dimension depending on the context. The *central trajectory* is the set of points (x, s, λ) satisfying, for a certain $\mu > 0$, the relations

$$\begin{cases} xs = \mu\mathbf{1}, \\ Ax = b, \quad c + Hx + A^\top\lambda = s, \\ x \geq 0, \quad s \geq 0. \end{cases} \qquad (CT)$$

For $\mu = 0$ we find back the optimality system (OS). Central-path algorithms (also said "path-following" algorithms) generate a sequence of points satisfying approximately the equations of the central path, for a sequence of μ-values tending to 0. To compute the directions of move of x and s, Newton's method is applied to the central-path equations. One speaks of affine direction if one takes $\mu = 0$ in the linearized equation of the central trajectory, and of centralization (or restoration) if μ is set to its current value.

In order to control the effects of the nonlinear term xs, the successive iterates are imposed to stay in a neighborhood of the central path. A point (x, s, λ) is said to be admissible if it satisfies the linear constraints in (CT). The *small neighborhood*, parameterized by $\alpha \in (0, 1)$, and the *large neighborhood*, parameterized by $\epsilon \in (0, 1)$, are defined respectively by

$$\mathcal{V}_\alpha := \left\{ (x, s, \lambda, \mu); \quad \mu > 0; \quad \left\| \frac{xs}{\mu} - \mathbf{1} \right\| \leq \alpha; \quad (x, s, \lambda) \text{ admissible} \right\},$$

$$\mathcal{N}_\epsilon := \left\{ (x, s, \lambda, \mu); \quad \mu > 0; \quad \epsilon\mathbf{1} \leq \frac{xs}{\mu} \leq \epsilon^{-1}\mathbf{1}; \quad (x, s, \lambda) \text{ admissible} \right\}.$$

Let $(x^0, s^0, \lambda_0, \mu_0) \in \mathcal{V}_\alpha$ be the initial point. The small neighborhood algorithm stops when a point (x, s, μ) is found in $\mathcal{V}_\alpha$ such that $\mu \leq \mu_\infty$, where μ_∞ is a given parameter. All interior-point algorithms presented here require $O(n^3)$ operations per iteration (if the data are dense). Given an initial point in the current neighborhood, with the initial value μ_0 of the parameter μ, the *complexity* of each algorithm is measured by the number of iterations necessary to compute a point in the neighborhood associated with $\mu \leq \mu_\infty$. Set $\bar{L} := \log(\mu_0/\mu_\infty)$.

The predictor-corrector method studied in Chap. 21 performs at each iteration a restoration step (centralization), followed by an affine move. In the small neighborhood case, used with $\alpha \leq 1/2$, the restoration step sends a point of $\mathcal{V}_\alpha$ into $\mathcal{V}_{\alpha/2}$, while the affine step reduces the optimality measure by a term of the order $1/\sqrt{n}$. The complexity is therefore $O(\sqrt{n}\bar{L})$. Besides, the asymptotic analysis shows that $\limsup \mu_{k+1}/\mu_k \leq 1 - \alpha^{1/2}|T|^{-1/4}/2$, where T is the set of indices for which strict complementarity is not satisfied. Finally, under the strict complementarity assumption, μ_k converges quadratically to 0.

A variant of this algorithm, based on the modified field introduced in Chap. 20, preserves the same complexity while converging superlinearly without the strict complementarity assumption. More precisely, there holds $\mu_{k+1} = O(\mu_k^{5/4})$.

Then, we study a large neighborhood algorithm, theoretically less efficient, but numerically better and closer to present implementations. Its theoretical complexity is as follows: a centralization step of order $1/n$ allows a relative reduction of the optimality measure μ, in the affine move, of the order $1/n$. Convergence therefore occurs in $O(n\bar{L})$ iterations. The asymptotic analysis is similar to that of the small neighborhood algorithm: convergence is quadratic under the strict complementarity hypothesis, and a modification of the algorithm using the modified field results in an algorithm enjoying the same complexity, satisfying $\mu_{k+1} = O(\mu_k^{5/4})$ without strict complementarity hypothesis.

A drawback of the above methods is to require the knowledge of an initial point in the small or large neighborhood of the central path. This assumption, made to facilitate the statement of the algorithms only, is not satisfied in general. We present two classes of algorithms alleviating this requirement.

The small neighborhood non-feasible algorithm, presented in Chap. 22, is very similar to the small neighborhood feasible algorithm: it computes the directions of move by applying Newton's method to the central-path equation. Its complexity is $O(n\bar{L})$ (compared to $O(\sqrt{n}\bar{L})$ in the feasible case). The asymptotic analysis is close to that of the feasible algorithm.

The second approach, presented mainly in the framework of linear programming, is based on the properties of *self-dual* problems, i.e. those such that the primal and dual formulations coincide. Any linear program can be cast into a self-dual problem, for which one knows a point on the central path. Solving the self-dual problem by an interior-point (feasible) algorithm yields a strictly complementary solution (of the self-dual problem). Once this solution is computed, it is easy (in exact arithmetic) to compute a solution of the original problem, or to conclude infeasibility of the latter.

Chapter 24 states the long-step algorithm, which is of the path-following type with only one displacement per iteration (while the predictor-corrector algorithm performs two displacements). The direction of displacement is a combination of the affine and centralization directions. The analysis, limited to the small neighborhood, shows that complexity is $O(\sqrt{n}\bar{L})$. The convergence of μ_k is superlinear in the neighborhood $\mathcal{V}_{1/4}$. A variant of the algorithm, with restoration steps inserted at some iterations, enjoys similar properties in the neighborhood $\mathcal{V}_{1/2}$.

Complexity Theory and Karmarkar's Algorithm

For problems with integer data, complexity theory gives estimates of the number of operations allowing the computation of a solution. Chapter 25 states this theory in the case of linear programming. Let L be the *size* of the problem; it can be defined as the amount of memory necessary to store (A, b, c). The key result is that, if the feasible set is nonempty and bounded, and if x is a basic point, then: either x is optimal, or there exists an optimal $\bar{x}$ such that $c^\top \bar{x} < c^\top x - 2^{-2L}$. Now, interior-point methods compute rapidly an approximate solution of the problem. Combined with a *purification* algorithm which computes a basic point via simplex-type computations, they yield the best complexity estimate known so far for linear programming problems.

Finally, we present Karmarkar's algorithm. Even though it is no longer much used, its statement shows how original it is. Motivated by the complexity of problems with integer data, this algorithm uses a projective transformation to minimize efficiently a potential function having a singularity at its optimum point. Undoubtedly, the works that followed it owe a lot to the impulse given by Karmarkar's pioneering contribution.

Other Monographs on Interior-Point Methods

Among other works devoted to interior-point algorithms, let us mention those of Saigal [319], who analyzes in detail affine algorithms, Den Hertog [103], who also discusses convex nonlinear or quadratic problems, Terlaky, Vial, and Roos [349], Vanderbei [353], Wright [366], and Ye [370]. A technical discussion of interior-point algorithms in the framework of monotone linear complementarity is given in Kojima et al. [217]. Nesterov and Nemirovski [273] present a general complexity theory of interior-point algorithms for nonlinear problems. One finds in Terlaky [348] a useful synthesis of various points of view and applications. The classical aspects of linear programming are discussed by Goldfarb and Todd [166]. Extensions to semidefinite programming problems are presented in Saigal, Vandenberghe, and Wolkowicz [363] and Ben-Tal and Nemirovski [23].

Notation

B, N, T	Partition of $\{1, \cdots, n\}$ associated with solutions of (LCP).
$\text{dist}(x, X)$	Distance from the point x to the set X.
$F(P)$	Feasible or admissible set of problem (P).
$\mathcal{L}(x, \lambda, s)$	Lagrangian of problem (P).
(LCP)	Monotone linear complementarity problem.
$(LP), (LD)$	Linear problem in standard form and its dual.
$\mathcal{N}(A), \mathcal{R}(A)$	Kernel and Range of A.
$\nabla f(x), \nabla^2 f(x)$	Gradient and Hessian of $f(x)$.
$P_{A,q}, P_A$	Orthogonal projection onto $\{z : Az + q = 0\}$ and $\mathcal{N}(A)$.
$\pi(x)$	Logarithmic potential: $\pi(x) := -\sum_i \log x_i$.
$(QP), (QD)$	Quadratic problem in standard form and its dual.
$\mathbb{R}, \mathbb{R}_+, \mathbb{R}_{++}$	Sets of real, nonnegative, positive real numbers.
$\mathbb{R}_-, \mathbb{R}_{--}$	Sets of nonpositive, negative real numbers.
$S(P), v(P)$	Solution set, value (optimal cost) of (P).
	If $F(P) = \emptyset$, then $v(P) = +\infty$ and $S(P) = \emptyset$.
u^c, v^c, u^a, v^a	Centralization and affine displacements.
w	Triple (x, s, μ).
$\delta(w)$	$\left\| \dfrac{xs}{\mu} - 1 \right\|$.
$\mathbf{1}$	Vector whose coordinates are all 1.
$\|\cdot\|, \|\cdot\|_\infty$	Euclidean norm, max-norm.

19 Linearly Constrained Optimization and Simplex Algorithm

Overview

This chapter recalls some theoretical results on linearly constrained optimization with convex objective function. In the case of a linear or quadratic objective, we show existence of an optimal solution whenever the value of the problem is finite, as well as existence of a basic solution in the linear case. Lagrangian duality theory is presented. In the linear case, existence of a strictly complementary solutions is obtained whenever the optimal value of the problem is finite. Finally, the simplex algorithm is introduced in the last part of the chapter.

19.1 Existence of Solutions

19.1.1 Existence Result

Consider a linearly constrained optimization problem of the type

$$\underset{x \in \mathbb{R}^n}{\text{Min}} \, f(x); \quad Ax = b, \quad x \geq 0, \tag{P}$$

where f is a convex function $\mathbb{R}^n \to \mathbb{R}$, A is a $p \times n$ matrix and $b \in \mathbb{R}^p$. Constraints written in this way are said to be in *standard form*. Any linearly constrained problem can be cast into an equivalent problem in standard form. This transformation is rarely advantageous from a numerical point of view. The interest of the standard form is to allow a simple statement of algorithms, which can be formulated in a more general format when coming to implementations.

We denote the *feasible set*, the *value* and the *solution set* of problem (P) respectively by

$$F(P) := \{x \in \mathbb{R}^n; \quad Ax = b, \quad x \geq 0\},$$
$$v(P) := \inf\{f(x); \quad x \in F(P)\},$$
$$S(P) := \{x \in F(P); \quad f(x) = v(P)\}.$$

By convention, the infimum on the empty set is $+\infty$. Thus, if $F(P)$ is empty, then $v(P) = +\infty$ and $S(P)$ is empty. A linearly constrained optimization problem is said to be *linear* (resp. *quadratic*) if the objective function f is linear (resp. quadratic). Linear and convex quadratic problems in standard form are therefore written respectively as

$$\underset{x \in \mathbb{R}^n}{\text{Min}} \; c^\top x; \quad Ax = b, \quad x \geq 0, \tag{LP}$$

$$\underset{x \in \mathbb{R}^n}{\text{Min}} \; c^\top x + \tfrac{1}{2} x^\top H x; \quad Ax = b; \quad x \geq 0, \tag{QP}$$

where H is an $n \times n$ symmetric positive semidefinite matrix. Of course, linear problems are particular cases of quadratic problems.

Theorem 19.1. *The solution set $S(P)$ is convex. When f is convex quadratic, this set is nonempty iff $v(P) \in \mathbb{R}$.*

Proof. Convexity of $S(P)$ follows from that of $F(P)$ and f. Clearly, if $S(P)$ is nonempty, then $v(P) \in \mathbb{R}$. Let us show the converse inclusion, assuming that f is quadratic.

(a) First, we prove the property for a linear objective. Denote the active constraint set at x by

$$N(x) := \{1 \leq i \leq n; \quad x_i = 0\}.$$

If $v(LP) \in \mathbb{R}$, consider a minimizing sequence; extracting a subsequence, there exists a minimizing sequence $\{x^k\}$ such that $N(x^k)$ is constant. Let $\{x^k\}$ be a minimizing sequence such that $N(x^k)$ is constant and maximal (i.e., there exists no other minimizing sequence whose active constraint set is constant and contains $N(x^k)$ strictly). If $\{x^k\}$ stays out of $S(LP)$, extract again a subsequence: we can assume that $c^\top x^{k+1} < c^\top x^k$. Set $d^k := x^{k+1} - x^k$. Then $c^\top d^k = c^\top (x^{k+1} - x^k) < 0$.

Set $t^k := \sup\{t \geq 0; x^k + td^k \in F(P)\}$. Since $N(x^{k+1}) = N(x^k)$, we have $t^k > 0$. Since $v(LP) \in \mathbb{R}$ and $c^\top d^k < 0$, we have $t^k < +\infty$ and a nonnegativity constraint becomes active at t^k. Then $\hat{x}^k := x^k + t^k d^k$ satisfies $N(x^k) \subset N(\hat{x}^k)$, with strict inclusion. Since $c^\top \hat{x}^k = c^\top (x^k + t^k d^k) < c^\top x^k$, the sequence $\{\hat{x}^k\}$ is minimizing. We can extract a minimizing sequence, whose active constraint set is constant, which contradicts the definition of $\{x^k\}$. An element of the sequence $\{x^k\}$ therefore lies in $S(LP)$. In particular, $S(LP) \neq \emptyset$.

(b) In the case of a quadratic objective, let us check first the property for an unconstrained problem of the type

$$\underset{x \in \mathbb{R}^n}{\text{Min}} \; f(x) := c^\top x + \tfrac{1}{2} x^\top H x.$$

If this problem has a finite value, then $c \perp \mathcal{N}(H)$ (otherwise, let $\hat{c}$ be the projection of c onto $\mathcal{N}(H)$; then the sequence $x^k := -k\hat{c}$ satisfies $f(x^k) = -k\|\hat{c}\|^2 \downarrow -\infty$). Decomposing x as $x = x^1 + x^2$, $x^1 \in \mathcal{N}(H)$, $x^2 \in \mathcal{N}(H)^\perp$, and likewise for c, we check that

$$f(x) = c^\top x + \tfrac{1}{2} x^\top H x = (c^2)^\top x^2 + \tfrac{1}{2}(x^2)^\top H^{22} x^2,$$

where H^{22} is positive definite. The set of minima of f is therefore nonempty and is given by

$$\mathcal{N}(H) \times \{-(H^{22})^{-1} c^2\}.$$

Consider now a constrained quadratic problem (QP). As in the case of linear problems, we form a minimizing sequence $\{x^k\}$ such that $N(x^k)$ is constant and maximal. Set $N^* := N(x^k)$, and consider the problem

$$\text{Min } f(x); \quad Ax = b; \quad x_i = 0, \quad i \in N^*. \qquad (P^*)$$

Since $\{x^k\} \subset F(P^*)$, we have

$$v(P) = \lim f(x^k) \geq v(P^*).$$

Let us show that $v(P) = v(P^*)$. Otherwise there exists $x^* \in F(P^*)$ such that $f(x^*) < v(P)$, and hence, $x^* \notin F(P)$. Move from x^k towards x^* on the segment

$$[x^k, x^*] := \{\alpha x^k + (1-\alpha)x^*; \quad \alpha \in [0,1]\}.$$

Call $\hat{x}^k$ the point obtained when a first nonnegativity constraint is hit. Then $\hat{x}^k \in F(P)$, $N^* \subset N(\hat{x}^k)$ with strict inclusion, and by convexity of f,

$$f(\hat{x}^k) \leq \max\{f(x^k), f(x^*)\} = f(x^k),$$

and hence, $\{\hat{x}^k\}$ is a minimizing sequence of (P) contradicting the definition of $\{x^k\}$ (extract from it a subsequence whose active set is constant), so that $v(P) = v(P^*)$.

Note that (P^*) reduces to an unconstrained problem, expressing x in a basis of

$$\mathcal{N}(A) \cap \{x \in \mathbb{R}^n; \quad x_i = 0, \quad i \in N^*\}.$$

Since $v(P^*) = v(P) \in \mathbb{R}$, problem (P^*) has a solution $\bar{x}$. If $\bar{x} \in F(P)$, this is the desired solution since $f(\bar{x}) = v(P^*) = v(P)$. Otherwise, a contradiction is obtained by constructing a sequence $\{\hat{x}^k\}$ as before, replacing x^* by $\bar{x}$. $\qquad \square$

19.1.2 Basic Points and Extensions

Let us introduce a fundamental concept for the analysis of linear problems. We will say that x is a *basic point* if $x \in F(LP)$ and $x_i \neq 0$ for at most p components.

Proposition 19.2. *If (LP) has solutions at all, one of them is basic.*

Proof. Let $x \in S(LP)$, having a minimal number of nonzero components, say q. We prove a contradiction if $q > p$.

We can assume $x_i > 0$, $i \in B := \{1, \dots, q\}$. Since A is $p \times n$, and hence, of rank at most p, the kernel of A_B has dimension at least $q - p$. Thus, let $d \in \mathbb{R}^n$ be such that $d \neq 0$, $Ad = 0$ and $d_i = 0$, $i > q$. Note that, for $|\rho|$ small enough, $x + \rho d \in F(LP)$, hence $c^\top(x + \rho d) \geq v(LP) = c^\top x$ and therefore $c^\top d = 0$. Changing d in $-d$ if necessary, we can assume $\min\{d_i\} < 0$. Then there exists a maximal value of ρ such that $x + \rho d \in F(LP)$, call it $\bar{\rho} := \min\{x_i/|d_i|; \ d_i < 0\}$. The point $x^\sharp := x + \bar{\rho}d$ is feasible, lies in $S(LP)$ and has at most $q - 1$ nonzero components. This contradicts the definition of x.

$\square$

This type of result extends to problems depending linearly only on some variables. In fact, if f is linear with respect to the first n_1 variables, and if $\bar{x}$ minimizes f on $F(P)$, we can consider the minimization with respect to the first n_1 variables, the others being fixed. The following result is a consequence of Proposition 19.2:

Corollary 19.3. *If f is linear with respect to the first n_1 variables, with $n_1 > p$, then the set $S(P)$, if nonempty, contains a point with at most p among its first n_1 components that are nonzero.*

Here is another extension of Proposition 19.2 to the case of a quadratic objective.

Lemma 19.4. *Suppose that f is quadratic with Hessian H positive semidefinite, and $n^a := |\mathcal{N}(H)| - p$ is positive. If nonempty, $S(QP)$ contains a solution with at least n^a zero components.*

Proof. Let $\bar{x} \in S(QP)$ have the maximum number of zero components, say n^b. Suppose $n^b < n^a$. Let X be the space of dimension $n - n^b$ spanned by the nonzero components of $\bar{x}$. Then $X \cap \mathcal{N}(A)$ has dimension at least $n-(n^b+p)$; therefore, the quadratic form $d \to d^\top Hd$ has on $X \cap \mathcal{N}(A)$ a kernel of dimension at least $|\mathcal{N}(H)| - (n^b + p) = n^a - n^b > 0$. Consequently, there exists $d \neq 0$, $d \in X \cap \mathcal{N}(A) \cap \mathcal{N}(H)$. We can assume $\min\{d_i; 1 \leq i \leq n\} < 0$, which gives $\hat{x} \in S(QP)$ in the form $\hat{x} = \bar{x} + \sigma d$, where $\sigma > 0$ has at least one more nonzero component than $\bar{x}$; this is the desired contradiction.

$\square$

19.2 Duality

Let us now introduce some elements of Lagrangian duality theory.

19.2.1 Introducing the Dual Problem

With problem (P) is associated the *Lagrangian*

$$\mathcal{L}(x, \lambda, s) := f(x) + \lambda^\top (Ax - b) - s^\top x.$$

Since

$$\underset{\lambda \in \mathbb{R}^p, s \in \mathbb{R}_+^n}{\text{Max}} \mathcal{L}(x, \lambda, s) = \begin{cases} f(x) & \text{if} \quad x \in F(LP), \\ +\infty & \text{otherwise,} \end{cases}$$

problem (P) can be interpreted as

$$\underset{x \in \mathbb{R}^n}{\text{Min}} \quad \underset{\lambda \in \mathbb{R}^p, s \in \mathbb{R}_+^n}{\sup} \mathcal{L}(x, \lambda, s).$$

The *dual problem* is obtained by inverting the min and max operations:

$$\underset{\lambda \in \mathbb{R}^p, s \in \mathbb{R}_+^n}{\text{Max}} \quad \underset{x \in \mathbb{R}^n}{\inf} \mathcal{L}(x, \lambda, s).$$

Example 19.5. Linear Optimization. In the linear case $f(x) = c^\top x$, the dual problem appears after some computations to be linear and to have the expression

$$\underset{\lambda, s}{\text{Max}} -b^\top \lambda; \quad c + A^\top \lambda = s, \quad s \geq 0. \tag{LD}$$

Remark 19.6. The same dual is obtained if only equality constraints are dualized. The Lagrangian is then $c^\top x + \lambda^\top (Ax - b)$, and (LP) is interpreted as

$$\underset{x \in \mathbb{R}_+^n}{\text{Min}} \sup_{\lambda \in \mathbb{R}^p} c^\top x + \lambda^\top (Ax - b).$$

The dual problem, obtained by inverting the min and max, is

$$\underset{\lambda \in \mathbb{R}^p}{\text{Max}} -b^\top \lambda; \quad c + A^\top \lambda \geq 0.$$

This problem is equivalent to (LD).   □

Example 19.7. Quadratic Optimization. When $f(x) = c^\top x + \frac{1}{2} x^\top H x$, minimizing the Lagrangian with respect to x is an unconstrained optimization problem. From Theorem 19.1, the minimum is attained iff there exists a solution, otherwise it is $-\infty$. The Lagrangian is convex, its minima are the zeros of its gradient, and are therefore characterized by

$$c + Hx + A^\top \lambda = s,$$

where the Lagrangian has the value

$$\begin{aligned} \mathcal{L}(x, \lambda, s) &= f(x) + \lambda^\top (Ax - b) - s^\top x, \\ &= (c + Hx + A^\top \lambda - s)^\top x - b^\top \lambda - \tfrac{1}{2} x^\top H x, \\ &= -b^\top \lambda - \tfrac{1}{2} x^\top H x. \end{aligned}$$

An expression of the dual problem is therefore

$$\text{Max}_{x,\lambda,s} -b^\top\lambda - \tfrac{1}{2}x^\top Hx; \quad c + Hx + A^\top\lambda = s; \quad s \geq 0. \qquad (QD)$$

In the case $H = 0$, the variable x disappears from this problem and we find back the expression of the dual problem (LD).

An Interpretation Extracted from Economics Let $b \in \mathbb{R}^p$ be a stock of nuclear waste (b_i is the quantity of product i). One wants to eliminate them by n different processes. Process j consumes a quantity a_{ij} of waste i, and has a unit cost c_j. Let x_j measure the use of process j. The least-cost solution is obtained by solving the linear problem in standard form. To interpret the dual problem, write it as (with $\hat{\lambda} = -\lambda$)

$$\text{Max}_{\hat{\lambda} \in \mathbb{R}^p} b^\top\hat{\lambda}; \quad A^\top\hat{\lambda} \leq c.$$

Then $\hat{\lambda}_i$ represents a price associated with refusal i, while $A^\top\hat{\lambda}$ is the value associated with the process $\{1, \cdots, n\}$. The dual problem is interpreted as the optimal assessment of the stock, under the constraint that c is an upper bound for the process values.

As suggested by this example, the dual problem has in general a practical interpretation, which is often useful.

19.2.2 Concept of Saddle-Point

The above duality scheme fits with the following abstract framework. Consider $\varphi : X \times Y \to \mathbb{R}$, where X and Y are two arbitrary sets. Introduce the *primal* and *dual* problems

$$\text{Min}_{x \in X} \sup_{y \in Y} \varphi(x, y), \qquad (P^\varphi)$$

$$\text{Max}_{y \in Y} \inf_{x \in X} \varphi(x, y), \qquad (D^\varphi)$$

as well as the *saddle-point* problem: find $(\bar{x}, \bar{y}) \in X \times Y$ satisfying

$$\forall(x, y) \in X \times Y, \quad \varphi(\bar{x}, y) \leq \varphi(\bar{x}, \bar{y}) \leq \varphi(x, \bar{y}).$$

Lemma 19.8. *There always holds that $v(D^\varphi) \leq v(P^\varphi)$. A saddle-point exists iff problems (D^φ) and (P^φ) have the same value and admit an optimal solution (at least); the set of saddle-points is then $S(P^\varphi) \times S(D^\varphi)$. In this case, we denote by $v(\varphi)$ the common value of (D^φ) and (P^φ), and the set of saddle-points is equal to $S(P^\varphi) \times S(D^\varphi)$.*

Proof. Let $(\hat{x}, \hat{y}) \in X \times Y$. Then

$$\inf_{x \in X} \varphi(x, \hat{y}) \leq \varphi(\hat{x}, \hat{y}) \leq \sup_{y \in Y} \varphi(\hat{x}, y),$$

and thus

$$\sup_{\hat{y} \in Y} \inf_{x \in X} \varphi(x, \hat{y}) \leq \inf_{\hat{x} \in X} \sup_{y \in Y} \varphi(\hat{x}, y),$$

which establishes the inequality $v(D^\varphi) \leq v(P^\varphi)$. Let us show that existence of a saddle-point $(\bar{x}, \bar{y})$ implies the condition in the Lemma. By definition of a saddle-point, we have

$$\sup_{y \in Y} \varphi(\bar{x}, y) \leq \varphi(\bar{x}, \bar{y}) \leq \inf_{x \in X} \varphi(x, \bar{y})$$

and the above inequalities actually hold as equalities, since the sup and inf are obtained respectively for $\bar{y}$ and $\bar{x}$, and hence,

$$v(P^\varphi) \leq \sup_{y \in Y} \varphi(\bar{x}, y) = \varphi(\bar{x}, \bar{y}) = \inf_{x \in X} \varphi(x, \bar{y}) \leq v(D^\varphi).$$

But $v(D^\varphi) \leq v(P^\varphi)$, and hence, $v(D^\varphi) = \varphi(\bar{x}, \bar{y}) = v(P^\varphi) = v(\varphi)$. Moreover

$$v(P^\varphi) = \varphi(\bar{x}, \bar{y}) = \sup_{y \in Y} \varphi(\bar{x}, y)$$

which indicates that $\bar{x} \in S(P^\varphi)$, and likewise $\bar{y} \in S(D^\varphi)$.

To finish the proof we check that, if $v(P^\varphi) = v(D^\varphi)$, and if $\bar{x} \in S(P^\varphi)$ and $\bar{y} \in S(D^\varphi)$, then $(\bar{x}, \bar{y})$ is a saddle-point of φ: this will show that the condition is sufficient and will imply that the set of saddle-points is $S(P^\varphi) \times S(D^\varphi)$. We have

$$v(\varphi) = \inf_{x \in X} \varphi(x, \bar{y}) \leq \varphi(\bar{x}, \bar{y}) \leq \sup_{y \in Y} \varphi(\bar{x}, y) = v(\varphi),$$

so that $\varphi(\bar{x}, \bar{y}) = \inf_{x \in X} \varphi(x, \bar{y}) \leq \varphi(x, \bar{y}), \forall x \in X$, the other saddle-point inequality being proved likewise.

$\square$

Now suppose that X, Y are convex subsets of $\mathbb{R}^n$ and $\mathbb{R}^m$, respectively, and that φ is of class C^1. We will say that $(\bar{x}, \bar{y})$ is a *critical point* of φ in (X, Y) when

$$\begin{cases} (\bar{x}, \bar{y}) \in X \times Y, \\ \varphi'_x(\bar{x}, \bar{y})(x - \bar{x}) \geq 0, & \text{for all } x \in X, \\ \varphi'_y(\bar{x}, \bar{y})(y - \bar{y}) \leq 0, & \text{for all } y \in Y. \end{cases}$$

The function φ is said to be *convex-concave* if $x \to \varphi(x, \hat{y})$ and $y \to \varphi(\hat{x}, y)$ are convex and concave, respectively, for all $(\hat{x}, \hat{y}) \in X \times Y$.

Lemma 19.9. *Suppose X and Y are convex. Then the set of saddle-points of φ is contained in the set of its critical points, and the two sets coincide if φ is convex-concave.*

Proof. Let $(\bar{x}, \bar{y})$ be a saddle-point of φ, $x \in X$, $y \in Y$ and $\sigma \in]0, 1[$. Since X and Y are convex, we have $\bar{x} + \sigma(x - \bar{x}) = (1 - \sigma)\bar{x} + \sigma x \in X$, and likewise $\bar{y} + \sigma(y - \bar{y}) \in Y$. Therefore

$$\frac{\varphi(\bar{x}, \bar{y} + \sigma(y - \bar{y})) - \varphi(\bar{x}, \bar{y})}{\sigma} \leq 0 \leq \frac{\varphi(\bar{x} + \sigma(x - \bar{x}), \bar{y}) - \varphi(\bar{x}, \bar{y})}{\sigma}.$$

Letting σ tend to 0, we obtain the characterization of a critical point.

If $(\bar{x}, \bar{y})$ is a critical point of φ, and if φ is convex-concave, use the fact that a convex (resp. concave) function bounds its affine approximation from above (resp. from below); we have for all $(x, y) \in X \times Y$

$$\varphi(\bar{x}, y) \leq \varphi(\bar{x}, \bar{y}) + \varphi'_y(\bar{x}, \bar{y})(y - \bar{y}) \leq \varphi(\bar{x}, \bar{y})$$
$$\leq \varphi(\bar{x}, \bar{y}) + \varphi'_x(\bar{x}, \bar{y})(x - \bar{x}) \leq \varphi(x, \bar{y})$$

which proves that $(\bar{x}, \bar{y})$ is a saddle-point of φ. The conclusion follows.

□

In our present linearly constrained optimization of a convex objective, we have $X = \mathbb{R}^n$, $Y = \mathbb{R}^p \times \mathbb{R}^n_+$, $\varphi = \mathcal{L}$, and the Lagrangian $\mathcal{L}$ is a convex-concave function of (x, λ, s). A saddle-point $(\bar{x}, \bar{\lambda}, \bar{s})$ is therefore characterized by

$$\begin{cases} \mathcal{L}'_x(\bar{x}, \bar{\lambda}, \bar{s})(x - \bar{x}) \geq 0, & \forall x \in \mathbb{R}^n, \\ \mathcal{L}'_{\lambda,s}(\bar{x}, \bar{\lambda}, \bar{s})(\lambda - \bar{\lambda}, s - \bar{s}) \leq 0, & \forall \lambda \in \mathbb{R}^p, \ s \in \mathbb{R}^n_+, \end{cases}$$

which, after some computations, appears to be equivalent to

$$\begin{cases} \mathcal{L}'_x(\bar{x}, \bar{\lambda}, \bar{s}) = \nabla f(\bar{x}) + A^\top \bar{\lambda} - \bar{s} = 0, \\ \mathcal{L}'_\lambda(\bar{x}, \bar{\lambda}, \bar{s}) = A\bar{x} - b = 0, \\ \bar{x} \geq 0 \text{ and } \bar{x}\bar{s} = 0; \end{cases}$$

here again, the product of two vectors is understood as componentwise. Altogether, we have shown

Theorem 19.10. *A saddle-point of $\mathcal{L}$ is characterized by what we will call the* optimality system

$$\begin{cases} \nabla f(\bar{x}) + A^\top \bar{\lambda} = \bar{s}, \\ A\bar{x} = b, \\ \bar{x} \geq 0, \ \bar{s} \geq 0, \ \bar{x}\bar{s} = 0. \end{cases}$$

If $(\bar{x}, \bar{\lambda}, \bar{s})$ is a saddle-point of the Lagrangian, we will say that $(\bar{\lambda}, \bar{s})$ is a *Lagrange multiplier* associated with $\bar{x}$. This terminology suggests that the multiplier appears in the Lagrangian through its product with the constraint. The set of Lagrange multipliers associated with a solution of (P) is the same for all solutions, since it is just $S(D)$.

Example 19.11. Consider the convex problem $\text{Min}\{e^{-x}; x \geq 0\}$. It has no solution. The primal and dual values are equal and the dual problem has the unique solution $\bar{\lambda} = 0$, which is therefore not a Lagrange multiplier.

Theorem 19.12. *If $S(P) \neq \emptyset$, the set of Lagrange multipliers is nonempty.*

Proof. Take $\bar{x} \in S(P)$ and $x \in F(P)$. Then, for $\sigma \in\,]0, 1[$

$$0 \leq \frac{f(\bar{x} + \sigma(x - \bar{x})) - f(\bar{x})}{\sigma}.$$

Pass to the limit; setting $\bar{c} := \nabla f(\bar{x})$, we obtain $\bar{c}^\top(x - \bar{x}) \geq 0$, for all $x \in F(P)$. This means that $\bar{x}$ solves the *linear* problem

$$\operatorname*{Min}_{x} \bar{c}^\top x; \quad Ax = b; \quad x \geq 0. \tag{$*$}$$

Thus, it suffices to show the existence of Lagrange multipliers for a linear problem. At $x \in F(P)$, denote the set of active constraints and the tangent cone to $F(P)$ respectively by

$$N(x) := \{i = 1, \cdots, n; \quad x_i = 0\},$$
$$T(x) := \{d \in \mathcal{N}(A); \quad d_i \geq 0, \ \forall i \in N(x)\}.$$

Let $d \in T(\bar{x})$. Then $\bar{x} + \sigma d \in F(P)$ for $\sigma > 0$ small enough, so that

$$\bar{c}^\top d \geq 0, \ \forall d \in T(\bar{x}).$$

We will obtain the multiplier as solving the quadratic problem

$$\operatorname*{Min}_{\lambda, s} g(\lambda, s) := \tfrac{1}{2}\|\bar{c} + A^\top \lambda - s\|^2; \quad s \geq 0, \ s_i = 0, \ i \notin N(\bar{x}).$$

This problem is feasible. Its objective function is everywhere nonnegative, so its value is finite. Being quadratic and convex, the problem has an optimal solution $(\bar{\lambda}, \bar{s})$ (Theorem 19.1). Let us show that

$$\bar{r} := \bar{c} + A^\top \bar{\lambda} - \bar{s}$$

is zero. The point $(\bar{\lambda}, \bar{s})$ minimizing g, it satisfies the optimality system

$$\begin{cases} 0 = g'_\lambda(\bar{\lambda}, \bar{s}) = A\bar{r}, \\ 0 \leq g'_s(\bar{\lambda}, \bar{s})(s - \bar{s}), \ \forall s \geq 0, \quad s_i = 0, i \notin N(x). \end{cases} \tag{$**$}$$

The second relation is equivalent to $\bar{r}_i \leq 0, i \in N(\bar{x})$ and $\bar{r}^\top \bar{s} = 0$, and hence,

$$0 \leq \|\bar{r}\|^2 = \bar{r}^\top \bar{r} = \bar{c}^\top \bar{r} + \bar{\lambda}^\top A\bar{r} - \bar{s}^\top \bar{r} = \bar{c}^\top \bar{r}. \tag{$***$}$$

Set $x(\sigma) := \bar{x} - \sigma\bar{r}$. Using $(**)$, we get

$$Ax(\sigma) = A\bar{x} - \sigma A\bar{r} = b,$$

and for $\sigma > 0$ small enough we have $x_i(\sigma) > 0$, $i \notin N(x)$, and $x_i(\sigma) \geq 0$, $i \in N(x)$ (since $\bar{r}_i \leq 0$, $i \in N(x)$), and hence, $x(\sigma) \in F(P)$. By $(***)$,

$$\bar{c}^\top x(\sigma) = \bar{c}^\top \bar{x} - \sigma \bar{c}^\top \bar{r} = \bar{c}^\top \bar{x} - \sigma \|\bar{r}\|^2.$$

Since $\bar{x}$ solves $(*)$, this implies $\bar{r} = 0$. The point $(\bar{\lambda}, \bar{s})$ therefore satisfies

$$\bar{c} + A^\top \bar{\lambda} = \bar{s} \ge 0,$$

and if $\bar{x}_i \ne 0$, then $i \notin N(x) \Rightarrow \bar{s}_i = 0$, and hence, $\bar{x}\bar{s} = 0$, so that $(\bar{\lambda}, \bar{s})$ is a Lagrange multiplier.

$\square$

If $S(P)$ is nonempty, the above result tells us that there exists a Lagrange multiplier; hence the Lagrangian has a saddle-point and the primal and dual values are equal. In the case of linear or quadratic optimization, we know that $S(QP)$ is nonempty iff $v(QP) \in \mathbb{R}$. Being a concave quadratic problem, the dual (QD) is equivalent to a convex quadratic problem whose dual is (P). Hence $S(QD) \ne \emptyset$ iff $v(QD) \in \mathbb{R}$; in this case it can be shown that, with the solutions of (D) are associated Lagrange multipliers, which solve (P). Since we always have $v(D) \le v(P)$, we deduce

Corollary 19.13. *There holds* $v(QP) = v(QD) \in \bar{\mathbb{R}}$, *except when*

$$v(QP) = +\infty \quad and \quad v(QD) = -\infty.$$

Example 19.14. The linear problem $\text{Min}_x\{-x; 0 \times x = -1; x \in \mathbb{R}_+\}$ is not feasible; neither is its dual $\text{Max}_{\lambda,s}\{\lambda; -1 + 0 \times \lambda = s; s \in \mathbb{R}_+\}$. We are in the situation where $-\infty = v(LD) < v(LP) = \infty$.

19.2.3 Other Formulations

With the primal and dual problems, can be associated the *primal-dual* or *mixed formulation*

$$\underset{x,\lambda,s}{\text{Min}}\, x^\top s; \quad Ax = b,\ x \ge 0,\ \nabla f(x) + A^\top \lambda = s,\ s \ge 0. \qquad (MP)$$

Its constraints are those appearing in the optimality conditions, and the complementarity condition is taken care of by the objective function. Since $x \ge 0$ and $s \ge 0$, the objective is nonnegative and therefore $v(MP) \ge 0$. If $(\bar{x}, \bar{\lambda}, \bar{s}) \in S(MP)$, then $\bar{x}$, $(\bar{\lambda}, \bar{s})$ is a primal-dual solution iff $v(MP) = 0$: indeed, the optimality system is satisfied iff $x^\top s = 0$.

In the linear or convex quadratic case, problem (MP) writes

$$\underset{x,\lambda,s}{\text{Min}}\, x^\top s; \quad Ax = b,\ x \ge 0,\ c + Hx + A^\top \lambda = s,\ s \ge 0, \qquad (MQ)$$

and we have: (i) the constraints of (MQ) are linear, and (MQ) is therefore a quadratic problem; (ii) if (MQ) is feasible, then $v(MQ) = 0$ and $S(MQ) \ne \emptyset$ by virtue of Corollary 19.13.

Another approach to duality, called *Wolfe duality* [360], consists in introducing the dual problem

$$\max_{x,\lambda,s} \mathcal{L}(x, \lambda, s); \quad \nabla f(x) + A^\top \lambda = s, \ s \geq 0. \tag{D^*}$$

Said otherwise, the Lagrangian is maximized under the "dual constraint". For a quadratic problem, the classical dual (QD) is recovered. More generally, $v(D^*) \leq v(D)$, and $v(D^*) = v(D) = v(P)$ if $S(P)$ is nonempty (since in this case, $\mathcal{L}$ has a solution which solves (D^*)).

Remark 19.15. Let (x, λ, s), feasible for Wolfe's dual, be such that x is feasible for (P). The difference between the associated costs is $f(x) - \mathcal{L}(x, \lambda, s) = x^\top s$. Problem (MP) is therefore interpreted as the minimization of the difference between primal and dual costs, under the primal feasibility and (Wolfe) dual constraints. □

19.2.4 Strict Complementarity

A primal-dual solution (x, λ, s) is called *strictly complementary* when $x + s > 0$, or equivalently $\max(x_i, s_i) > 0, \forall i \in \{1, \cdots, n\}$. This property plays an important role in the analysis of the asymptotic behavior of interior-point algorithms. A quadratic problem need not have any complementary solution, even when a solution exists; a counter-example is $\min\{x^2; \ x \geq 0\}$. The situation is different for linear optimization.

Proposition 19.16. *(Goldman-Tucker, [167]) If (LP) has a finite value, then it has a strictly complementary solution. More precisely, there exist a partition $\bar{B}$ and $\bar{N}$ of $\{1, \cdots, n\}$, and a primal-dual solution $(\bar{x}, \bar{\lambda}, \bar{s})$, such that $\bar{x}_{\bar{B}} > 0$ and $\bar{s}_{\bar{N}} > 0$, and if (x, λ, s) is a primal-dual solution, then $x_{\bar{N}} = 0$ and $s_{\bar{B}} = 0$.*

Proof. Define

$$\bar{B} := \{i \in \{1, \cdots, n\}; \ \exists x^i \in S(LP); \ x_i^i > 0\}; \quad \bar{N} := \{1, \cdots, n\} \backslash \bar{B}.$$

Since $S(LP)$ is convex, the point $\bar{x} := |\bar{B}|^{-1} \sum_{i \in \bar{B}} x^i$ lies in $S(LP)$ and satisfies $\bar{x}_{\bar{B}} > 0$. Consider the linear problem

$$\min_x \left(-\sum_{i \in \bar{N}} x_i \right); \ Ax = b; \ c^\top x \leq v(LP); \ x \geq 0. \tag{$*$}$$

Its feasible set is $F(*) = S(LP)$. By definition of $\bar{N}$, if $x \in S(LP)$, then $x_{\bar{N}} = 0$, hence $v(*) = 0$ and $S(*) = F(*) = S(LP)$.

The dual problem writes

$$\max_{(\lambda,\beta,s)} -\lambda^\top b - \beta v(LP); \quad -\sum_{i \in \bar{N}} e^i + A^\top \lambda + \beta c = s \geq 0, \quad \beta \geq 0,$$

where e^i is the ith basis vector and $(\lambda, \beta, s) \in \mathbb{R}^p \times \mathbb{R}_+ \times \mathbb{R}_+^n$ is the multiplier. Since $v(*) = 0$, there exists a dual solution $(\hat{\lambda}, \hat{\beta}, \hat{s})$. Take $(\lambda, s) \in S(LD)$; then

$$(1 + \hat{\beta})c + A^\top(\hat{\lambda} + \lambda) = \hat{s} + s + \sum_{i \in \bar{N}} e^i.$$

Set

$$\bar{\lambda} := (1 + \hat{\beta})^{-1}(\hat{\lambda} + \lambda) ; \quad \bar{s} := (1 + \hat{\beta})^{-1}(\hat{s} + s + \sum_{i \in \bar{N}} e^i).$$

Then

$$c + A^\top \bar{\lambda} = \bar{s} \geq 0; \quad \bar{s}_{\bar{N}} \geq (1 + \hat{\beta})^{-1} \sum_{i \in \bar{N}} e^i > 0.$$

Besides, $\bar{s} \geq 0$ and $\bar{x}\bar{s} = 0$ (indeed $\bar{x} \in S(*)$, hence $\bar{x}^\top \hat{s} = 0$); we deduce $(\bar{\lambda}, \bar{s}) \in S(LD)$. Finally, if (x, λ, s) is a primal-dual solution of (LP), the complementarity relation with $\bar{x}$ and $\bar{s}$ imply $x_{\bar{N}} = 0$ and $s_{\bar{B}} = 0$.

$\square$

We will call $(\bar{B}, \bar{N})$ given above the *optimal partition* of the problem.

19.3 The Simplex Algorithm

The classical method for solving linear problems is the so-called simplex method, which we now present. Despite the importance of interior-point algorithms, a study of the simplex method is useful for three reasons. First, this method is competitive in many cases, and gives a benchmark to assess the performances of other methods. Second, to compute an exact solution – which is on the boundary of the domain – interior-point methods need a so-called purification process, which is a variant of the simplex method. Finally, stating the simplex algorithm allows an illustration of various concepts intrinsically linked to linear problems themselves, and not to the techniques used for solving them, as is the case for the existence of basic points.

19.3.1 Computing the Descent Direction

Let $\tilde{x}$ be a basic point. Denote by

$$B(\tilde{x}) := \{1 \leq i \leq n; \tilde{x}_i > 0\}$$

the set of non-active positivity constraints; $B(\tilde{x})$ is called the basis. The set of columns of A, indexed in a set I, will be denoted by A_I. Set

$$B := B(x); \quad N := \{1, \cdots, n\} \backslash B.$$

We assume $|B| = p$. Without loss of generality, we can assume $B = \{1, \cdots, p\}$. Partition $x \in \mathbb{R}^n$ and A according to their indices and columns, so that

$$x = (x_B, x_N); \quad A = (A_B, A_N); \quad Ax = A_B x_B + A_N x_N.$$

We call A_B the *basis matrix*. Suppose A_B invertible. Then, from the relation $Ax = b$, we can extract x_B as a function of x_N:

$$x_B(x_N) := A_B^{-1}(b - A_N x_N).$$

To compute a descent direction, the constraints $x_B \geq 0$ which are not active at the current point $\tilde{x}$ can be ignored. Locally, our problem therefore writes (using $c^\top x = c_B^\top x_B + c_N^\top x_N$):

$$\operatorname*{Min}_{x_N} c_B^\top A_B^{-1}(b - A_N x_N) + c_N^\top x_N; \quad x_N \geq 0,$$

or equivalently

$$\operatorname*{Min}_{x_N} (c_N - A_N^\top A_B^{-t} c_B)^\top x_N; \quad x_N \geq 0.$$

We call *reduced cost* the quantity $r := c_N - A_N^\top A_B^{-t} c_B$. Decomposing its expression so as to disclose the *multiplier estimate* λ, we obtain

$$A_B^\top \lambda = -c_B; \quad r = c_N + A_N^\top \lambda.$$

Combining the above two equalities, we observe that

$$c + A^\top \lambda = s \quad \text{where} \quad s := \begin{pmatrix} 0 \\ r \end{pmatrix}, \quad \text{and} \quad x^\top s = 0.$$

In case $r \geq 0$, the optimality system is satisfied: (x, λ, s) is therefore a solution of (MP). From Theorem 19.10, x solves (LP). We can thus detect optimality.

By contrast, if there exists j such that $r_j < 0$, let e^j be the jth basis vector, and let $d \in \mathbb{R}^n$ be given by

$$d_N = e^j; \quad d_B = -A_B^{-1} A_N d_N = -A_B^{-1} A_j.$$

Then

$$Ad = 0; \quad d_N \geq 0; \quad d^\top c = d^\top r = r_j < 0,$$

and thus, for small $\rho > 0$, the point $x(\rho) := x + \rho d$ is feasible and $c^\top x(\rho) := c^\top x + \rho c^\top d < c^\top x$. Since x has p nonzero components, and since $d_N = e^j$, $x(\rho)$ has at most $p + 1$ nonzero components. Consider the largest feasible step. If it is $+\infty$, we deduce $v(LP) = -\infty$ since $c^\top d < 0$. If this step is finite, it cancels one component (at least). A new basic point is thus obtained.

19.3.2 Stating the algorithm

Summing up the set of operations, we obtain the simplex algorithm:

Algorithm 19.17. (Simplex algorithm)

Data: choose a basic point x^0; $k \leftarrow 0$.

REPEAT

- Compute the multiplier, solving $A_B^\top \lambda = -c_B$.
- Compute the reduced cost $r \leftarrow c_N + A_N^\top \lambda$.
- If $r \geq 0$, stop: x solves (LP).
- Direction of move: j such that $r_j < 0$, $d_N \leftarrow e^j$; $d_B \leftarrow -A_B^{-1} A_N d_N$.
- If $d \geq 0$, stop: $v(LP) = -\infty$.
- Stepsize: $i \leftarrow \operatorname{argmin}\{\frac{x_i}{|d_i|}; \ d_i < 0\}$, and $\rho \leftarrow \dfrac{x_i}{|d_i|}$.
- New point $x^\sharp \leftarrow x + \rho d$. $B \leftarrow (B \cup \{j\}) \backslash \{i\}$; $N \leftarrow (N \cup \{i\}) \backslash \{j\}$.
- $k \leftarrow k + 1$.

Remark 19.18. Call j the variable index entering the basis, and i the one exiting the basis. A classical choice is $j = \operatorname{argmin}\{r_j\}$. Note that this choice depends on the scaling of the variables. $\qquad\square$

The basic point x will be called a *regular basic point* when

$$|B(x)| = p \text{ and } A_{B(x)} \text{ has rank } p.$$

As long as the basic points generated by the algorithm are regular, this algorithm is well-defined. A singularity may appear if several basic variables reach their bound simultaneously. On the other hand, the following lemma shows that the basis matrix never becomes singular (at least in exact arithmetic; if the computations are performed in finite arithmetic, a singularity may of course appear).

Lemma 19.19. *The basis matrix cannot become singular along the algorithm iterations.*

Proof. Let $A^\flat$ and $A^\sharp$ be the basis matrix at the beginning and at the end of an iteration. We have to show that $A^\sharp$ is invertible if $A^\flat$ is such. We can assume that the changed column is the last one. Let $A_n^\sharp$ bet the last column of $A^\sharp$, and $d \neq 0$ be such that $A^\sharp d = 0$. Set $\tilde{d} := (d_1, \cdots, d_{n-1}, 0)$. Then

$$A^\sharp d = A^\flat \tilde{d} + A_n^\sharp d_n = 0.$$

Since $A^\flat$ is invertible, this implies $d_n \neq 0$; we can assume $d_n = 1$, so

$$0 = A^\sharp d = A^\flat \tilde{d} + A_n^\sharp.$$

During one simplex iteration, we have computed a direction of move $\bar{d}$ solving $A^\flat \bar{d} = -A_n^\sharp$. Since $A^\flat$ is invertible, this implies $\bar{d} = \tilde{d}$, hence $\bar{d}_n = 0$, which is impossible: the component of displacement associated with the variable exiting the basis cannot be zero.

$\qquad\square$

Theorem 19.20. *If the basic point x^0 is regular, and if the variable exiting the basis is defined unambiguously, then the simplex algorithm stops after finitely many iterations.*

Proof. In view of the above lemma, $\{x^k\}$ is a sequence of regular basic points. Then it suffices to show that the number of regular basic points is finite. If they are infinitely many, two of them are different, say $x^\sharp$ and $x^\flat$, associated with the same basis B. Hence $A_B x_B^\sharp = A x^\sharp = b = A x^\flat = A_B x_B^\flat$, so that $A_B(x_B^\sharp - x_B^\flat) = 0$, hence $x_B^\sharp = x_B^\flat$. Set $N := \{1, \cdots, n\} \backslash B$. Since $x_N^\sharp = x_N^\flat = 0$, it follows that $x^\sharp = x^\flat$, a contradiction with the initial hypothesis. $\qquad \square$

In general, no basic point of the problem to solve is available. Actually, it is not even known whether this problem is feasible! Nevertheless, computing a basic point can be formulated as an auxiliary problem, of which a basic point is known. For example, consider the linear problem

$$\operatorname*{Min}_{x,z} \sum_{i=1}^{p} z_i; \quad Ax + wz = b; \quad x \geq 0; \quad z \geq 0, \tag{$*$}$$

and take

$$x^0 \in \mathbb{R}_+^n; \quad w := b - A x^0; \quad z^0 := \mathbb{1} \quad \text{(in } \mathbb{R}^p).$$

Then the point (x^0, z^0) is feasible for $(*)$. In particular, chosing $x^0 = 0$, this point is basic. Finally, it is clear that $v(*) \geq 0$, and that $v(*) = 0$ iff (P) is feasible. Moreover, the simplex algorithm yields a basic point for (P), which allows the initialization of a simplex algorithm to solve (LP). We call Phase I and Phase II the successive steps, seeking a feasible point, and then minimizing the objective function.

19.3.3 Dual simplex

Sometimes one has to solve a sequence of linear programs, the kth problem being identical to the $(k-1)$th except for some additional linear inequalities. This happens for instance when integer linear programming problems are solved by using continuous relaxations and integrality cuts. Then a (primal-dual) solution of the $(k-1)$th problem is dual feasible for the kth problem. It is therefore useful to have a version of the simplex algorithm for the dual problem, since we already know a feasible starting point. We briefly give the main ideas of the dual simplex algorithm, and show that the implementation is again based on a primal (B, N) partition. The dual program

$$\operatorname*{Max}_{\lambda,s} -b^\top \lambda; \quad c + A^\top \lambda = s; \quad s \geq 0$$

has $n + p$ variables and n constraints. The multiplier λ will be in the dual basis, and cannot leave it since it is unbounded. There are $n - p$ components of s in the dual basis. For reasons that will appear soon, denote by N (resp. B) the set of *basic* (resp. *nonbasic*) components of s. Observe that $|B| = p$ and $|N| = n - p$. With a dual feasible (λ, s), the dual simplex algorithm associates the estimate x of the multiplier of the affine constraint, obtained by expressing that the Lagrangian

$$L(x, \lambda, s) = -b \cdot \lambda + x \cdot (c + A^\top \lambda - s) = c \cdot x + \lambda \cdot (Ax - b) - s \cdot x$$

has zero partial derivatives with respect to basic variables, i.e., $Ax = b$ and $x_N = 0$. In other words, x is the (primal, not necessarily feasible) basic solution of $A_B x_B = b$. The reduced gradient with respect to the nonbasic variables s_B is nothing but $-x_B$. So the entering variable may be the most negative component of x_B.

The next step is to compute the direction of move of basic variables when (nonbasic) component $i \in B$ enters the basis. We have to solve in (d^λ, d_N^s) the equations $A^\top d^\lambda = d_N^s + e_i$, where here we identify d_N^s and its extension by zero over $\mathbb{R}^n$, and e_i is the ith basis vector in $\mathbb{R}^n$. For that, compute first d^λ solution of $A_B^\top d^\lambda = e_i$, from which $d_N^s = A_N^\top d^\lambda$ follows. The algorithm performs then the greatest move in direction (d^λ, d_N^s); it stops when the primal estimate is feasible.

We see that, as for the primal simplex method, the main operations are the factorization and the update of the (primal) basis matrix A_B, and that the dual simplex method can be interpreted as computing a sequence of "non feasible basic points", and trying to reduce the primal infeasibility while keeping dual feasibility.

19.4 Comments

The (small) part of Lagrangian duality theory exposed here, also called min-max duality, was developed by Von Neumann (see von Neumann and Morgenstein [274]). Another approach, based on perturbations of the optimization problem can be found in Rockafellar [309] and Ekeland and Temam [117]. Bonnans and Shapiro [50] give an overview of some generalizations of linear programming to infinite dimensional spaces.

The simplex algorithm is due to Dantzig [94]. Often, A is sparse, and a sparse LU decomposition (Gauss method) of the basis matrix is used to compute λ and d_B. The operation of updating the factors from one iteration to the next one is called *pivoting*. As only one column is changed at each iteration, this can be done in $O(p^2)$ operations (compared to $O(p^3)$ for the factorization). A whole literature is devoted to this subject (Reid [304]).

The process of updating the LU factors of the basis has two shortcomings: an increase of the memory requirement to store the factorization of the basis matrix, and a degradation of their numerical stability. After some tens or hundreds of pivotings, it may therefore become necessary to re-factorize the basis matrix. It is important to control at each iteration the accuracy of the direction and multipliers, in order to re-factor at appropriate times.

Most softwares perform their computations with floating-point numbers, of length *apriori* fixed in the computer. In these conditions, everything can happen! In particular, a certain basic variable (other than the one exiting the basis) may take a slightly negative value. To remedy this situation, one can either decide to get this variable off the basis, or to continue the algorithm, after including in the objective a penalty term of the negative part of the basic variables.

A natural extension of the simplex algorithm to quadratic problems, or more generally linearly constrained problems, is the reduced-gradient method (Luenberger [239]), which partitions the variables in three sets: a basis, guaranteeing satisfaction of the linear constraints, the variables called nonbasic, stuck to their bound, and the remaining variables, called superbasic. Just as the simplex method, this type of algorithm has the advantage of factoring the basis matrix only, and the drawback of requiring many iterations, in order to identify the active constraints at the optimum. In applications, one may have to solve a sequence of optimization problems slightly different from each other: such is the case, for example, in the study of the sensitivity of the solution with respect to a small variation of the data. Then, reduced-gradient methods can be very efficient.

20 Linear Monotone Complementarity and Associated Vector Fields

Overview

We develop the theoretical tools necessary for the algorithms to follow. The logarithmic penalty technique allows the introduction of the central path. In the case of linear or quadratic optimization, the optimality system and the central path are suitably cast into the framework of linear monotone complementarity problems.

The analysis of linear monotone complementarity problems starts with some results of global nature, involving the partition of variables, standard and canonical forms. Then comes a discussion on the magnitude of the variables in a neighborhood of the central path. We introduce two families of vector fields associated with the central path: the affine and centralization directions. The magnitude of the components of these fields are analyzed in detail. We also discuss the convergence of the differential system obtained by a convex combination of the affine and centralization directions.

Since the results stated here are motivated by the analysis of algorithms presented afterwards, a quick reading is sufficient for a first step. Besides, a large part of the technical difficulties are due to the modified field theory, useful for problems without strict complementarity. A reader interested mainly by linear optimization (where strict complementarity always holds) can therefore skip this part.

20.1 Logarithmic Penalty and Central Path

20.1.1 Logarithmic Penalty

We call *logarithmic potential* the strictly convex function

$$\pi(x) := -\sum_{i=1}^{n} \log x_i$$

defined on $\{x \in \mathbb{R}^n;\ x > 0\}$. Consider the linearly constrained minimization problem

$$\operatorname*{Min}_{x} f(x); \quad Ax = b,\ x \geq 0, \tag{P}$$

with $f : \mathbb{R}^n \to \mathbb{R}$ convex and C^1, and A a $p \times n$ matrix of rank p. When $f(x) = c^\top x + \frac{1}{2} x^\top H x$, we find again a quadratic program in standard form. The problem with logarithmic penalty associated with (P) is

$$\operatorname*{Min}_x f(x) + \mu \pi(x); \quad Ax = b, \ x > 0, \qquad (\mathcal{P}_\mu)$$

where $\mu > 0$ is the penalty parameter. Under certain assumptions (not specified here), problem (P_μ) can be shown to have a unique solution, which converges to a solution of (P) when $\mu \downarrow 0$. Solving approximately a sequence of problems (P_μ), with $\mu \downarrow 0$, therefore yields a means of solving (P) approximately.

20.1.2 Central Path

Since (P_μ) is a linearly constrained convex problem, x solves (P_μ) iff there exists $\lambda \in \mathbb{R}^p$ such that (x, λ) solves the first-order optimality system

$$\begin{cases} \nabla f(x) - \mu x^{-1} + A^\top \lambda = 0, \\ Ax = b, \ x > 0, \end{cases} \qquad (20.1)$$

with $x^{-1} := (x_1^{-1}, \cdots, x_n^{-1})^\top$. Set $s := \mu x^{-1}$, then $xs = \mu \mathbf{1}$ and the optimality system is equivalent to

$$\begin{cases} xs = \mu \mathbf{1}, \\ \nabla f(x) + A^\top \lambda = s, \\ Ax = b, \\ x \geq 0, \ s \geq 0. \end{cases} \qquad (20.2)$$

For $\mu = 0$ we obtain the optimality system associated with (P). Note that $\pi(x)$ is strictly convex: (P_μ) has therefore one solution x^μ at most. The solution set of (20.2) has the form $(x^\mu, s^\mu, \lambda^\mu)$ with $s^\mu = \mu (x^\mu)^{-1}$. Moreover A has rank p, is therefore onto and $\mathcal{N}(A^\top) = \mathcal{R}(A)^\perp = \{0\}$, so (20.2) defines λ^μ in a unique way.

In the sequel, our analysis will involve (x^μ, s^μ). A way of eliminating λ consists in noting that the second relation of (20.2) is equivalent to

$$\nabla f(x) - s \in \mathcal{R}(A^\top) = \mathcal{N}(A)^\perp.$$

Since A has rank p, $\mathcal{N}(A)$ has dimension $n - p$. Let M be an $n \times (n - p)$ matrix, whose columns form a basis of $\mathcal{N}(A)$; then

$$\nabla f(x) - s \in \mathcal{R}(A^\top) \Leftrightarrow M^\top (\nabla f(x) - s) = 0.$$

Relation (20.2) is therefore equivalent to

$$\begin{cases} xs = \mu \mathbf{1}, \\ M^\top (\nabla f(x) - s) = 0, \\ Ax = b, \\ x \geq 0, \ s \geq 0. \end{cases} \qquad (20.3)$$

Note that, in the case of a quadratic objective, the second relation of (20.2) writes $c + Hx + A^\top \lambda = s$. Then (20.3) reduces to

$$
\begin{cases}
xs = \mu\mathbf{1}, \\
M^\top(c + Hx - s) = 0, \\
Ax = b, \\
x \geq 0, \quad s \geq 0.
\end{cases}
\tag{20.4}
$$

The lemma below makes precise to which extent a solution of (20.3) approximates a solution of the associated minimization problem.

Lemma 20.1. *If (x, s) solves (20.3), then $f(x) \leq v(P) + n\mu$. More generally, if (x, s) only satisfies the last three relations of (20.3), then $f(x) \leq v(P) + x^\top s$.*

Proof. The dual problem of (P) is written in terms of the Lagrangian

$$
\operatorname*{Max}_{\substack{\lambda \\ s \geq 0}} \inf_x \; f(x) + \lambda^\top (Ax - b) - s^\top x.
\tag{D}
$$

This Lagrangian is convex, it is minimal when $\nabla f(x) + A^\top \lambda = s$, a relation satisfied by any triple (x, s, λ) for which the last three relations of (20.3) hold; so

$$
f(x) + (\lambda)^\top (Ax - b) - (s)^\top x \leq v(D) \leq v(P).
$$

Since $Ax - b = 0$, we have that $f(x) - v(P) \leq x^\top s$. If, in addition, the first relation of (20.3) is satisfied, then $x^\top s = n\mu$, and the conclusion follows. □

We call *primal central path* (resp. *primal-dual, dual*), the set of primal (resp. primal-dual, dual) solutions x^μ (resp. (x^μ, s^μ), s^μ) to (20.3).

In figure 20.1 we represent the primal central path for the problem of minimizing $x_2 + \varepsilon x_1$ subject to $x_1 \in [0, 1]$ and $x_2 \geq 0$, both for $\varepsilon = 0$ (half-line $x_1 = 1/2$ and $x_2 \geq 0$) and for $\varepsilon = 0.04$. When $\varepsilon > 0$ the primal central path converges to the point $(0, 0)$. Therefore, when $\varepsilon > 0$ is small, there is a *sharp turn* close to the point $(\frac{1}{2}, 0)$. We will see that efficient (central) path-following algorithms exist, despite the existence of such sharp turns.

20.2 Linear Monotone Complementarity

We introduce a structure generalizing optimality systems of quadratic problems. It will ease our study of the algorithms to come.

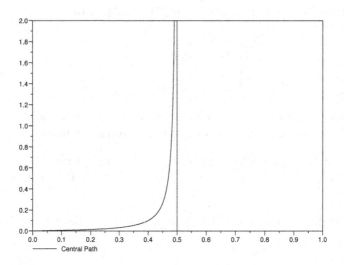

Fig. 20.1. A central path with sharp turn

20.2.1 General Framework

We call *linear complementarity problem* the following set of relations:

$$\begin{cases} xs = 0, \\ Qx + Rs = h, \\ x \geq 0, \quad s \geq 0, \end{cases} \qquad (LCP)$$

with Q and R matrices $n \times n$ and $h \in \mathbb{R}^n$. The problem is said to be *monotone* if

$$Qu + Rv = 0 \Rightarrow u^\top v \geq 0.$$

In what follows, we will always suppose that (Q, R) satisfies this property. The *feasible set* of (LCP) is

$$F(LCP) := \{x \in \mathbb{R}^n_+, \ s \in \mathbb{R}^n_+; \quad Qx + Rs = h\},$$

and its *solution set* is

$$S(LCP) := \{x \in F(LCP); \ xs = 0\}.$$

We will say that (x, s) is an *interior point* if $(x, s) \in F(LCP)$, $x \in \mathbb{R}_{++}$ and $s \in \mathbb{R}_{++}$.

Let us show that a convex quadratic problem is a particular case of linear monotone complementarity. Relation (20.4) does include n inequality relations. Moreover, the associated homogeneous relation

$$M^\top(Hu - v) = 0; \quad Au = 0$$

is equivalent to $(Hu - v \perp \mathcal{N}(A); \quad Au = 0)$, hence it implies $0 = u^\top(Hu - v)$, i.e. $u^\top v = u^\top Hu \geq 0$, and monotonicity follows. Note that in the case of linear optimization, u and v are orthogonal: indeed $u \in \mathcal{N}(A)$ and $v \in \mathcal{R}(A^\top) = \mathcal{N}(A)^\perp$.

Lemma 20.2. *The solution set of a linear monotone complementarity problem is convex.*

Proof. Let $(x^\sharp, s^\sharp)$ and $(x^\flat, s^\flat)$ be two solutions, $\alpha \in]0, 1[$, and

$$(x, s) := \alpha(x^\sharp, s^\sharp) + (1 - \alpha)(x^\flat, s^\flat).$$

We have to show $(x, s) \in S(LCP)$. We obtain easily $Qx + Rs = h$, $x \geq 0$ and $s \geq 0$, there remains to show $xs = 0$. From

$$Q(x^\sharp - x^\flat) + R(s^\sharp - s^\flat) = 0,$$

we deduce $(x^\sharp - x^\flat)^\top(s^\sharp - s^\flat) \geq 0$, or equivalently

$$(x^\sharp)^\top s^\flat + (x^\flat)^\top s^\sharp \leq (x^\sharp)^\top s^\sharp + (x^\flat)^\top s^\flat.$$

Since $(x^\sharp, s^\sharp)$ and $(x^\flat, s^\flat)$ solve (LCP), the right-hand side is zero. Because the left-hand side is a sum of nonnegative terms, each term is zero; thus $x^\sharp s^\flat = x^\flat s^\sharp = 0$. Developing xs and using the expressions of x and s, we do obtain $xs = 0$.

$\square$

The *central path* for (LCP) is defined as the set of (x, s) such that, for some $\mu > 0$,

$$\begin{cases} xs = \mu \mathbf{1}, \\ Qx + Rs = h, \\ x \geq 0, \ s \geq 0. \end{cases}$$

It is easy to check that this definition extends the one given previously for convex optimization problems. We define the *proximity or centrality measure* as being

$$\delta(x, s, \mu) := \left\| \frac{xs}{\mu} - \mathbf{1} \right\|$$

where $\| \ \|$ is the Euclidean norm. The central path is therefore the set of points (x, s) feasible for (LCP), of zero proximity for some μ. The (Euclidean) *small neighborhood* of size $\alpha > 0$ is defined as being

$$\mathcal{V}_\alpha := \left\{ (x, s, \mu); \quad (x, s) \in F(LCP), \ \mu > 0; \quad \left\| \frac{xs}{\mu} - \mathbf{1} \right\| \leq \alpha \right\}. \quad (20.5)$$

It is also useful to consider *large neighborhoods* of the type

$$\mathcal{N}_\epsilon := \left\{ (x, s, \mu); \quad (x, s) \in F(LCP); \quad \mu > 0; \quad \epsilon \mathbf{1} \le \frac{xs}{\mu} \le \epsilon^{-1}\mathbf{1} \right\}, \quad (20.6)$$

with $0 < \epsilon \le 1/2$. The next proposition shows that the central path of a linear complementarity problem can be interpreted as the locus of minima of the mixed potential

$$\textstyle\prod(x, s) := x^\top s - \mu \sum_{i=1}^{n} \log x_i s_i = x^\top s + \mu\pi(x) + \mu\pi(s)$$

on the interior of the feasible set.

Proposition 20.3. *Every point (x^μ, s^μ, μ) of the central path is the unique solution of the mixed centralization problem*

$$\operatorname*{Min}_{x,s} \textstyle\prod(x, s); \quad Qx + Rs = h, \ x > 0, \ s > 0.$$

Proof. a) We claim that the above mixed centralization problem has at least one solution whenever it is feasible (if not, then obviously the central path is empty). Indeed, we have that $\prod(x, s) = \sum_{i=1}^{n}(x_i s_i - \mu \log(x_i s_i))$, while the function $t \to t - \mu \log t$ is bounded from below, and goes to $+\infty$ when $t \downarrow 0$ or $t \uparrow +\infty$. It is then a simple exercise to prove that a minimizing sequence (x^k, s^k) is bounded and none of the components of x_k or s_k has a zero limit-point. Therefore, by standard arguments, (x^k, s^k) has at least one limit-point, and the latter is solution of the mixed centralization problem.
b) Let (x, s) be solution of the mixed centralization problem. We claim that $xs = \mu\mathbf{1}$. Indeed, there exists λ such that the following optimality system is satisfied:

$$\nabla \textstyle\prod(x, s) + (Q\ R)^\top \lambda = 0; \quad Qx + Rs = h, \ x > 0, \ s > 0,$$

and therefore

$$(*) \qquad \begin{cases} s - \mu x^{-1} + Q^\top \lambda = 0, \\ x - \mu s^{-1} + R^\top \lambda = 0. \end{cases}$$

If $\lambda = 0$, we have $s = \mu x^{-1}$, i.e. $xs = \mu\mathbf{1}$, and we are done. We now show that $\lambda = 0$. Multiplying the first (resp. second) relation of $(*)$ by $d = \sqrt{x/s}$ (resp. $d^{-1} = \sqrt{s/x}$) and setting $D = \operatorname{diag}(d)$, we get:

$$\left.\begin{matrix} \sqrt{xs} - \mu/\sqrt{xs} + DQ^\top \lambda = 0 \\ \sqrt{xs} - \mu/\sqrt{xs} + D^{-1}R^\top \lambda = 0 \end{matrix}\right\} \Rightarrow (QD - RD^{-1})^\top \lambda = 0.$$

Then it suffices to show that $(QD - RD^{-1})$ is invertible. Take $\bar{u} \in \mathcal{N}(QD - RD^{-1})$. Set $u := d\bar{u}$, $v := -d^{-1}\bar{u}$. Then $Qu + Rv = 0$, hence $0 \le u^\top v = -\|\bar{u}\|^2$, so $\bar{u} = 0$; the claim follows.

c) We know that there exists $(\hat{x}, \hat{s})$, solution of the mixed centralization problem, that satisfies $\hat{x}\hat{s} = \mathbf{1}$. We end the proof of the lemma by checking that, if (x, s, μ) satisfies the equation of the central path, then $x = \hat{x}$ and $s = \hat{s}$. Indeed, we have that

$$0 \le (\hat{x} - x)^\top (\hat{s} - s) = 2n\mu - x^\top \hat{s} - s^\top \hat{x},$$

or equivalently, eliminating $\hat{s}$ and s from the central path relations,

$$\sum_{i=1}^{n} \left(\frac{x_i}{\hat{x}_i} + \frac{\hat{x}_i}{x_i} \right) = 2n.$$

Since $(t, t') \to t/t' + t'/t$ is, if t and t' are positive, always greater or equal 2, with equality only if $t = t'$, we deduce that $x = \hat{x}$, and hence, $s = \hat{s}$.   □

Remark 20.4. Solving a linear complementarity problem amounts to solving the quadratic problem

$$\operatorname*{Min}_{x,s} x^\top s; \quad Qx + Rs = h, \ x \ge 0, \ s \ge 0.$$

The potential $\prod$ is interpreted as the objective with logarithmic penalty associated with the above problem.   □

Remark 20.5. If a linear complementarity problem corresponds to the optimality conditions of the quadratic problem

$$\operatorname{Min} c^\top x + \tfrac{1}{2} x^\top H x; \quad x = b; \quad x \ge 0,$$

then from Remark 19.15,

$$\prod(x, s) = c^\top x + \tfrac{1}{2} x^\top H x - (-b^\top \lambda - \tfrac{1}{2} x^\top H x) + \mu\pi(x) + \mu\pi(s)$$
$$\left(c^\top x + \tfrac{1}{2} x^\top H x + \mu\pi(x) \right) + \left(b^\top \lambda + \tfrac{1}{2} x^\top H x + \mu\pi(s) \right)$$

is interpreted as the difference between the primal and dual objective with logarithmic penalty. In the linear case $(H = 0)$, the primal and dual penalized objectives are uncoupled.   □

20.2.2 A Group of Transformations

Let M be an invertible matrix. Clearly, (LCP) is not changed if it is rewritten as

$$\begin{cases} xs = 0, \\ MQx + MRs = Mh, \\ x, s \ge 0. \end{cases}$$

Nothing is changed either when permuting rows (same permutation applied to the rows of Q and R, and to the components of h) or columns (same permutation applied to the columns of Q and R, and to the components of x and s). Let us introduce another family of transformations, consisting in permuting some components of x and s. More precisely, let (I, J) be a partition of $\{1, \cdots, n\}$. We study the mapping $(x, s) \rightarrow (x', s')$, defined by permuting the components of x and s indexed in J; said otherwise,

$$x'_I = x_I, \; x'_J = s_J; \qquad s'_I = s_I, \; s'_J = x_J.$$

Partition Q and R according to their columns in Q_I, Q_J and R_I, R_J, so that

$$Qx = Q_I x_I + Q_J x_J; \quad Rs = R_I s_I + R_J s_J.$$

Let Q' and R' be two $n \times n$ matrices, defined by the relation

$$Q'x' = Q_I x'_I + R_J x'_J; \quad R's' = R_I s'_I + Q_J s'_J.$$

Consider the problem

$$\begin{cases} x's' = 0, \\ Q'x' + R's' = h, \\ x', s' \geq 0. \end{cases} \qquad (LCP')$$

Then (LCP') is the image of (LCP) under the above component permutation, in the sense that (x, s) is feasible (resp. solution) of (LCP) iff its image is feasible (resp. solution) for (LCP'). Besides, the small and large neighborhoods of the central path (LCP') are the images of the neighborhood of size $\alpha > 0$ of (LCP).

On the set of linear monotone complementarity problems of dimension n, parameterized by (Q, R, h), consider the transformations obtained by permuting rows, columns, component-exchanges of x and s, and constraint-composition by an invertible matrix. These transformations form a *group*, which we will denote by $\mathcal{T}$.

In Chapter 21, we study algorithm families that are *invariant* under this type of transformation: more precisely, the algorithms applied to one form or the other generate directions of move and sequences of points which are invariant. These algorithms are therefore *invariants* of the group, and they can be seen as defined on *equivalence classes* of the group (sets of problems in mutual correspondence through a transformation).

20.2.3 Standard Form

The *standard form* of a linear complementarity problem (which has nothing to do with the standard form of linear constraints) is defined as follows:

$$\begin{cases} xs = 0, \\ s = Mx + q, \\ x, s \geq 0. \end{cases} \qquad (SLCP)$$

We will say that a transformation $\tau \in \mathcal{T}$ reduces an (LCP) problem in standard form if the image of (LCP) by τ is in standard form. We will check that any problem of the type (LCP) can be reduced to standard form.

Lemma 20.6. *With any problem of the type (LCP) is associated a transformation $\tau \in \mathcal{T}$ which reduces this problem to standard form.*

Proof. If R is invertible, reducing (LCP) to standard form is done by taking $M := -R^{-1}Q$ and $q := R^{-1}h$. To obtain the conclusion, it therefore suffices to prove that appropriate transformations in $\mathcal{T}$ can make R invertible.

Let r be the rank of Q. Reordering row and columns if necessary, Q can be factored to $Q = LU$, with L invertible and U made as follows: its last $n - r$ rows are identically zero, and its r first contain an invertible block U_1 of size $r \times r$:

$$Q = L \begin{pmatrix} U_1 & U_2 \\ 0 & 0 \end{pmatrix}.$$

Set $D := L^{-1}R$. We denote by u^1 and u^2 the first r and last $n-r$ components of u, and likewise for v. The relation $Qu + Rv = 0$ is equivalent to:

$$\begin{cases} U_1 u^1 + U_2 u^2 + D_{11} v^1 + D_{12} v^2 = 0, \\ \qquad\qquad\qquad\quad D_{21} v^1 + D_{22} v^2 = 0, \end{cases}$$

where D has been partitioned in blocks. Let us check that D_{22} is invertible. Take arbitrary $v^1 = 0$, v^2 in $\mathcal{N}(D_{22})$, and $u^2 = -v^2$. Since U_1 is invertible, there exists u^1 such that $Qu + Rv = 0$. The monotonicity relation gives $0 \leq u^\top v = -\|v^2\|^2$, hence $v^2 = 0$. Because the square matrix D_{22} is invertible, the relation $Qu + Rv = 0$ is equivalent to

$$\begin{pmatrix} u^1 \\ v^2 \end{pmatrix} = - \begin{pmatrix} U_1 & D_{12} \\ 0 & D_{22} \end{pmatrix}^{-1} \begin{pmatrix} U_2 u^2 + D_{11} v^1 \\ D_{21} v^1 \end{pmatrix}.$$

Taking the new s as (x^1, s^2), the result is obtained. $\qquad\qquad\square$

If an algorithm is invariant with respect to the transformations of $\mathcal{T}$, its study can therefore be limited to the standard form.

20.2.4 Partition of Variables and Canonical Form

We now introduce the notion of optimal partition for a monotone linear complementarity problem.

Lemma 20.7. *If $S(LCP) \neq \emptyset$, then there exists a partition (B, N, T) of $\{1, \cdots, n\}$ such that, for all $(x, s) \in S(LCP)$, the following holds:*

$$x_i = 0, i \in N \cup T, \text{ and } s_j = 0, j \in B \cup T.$$

In addition, there exists $(\bar{x}, \bar{s}) \in S(LCP)$ such that $\bar{x}_B > 0$ and $\bar{s}_N > 0$.

Proof. Define (B, N, T) by

$$
\begin{aligned}
B &:= \{i \in \{1, \cdots, n\}; \quad \exists (x, s) \in S(LCP); \quad x_i > 0\}, \\
N &:= \{j \in \{1, \cdots, n\}; \quad \exists (x, s) \in S(LCP); \quad s_j > 0\}, \\
T &:= \{1, \cdots, n\} \backslash (B \cup N).
\end{aligned}
$$

With each $i \in B$ is associated $(x^i, s^i) \in S(LCP)$ such that $x_i^i > 0$. With each $j \in N$ is associated likewise $(x^j, s^j) \in S(LCP)$ such that $s_j^j > 0$. Since $S(LCP)$ is convex (Lemma 20.2), it contains the point

$$
(\bar{x}, \bar{s}) := \frac{1}{|B| + |N|} \left(\sum_{i \in B} (x^i, s^i) + \sum_{j \in N} (x^j, s^j) \right).
$$

This point satisfies $\bar{x}_B > 0$ and $\bar{s}_N > 0$. From $\bar{x}\bar{s} = 0$ we deduce $B \cap N = \emptyset$, which implies that (B, N, T) is a partition of $\{1, \cdots, n\}$. If $(x, s) \in S(LCP)$, we have by definition of B, N and T, that $x_i = 0, i \notin B$, and $s_j = 0, j \notin N$; the result follows.

$\square$

We will say that the problem is in *canonical form* if $N = \emptyset$. A permutation of components of x and s in N can reduce the problem to this form. Note that a problem in canonical form need not be in standard form!

We say that (LCP) satisfies the *strict complementarity* assumption if $T = \emptyset$. In view or Lemma 20.7, this is equivalent to the existence of a *strictly complementary* solution, i.e. an $(\bar{x}, \bar{s}) \in S(LCP)$ such that $\bar{x}_i > 0$ or $\bar{s}_i > 0$, for all $i \in \{1, \cdots, n\}$, or also $\bar{x} + \bar{s} > 0$. In the case of linear optimization, this hypothesis always holds (Goldman-Tucker Theorem 19.16).

20.2.5 Magnitudes in a Neighborhood of the Central Path

Let us start with a compactness result. Recall that (x, s) is an *interior point* of (LCP) when $(x, s) \in F(LCP)$, $x > 0$ and $s > 0$.

Theorem 20.8. *Suppose (LCP) has an interior point. Then*
(i) for all $\eta \geq 0$, the set $\{(x, s) \in F(LCP); x^\top s \leq \eta\}$ is bounded;
(ii) $S(LCP)$ is nonempty and bounded.

Proof. (i) Call $(\hat{x}, \hat{s})$ the interior point, and

$$
\beta := \min\{\hat{x}_i, \hat{s}_j, 1 \leq i, j \leq n\}.
$$

If $(x, s) \in F(LCP)$, we deduce from $(x - \hat{x})^\top (s - \hat{s}) \geq 0$ that

$$\beta \left(\sum_{i=1}^{n} x_i + \sum_{j=1}^{n} s_j \right) \le \hat{x}^\top s + \hat{s}^\top x \le x^\top s + \hat{x}^\top \hat{s}.$$

But $x \ge 0$ and $s \ge 0$, and hence, the set $\{(x, s) \in F(LCP); \; x^\top s \le \eta\}$ is bounded.

(ii) Taking $\eta = 0$, it follows from (i) that $S(LCP)$ is bounded. Let us show that it is nonempty. Set

$$\bar{\mu} := \inf\{\mu \ge 0; \quad \exists (x, s) \in F(LCP); \quad xs = \mu \hat{x} \hat{s}\}.$$

Then $0 \le \bar{\mu} \le 1$, since from $\bar{\mu} = 1$, the point $(\hat{x}, \hat{s})$ lies in the above set. From (i), if $\mu_k \downarrow \bar{\mu}$ and $(x^k, s^k) \in F(LCP)$ satisfies $x^k s^k = \mu_k \hat{x} \hat{s}$, then (x^k, s^k) is bounded, and therefore has a cluster point $(\bar{x}, \bar{s}) \in F(LCP)$ satisfying $\bar{x}\bar{s} = \bar{\mu} \hat{x} \hat{s}$.

If $\bar{\mu} = 0$, then $\bar{x}\bar{s} = 0$, hence $(\bar{x}, \bar{s}) \in S(LCP)$. To show nonemptiness of $S(LCP)$ it therefore suffices to check that $\bar{\mu}$ cannot be positive. If so, we would have $\bar{x} > 0$ and $\bar{s} > 0$. Apply the implicit function theorem to the system

$$\begin{cases} xs = \mu \hat{x} \hat{s}, \\ Qx + Rs = h. \end{cases}$$

Its Jacobian with respect to (x, s) is invertible. Indeed, let (u, v) be a nonzero element of the kernel:

$$\begin{cases} su + xv = 0, \\ Qu + Rv = 0. \end{cases}$$

Dividing the first relation by $\sqrt{xs}$, and because $(u, v) \ne 0$, it follows that

$$0 = \left\| \sqrt{\frac{s}{x}} u + \sqrt{\frac{x}{s}} v \right\|^2 > 2 \left(\sqrt{\frac{s}{x}} u \right)^\top \left(\sqrt{\frac{x}{s}} v \right) = 2u^\top v;$$

this contradicts the monotonicity property implied by the second relation. The implicit function theorem thus allows the local expression of (x, s) as a function of μ. In particular, the system has a solution in $F(LCP)$ for μ-values smaller than $\bar{\mu}$, which is a contradiction.

$\square$

We will say that $\eta \approx \delta$, where η and δ are two scalars, when $\eta = O(\delta)$ and $\delta = O(\eta)$. If $x \in \mathbb{R}^n$, we will say that $x \approx \delta$ when $x_i \approx \delta$, for all $i \in \{1, \cdots, n\}$. The estimates of order of magnitude are understood for μ close to 0. If y is a vector, $O(y)$ means $O(\|y\|)$.

Lemma 20.9. *Suppose that* (LCP) *(in canonical form:* $N = \emptyset$*) has an interior point. Let* $\varepsilon > 0$. *If* $(x, s, \mu) \in \mathcal{N}_\varepsilon$, *then*

$$x_B \approx 1, \quad s_B \approx \mu, \quad x_T \approx \sqrt{\mu}, \quad s_T \approx \sqrt{\mu}.$$

Proof. From Theorem 20.8 and Lemma 20.7, there exists $(\bar{x}, \bar{s}) \in S(LCP)$ with $\bar{x}_B > 0$. Let $(x, s) \in \mathcal{N}_\epsilon$; then $(x - \bar{x})^\top (s - \bar{s}) \geq 0$. Since $\bar{x}_T$ and $\bar{s}$ are zero, we deduce $\bar{x}_B^\top s_B \leq x^\top s \leq \epsilon^{-1} n\mu$. Since $\bar{x}_B > 0$, this implies $s_B = O(\mu)$. But $xs \approx \mu$ by definition of $\mathcal{N}_\epsilon$, and $x = O(1)$ by Theorem 20.8, hence $x_B \approx 1$ and $s_B \approx \mu$.

There remains to show that $x_T \approx \sqrt{\mu}$ and $s_T \approx \sqrt{\mu}$. From Theorem 20.8, there exists $(\bar{x}, \bar{s}) \in S(LCP)$. Then, since $N = \emptyset$,

$$Q_B x_B + Q_T x_T + Rs = h = Q_B \bar{x}_B,$$

hence

$$Q_T x_T + Rs \in \mathcal{R}(Q_B).$$

Set $\eta := \|x_T\| + \|s\|$. If $\eta = O(\sqrt{\mu})$, then $x_T = O(\sqrt{\mu})$ and $s_T = O(\sqrt{\mu})$. Since $x_T s_T \approx \mu$, the conclusion follows.

Otherwise, since $x_T s_T \approx \mu$, we have that the upper limit of $\{\eta/\sqrt{\mu}\}$, when $\mu \downarrow 0$, is $+\infty$, the supremum being attained when $\mu \to 0$. Let (u_T, v) be a cluster point of $\eta^{-1}(x_T, s)$ corresponding to a sequence $\mu_k \downarrow 0$ such that $\lim \eta_k/\mu_k = +\infty$. Then $(u_T, v) \neq 0$ and $s_B \approx \mu$ implies $v_B = 0$, hence $(u_T, v_T) \neq 0$. Passing to the limit in the relation

$$\eta^{-1}(Q_T x_T + Rs) \in \mathcal{R}(Q_B),$$

and since $\mathcal{R}(Q_B)$ is closed, we obtain the relation

$$Q_T u_T + Rv = -Q u_B$$

for a certain u_B. Let u be the vector formed by u_B and u_T. Let us show that $(\bar{x}^\varepsilon, \bar{s}^\varepsilon) := (\bar{x}, \bar{s}) + \varepsilon(u, v) \in S(LCP)$ if $\varepsilon > 0$ is small. Since (u_T, v) is nonnegative, this point lies in $\mathbb{R}_+^n \times \mathbb{R}_+^n$ (for $\varepsilon > 0$ small enough) and it satisfies the linear constraints. Moreover, $\bar{s}_B^\varepsilon = 0$. Finally, if $i \in T$, then $u_i > 0 \Rightarrow v_i = 0$ (because $x_T s_T \approx \mu$ and $\lim \eta_k/\sqrt{\mu_k} = \infty$); the converse is also true, therefore $u_T v_T = 0$, so

$$\bar{x}^\varepsilon \bar{s}^\varepsilon = \varepsilon(\bar{x} + \varepsilon u)v = 0.$$

We have shown $(\bar{x}^\varepsilon, \bar{s}^\varepsilon) \in S(LCP)$. However, $(\bar{x}_T^\varepsilon, \bar{s}_T^\varepsilon) = \varepsilon^2(u_T, v_T) \neq 0$, which contradicts the definition of T.

$\square$

20.3 Vector Fields Associated with the Central Path

In what follows, we study algorithms solving the central path equations by Newton's method. We analyze here the vector fields associated with the displacements in Newton's method.

20.3.1 General Framework

Let us analyze the directions obtained by linearizing the central path equation at a point (x, s, μ) of $\mathbb{R}^n_{++} \times \mathbb{R}^n_{++} \times \mathbb{R}_{++}$:

$$\begin{cases} xs = \mu\mathbf{1}, \\ Qx + Rs = h. \end{cases}$$

Said otherwise, let us compute the displacement used in Newton's method. We will call

$$w := (x, s, \mu)$$

the current point, supposed feasible. We aim at a new value of μ with value $\gamma\mu$, $\gamma \in [0, 1]$. Denoting by (u, v) the displacement, the linearization at $(x, s) \in F(LCP)$ writes:

$$\begin{cases} su + xv = \gamma\mu\mathbf{1} - xs, \\ Qu + Rs = 0. \end{cases} \tag{20.7}$$

We will speak of *centralization direction* when we aim at the same μ (then $\gamma = 1$) and of *affine direction* when $\gamma = 0$ (maximal reduction of μ). These directions are therefore defined by

$$\begin{cases} su^c + xv^c = \mu\mathbf{1} - xs, \\ Qu^c + Rv^c = 0, \end{cases} \qquad \begin{cases} su^a + xv^a = -xs, \\ Qu^a + Rv^a = 0. \end{cases}$$

We note the relation

$$\begin{cases} u = \gamma u^c + (1 - \gamma)u^a, \\ v = \gamma v^c + (1 - \gamma)v^a. \end{cases}$$

20.3.2 Scaling the Problem

In the remainder of this chapter, except otherwise stated, we use the canonical form $N = \emptyset$. We assume $(x, s) \in \mathcal{N}_\epsilon$. Set

$$d := \sqrt{\frac{x}{s}}; \quad \phi := \sqrt{\frac{xs}{\mu}}.$$

Note that $d_B \approx 1/\sqrt{\mu}$ and $d_T \approx 1$ (Lemma 20.9) and $\phi \approx 1$ ($\phi = 1$ on the central path). We scale the problem by multiplying the first equation of (20.7) by $1/\sqrt{xs}$, hence

$$\begin{cases} d^{-1}u + dv = \sqrt{\mu}(\gamma\phi^{-1} - \phi), \\ Qu + Rv = 0. \end{cases} \tag{20.8}$$

Set $D = \operatorname{diag}(d)$. The first equation suggests to consider the change of variables

$$\bar{u} := d^{-1}u, \quad \bar{v} := dv, \quad \bar{Q} := QD, \quad \bar{R} := RD^{-1}, \tag{20.9}$$

which corresponds to a change of variables in the (x, s)-space, mapping the current points x and s onto the same image

$$\bar{x} = d^{-1}x = \sqrt{xs} = ds = \bar{s}.$$

We obtain the scaled equations

$$\begin{cases} \bar{u} + \bar{v} = \sqrt{\mu}(\gamma\phi^{-1} - \phi), \\ \bar{Q}\bar{u} + \bar{R}\bar{v} = 0, \end{cases} \tag{20.10}$$

and $\bar{u}^\top \bar{v} \geq 0$ when $\bar{u}, \bar{v}$ satisfies the second relation of (20.10). This formulation allows a first estimate of the magnitude of the components of the directions:

Lemma 20.10. *Let* $(x, s, \mu) \in \mathcal{N}_\epsilon$. *The solution of equation (20.10) satisfies*

$$\bar{u} = O(\sqrt{\mu}); \quad u_B = O(1); \quad u_T = O(\sqrt{\mu});$$
$$\bar{v} = O(\sqrt{\mu}); \quad v_B = O(\mu); \quad v_T = O(\sqrt{\mu}).$$

Proof. From (20.10), we deduce

$$\|\bar{u}\|^2 + \|\bar{v}\|^2 + 2\bar{u}^\top \bar{v} = \mu\|\gamma\phi^{-1} - \phi\|^2 = O(\mu).$$

Since $\bar{u}^\top \bar{v} = u^\top v \geq 0$, it follows $\|\bar{u}\| = O(\sqrt{\mu})$ and $\|\bar{v}\| = O(\sqrt{\mu})$. Using $d_B \approx 1/\sqrt{\mu}$ and $d_T \approx 1$, as well as $u = d\bar{u}$ and $v = d^{-1}\bar{v}$, we obtain the conclusion.

$\square$

20.3.3 Analysis of the Directions

The following lemmas will be useful in the asymptotic analysis of the algorithms, and will allow an accurate analysis of the affine direction. First, let us show that the move of the large variables is interpreted as perturbing a certain projection. This requires a preliminary lemma.

Lemma 20.11. *If* (u, v) *is such that* $Qu + Rv = 0$, *then* $v \in \mathcal{N}(Q)^\perp = \mathcal{R}(Q^\top)$. *Moreover,* Q_B *denoting the columns of* Q *indexed in* B:

$$v_B \in \mathcal{N}(Q_B)^\perp = \mathcal{R}(Q_B^\top).$$

Proof. Take $u' \in \mathcal{N}(Q)$. Then $Q(u+u')+Rv = 0$, hence $(u+u')^\top v \geq 0$. Since u' is arbitrary in $\mathcal{N}(Q)$, this implies $v^\top u' = 0$, i.e. $v \in \mathcal{N}(Q)^\perp = \mathcal{R}(Q^\top)$. In particular, let u'_B be arbitrary in $\mathcal{N}(Q_B)$. Set $u'_T = 0$. Then $0 = v^\top u' = v_B^\top u'_B$, hence $v_B \in \mathcal{N}(Q_B)^\perp = \mathcal{R}(Q_B^\top)$.

$\square$

Define $P_{A,q}x$ to be the solution of the projection problem

$$\underset{z}{\text{Min}}\, \|z - x\|^2;\ Az + q = 0,$$

and set $P_A x := P_{A,0}x$. It is easily checked (for example by writing the optimality system of the projection problem) that

$$P_{A,q}x = P_A x + P_{A,q}0.$$

Lemma 20.12. *Let $(x, s, \mu) \in \mathcal{N}_\epsilon$ and (u, v) solve (20.10). Set $z := Rv + Q_T u_T$. Then*

$$\bar{u}_B = P_{\bar{Q}_B,z}(\gamma\sqrt{\mu}\phi_B^{-1}) = \gamma\sqrt{\mu}P_{\bar{Q}_B}\phi_B^{-1} + P_{\bar{Q}_B,z}0, \qquad (20.11)$$

$$u_B = \gamma\sqrt{\mu}d_B P_{\bar{Q}_B}\phi_B^{-1} + O(\sqrt{\mu}) = O(\gamma) + O(\sqrt{\mu}). \qquad (20.12)$$

If, in addition, $T = \emptyset$, then

$$u = \gamma\sqrt{\mu}d P_{\bar{Q}}\phi^{-1} + O(\mu) = O(\gamma) + O(\mu). \qquad (20.13)$$

In particular, $u^a = O(\mu)$ if $T = \emptyset$, and $u^a = O(\sqrt{\mu})$ in the general case.

Proof. The optimality conditions of the least-square problem

$$\underset{\bar{u}_B}{\text{Min}}\, \tfrac{1}{2}\|\bar{u}_B - \gamma\sqrt{\mu}\phi_B^{-1}\|^2;\quad \bar{Q}_B\bar{u}_B + z = 0, \qquad (*)$$

are

$$\bar{u}_B - \gamma\sqrt{\mu}\phi_B^{-1} \in \mathcal{R}(\bar{Q}_B^\top);\quad \bar{Q}_B\bar{u}_B + z = 0.$$

The second relation is obviously satisfied. In order to check the first one, note that from (20.10),

$$\bar{u}_B - \gamma\sqrt{\mu}\phi_B^{-1} = -\sqrt{\mu}\phi_B - \bar{v}_B.$$

By the previous lemma, $v_B \in \mathcal{R}(Q_B^\top)$; therefore $\bar{v}_B \in \mathcal{R}(\bar{Q}_B^\top)$. Moreover, let $(x^*, s^*) \in S(LCP)$. Then $0 = Q(x - x^*) + R(s - s^*)$, hence $\mathcal{R}(Q_B^\top) \ni (s - s^*)_B = s_B$. It follows $\phi_B = d_B s_B/\mu \in \mathcal{R}(\bar{Q}_B^\top)$. Problem $(*)$ is therefore solved by $\bar{u}_B$, and (20.11) follows.

Now let us show that $d_B P_{\bar{Q}_B,z}0 = O(\|z\|)$. Indeed, let Q_B^- be a right inverse of Q_B: a linear mapping from $\mathcal{R}(Q_B)$ to $\mathbb{R}^n$ such that $Q_B Q_B^- y = y$, $\forall\, y \in \mathcal{R}(Q_B)$. Then $\bar{Q}_B(D_B^{-1}Q_B^- z) = z$, hence

$$\|P_{\bar{Q}_B,z}0\| \le \| - D_B^{-1}Q_B^- z\| \le \|D_B^{-1}\|\|Q_B^-\|\|z\| = O(\sqrt{\mu}\|z\|). \qquad (**)$$

Since $d_B = O(1/\sqrt{\mu})$, we do have $d_B P_{\bar{Q}_B,z}0 = O(\|z\|)$. Combining with (20.11), we obtain

$$u_B = d_B \bar{u}_B = \gamma\sqrt{\mu}d_B P_{\bar{Q}_B}\phi_B^{-1} + O(\|z\|).$$

The conclusion is obtained by noting that, from Lemma 20.10, $z = Rv = O(\mu)$ if $T = \emptyset$, and $z = Rv + Q_T u_T = O(\sqrt{\mu})$ otherwise.

$\square$

Lemma 20.13. *Take $(x, s, \mu) \in \mathcal{N}_\epsilon$ and let (x^*, s^*) be the element of $S(LCP)$ closest to (x, s) (for the Euclidean norm). Then*

$$(x^*, s^*) = (x, s) + O(\sqrt{\mu}). \tag{20.14}$$

If strict complementarity holds, then

$$(x^*, s^*) = (x, s) + O(\mu). \tag{20.15}$$

Proof. One can check that, for μ small enough, the projection of (x, s) onto $S(LCP)$ coincides with the projection onto

$$\mathcal{V} := \{(x, s) \in \mathbb{R}^n; \quad x_T = 0; \quad s = 0; \quad Qx = h\}.$$

Indeed, let (x^*, s^*) be the projection of (x, s) onto $\mathcal{V}$. Then the amount

$$(x^*, s^*) - (x, s) = O(\|x_T\| + \|s\|) \tag{$*$}$$

converges towards 0 when $\mu \downarrow 0$, while $x_B \approx 1$; hence $x_B^* \approx 1$, which allows to check $(x^*, s^*) \in S(LCP)$. Because $S(LCP) \subset \mathcal{V}$ it is clear that (x^*, s^*) is the projection of (x, s) onto $S(LCP)$. The conclusion is deduced from $(*)$ and from Lemma 20.9. $\qquad\square$

The next theorem estimates the result of an affine move.

Theorem 20.14. *Let $(x, s, \mu) \in \mathcal{N}_\epsilon$. Then*
(i) *If strict complementarity holds, then*

$$s + v^a = O(\mu^2). \tag{20.16}$$

(ii) *If, on the contrary, $T \neq \emptyset$, then*

$$s_B + v_B^a = O(\mu^{3/2}) \tag{20.17}$$

and
$$x_T + 2u_T^a = O(\mu^{3/4}); \quad s_T + 2v_T^a = O(\mu^{3/4}). \tag{20.18}$$

Proof. From the affine step equation, we deduce $s + v^a = -sx^{-1}u^a$. Combining with Lemma 20.9, it follows that $s_B + v_B^a = O(\mu\|u^a\|)$. We then obtain (20.16)–(20.17) with Lemma 20.12.

Now let us establish (20.18). Call (x^I, s^I) the point obtained after a move *twice as long* as the affine direction:

$$x^I := x + 2u^a; \quad s^I := s + 2v^a.$$

Let (x^*, s^*) be the point of $S(LCP)$ closest to (x, s). Then, from Lemmas 20.10, 20.12 and 20.13,

$$\|(x^*, s^*) - (x^I, s^I)\| \leq \|(x^*, s^*) - (x, s)\| + 2\|(u^a, v^a)\| = O(\sqrt{\mu}).$$

But $Q(x^I - x^*) + Rs^I = 0$, hence $(x^I - x^*)^\top s^I \geq 0$, i.e.

$$(x_T^I)^\top (s_T^I) \geq (x_B^* - x_B^I)^\top s_B^I = O(\mu^{3/2}). \qquad (*)$$

Now

$$d^{-1}x^I + ds^I = d^{-1}x + ds + 2(d^{-1}u^a + dv^a) = 2\sqrt{xs} - 2\sqrt{xs} = 0.$$

As a result, for all $i \in \{1, \cdots, n\}$ we have

$$0 = (d_i^{-1}x_i^I + d_i s_i^I)^2 \geq 2x_i^I s_i^I, \qquad (**)$$

and the relation $0 \geq x_i^I s_i^I \geq (x_T^I)^\top (s_T^I) \geq O(\mu^{3/2})$, for all $i \in T$, follows from $(*)$. Combining with $(**)$, we have for all $i \in T$

$$(d_i^{-1}x_i^I)^2 + (d_i s_i^I)^2 = -2x_i^I s_i^I = O(\mu^{3/2}).$$

Since $d_T \approx 1$, we deduce $(x_i^I)^2 + (s_i^I)^2 = O(\mu^{3/2})$, and the conclusion follows.

$\square$

Remark 20.15. The affine move does not preserve positivity of variables in general: $(x + u^a)(s + v^a) = u^a v^a$ has no reason to be positive. $\qquad \square$

20.3.4 Modified Field

Theorem 20.14 shows that, if T is empty, the affine move yields a new point very close to $S(LCP)$. In fact, take again the proof of Lemma 20.13: we see that the estimate of the distance to $S(LCP)$ passes from $O(\mu)$ to $O(\|s + v^a\|) = O(\mu^2)$. On the other hand, if $T \neq \emptyset$, the estimate of the distance remains $\sqrt{\mu}$: this is due to the components of x and s in T which, because $(x_T, s_T) \approx \sqrt{\mu}$, are roughly divided by 2 (for small μ). We will see how to construct a *modified field* which, added to the affine field, yields a point closer to $S(LCP)$. This field uses explicitly the set T, which is of course unknown. Let us check that it is possible to construct an estimate of T, which is exact for μ small enough. The following lemma does not require $N = \emptyset$.

Lemma 20.16. (i) *For* $(x, s, \mu) \in \mathcal{N}_\epsilon$ *and* $\mu \downarrow 0$, *we have*

$$\begin{array}{llll} u_i^a/x_i \to 0 & and & v_i^a/s_i \to -1, & \forall i \in B, \\ u_i^a/x_i \to -1 & and & v_i^a/s_i \to 0, & \forall i \in N, \\ u_i^a/x_i \to -1/2 & and & v_i^a/s_i \to -1/2, & \forall i \in T. \end{array}$$

(ii) *Let* $\hat{T}(x, s, \mu)$ *be the estimate of* T *defined as follows:*

$$\hat{T}(x, s, \mu) := \left\{ i = 1, \cdots, n; \quad u_i^a/x_i \in \left[-\tfrac{3}{4}, -\tfrac{1}{4}\right] \text{ and } v_i^a/s_i \in \left[-\tfrac{3}{4}, -\tfrac{1}{4}\right] \right\}.$$

Then $\hat{T}(x, s, \mu) = T$ *if* $(x, s, \mu) \in \mathcal{N}_\epsilon$ *and* μ *is small enough.*

Proof. It suffices to consider the case $N = \emptyset$, $T \neq \emptyset$. Combining Lemmas 20.9, 20.12 and Theorem 20.14, we get

$$x_B \approx 1 \text{ and } u_B^a = O(\sqrt{\mu}) \Rightarrow u_B^a/x_B = O(\sqrt{\mu}),$$
$$s_B \approx \mu \text{ and } s_B + v_B^a = O(\mu^{3/2}) \Rightarrow v_B^a/s_B = -1 + O(\sqrt{\mu}),$$
$$x_T \approx \sqrt{\mu} \text{ and } x_T + 2u_T^a = O(\mu^{3/4}) \Rightarrow u_T^a/x_T = -\tfrac{1}{2} + O(\mu^{1/4}),$$

and an estimate similar to the last one holds for v_T^a/s_T, hence (i); as for (ii), it is an immediate consequence of (i). $\qquad\square$

For $M \subset \{1, \cdots, n\}$, let us introduce the *modified field* (u^M, v^M) defined as solving

$$\begin{cases} su^M + xv^M = -x_M s_M, \\ Qu^M + Rv^M = 0. \end{cases}$$

Here x_M (resp. s_M) is the extension to $\mathbb{R}^n$ of the restriction of x (resp. s) to M; said otherwise

$$(x_M)_i = (s_M)_i = 0, \ i \notin M; \quad (x_M)_i = x_i, \ (s_M)_i = s_i, \ i \in M.$$

If $M = \emptyset$, then $x_M = s_M = 0$ and the modified field is zero. In the general case, we will see that, when $M = T$, then (u_T^M, v_T^M) is close to $-\frac{1}{2}(x_T, s_T)$, while u_B^M and v_B^M are small enough to guarantee that the sum of the affine and modified fields yield a move substantially closer to $S(LCP)$.

Theorem 20.17. *Let $(x, s, \mu) \in \mathcal{N}_\epsilon$ and suppose $M = T$. Then*
(i) *the modified field has the following magnitude:*

$$u_B^M = O(\sqrt{\mu}), \ v_B^M = O(\mu^{3/2}), \ u_T^M = O(\sqrt{\mu}), \ v_T^M = O(\sqrt{\mu});$$

(ii) *there holds that*

$$x_T + 2u_T^M = O(\mu^{3/4}); \quad s_T + 2v_T^M = O(\mu^{3/4}).$$

Proof. Set $\bar{u}^M := d^{-1}u^M$ and $\bar{s}^M := dv^M$. Proceeding as in the proof of Lemma 20.10, we first check that

$$\bar{u}^M = O(\sqrt{\mu}); \quad u_B = O(1); \quad u_T = O(\sqrt{\mu});$$
$$\bar{v}^M = O(\sqrt{\mu}); \quad v_B = O(\mu); \quad v_T = O(\sqrt{\mu}).$$

For $M = T$, we have with Lemma 20.11

$$\bar{u}_B^M = -\bar{v}_B^M \in D_B \mathcal{R}(Q_B^\top) = \mathcal{R}(\bar{Q}_B^\top).$$

Using the technique of proof of Lemma 20.10, we deduce that $\bar{u}_B = P_{\bar{Q}_{B,z^M}}0$, where $z^M := Q_T u_T^M + Rv^M = O(\sqrt{\mu})$, and that $\|u_B^M\| = \|d_B \bar{u}_B^M\| = O(\sqrt{\mu})$. Finally, from $(su^M + xv^M)_B = 0$ we deduce with Lemma 20.10

$v_B^M = -s_B u_B^M / x_B = O(\mu^{3/2})$; (i) follows. To prove (ii), we follow the proof of Theorem 20.14. Set

$$x^I := x + 2u^M; \quad s^I := s + 2v^M,$$

and let (x^*, s^*) be the point of $S(LCP)$ closest to (x, s). By Lemma 20.13, we have $x^* = x + O(\sqrt{\mu})$. From monotonicity,

$$0 \leq (x^I - x^*)^\top (s^I - s^*) = (x^I - x^*)^\top s^I,$$

hence, with (20.14): $(x_T^I)^\top s_T^I \geq (x_B^* - x_B^I)^\top s_B^I \geq O(\mu^{3/2})$. Because $d_T^{-1} x_T^I + d_T s_T^I = 0$, we have

$$\|d_T^{-1} x_T^I\|^2 + \|d_T s_T^I\|^2 \leq -2(x_T^I)^\top s_T^I = O(\mu^{3/2}).$$

With $d_T \approx 1$, we deduce that x_T^I and s_T^I are of order $\mu^{3/4}$, hence the conclusion.

$\square$

Combining Theorems 20.14 and 20.17, we deduce

Corollary 20.18. *Let* $(x, s, \mu) \in \mathcal{N}_\epsilon$. *The point*

$$(x^{\sharp\sharp}, s^{\sharp\sharp}) := (x, s) + (u^a, v^a) + (u^M, v^M)$$

satisfies

$$\text{dist}\left((x^{\sharp\sharp}, s^{\sharp\sharp}), S(LCP)\right) = O(\mu^{3/4}) \quad \text{and} \quad (x^{\sharp\sharp})^\top s^{\sharp\sharp} = O(\mu^{3/2}).$$

Remark 20.19. The point thus obtained need not be feasible. In Chap. 21, we will construct algorithms using the modified field, in which the sequence of points is kept in a neighborhood of the central path. The new μ is then of order $\mu^{5/4}$ only. $\square$

20.4 Continuous Trajectories

20.4.1 Limit Points of Continuous Trajectories

This section is devoted to the differential equations associated with the vector fields studied previously. We will see that it "suffices" to integrate these differential equations to obtain, depending on the value of a control parameter $\gamma(t)$, either a point on the central path, or a solution of (LCP) whose distance to the central path can be controlled. Of course, the numerical resolution of the differential equation induces discretization errors which will have to be coped with, to obtain an implementable algorithm; especially as one is interested by long steps, so as to converge more rapidly to a solution of (LCP).

The whole difficulty of "discretized" algorithms studied below is precisely to take these various aspects into account. Studying continuous trajectories is therefore only a first approach, which gives a hint of what can be hoped for from discrete algorithms.

So we study the differential equation

$$\frac{d}{dt}(x, s, \mu) = (u(x, s), v(x, s), -(1 - \gamma)\mu),$$

where $(u(x, s), v(x, s))$ solves

$$\begin{cases} su + xv = \gamma\mu\mathbf{1} - xs, \\ Qu + Rv = 0, \end{cases}$$

and γ takes its values in $[0, 1]$, as a function of (x, s, μ, t) (smooth enough so that the differential equation is well-defined). We will use the notation $\gamma(t) = \gamma(x(t), s(t), \mu(t), t)$.

Proposition 20.20. *There holds*

$$(xs - \mu\mathbf{1})(t) = e^{-t}(xs - \mu\mathbf{1})(0),$$
$$(\frac{xs}{\mu} - 1)(t) = e^{-\int_0^t \gamma(\sigma)d\sigma}(\frac{xs}{\mu} - 1)(0).$$

Proof. Using the derivative formulae of (x, s, μ), we obtain

$$\frac{d}{dt}(xs - \mu\mathbf{1})(t) = xv + su + (1 - \gamma)\mu = -(xs - \mu\mathbf{1}),$$
$$\frac{d}{dt}(\frac{xs}{\mu} - 1)(t) = -\frac{1}{\mu}(xs - \mu\mathbf{1}) - \frac{xs - \mu\mathbf{1}}{\mu^2}\frac{d\mu}{dt} = -\gamma(t)(\frac{xs}{\mu} - 1)(t),$$

and the result follows.

□

Set

$$c_1 := \int_0^\infty (1 - \gamma(t))dt \quad \text{and} \quad c_2 := \int_0^\infty \gamma(t)dt. \tag{20.19}$$

Corollary 20.21. *If $c_1 < \infty$, then $\mu(t) \downarrow \mu_\infty := e^{-c_1}\mu(0)$ and $(x(t), s(t))$ converges to the point of the central path associated with μ_∞. If $c_1 = \infty$, then $\mu(t) \downarrow 0$.*

The trajectories obtained by integrating the above system when $\gamma = 0$ are called affine. When $\gamma = 1$ we will speak of the centering trajectory. For the affine trajectory, we obtain

$$x(t)s(t) = e^{-t}x(0)s(0) \quad \text{and} \quad \mu(t) = e^{-t}\mu(0).$$

Let $\mathbb{R}^B$ be the vector space spanned by the components in B of the elements of $\mathbb{R}^n$ ($\mathbb{R}^B$ is isomorphic to $\mathbb{R}^{|B|}$). The logarithmic potential in $\mathbb{R}^B$ is $\pi(x_B) :=$

$-\sum_{i\in B}\log x_i$. We call *analytic center* of $S(LCP)$, denoted by $(\bar{x}^*,\bar{s}^*)$, the point of components $\bar{s}_B^* = 0$, $\bar{x}_T^* = \bar{s}_T^* = 0$, and $\bar{x}_B^*$ solving

$$\operatorname*{Min}_{x_B>0} \pi(x_B); \quad Q_B x_B = h.$$

The above problem is convex, with a strictly convex objective; it has at most one solution, characterized by the optimality system

$$\bar{x}_B^* \in \mathbb{R}_{++}^B; \quad Q_B\bar{x}_B^* = h; \quad \nabla\pi(\bar{x}_B^*) = -(\bar{x}_B^*)^{-1} \in \mathcal{R}(Q_B^\top).$$

More generally, let $\eta \in \mathbb{R}_{++}^n$ be a weighting vector; the associated weighted analytic center, solving

$$\operatorname*{Min}_{x_B>0} -\sum_{i\in B}^{n} \eta_i \log x_i; \quad Q_B x_B = h,$$

is characterized by

$$x_B^\eta \in \mathbb{R}_{++}^B; \quad Q_B x_B^\eta = h; \quad \eta_B(x_B^\eta)^{-1} \in \mathcal{R}(Q_B^\top).$$

Theorem 20.22. *Let c_2 be defined by (20.19). Set*

$$\eta := e^{-c_2}\left(\frac{xs}{\mu} - 1\right)(0) + 1.$$

Then $x_B(t) \to x_B^\eta$, the weighted analytic center associated with η. In particular, if $c_2 = \infty$, then $(x(t), s(t))$ converges to the analytic center of $S(LCP)$.

Proof. From Theorem 20.8, $x(t)$ is bounded. Take a cluster point x^* of $x(t)$ when $\mu \downarrow 0$: we know that $x_B^* > 0$ (Lemma 20.9). Let $d_B \in \mathcal{N}(Q_B)$. Set $d := (d_B, d_T = 0)$. Then, for $\theta > 0$ small enough, the point $x(\theta) := x^* + \theta d$ lies in $S(LCP)$ and $Q(x - x(\theta)) + Rs = 0$, hence $(x - x(\theta))^\top s \geq 0$. By Proposition 20.20, we have that $(xs/\mu - 1)(t) \to e^{-c_2}(xs/\mu - 1)(0)$, and hence, $xs/\mu \to \eta$, so that $(x_B^*)^{-1}\eta$ is a cluster point of s_B/μ. As a result, for the subsequence corresponding to this cluster point x^*,

$$0 \leq \lim (x - x(\theta))^\top s/\mu = \lim (x_B - x_B(\theta))^\top s_B/\mu = \theta d^\top (x_B^*)^{-1}\eta.$$

Since d_B is arbitrary in $\mathcal{N}(Q_B)$, we deduce $(x_B^*)^{-1}\eta \in \mathcal{N}(Q_B)^\perp = \mathcal{R}(Q_B^\top)$, hence $x_B^* = x_B^\eta$. If $c_2 = \infty$, then $\eta = 1$, hence the result. □

20.4.2 Developing Affine Trajectories and Directions

The next theorem expresses the development of the small variables and of the affine field along an affine trajectory. It gives a motivation of the form of line-search for modified-field algorithms to follow.

Theorem 20.23. *Let (x, s, μ) lie on the affine trajectory of weight $y = xs/\mu$ and with end point $(x^y, 0, 0)$; let (u^a, v^a) be the associated affine field. Then*
(i) $s_B = \mu y_B(\bar{x}_B^y)^{-1} + o(\mu)$,
(ii) $v_B^a = -\mu y_B(\bar{x}_B^y)^{-1} + o(\mu)$,
(iii) *We have that*
$$\begin{cases} x_T = \sqrt{\mu}\hat{x}_T + o(\sqrt{\mu}), \\ s_T = \sqrt{\mu}\hat{s}_T + o(\sqrt{\mu}), \end{cases}$$
where $(\hat{x}_T, \hat{s}_T)$ is the unique solution of

$$\hat{x}_T \in \mathbb{R}^n_{++}, \quad \hat{s}_T \in \mathbb{R}^N_{++}; \quad \hat{x}_T\hat{s}_T = \mu y_T; \quad Q_T\hat{x}_T + R_T\hat{s}_T \in \mathcal{R}(Q_B). \quad (20.20)$$

Moreover, set $\hat{z} := Q_T\hat{x}_T + R_T\hat{s}_T$ and $\check{Q} := Q_B Y_B^{-1/2}\bar{X}_B^y$. Then

$$u_B^a = \sqrt{\mu}P_{\check{Q},\hat{z}}0 + o(\sqrt{\mu}).$$

(iv) *If $T = \emptyset$, set $\tilde{z} := RY(X^y)^{-1}$. Then $u^a = \mu Y^{-1/2}X^y P_{\check{Q},\hat{z}}0 + o(\mu)$.*

Proof. Relation (i) is a consequence of the previous theorem and of the relation $s_B x_B = \mu y_B$; combined with Theorem 20.14, we deduce (ii).

Another consequence of (i) is that

$$d_B = \sqrt{x_B/s_B} = (Y_B^{-1/2}\bar{x}_B^y + o(1))/\sqrt{\mu}.$$

Set $\bar{d} := \sqrt{\mu x_B/s_B} = \sqrt{\mu}d_B$ and $z := Rs + Q_T x_T$. Then $\bar{d} = Y_B^{-1/2}x_B^y + o(1)$. From Lemma 20.12, we have

$$u_B^a = D_B P_{Q_B D_B, z}0 = \bar{D}P_{Q_B\bar{D}, z}0 = Y_B^{-1/2}X_B^y P_{Q_B Y_B^{-1/2}\bar{X}_B^y, z}0 + o(\|z\|). \quad (*)$$

If $T = \emptyset$, then $z = Rs = \mu RY_B(x^y)^{-1} + o(\mu)$, hence (iv).

Let us prove (iii). From Lemma 20.9 we know that $x_T \approx \sqrt{\mu}$ and $s_T \approx \sqrt{\mu}$, hence $(x_T, s_T)/\sqrt{\mu}$ has at least one cluster point $(\hat{x}_T, \hat{s}_T)$ and $\hat{x}_T > 0$, $\hat{s}_T > 0$. Passing to the limit in

$$\frac{x_T}{\sqrt{\mu}} \times \frac{s_T}{\sqrt{\mu}} = y; \quad Q_T x_T + R_T s_T = -Q_B x_B - R_B s_B \in \mathcal{R}(Q_B) + O(\mu),$$

we deduce that $(\hat{x}_T, \hat{s}_T)$ satisfies (20.20). There remains to show that the solution of (20.20) is unique. Let $(\tilde{x}_T, \tilde{s}_T)$ be another solution; there exists $\hat{x}_B, \tilde{x}_B$ such that, completing the vectors $\hat{s}$ and $\tilde{s}$ by 0 on s_B, we have

$$Q\hat{x} + R\hat{s} = 0 = Q\tilde{x} + R\tilde{s},$$

and therefore $Q(\hat{x} - \tilde{x}) + R(\hat{x} - \tilde{x}) = 0$, hence $(\hat{x} - \tilde{x})^\top(\hat{s} - \tilde{s}) \geq 0$. Since $(\hat{s} - \tilde{s})_B = 0$, this reduces to $(\hat{x}_T - \tilde{x}_T)^\top(\hat{s}_T - \tilde{s}_T) \geq 0$. Using $\hat{x}\hat{s} = \tilde{x}\tilde{s} = \mu y$, get $\hat{x}_T^\top\tilde{s}_T + \tilde{x}_T^\top\hat{s}_T \leq 2\mu\|y_T\|_1$. Substituting $\tilde{s}_T = \mu y(\tilde{x}_T^{-1})$ and $\hat{s}_T = \mu y\hat{x}_T^{-1}$, we obtain

$$\hat{x}_T^\top Y_T \tilde{x}_T^{-1} + \tilde{x}_T^\top Y_T \hat{x}_T^{-1} \leq 2 \sum_{i \in T} y_i.$$

In particular, there exists $i \in T$ such that $\hat{x}_i y_i / \tilde{x}_i + \tilde{x}_i y_i / \hat{x}_i \leq 2y_i$, hence $\hat{x}_i / \tilde{x}_i + \tilde{x}_i / \hat{x}_i \leq 2$.

Since, for positive α and β, the function $(\alpha, \beta) \to \alpha/\beta + \beta/\alpha$ is larger than 2, except if $\alpha = \beta$, we deduce recursively that $\hat{x}_T = \tilde{x}_T$, whence $\hat{s}_T = \tilde{s}_T$, which proves (20.20). Combining with (∗), we obtain the formula for u_B^a. ☐

20.4.3 Mizuno's Lemma

The Lemma below will be used many times.

Lemma 20.24. *(Mizuno [259]). Let u and v of $\mathbb{R}^n$ be such that $u^\top v \geq 0$. Then*

$$\|uv\| \leq \frac{1}{\sqrt{8}} \|u + v\|^2.$$

Proof. Take β and γ in $\mathbb{R}$. From $(\beta + \gamma)^2 = \beta^2 + 2\beta\gamma + \gamma^2 = (\beta - \gamma)^2 + 4\beta\gamma$ we deduce $\beta\gamma \leq \frac{1}{4}(\beta + \gamma)^2$. Besides,

$$\|uv\|^2 = \sum_{i=1}^n (u_i v_i)^2 \leq \left(\sum_{u_i v_i > 0} u_i v_i \right)^2 + \left(\sum_{u_i v_i < 0} u_i v_i \right)^2.$$

Since $u^\top v \geq 0$, we have $\left| \sum_{u_i v_i < 0} u_i v_i \right| \leq \sum_{u_i v_i > 0} u_i v_i$ and therefore

$$\|uv\|^2 \leq 2 \left(\sum_{u_i v_i > 0} u_i v_i \right)^2 \leq 2 \left(\sum_{u_i v_i > 0} \frac{1}{4}(u_i + v_i)^2 \right)^2,$$

$$\leq \frac{2}{4^2}(\|u + v\|^2)^2 = \left(\frac{1}{\sqrt{8}} \|u + v\| \right)^2,$$

hence the result. ☐

20.5 Comments

With the book by Fiacco and McCormick [122], one can follow the history of the approach by logarithmic penalty. The first attempts to minimize the objective with a log-penalty used quasi-Newton methods (for unconstrained problems), that were available in the sixties. In this approach, it is difficult to control the accuracy with which the problem with μ fixed must be solved. Log-penalty methods were then abandoned, to the advantage of augmented

Lagrangian algorithms (see Bertsekas [25]), and then of Newton-type methods, applied to the optimality system, that are presented in part III of this book.

Linear complementarity problems, monotone or not, are discussed in Cottle, Pang and Stone [91]. The asymptotic analysis of points and vector fields in the large neighborhood is taken mainly from Monteiro and Tsuchiya [263], Mizuno, Jarre and Stoer [260, 261]. Gonzaga [172] gives a good introduction to the algorithmic consequences of the central path concept.

21 Predictor-Corrector Algorithms

21.1 Overview

Path-following algorithms take as direction of displacement the Newton direction associated with the central-path equation. They use a parameter μ, measuring the violation of the complementarity conditions. The algorithm's complexity is evaluated by the number of operations necessary to compute a point associated with the target measure μ_∞. The measure of the initial point is μ_0, and we set $\bar{L} := \log(\mu_0/\mu_\infty)$. The predictor-corrector method performs at each iteration a restoration (centralization) step, followed by an affine step.

In the case of the small neighborhood, the restoration step reduces by at least a half the proximity measure, while the affine step reduces by a term of the order $1/\sqrt{n}$ the optimality measure, thus yielding a convergence in $O(\sqrt{n}\bar{L})$ iterations. The asymptotic analysis shows that $\limsup \mu_{k+1}/\mu_k \leq 1 - \alpha^{1/2}|T|^{-1/4}/2$, where T is the set of indices for which strict complementarity is not satisfied, and α is the size of the neighborhood. Besides, under the strict complementarity assumption, μ_k converges quadratically to 0. To obtain superlinear convergence in the general case, the modified field of Chap. 20 can be used; the algorithm thus obtained satisfies $\mu_{k+1} = O(\mu_k^{5/4})$.

Then we study a large-neighborhood algorithm. A centralization step of order $1/n$ allows a reduction of the optimality measure μ, during the affine step, of a term of order $1/n$. Convergence therefore holds in $O(nL)$ iterations. The asymptotic analysis is similar to that of the small-neighborhood algorithm: convergence is quadratic under strict complementarity assumption, and a modification of the algorithm using the modified field allows the statement of an algorithm with the same complexity and superlinear convergence of order $5/4$.

It is worth mentioning that these algorithms are *feasible*: they assume the knowledge of a feasible starting point, close to the central path. This assumption is of course excessive, and we will see later how to cope with it. This will involve technical complications, and it is preferable to avoid them in a first step.

21.2 Statement of the Methods

21.2.1 General Framework for Primal-Dual Algorithms

We consider the framework of the monotone complementarity problem

$$\begin{cases} xs = 0, \\ Qx + Rs = h, \\ x \geq 0, \ s \geq 0, \end{cases} \qquad (LCP)$$

with $u^\top v \geq 0$ if $Qu + Rv = 0$. Newton's method applied to the resolution of the central-path equation

$$\begin{cases} xs = \gamma\mu\mathbf{1}, \\ Qx + Rs = h, \end{cases}$$

where $\gamma \in [0, 1]$ is the reduction factor of μ, and the current point is assumed feasible, is, denoting by (u, v) the displacement:

$$\begin{cases} su + xv = \gamma\mu\mathbf{1} - xs, \\ Qu + Rs = 0. \end{cases} \qquad (21.1)$$

We will use the notation $w := (x, s, \mu)$ and $\theta > 0$ will be the step in the direction (u, v). We choose for the new value of μ the quantity

$$\mu_\sharp := (1 - \theta)\mu + \theta\gamma\mu = \mu + \theta(\gamma - 1)\mu = (1 - \theta + \theta\gamma)\mu,$$

so that the new value of μ equals the old one if $\gamma = 1$, and equals $\gamma\mu$ if $\theta = 1$. The new point is denoted by

$$w^\sharp := (x^\sharp, s^\sharp, \mu_\sharp),$$

with

$$x^\sharp := x + \theta u; \qquad s^\sharp := s + \theta v.$$

For $\gamma = 1$ (resp. $\gamma = 0$), we obtain the centralization (resp. affine) displacement, of equation

$$\begin{cases} su^c + xv^c = \mu\mathbf{1} - xs, \\ Qu^c + Rv^c = 0, \end{cases} \qquad (21.2)$$

and

$$\begin{cases} su^a + xv^a = -xs, \\ Qu^a + Rv^a = 0, \end{cases} \qquad (21.3)$$

and we denote by w^c, w^a the centralized and affine points obtained with $\theta = 1$:

$$w^c := (x^c, s^c, \mu_c) = (x + u^c, s + v^c, \mu),$$
$$w^a := (x^a, s^a, \mu_a) = (x + u^a, s + v^a, 0).$$

The formula for $\mu_\sharp$ can be written as

$$\mu_\sharp = (1 - \theta)\mu + \theta(\gamma\mu_c + (1 - \gamma)\mu_a),$$

so that

$$w^\sharp = (1 - \theta)w + \theta(\gamma w^c + (1 - \gamma)w^a).$$

21.2.2 Weighting After Displacement

The following formulae allow the evaluation of the centralization of the point obtained after a Newton displacement, independently of the algorithm. We have

$$(x + \theta u)(s + \theta v) = xs + \theta(su + xv) + \theta^2 uv,$$
$$= (1 - \theta)xs + \theta\gamma\mu\mathbf{1} + \theta^2 uv,$$

and therefore

$$\frac{x^\sharp s^\sharp}{\mu_\sharp} - \mathbf{1} = \frac{(1 - \theta)xs + \theta\gamma\mu\mathbf{1} + \theta^2 uv}{(1 - \theta + \theta\gamma)\mu} - \mathbf{1} = \frac{1 - \theta}{1 - \theta + \theta\gamma}\left(\frac{xs}{\mu} - \mathbf{1}\right) + \theta^2\frac{uv}{\mu_\sharp}.$$

In particular, for the centralization step ($\gamma = 1$, $\mu_\sharp = \mu$), and for the affine step ($\gamma = 0$, $\mu_\sharp = (1 - \theta)\mu$), it gives

$$\frac{x^c s^c}{\mu} - \mathbf{1} = (1 - \theta)\left(\frac{xs}{\mu} - \mathbf{1}\right) + \theta^2\frac{u^c v^c}{\mu}\ ;$$

$$\frac{x^a s^a}{\mu_\sharp} - \mathbf{1} = \frac{xs}{\mu} - \mathbf{1} + \frac{\theta^2}{1 - \theta}\frac{u^a v^a}{\mu}.$$

21.2.3 The Predictor-Corrector Method

The *generic predictor-corrector algorithm* uses a neighborhood of the central path $\mathcal{G}$, which can be for example $\mathcal{V}_\alpha$ or $\mathcal{N}_\epsilon$. It simply consists in alternating the centralization steps, so as to get closer to the central path, and the affine steps in which θ^a is the largest step allowing to stay in $\mathcal{G}$.

Algorithm 21.1. GPC (Generic Predictor-Corrector Algorithm)
    Data: $\mu_\infty > 0$, $(x^0, s^0, \mu_0) \in \mathcal{G}$. $k \leftarrow 0$.
    REPEAT

- $x \leftarrow x^k$, $s \leftarrow s^k$, $\mu \leftarrow \mu_k$;
- Centralization: compute (u^c, v^c) solving (21.2);
  $x(\theta) := x + \theta u^c$, $s(\theta) := s + \theta v^c$.
  Compute $\theta^c \in]0, 1]$ such that $(x(\theta^c), s(\theta^c), \mu) \in \mathcal{G}$.
  $x \leftarrow x(\theta^c)$, $s \leftarrow s(\theta^c)$.
- Affine displacement: compute (u^a, v^a) solving (21.3);
  $x(\theta) := x + \theta u^a$, $s(\theta) := s + \theta v^a$, $\mu(\theta) := (1 - \theta)\mu$.
  Compute θ^a, the largest value in $]0, 1[$ such that
  $(x(\theta), s(\theta), \mu(\theta)) \in \mathcal{G}$, $\forall\, \theta \in [0, \theta^a]$.
  $x^{k+1} \leftarrow x(\theta^a)$, $s^{k+1} \leftarrow s(\theta^a)$, $\mu_{k+1} \leftarrow (1 - \theta^a)\mu_k$; $k \leftarrow k + 1$.
    UNTIL $\mu_k < \mu_\infty$.

Remark 21.2. We stress the fact that the affine step is obtained by computing the affine direction at the point obtained after centering. Each iteration therefore requires solving two linear systems, whose associated matrices are different.   □

21.3 A small-Neighborhood Algorithm

21.3.1 Statement of the Algorithm. Main Result

The *small-neighborhood algorithm* is a particular instance of the above general algorithm, in which we choose $\mathcal{G} = \mathcal{V}_\alpha$, with $\alpha \leq 1/2$, and a centering step equal to 1. Said otherwise:

Algorithm 21.3. PC (Predictor-Corrector Algorithm)
  Data $\mu_\infty > 0$, $\alpha \in (0, 1/2]$; $(x^0, s^0, \mu_0) \in \mathcal{V}_\alpha$; $k \leftarrow 0$.
  REPEAT
  - $x \leftarrow x^k$, $s \leftarrow s^k$, $\mu \leftarrow \mu_k$;
  - Centralization: compute (u^c, v^c) solving (21.2);
    $x \leftarrow x + u^c$, $s \leftarrow s + v^c$.
  - Affine displacement: compute (u^a, v^a) solving (21.3);
    $x(\theta) := x + \theta u^a$, $s(\theta) := s + \theta v^a$, $\mu(\theta) := (1 - \theta)\mu$.
    Compute θ^a, the largest value in $]0, 1[$ such that
    $(x(\theta), s(\theta), \mu(\theta)) \in \mathcal{V}_\alpha$, $\forall\, \theta \in [0, \theta^a]$.
    $x^{k+1} \leftarrow x(\theta^a)$, $s^{k+1} \leftarrow s(\theta^a)$, $\mu_{k+1} \leftarrow (1 - \theta^a)\mu_k$; $k \leftarrow k + 1$.
  UNTIL $\mu_k < \mu_\infty$.

Theorem 21.4. (i) *If $\mu_\infty > 0$, set $\bar{L} := \log(\mu_0/\mu_\infty)$. Then Algorithm* **PC** *stops after at most $O(\sqrt{n}\bar{L})$ iterations (more precisely $\sqrt{2n/\alpha}\bar{L}$ iterations).*
(ii) *If $\mu_\infty = 0$, then $\limsup \mu_{k+1}/\mu_k \leq 1 - \alpha^{1/2}|T|^{-1/4}/2$.*
(iii) *Suppose (LCP) has a strictly complementary solution. If $\mu_\infty = 0$, then $\{\mu_k\}$ converges quadratically to 0.*

This theorem will be proved later. The complexity estimate $O(\sqrt{n}\bar{L})$ is the best known at present, even for linear programming. According to (ii), the asymptotic speed can be slow if $|T|$ is large. In case of linear programming, T is empty: convergence is therefore quadratic.

21.3.2 Analysis of the Centralization Move

Combining formulae from §21.2.2 and Mizuno's Lemma 20.24, an estimate of the proximity measure after a centralization move is obtained. This estimate shows that, if $\alpha \leq 1/2$, the centralization move with $\theta^c = 1$ reduces the proximity measure from α to $\alpha/\sqrt{8}$.

Lemma 21.5. (i) *Let $w := (x, s, \mu)$ be such that (x, s) lies in the interior of (LCP). Then*

$$\frac{\|u^c v^c\|}{\mu} \leq \frac{1}{\sqrt{8}} \left\| \frac{\mu}{xs} \right\|_\infty \delta(w)^2.$$

(ii) *Set $\alpha^c := \alpha/\sqrt{8}$. If $\delta(w) \leq \alpha \leq 1/2$, then $w^c \in \mathcal{V}_{\alpha^c}$ and*

$$\delta(w^c) \le \frac{1}{\sqrt{2}}\delta(w)^2 \le \frac{1}{\sqrt{8}}\delta(w).$$

Moreover, $(x + \theta u^c, s + \theta v^c, \mu) \in \mathcal{V}_\alpha$ for all $\theta \in [0, 1]$.

Proof. Dividing the equation $su^c + xv^c = \mu\mathbf{1} - xs$ by $\sqrt{xs}$, get

$$\sqrt{s/x}\,u^c + \sqrt{x/s}\,v^c = \frac{\mu}{\sqrt{xs}} - \sqrt{xs}.$$

But $\left(\sqrt{\frac{s}{x}}u^c\right)^\top \left(\sqrt{\frac{x}{s}}v^c\right) = (u^c)^\top v^c \ge 0$. Hence, with Lemma 20.24,

$$\frac{\|u^c v^c\|}{\mu} = \frac{1}{\mu}\left\|\left(\sqrt{\frac{s}{x}}u^c\right)\left(\sqrt{\frac{x}{s}}v^c\right)\right\| \le \frac{1}{\sqrt{8}}\left\|\sqrt{\frac{\mu}{xs}} - \sqrt{\frac{xs}{\mu}}\right\|^2$$

$$\le \frac{1}{\sqrt{8}}\left\|\frac{\mu}{xs}\right\|_\infty \left\|1 - \frac{xs}{\mu}\right\|^2$$

which proves (i). If $\delta(w) \le 1/2$, then $\frac{x_i s_i}{\mu} \ge 1 - \delta(w) \ge 1/2$, hence $\left\|\frac{\mu}{xs}\right\|_\infty \le 2$
and $\delta(w^c) \le \delta(w)^2/\sqrt{2} \le \delta(w)/\sqrt{8}$.

It remains to prove that, if $\delta(w) \le 1/2$, then $x^c > 0$ and $s^c > 0$. For this, it suffices to check the last relation of the lemma. The centrality measure at the new point $w + \theta(w^c - w)$ is, in view of 21.2.2 and the relation just proved,

$$\delta(w^c) = \left\|(1 - \theta)\left(\frac{xs}{\mu} - 1\right) + \theta^2 \frac{u^c v^c}{\mu}\right\|,$$

$$\le (1 - \theta)\delta(w) + \theta^2 \frac{\|u^c v^c\|}{\mu} \le (1 - \theta)\delta(w) + \frac{\theta^2}{\sqrt{8}}\delta(w).$$

Bounding $\theta^2/\sqrt{8}$ from above by θ, we deduce that $\delta(w^c) \le \delta(w) \le 1/2$. Therefore

$$(x + \theta u^c)(s + \theta v^c) \ge \frac{\mu}{2}\mathbf{1}, \quad \forall \theta \in [0, 1].$$

The left-hand side is a continuous function of θ which, for $\theta = 0$, belongs to $\mathbb{R}^n_{++}$. The inequality prevents each component from vanishing; we deduce that, when θ increases from 0 to 1, $(x + \theta u^c)(s + \theta v^c)$ stays strictly positive. Each term of the product therefore keeps the same $+$ sign, as was to be proved.

□

21.3.3 Analysis of the Affine Step and Global Convergence

The problem is now to determine a value as large as possible for the affine step θ. The condition to satisfy is $\delta^a \le \alpha$, knowing that the current step obtained after centering has, in view of the above lemma, a proximity measure of at most $\alpha/\sqrt{8}$. The next lemma gives an estimate of θ^a allowing the polynomial complexity of Theorem 21.4. (The complexity estimate comes from item (iii); items (i) and (ii) will be useful for asymptotic analysis.)

Lemma 21.6. *Take* $(x, s, \mu) \in \mathcal{V}_{\alpha/2}$, *with* $\alpha \leq 1/2$. *Then*

(i) $\|u^a v^a\| \leq x^\top s / \sqrt{8}$,   (ii) $\dfrac{(\theta^a)^2}{1 - \theta^a} \geq \dfrac{\alpha\mu}{2\|u^a v^a\|}$,   (iii) $\theta^a \geq \sqrt{\dfrac{\alpha}{2n}}$.

Proof. Apply Mizuno's Lemma 20.24 to the inequality $su^a + xv^a = -xs$, after the scaling obtained by dividing the preceding equation by $\sqrt{xs}$. It follows that

$$\sqrt{\frac{s}{x}}u^a + \sqrt{\frac{x}{s}}v^a = -\sqrt{xs}.$$

Since $\left(\sqrt{\dfrac{s}{x}}u^a\right)^\top \left(\sqrt{\dfrac{x}{s}}v^a\right) = (u^a)^\top v^a \geq 0$, with Lemma 20.24, we obtain:

$$\|u^a v^a\| = \left\|\left(\sqrt{\frac{s}{x}}u^a\right)\left(\sqrt{\frac{x}{s}}v^a\right)\right\| \leq \frac{1}{\sqrt{8}}\|\sqrt{xs}\|^2 = \frac{1}{\sqrt{8}}x^\top s.$$

Item (i) follows. Using the expression of δ^a given in 21.2.2, we have

$$\delta^a = \left\|\frac{xs}{\mu} - 1 + \frac{\theta^2}{(1-\theta)}\frac{u^a v^a}{\mu}\right\| \leq \left\|\frac{xs}{\mu} - 1\right\| + \frac{\theta^2}{1-\theta}\frac{\|u^a v^a\|}{\mu}$$
$$\leq \frac{\alpha}{2} + \frac{\theta^2}{1-\theta}\frac{\|u^a v^a\|}{\mu}.$$

The function $\theta^2/(1 - \theta)$ is nondecreasing in $(0, 1)$, and the condition for acceptation is $\delta^a \leq \alpha$, we therefore have

$$\frac{(\theta^a)^2}{1 - \theta^a}\frac{\|u^a v^a\|}{\mu} \geq \frac{\alpha}{2},$$

hence (ii). We have $\left\|\dfrac{xs}{\mu} - 1\right\| \leq \dfrac{\alpha}{2} \leq \dfrac{1}{4}$, hence $x_i s_i \leq \dfrac{5}{4}\mu$ and with (i)

$$\|u^a v^a\| \leq \frac{5/4}{\sqrt{8}}n\mu \leq \frac{n}{2}\mu,$$

hence $\dfrac{\mu}{\|u^a v^a\|} \geq \dfrac{2}{n}$. Combining with (ii), we obtain $\dfrac{(\theta^a)^2}{1 - \theta^a} \geq \dfrac{\alpha}{n}$. If $\theta^a \geq 1/2$, the conclusion follows (for $\alpha \leq 1/2$ and $n \geq 1$). Otherwise, $1 - \theta^a \geq 1/2$, hence $(\theta^a)^2 \geq (1 - \theta^a)\dfrac{\alpha}{n} \geq \dfrac{\alpha}{2n}$, hence (iii). $\qquad\square$

Complexity Estimate of the Algorithm Lemma 21.5 shows that the centering step divides the proximity step by at least $1/2$, which allows us to check (Lemma 21.6) that the affine step is at least $\sqrt{\alpha/2n}$. From the formula of $\mu_\sharp$, we have

$$\mu_k \leq \left(1 - \sqrt{\frac{\alpha}{2n}}\right)^k \mu_0.$$

We will have that $\mu_k \leq \mu_\infty$ as soon as

$$k \log\left(1 - \sqrt{\frac{\alpha}{2n}}\right) + \log \mu_0 \leq \log \mu_\infty,$$

i.e.

$$k \left|\log\left(1 - \sqrt{\frac{\alpha}{2n}}\right)\right| \geq \log \frac{\mu_0}{\mu_\infty} = \bar{L}.$$

But $|\log(1 - \beta)| \geq \beta$ if $\beta \in [0, 1[$, hence $k = \sqrt{\frac{2n}{\alpha}} \bar{L} = O(\sqrt{n}\bar{L})$ is (rounding off to the least upper integer) an estimate larger to the number of iterations of the algorithm.

21.3.4 Asymptotic Speed of Convergence

a) Consider first the case where strict complementarity holds. Lemmas 20.10 and 20.12 show that, under this assumption, we have that $u^a = O(\mu)$ and $v^a = O(\mu)$. In view of Lemma 21.6(ii), since $\theta^a \leq 1$, we have that

$$\frac{1}{1 - \theta^a} \geq \frac{\alpha\mu}{2\|u^a v^a\|},$$

and therefore

$$\mu_\sharp = (1 - \theta^a)\mu \leq \frac{2}{\alpha}\|u^a v^a\| = O(\mu^2),$$

which is the required result.

b) Consider now the case without strict complementarity. From Theorem 20.14 and Lemmas 20.10 and 20.12, we have that

$$x_T \approx O(\sqrt{\mu}); \quad s_T \approx O(\sqrt{\mu}); \quad u_B^a = O(\sqrt{\mu}); \quad v_B^a = O(\mu);$$

$$u_T^a = -\frac{1}{2}x_T + O(\mu^{3/4}); \quad v_T^a = -\frac{1}{2}s_T + O(\mu^{3/4}).$$

As a result,

$$\|u^a v^a\|^2 = \|u_T^a v_T^a\|^2 + \|u_B^a v_B^a\|^2 = \frac{1}{16}\|x_T s_T\|^2 + O(\mu^{5/2}),$$

and therefore

$$\|u^a v^a\| = \frac{1}{4}\|x_T s_T\| \left(1 + \frac{O(\mu^{5/2})}{\|x_T s_T\|^2}\right) = \frac{1}{4}\|x_T s_T\| + O(\mu^{3/2}).$$

Since $(x, s, \mu) \in V_\alpha$:

$$\frac{\|x_T s_T\|}{\mu} \leq \frac{\|x_T s_T\|_\infty}{\mu} \sqrt{|T|} \leq (1 + \alpha/2)\sqrt{|T|}.$$

Combining Lemma 21.6(ii) with the above two inequalities, we obtain

$$\frac{(\theta^a)^2}{1 - \theta^a} \geq \frac{\alpha\mu}{2\|u^a v^a\|} \geq \frac{2\alpha|T|^{-1/2}}{(1 + \alpha/2)} + O(\mu^{1/2}) \geq \alpha|T|^{-1/2} + O(\mu^{1/2}). \quad (21.4)$$

If $\theta^a \geq 1/2$, the conclusion follows, otherwise $(1 - \theta^a) \geq 2/3$, hence

$$(\theta^a)^2 \geq (1 - \theta^a)\alpha|T|^{-1/2} + O(\mu^{1/2}) \geq \frac{\alpha}{2}|T|^{-1/2} + O(\mu^{1/2}).$$

Then $\mu_{k+1} = (1 - \theta^a)\mu_k \leq (1 - \alpha^{1/2}|T|^{-1/4}/2)\mu_k + O(\mu_k^{3/2})$, and the result follows.

21.4 A Predictor-Corrector Algorithm with Modified Field

21.4.1 Principle

If strict complementarity does not hold, the asymptotic speed of the small-neighborhood predictor-corrector algorithm presented in the previous section is after all rather low, especially if $|T|$ is large (we have given only an upper estimate of the convergence speed, but a lower estimate of the same order can be found in Mizuno [260]). However, we have studied in Chap. 20 a modified field (u^M, v^M) solving the linear system

$$\begin{cases} su^M + xv^M = -x_M s_M, \\ Qu^M + Rv^M = 0. \end{cases} \quad (21.5)$$

In this equation $M \subset \{1, \cdots, n\}$ is an estimate of T computed by the algorithm, and x_M, s_M is the restriction of x and s to M, identified to its extension by 0 on the whole of $\mathbb{R}^n$. It has been established that, if μ is small and $M = T$, then the sum of the affine field and of the modified field allows the computation of a point whose distance to $S(LCP)$ is of order $\mu^{3/4}$ (while the distance from the current point to $S(LCP)$ is of order $\mu^{1/2}$). However, this modified field is useful only in a neighborhood of $S(LCP)$. Close to the starting point of the algorithm, it can produce a useless direction, and impair convergence. To obtain an efficient implementable algorithm, it is therefore necessary to conceive a mechanism to eliminate the influence of the modified field if it does not produce an interesting direction, so as to preserve global convergence, while ensuring a step close to 1 when approaching $S(LCP)$. A classical mechanism in such a situation consists in seeking the point along a trajectory of the type

$$(x(\theta), s(\theta)) := (x, s) + \theta(u^a, v^a) + \varphi(\theta)(u^M, v^M), \tag{21.6}$$

where φ is a continuous function satisfying $\varphi(0) = \varphi'(0) = 0$ and $\varphi(1) = 1$. The usual choice is $\varphi(\theta) = \theta^2$; however, the convergence analysis to follow works with a specific choice of φ only, suggested by Theorem 20.23. One checks in this latter result that, along the affine trajectory,

$$x_T = \sqrt{\mu}\hat{x}_T + o(\sqrt{\mu}) \quad \text{and} \quad s_T = \sqrt{\mu}\hat{s}_T + o(\sqrt{\mu}),$$

where $\hat{x}_T$ and $\hat{s}_T$ are defined in this theorem. Let θ be the step-value. Denote by x^θ and s^θ the values of x and s *along the affine trajectory*, associated with the parameter $\mu_\theta := (1-\theta)\mu$ (not to be confused with $x(\theta)$ and $s(\theta)$; likewise, the notations u^θ, v^θ should not be confused with the notations $u(\theta)$ and $v(\theta)$ defined in (21.10)). Set

$$u(\theta) := x^\theta - x; \quad v(\theta) := s^\theta - s.$$

It would be ideal to compute a move (u, v) close to $(u(\theta), v(\theta))$. For this, we will evaluate the quantity $su(\theta) + xv(\theta)$. Note first that

$$u_T(\theta) := x_T^\theta - x_T = (\sqrt{1-\theta} - 1)\sqrt{\mu}\hat{x}_T + o(\sqrt{\mu}) = (\sqrt{1-\theta} - 1)x_T + o(\sqrt{\mu}),$$

and likewise

$$v_T(\theta) := (\sqrt{1-\theta} - 1)s_T + o(\sqrt{\mu}),$$

so that

$$u_T(\theta)v_T(\theta) = (\sqrt{1-\theta} - 1)^2 x_T s_T + o(\mu).$$

Besides

$$su(\theta) + xv(\theta) = x^\theta s^\theta - xs - (x^\theta - x)(s^\theta - s) = -\theta xs - u(\theta)v(\theta), \tag{21.7}$$

and therefore

$$\begin{cases} s_B u_B(\theta) + x_B v_B(\theta) = -\theta xs + o(\mu), \\ s_T u_T(\theta) + x_T v_T(\theta) = -\left(\theta + (\sqrt{1-\theta} - 1)^2\right) x_T s_T + o(\mu). \end{cases}$$

Said otherwise

$$su(\theta) + xv(\theta) = \theta(su^a + xv^a) + \varphi(\theta)(su^M + xv^M) + o(\mu),$$

where (u^M, v^M) solves (21.5) with $M = T$ and

$$\varphi(\theta) := (\sqrt{1-\theta} - 1)^2. \tag{21.8}$$

Note that $\varphi(0) = \varphi'(0) = 0$ and $\varphi(1) = 1$. This suggests to search a point along the path defined by (21.6), with φ as above. We set

$$f(\theta) = (1 + \sqrt{1-\theta})^{-2}. \tag{21.9}$$

Then $\varphi(\theta) := \theta^2 f(\theta)$, and (21.6) writes

$$\begin{cases} (x(\theta), s(\theta)) := (x, s) + \theta(u^\theta, v^\theta) \\ \\ (u^\theta, v^\theta) \quad := (u^a, v^a) + \theta f(\theta)(u^M, v^M). \end{cases} \tag{21.10}$$

21.4.2 Statement of the Algorithm. Main Result

The predictor-corrector algorithm with modified field goes as follows:

Algorithm 21.7. PCM (Predictor-Corrector Algorithm, Modified Field)
   Data: $\mu_\infty > 0$, $\alpha \in (0, 1/2]$; $(x^0, s^0, \mu_0) \in \mathcal{V}_\alpha$; $k \leftarrow 0$.
   REPEAT
   - $w \leftarrow w^k$.
   - Centralization: compute (w^c) solving (21.2); $w \leftarrow w^c$.
   - Affine move: compute M. Compute (u^a, v^a) and (u^M, v^M) solving
     (21.3) and (21.5). Define $x(\theta), s(\theta)$ by (21.10); $\mu(\theta) := (1 - \theta)\mu$.
     Compute θ^a, the largest value in $]0, 1[$ such that
     $(x(\theta), s(\theta), \mu(\theta)) \in \mathcal{V}_\alpha$, $\forall \, \theta \in [0, \theta^a]$.
     $x^{k+1} \leftarrow x(\theta^a)$, $s^{k+1} \leftarrow s(\theta^a)$, $\mu_{k+1} \leftarrow (1 - \theta^a)\mu_k$; $k \leftarrow k + 1$.
   UNTIL $\mu_k < \mu_\infty$.

Theorem 21.8. *Let (x^k, s^k, μ_k) be computed by Algorithm* **PCM**, *with M obtained by the estimate in Lemma 20.16. Then*
(i) *If $\mu_\infty > 0$, set $\bar{L} := \log(\mu_0/\mu_\infty)$. Algorithm* **PCM** *stops after at most $O(\sqrt{n}\bar{L})$ iterations (more precisely after at most $4\sqrt{n/\alpha}\bar{L}$ iterations).*
(ii) *If $\mu_\infty = 0$, then the algorithm identifies the set T (i.e. we have $M = T$) after finitely many iterations, and moreover $\mu_{k+1} = O(\mu_k^{5/4})$. If, in addition, $T = \emptyset$, convergence of the sequence $\{\mu_k\}$ is quadratic.*

   This theorem is proved below.

21.4.3 Complexity Analysis

We will show that the affine step θ^a is at least $\frac{1}{4}\sqrt{\alpha/n}$ at each iteration. Since the centering step is the same as in Algorithm 21.3, we know that the point obtained after centering lies in $\mathcal{V}_{\alpha/2}$. Accordingly, let us study the affine step. From (21.10)

$$x(\theta)s(\theta) = xs + \theta \left(su^\theta + xv^\theta \right) + \theta^2 u^\theta v^\theta = (1 - \theta)xs - \theta^2 f(\theta)x_M s_M + \theta^2 u^\theta v^\theta.$$

As a result, setting $\mu_\theta := (1 - \theta)\mu$:

$$\frac{x(\theta)s(\theta)}{\mu_\theta} - \mathbf{1} = \frac{xs}{\mu} - \mathbf{1} - \frac{\theta^2 f(\theta)}{(1 - \theta)} \frac{x_M s_M}{\mu} + \frac{\theta^2}{(1 - \theta)} \frac{u^\theta v^\theta}{\mu}. \qquad (21.11)$$

Using $f(\theta) \leq 1$, and setting $d := \sqrt{x/s}$, we get with Mizuno's Lemma 20.24

$$\|u^\theta v^\theta\| = \|d^{-1}u^\theta \, dv^\theta\| \leq \frac{1}{\sqrt{8}}\|d^{-1}u^\theta + dv^\theta\|^2,$$

$$= \frac{1}{\sqrt{8}} \left\| -\sqrt{xs} - \theta f(\theta)\sqrt{x_M s_M} \right\|^2 \leq \frac{1}{\sqrt{8}} \left(\|\sqrt{xs}\| + \|\sqrt{x_M s_M}\| \right)^2,$$

$$\leq \|\sqrt{xs}\|^2 \sqrt{2},$$

and hence,

$$\frac{\|u^\theta v^\theta\|}{\mu} \le \frac{x^\top s}{\mu}\sqrt{2}. \tag{21.12}$$

In particular, since $(x, s, \mu) \in \mathcal{V}_{\alpha/2}$, we have

$$\frac{\|u^\theta v^\theta\|}{\mu} \le n(1 + \alpha/2)\sqrt{2} \le 2n.$$

On the other hand,

$$\frac{\|x_M s_M\|}{\mu} \le \frac{\|x_M s_M\|_\infty}{\mu}\sqrt{|M|} \le (1 + \alpha/2)\sqrt{|M|} \le \frac{5}{4}\sqrt{n}.$$

If $\theta \ge 1/2$, we obtain the result; otherwise, using $(1 - \theta)^{-1} \le 2$ if $\theta \le 1/2$ and $f(\theta) \le 1$ for $\theta \in (0, 1)$, we get with (21.11)

$$\left\|\frac{x(\theta)s(\theta)}{\mu_\theta} - 1\right\| \le \frac{\alpha}{2} + \frac{\theta^2}{1 - \theta}\left(\frac{5}{4}\sqrt{n} + 2n\right) \le \frac{\alpha}{2} + 7n\theta^2.$$

As a result

$$\theta^a \ge \sqrt{\frac{\alpha}{14n}} \ge \frac{1}{4}\sqrt{\frac{\alpha}{n}},$$

which was to be proved.

21.4.4 Asymptotic Analysis

If $T = \emptyset$, then, after a finite number of iterations, $M = \emptyset$ by lemma 20.16, so that the iterates are the same as those computed by algorithm **PC**. Therefore, by theorem 21.4(iii), the quadratic convergence of $\{\mu_k\}$ towards 0 occurs.

It remains to deal with the case when $T \ne \emptyset$. Write f instead of $f(\theta)$, and note the key relation

$$f = \tfrac{1}{4}(1 + \theta f)^2.$$

In view of Lemma 20.10 and Theorems 20.14 and 20.17, when $M = T$ (which is true for M large enough) we have

$$u_T^a = u_T^M + O(\mu^{3/4}) = -\tfrac{1}{2}x_T + O(\mu^{3/4});$$
$$v_T^a = v_T^M + O(\mu^{3/4}) = -\tfrac{1}{2}s_T + O(\mu^{3/4}),$$
$$\|u_B^a\| + \|u_B^M\| = O(\sqrt{\mu}); \quad \|v_B^a\| = O(\mu); \quad \|v_B^M\| = O(\mu^{3/2});$$

Therefore

$$u^\theta v^\theta = (u^a + \theta f u^M)(v^a + \theta f v^M)$$
$$= \tfrac{1}{4}(1 + \theta f)^2 x_T s_T + O(\mu^{5/4}) = f x_T s_T + O(\mu^{5/4}).$$

Combining with (21.11), we get

$$\frac{x(\theta)s(\theta)}{\mu_\theta} - 1 = \frac{xs}{\mu} - 1 + \frac{1}{(1 - \theta)}O(\mu^{1/4}). \tag{21.13}$$

Since $\delta(x, s, \mu) \le \alpha/2$, we deduce that $(1 - \theta)^{-1}O(\mu^{1/4}) \ge \alpha/2$, and therefore $\mu_\sharp = (1 - \theta)\mu \le O(\mu^{5/4})$, which was to be proved.

21.5 A Large-Neighborhood Algorithm

21.5.1 Statement of the Algorithm. Main Result

As before, the algorithm alternates centering steps and affine moves. Unfortunately, using large neighborhoods is no longer a guarantee that a centering step with $\theta = 1$ does improve the centering. A stepsize must be computed, to sufficiently re-enter $\mathcal{N}_\epsilon$, so that the subsequent affine step is large enough. The algorithm leaves a choice between computing the optimal stepsize and estimating a simple suboptimal stepsize.

The statement of the following algorithm includes the choice at each iteration of a set $M \subset \{1, \cdots, n\}$ which approximates T. If $M = \emptyset$, the modified direction (u^M, v^M) vanishes and the move corresponds to the predictor-corrector method without modified field. Otherwise, the search for the new point is done on an arc as in the small-neighborhood algorithm. The distance function in this section is computed using the norm $\|x\|_\infty = \max_i |x_i|$. The algorithm is stated as follows:

Algorithm 21.9. PCL (Predictor-Corrector in Large Neighborhood)
  Data $\mu_\infty > 0$, $\varepsilon \in (0, 1/2]$; $(x^0, s^0, \mu_0) \in \mathcal{N}_\epsilon$. $\bar{\theta} := (1 - \varepsilon)\varepsilon^3 \sqrt{2}/n$. $k \leftarrow 0$.
  REPEAT
  - $w \leftarrow w^k$.
  - Centralization: compute w^c;
    take either θ equal to $\bar{\theta}$, or to the solution of

    $$\underset{\theta \in]0,1[}{\text{Max}} \ \text{dist}(\theta w^c + (1 - \theta)w, \partial \mathcal{N}_\epsilon); \quad \theta w^c + (1 - \theta)w \in \mathcal{N}_\epsilon,$$

    $w \leftarrow \theta w^c + (1 - \theta)w$.
  - Affine move: Choose M and compute (u^a, v^a) and (u^M, v^M) solving (21.3) and (21.5). Define $x(\theta), s(\theta)$ by (21.10).
    Compute θ^a, the largest value in $]0, 1[$ such that
    $(x(\theta), s(\theta), (1 - \theta)\mu) \in \mathcal{N}_\epsilon$, $\forall \theta \in [0, \theta^a]$.
    $x^{k+1} \leftarrow x(\theta^a)$, $s^{k+1} \leftarrow s(\theta^a)$, $\mu_{k+1} \leftarrow (1 - \theta^a)\mu_k$; $k \leftarrow k + 1$.
  UNTIL $\mu_k < \mu_\infty$.

Theorem 21.10. *Let (x^k, s^k, μ_k) be computed by Algorithm 21.7. Then*
(i) If $\mu_\infty > 0$, set $\bar{L} := \log(\mu_0/\mu_\infty)$. Algorithm **PCL** *stops after at most $O(n\bar{L})$ iterations (more precisely $7\varepsilon^{-2}n\bar{L}$ iterations).*
(ii) If (LCP) has a strictly complementary solution, $M = \emptyset$ at each iteration and $\mu_0 = 0$, then $\{\mu_k\}$ converges quadratically to 0.
(iii) If $\mu_\infty = 0$, and if M is the estimate of T of Lemma 20.16, then the algorithm identifies the set T (i.e. $M = T$) after finitely many iterations, and moreover $\mu_{k+1} = O(\mu_k^{5/4})$. If, in addition, $T = \emptyset$, then $\{\mu_k\}$ converges quadratically to 0.

Remark 21.11. If one chooses $M = \emptyset$ while $T \neq \emptyset$, an estimate of the asymptotic speed can be given, of the type of that obtained in Theorem 21.4.

□

21.5.2 Analysis of the Centering Step

We will prove

Lemma 21.12. *Let $w \in \mathcal{N}_\epsilon$ and w^c be the associated centering step. Set*

$$w^\theta := \theta w^c + (1-\theta)w \quad \text{and} \quad \bar\theta := (1-\varepsilon)\varepsilon^3 \frac{\sqrt 2}{n}.$$

Then $w^{\bar\theta} \in \mathcal{N}_\epsilon$ and $\mathrm{dist}(w^{\bar\theta}, \partial\mathcal{N}_\epsilon) \geq \dfrac{\varepsilon^3}{8n}$.

Proof. From 21.2.2, we know that

$$\frac{x^\sharp s^\sharp}{\mu^\sharp} - 1 = (1-\theta)\left(\frac{xs}{\mu} - 1\right) + \theta^2 \frac{u^c v^c}{\mu}.$$

Since $w \in \mathcal{N}_\epsilon$, we have

$$\left((1-\theta)(\varepsilon - 1) - \theta^2 \frac{\|u^c v^c\|_\infty}{\mu}\right)\mathbf{1} \leq \frac{x^\sharp s^\sharp}{\mu^\sharp} - 1$$
$$\leq \left((1-\theta)(\frac{1}{\varepsilon} - 1) + \theta^2 \frac{\|u^c v^c\|_\infty}{\mu}\right)\mathbf{1},$$

and therefore

$$\left(\varepsilon + (1-\varepsilon)\theta - \theta^2 \frac{\|u^c v^c\|_\infty}{\mu}\right)\mathbf{1} \leq \frac{x^\sharp s^\sharp}{\mu^\sharp} \leq \left(\theta + \frac{1-\theta}{\varepsilon} + \theta^2 \frac{\|u^c v^c\|_\infty}{\mu}\right)\mathbf{1},$$
$$(*)$$

According to Lemma 21.5, we have

$$\frac{\|u^c v^c\|}{\mu} \leq \frac{1}{\sqrt 8}\left\|\frac{\mu}{xs}\right\|_\infty \left\|1 - \frac{xs}{\mu}\right\|^2 \leq \frac{n}{\varepsilon\sqrt 8}\left\|1 - \frac{xs}{\mu}\right\|_\infty^2 \leq \frac{n}{\varepsilon^3\sqrt 8}.$$

Since $\|u^c v^c\|_\infty \leq \|u^c v^c\|$, it follows with $(*)$

$$\left(\varepsilon + (1-\varepsilon)\theta - \frac{n\theta^2}{\varepsilon^3\sqrt 8}\right)\mathbf{1} \leq \frac{x^\sharp s^\sharp}{\mu^\sharp} \leq \left(\theta + \frac{1-\theta}{\varepsilon} + \frac{n\theta^2}{\varepsilon^3\sqrt 8}\right)\mathbf{1}.$$

Set

$$\Delta := \min\left((1-\varepsilon)\theta - \frac{n\theta^2}{\varepsilon^3\sqrt 8}, (\varepsilon^{-1} - 1)\theta - \frac{n\theta^2}{\varepsilon^3\sqrt 8}\right) = (1-\varepsilon)\theta - \frac{n\theta^2}{\varepsilon^3\sqrt 8}.$$

If $\Delta \geq 0$, we have $w^\sharp \in \mathcal{N}_\epsilon$ and $\mathrm{dist}(w^\sharp, \partial\mathcal{N}_\epsilon) \geq \Delta$. The maximal value of Δ is obtained for $\theta = \bar\theta$. Using $\varepsilon \leq 1/2$, we obtain

$$\mathrm{dist}(w^\sharp, \partial\mathcal{N}_\epsilon) \geq \frac{1}{2}(1-\varepsilon)^2 \frac{\varepsilon^3\sqrt 2}{n} \geq \frac{\sqrt 2}{8}\frac{\varepsilon^3}{n} \geq \frac{\varepsilon^3}{8n}.$$

$\square$

21.5.3 Analysis of the Affine Step

During the affine step, we have from (21.11), the relation $f(\theta) \leq 1$, $\theta \in (0,1)$ and the preceding lemma:

$$\frac{\theta^2}{1-\theta}\left(\frac{\|x_M s_M\|_\infty}{\mu} + \frac{\|u^\theta v^\theta\|_\infty}{\mu}\right) \geq \frac{\varepsilon^3}{8n}.$$

Using (21.12) and the fact that $(x, s, \mu) \in \mathcal{N}_\epsilon$, we get

$$\frac{\|x_M s_M\|_\infty}{\mu} + \frac{\|u^\theta v^\theta\|_\infty}{\mu} \leq \frac{\|xs\|_\infty}{\mu} + \frac{x^\top s}{\mu}\sqrt{2} \leq \frac{1}{\varepsilon} + n\frac{\sqrt{2}}{\varepsilon} \leq \frac{3n}{\varepsilon}, \quad (21.14)$$

and therefore

$$\frac{\theta^2}{1-\theta} \geq \frac{\varepsilon^4}{24n^2}. \tag{$*$}$$

In particular, if $\theta \leq 1/2$, we will have $\theta^2 \geq \varepsilon^4/48n^2$, hence $\theta \leq \varepsilon^2/7n$ which is indeed smaller than $1/2$. We therefore have $\mu_k \leq (1 - \frac{\varepsilon^2}{7n})^k \mu_0$. The condition $\mu_k \leq \mu_\infty$ is satisfied as soon as $(1 - \frac{\varepsilon^2}{7n})^k \leq \mu_\infty/\mu_0$, which in turn is satisfied when (barring roundoff)

$$k \geq \frac{\log(\mu_\infty/\mu_0)}{|\log(1 - \varepsilon^2/7n)|} \geq \frac{\bar{L}}{\varepsilon^2/7n} = \frac{7}{\varepsilon^2}n\bar{L},$$

which was to be proved.

21.5.4 Asymptotic Convergence

The argument is completely similar to that of the predictor-corrector method with small neighborhood. In fact, the relation (21.13) is valid in a large neighborhood. With Lemma 21.12, we deduce that $(1 - \theta) = O(\mu^{1/4})$, hence $\mu_\sharp = (1 - \theta)\mu = O(\mu^{5/4})$.

21.6 Practical Aspects

The two non-trivial steps in the algorithm are the computation of the direction of move and the line-search. We have seen that the direction (u, v) solves (21.1). In the case of a quadratic problem, solving this problem amounts to computing (u, v, η), where η represents the variation of the multiplier, solving

$$\begin{cases} su + xv = \gamma\mu\mathbf{1} - xs, \\ Au = 0, \quad Hu + A^\top\eta = v. \end{cases} \tag{21.15}$$

After scaling, we obtain (in the feasible case) the system

$$\begin{cases} \bar{u} + \bar{v} = \bar{f}, \\ \bar{A}\bar{u} = 0, \quad \bar{H}\bar{u} + \bar{A}^{\top}\eta - \bar{v} = 0, \end{cases} \qquad (21.16)$$

with $\bar{f} := (\gamma\mu\mathbf{1} - xs)/\sqrt{xs}$, $\bar{A} := AD$ and $\bar{H} := DHD$. Eliminating $\bar{v}$, we get

$$\begin{cases} (I + \bar{H})\bar{u} + \bar{A}^{\top}\eta = \bar{f}, \\ \bar{A}\bar{u} = 0. \end{cases} \qquad (21.17)$$

Then, eliminating $\bar{u}$, we obtain the reduced system

$$\bar{A}(I + \bar{H})^{-1}\bar{A}^{\top}\eta = \bar{A}(I + \bar{H})^{-1}\bar{f}. \qquad (21.18)$$

One can thus solve the symmetric (but not positive definite) system (21.17), which is equivalent to the quadratic problem

$$\underset{\bar{u}}{\text{Min}} \, \bar{f}^{\top}\bar{u} + \frac{1}{2}\bar{u}^{\top}(I + \bar{H})\bar{u}; \quad \bar{A}\bar{u} = 0,$$

or the reduced system. If the data A and H are sparse, then the system (21.17) is sparse. Such is not always the case of the reduced system, and this for two reasons. First, $(I + \bar{H})^{-1}$ can be dense; however, in the case of linear programming, and more generally if H is diagonal, this matrix is diagonal. Then the matrix of the reduced system has the same sparsity as $AA^{\top}$. On the other hand, $AA^{\top}$ is dense as soon as A has at least one dense column.

The matrix $AA^{\top}$ is positive definite if A has full rank. In this case, it can be factorized by the Cholesky method. The factorization being stable independently of the order of the pivots, this order can be chosen so as to minimize the density of the Cholesky factor, by a "symbolic" factorization which will be performed only once.

If A has full (almost) columns, then $AA^{\top}$ is (almost) full. Therefore dense columns of A are treated separately, through a recursive use of the Morrison-Sherman-Woodbury formula, see Wright [366].

21.7 Comments

The first primal-dual methods, based on decreasing the primal-dual potential

$$q \log x^{\top}s + \pi(x) + \pi(s)$$

with $q > 0$ fixed, also have a complexity of $O(\sqrt{n}L)$ iterations, see Kojima et al. [217]. The advantage of methods based on the concept of neighborhood is to allow a fast asymptotic convergence.

The predictor-corrector method is due to Mizuno, Todd and Ye [262]. This is the first algorithm enjoying at the same time a complexity of $O(\sqrt{n}\bar{L})$ iterations and a quadratic convergence. The large-neighborhood algorithm of this chapter, including the modified field, seems to be new. An analysis of the large-neighborhood algorithm, without modified field, but with possible errors in the computation of directions, is presented in Bonnans, Pola and Rebaï [47]. The theory of modified field is stated in Mizuno [260], in the non-feasible framework that we present in the next chapter.

22 Non-Feasible Algorithms

22.1 Overview

Let us study the resolution of the linear monotone complementarity problem without assuming the knowledge of a starting point in a given neighborhood of the central path. The algorithm is based on solving by Newton's method the central path equations, starting from a non-feasible point. Implementing the algorithm is as simple as in the feasible case. Unfortunately, the complexity estimate in small neighborhoods in now $O(n\bar{L})$ iterations, compared to $O(\sqrt{n}\bar{L})$ in the feasible case.

A large part of the analysis is similar to that of the feasible case. The specific difficulties lie in the estimate of the affine step to obtain the desired complexity.

22.2 Principle of the Non-Feasible Path Following

22.2.1 Non-Feasible Central Path

Recall the format of the monotone complementarity problem

$$\begin{cases} xs & = & 0, \\ Qx + Rs = & h, \\ x \geq 0, & s \geq 0, \end{cases} \qquad (LCP)$$

where monotonicity is defined by the implication

$$Qu + Rv = 0 \Rightarrow u^\top v \geq 0.$$

While it is in general impossible to obtain simply a feasible starting point (and a fortiori close to the central path), we can always compute $(x^0, s^0) \in \mathbb{R}^n_{++} \times \mathbb{R}^n_{++}$ and $\mu_0 > 0$ such that $x^0 s^0 = \mu_0 \mathbf{1}$. One can for example take $x^0 = \mu_{01}\mathbf{1}$ and $s^0 = \mu_{02}\mathbf{1}$, with $\mu_{01} > 0$ and $\mu_{02} > 0$ such that $\mu_{01}\mu_{02} = \mu_0$. Set

$$g := \frac{1}{\mu_0}(h - Qx^0 - Rs^0).$$

Then (x^0, s^0, μ_0) lies in the *perturbed central path* associated with g, defined by

$$\mathcal{C}^g(LCP) := \{(x, s, \mu) \in \mathbb{R}^n_{++} \times \mathbb{R}^n_{++} \times \mathbb{R}_{++} \; ; \; xs = \mu\mathbf{1}, \quad Qx + Rs = h - \mu g\}.$$

If we can construct a sequence of points close enough to the perturbed central path, with $\mu_k \to 0$, then every cluster point of this sequence solves (LCP).

The algorithmic family to be studied computes displacements via Newton's method applied to the equation of the perturbed central path. Accordingly the linear equation

$$Qx + Rs = h - \mu g$$

will be satisfied at each iteration. Said otherwise, the sequence thus constructed will stay in the set

$$F^g(LCP) := \{(x, s, \mu) \in \mathbb{R}^n_{++} \times \mathbb{R}^n_{++} \times \mathbb{R}_{++} \; ; \quad Qx + Rs = h - \mu g\}$$

which plays the role of a *feasible set perturbed* by g. Then the proximity to the perturbed central path can be measured, as in the feasible case, by

$$\delta(x, s, \mu) = \left\| \frac{xs}{\mu} - \mathbf{1} \right\|,$$

and the small and large neighborhoods associated with the perturbed central path will be defined as

$$\mathcal{V}^g_\alpha := \left\{ (x, s, \mu) \in F^g(LCP); \quad \left\| \frac{xs}{\mu} - \mathbf{1} \right\| \leq \alpha \right\},$$

$$\mathcal{N}^g_\epsilon := \left\{ (x, s, \mu) \in F^g(LCP); \quad \epsilon\mathbf{1} \leq \frac{xs}{\mu} \leq \frac{1}{\epsilon}\mathbf{1} \right\}.$$

In this approach, infeasibility is reduced at the same speed as $x^\top s$, which is of course arbitrary, these two quantities being of different natures. This suggests an appropriate magnitude for μ_0: in fact, if μ_{01} and μ_{02} are of same order, then if $\mu_0 \downarrow 0$ we have $\|g\| \to \infty$ (in general), while if $\mu_0 \uparrow \infty$, we have $\|g\| \to 0$. It seems therefore reasonable to choose μ_{01} and μ_{02} so that μ_0 and $\|g\|$ are of same order.

22.2.2 Directions of Move

The problem is to solve the system

$$\begin{cases} xs = \mu\mathbf{1}, \\ Qx + Rs = h - \mu g, \end{cases}$$

by a Newton method. We will denote by $w := (x, s, \mu)$ the current point satisfying the linear constraint. We aim at a new value of μ equal to $\gamma\mu$,

$\gamma \in [0, 1]$. Newton's method writes, linearizing at $(x, s) \in F^g(LCP)$, and denoting by (u, v) the displacement:

$$\begin{cases} su + xv = \gamma\mu\mathbf{1} - xs, \\ Qu + Rs = (1 - \gamma)\mu g. \end{cases} \tag{22.1}$$

We call $\theta > 0$ the step in the direction (u, v). The new point is

$$w^\sharp := (x^\sharp, s^\sharp, \mu_\sharp); \quad x^\sharp := x + \theta u; \quad s^\sharp := s + \theta v, \quad \mu_\sharp := (1 - \theta + \theta\gamma)\mu.$$

The centering and affine steps are obtained by taking respectively γ equal to 1 and 0, and their equations are therefore

$$\begin{cases} su^c + xv^c = \mu\mathbf{1} - xs, \\ Qu^c + Rv^c = 0, \end{cases} \tag{22.2}$$

$$\begin{cases} su^a + xv^a = -xs, \\ Qu^a + Rv^a = \mu g. \end{cases} \tag{22.3}$$

The general case is obtained by convex combination of the preceding two values. As in the feasible case, we obtain

$$\begin{cases} w^\sharp & = (1 - \theta)w + \theta(\gamma w^c + (1 - \gamma)w^a), \\ \\ \dfrac{x^\sharp s^\sharp}{\mu_\sharp} - \mathbf{1} & = \dfrac{(1 - \theta)xs + \theta\gamma\mu\mathbf{1} + \theta^2 uv}{(1 - \theta + \theta\gamma)\mu} - \mathbf{1} \\ \\ & = \dfrac{1 - \theta}{1 - \theta + \theta\gamma}\left(\dfrac{xs}{\mu} - \mathbf{1}\right) + \theta^2\dfrac{uv}{\mu_\sharp}. \end{cases}$$

As a result, for the centering step ($\theta = 1$, $\gamma = 1$, $\mu_\sharp = \mu$) and for the affine step ($\gamma = 0$, $\mu_\sharp = (1 - \theta)\mu$), we obtain the same formulae as in the feasible case

$$\frac{x^c s^c}{\mu} - \mathbf{1} = \frac{u^c v^c}{\mu}; \quad \frac{x^a s^a}{(1 - \theta)\mu} - \mathbf{1} = \frac{xs}{\mu} - \mathbf{1} + \frac{\theta^2}{1 - \theta}\frac{u^a v^a}{\mu}.$$

The analysis of the centering steps is strictly identical to that of the feasible case. By contrast, the relation $(u^a)^\top v^a \geq 0$ is destroyed, which substantially complicates the estimates of the affine step.

Before stating the algorithm, let us establish some properties of neighborhoods of the perturbed central path.

22.2.3 Orders of Magnitude of Approximately Centered Points

We start with a compactness result. We will say that (x^0, s^0) *dominates* (x^*, s^*) if $x^0 \geq x^*$ and $s^0 \geq s^*$.

Lemma 22.1. *Take* (x, s, μ) *and* (x^0, s^0, μ_0) *in* $\mathcal{N}_\varepsilon^g$, *with* $\mu \leq \mu_0$, *and* $(x^*, s^*) \in S(LCP)$. *Then*

$$x^\top s^0 + s^\top x^0 \leq (\mu + \mu_0)\frac{n}{\varepsilon} + \frac{\mu_0 - \mu}{\mu_0}\left((x^0)^\top s^* + (s^0)^\top x^*\right).$$

If, in addition, (x^0, s^0) *dominates* (x^*, s^*), *then*

$$x^\top s^0 + s^\top x^0 \leq 2\mu_0\frac{n}{\varepsilon}.$$

Proof. Due to the invariance properties discussed in chapter 20, we can without loss of generality assume $N = \emptyset$, and therefore $s^* = 0$. Let $(x, s, \mu) \in \mathcal{N}_\varepsilon^g$. Then

$$Q(\mu_0 x - \mu x^0 - (\mu_0 - \mu)x^*) + R(\mu_0 s - \mu s^0) = 0.$$

The pair (Q, R) being monotone, we deduce that

$$(\mu_0 x - \mu x^0 - (\mu_0 - \mu)x^*)^\top (\mu_0 s - \mu s^0) \geq 0.$$

Using $x^\top s \leq n\mu/\varepsilon$, $(x^0)^\top s^0 \leq n\mu_0/\varepsilon$, $s^\top x^* \geq 0$ and $\mu \leq \mu_0$, we get

$$\mu\mu_0(x^\top s^0 + s^\top x^0) \leq \mu_0^2 x^\top s + \mu^2 (x^0)^\top s^0 - (\mu_0 - \mu)(\mu_0 s - \mu s^0)^\top x^*,$$
$$\leq \frac{n}{\varepsilon}\mu_0^2\mu + \frac{n}{\varepsilon}\mu^2\mu_0 + \mu(\mu_0 - \mu)(s^0)^\top x^*.$$

Dividing by $\mu\mu_0$, we obtain the first relation (taking $s^* = 0$ into account). If (x^0, s^0) dominates (x^*, s^*), then

$$(x^0)^\top s^* + (s^0)^\top x^* = (s^0)^\top x^* \leq (s^0)^\top x^0 \leq \mu_0\frac{n}{\varepsilon}.$$

Combining with the first part of the lemma, the conclusion follows.   □

It is convenient to disclose a couple $(\tilde{x}, \tilde{s})$ such that

$$Q\tilde{x} + R\tilde{s} = g. \qquad (22.4)$$

Lemma 22.2. *If* $w = (x, s, \mu) \in \mathcal{N}_\varepsilon^g$, *then*

$$x_B \approx 1, \quad s_B \approx \mu, \quad x_T \approx \sqrt{\mu}, \quad s_T \approx \sqrt{\mu}.$$

Proof. From Lemma 22.1, we know that $x = O(1)$ and $s = O(1)$. By definition of $\mathcal{N}_\varepsilon^g$, $xs \approx \mu$. Let (x^*, s^*) be a solution of (LCP) such that $x_B^* > 0$. From

$$Qx + Rs + \mu g = h = Qx^*$$

and (22.4), we deduce

$$Q(x - x^* + \mu\tilde{x}) + R(s + \mu\tilde{s}) = 0.$$

Therefore $0 \leq (x - x^* + \mu\tilde{x})^\top (s + \mu\tilde{s})$, i.e., using again $s = O(1)$ and $x = O(1)$:

$$(x_B^*)^\top s_B = (x^*)^\top s \leq x^\top s + \mu\tilde{x}^\top s + \mu(x - x^* + \mu\tilde{x})^\top \tilde{s}$$
$$= x^\top s + O(\mu) = O(\mu).$$

Since $x_B^* > 0$, we deduce that $s_B = O(\mu)$; but $x_B = O(1)$ and $x_B s_B = O(\mu)$, hence $s_B \approx \mu$ and $x_B \approx 1$.

Proving the relations $x_T \approx \sqrt{\mu}$ and $s_T \approx \sqrt{\mu}$ is done as in the feasible case (Lemma 20.9).

$\square$

We now study orders of magnitude of the Newton direction. As with feasible algorithms, the aim of the analysis is to obtain a formula allowing the asymptotic evaluation of the size of the move and, in particular, that of the affine move. Let us scale the system, as in the feasible case. For this, set

$$d = \sqrt{\frac{x}{s}}, \quad \phi = \sqrt{\frac{xs}{\mu}}, \quad \bar{u} := d^{-1}u, \quad \bar{v} := dv, \quad \bar{Q} := QD, \quad \bar{R} := RD^{-1}.$$

$$(22.5)$$

From the previous lemma and the definition of $\mathcal{N}_\epsilon^g$, we have $d_B \approx \mu^{-1/2}$, $d_T \approx 1$ and $\phi \approx 1$. Dividing the first equation by $\sqrt{xs}$, we obtain the equation of the scaled direction

$$\begin{cases} \bar{u} + \bar{v} = \sqrt{\mu}(\gamma\phi^{-1} - \phi), \\ \bar{Q}\bar{u} + \bar{R}\bar{v} = \mu(1 - \gamma)g. \end{cases}$$

$$(22.6)$$

22.2.4 Analysis of Directions

The next lemmas will allow us to analyze accurately the directions of displacement. The two lemmas below estimate the size of the displacement and interpret the displacement of large variables as a perturbation of a certain projection.

Lemma 22.3. *The Newton direction satisfies*

(i) $\qquad\qquad \bar{u} = O(\sqrt{\mu}); \quad \bar{v} = O(\sqrt{\mu}),$

(ii) $\qquad\qquad v_B = O(\mu); \quad u_T = O(\sqrt{\mu}), \quad v_T = O(\sqrt{\mu}).$

Proof. Take $(\tilde{x}, \tilde{s})$ satisfying (22.4). Set

$$\hat{u} := \bar{u} - \mu(1 - \gamma)d^{-1}\tilde{x}, \quad \hat{v} := \bar{v} - \mu(1 - \gamma)d\tilde{s}.$$

$$(22.7)$$

By (22.6), we have that

$$\begin{cases} \hat{u} + \hat{v} = \sqrt{\mu}(\gamma\phi^{-1} - \phi) - \mu(1 - \gamma)(d^{-1}\tilde{x} + d\tilde{s}) = O(\sqrt{\mu}), \\ \bar{Q}\hat{u} + \bar{R}\hat{v} = 0. \end{cases}$$

$$(22.8)$$

Mizuno's Lemma 20.24 implies $\hat{u} = O(\sqrt{\mu})$ and $\hat{v} = O(\sqrt{\mu})$, hence $\bar{u} = O(\sqrt{\mu})$, $\bar{v} = O(\sqrt{\mu})$ and

$$v_B = d_B^{-1}\bar{v}_B = O(\mu), \quad u_T = d_T\bar{u}_T = O(\sqrt{\mu}), \quad v_T = d_T^{-1}\bar{v}_T = O(\sqrt{\mu}),$$

which was to be proved.

$\square$

Lemma 22.4. *Let $(\tilde{x}, \tilde{s})$ satisfying (22.4). Set $z := Rv + Q_T u_T$. The solution (u, v) of (22.1) satisfies*

$$\bar{u}_B = \gamma\sqrt{\mu}P_{\bar{Q}_B}(\phi_B^{-1} + \sqrt{\mu}d_B\tilde{s}_B) + P_{\bar{Q}_B, z}0, \qquad (22.9)$$

$$u_B = \gamma\sqrt{\mu}d_B P_{\bar{Q}_B}(\phi_B^{-1} + \sqrt{\mu}d_B\tilde{s}_B) + O(\sqrt{\mu}). \qquad (22.10)$$

If, in addition, strict complementarity holds $(T = \emptyset)$, then

$$u = \gamma\sqrt{\mu}dP_{\bar{Q}}(\phi^{-1} + \sqrt{\mu}d\tilde{s}) + O(\mu). \qquad (22.11)$$

In particular, $u^a = O(\mu)$ if $T = \emptyset$, and $u^a = O(\sqrt{\mu})$ in the general case.

Proof. Let us show (22.9). The optimality conditions of the least-squares problem

$$\underset{\bar{u}_B}{\text{Min}} \tfrac{1}{2}\|\bar{u}_B - \gamma\sqrt{\mu}\phi_B^{-1} - \mu d_B\tilde{s}_B\|^2; \quad \bar{Q}_B\bar{u}_B + z = 0,$$

are

$$\bar{u}_B - \gamma\sqrt{\mu}\phi_B^{-1} - \mu d_B\tilde{s}_B \in \mathcal{R}(\bar{Q}_B^\top); \quad \bar{Q}_B\bar{u}_B + z = 0.$$

The second relation is obviously satisfied. To check the first, take again the variables $(\hat{u}, \hat{v})$ defined in (22.7). From (22.8) and Lemma 20.11, it follows that $\hat{v}_B \in \mathcal{R}(\bar{Q}_B^\top)$. From

$$\begin{cases} \bar{u} + \hat{v} = \sqrt{\mu}\gamma(\phi^{-1} + \sqrt{\mu}d\tilde{s}) - \sqrt{\mu}(\phi + \sqrt{\mu}d\tilde{s}), \\ \bar{Q}_B\bar{u}_B + z = 0, \end{cases}$$

we deduce that

$$u_B = \sqrt{\mu}d_B P_{\bar{Q}_B}\left(\gamma(\phi_B^{-1} + \sqrt{\mu}d_B\tilde{s}_B) - \sqrt{\mu}(\phi_B + \sqrt{\mu}d_B\tilde{s}_B)\right) + d_B P_{\bar{Q}_B, z}0.$$

Let us show that $P_{\bar{Q}_B}(\phi_B + \sqrt{\mu}d\tilde{s}_B) = 0$, i.e.

$$\phi_B + \sqrt{\mu}d_B\tilde{s}_B \in \mathcal{N}(\bar{Q}_B^\top)^\perp = \mathcal{R}(\bar{Q}_B^\top). \qquad (22.12)$$

Since $\phi/d = s/\sqrt{\mu}$, this is just $s_B + \mu\tilde{s}_B \in \mathcal{R}(Q_B^\top)$, which is a consequence of

$$Q(x + \mu\tilde{x}) + R(s + \mu\tilde{s}) = h \in \mathcal{R}(Q)$$

and of Lemma 20.11. This proves (22.9).

One can check that $d_B P_{\bar{Q}_B, z}0 = O(\|z\|)$ proceeding as in the feasible case (Lemma 20.12). The conclusion then follows by combining (22.9) with the above relation, and noting that, from Lemma 22.3, $z = Rv = O(\mu)$ if $T = \emptyset$, and $z = Rv + Q_T u_T = O(\sqrt{\mu})$ if not.

$\square$

Lemma 22.5. *Take $(x, s) \in \mathcal{N}_\epsilon^g$ and (x^*, s^*) be the element of $S(LCP)$ closest to (x, s) (for the Euclidean norm). If strict complementarity holds, then*

$$(x^*, s^*) = (x, s) + O(\mu).\tag{22.13}$$

In the general case, we have that

$$(x^*, s^*) = (x, s) + O(\sqrt{\mu}).\tag{22.14}$$

Proof. Proceeding as in the feasible case, check that for μ small enough, the projection of (x, s) onto (LCP) coincides with the projection onto

$$\mathcal{V} := \{(x, s) \in \mathbb{R}^n ; x_T = 0; \ s = 0; \ Qx = h\}.$$

We therefore have that $(x^*, s^*) - (x, s) = O(\|x_T\| + \|s\|)$, hence the conclusion follows with Lemma 22.2.

$\square$

The next theorem gives an accurate estimate of the result of an affine move.

Theorem 22.6. *If strict complementarity holds, then*

$$s + v^a = O(\mu^2).\tag{22.15}$$

In the general case, we have that

$$s_B + v_B^a = O(\mu^{3/2})\tag{22.16}$$

and

$$x_T + u_T^a = \tfrac{1}{2}x_T + O(\mu^{3/4}); \quad s_T + v_T^a = \tfrac{1}{2}s_T + O(\mu^{3/4}).\tag{22.17}$$

Proof. From the equation of the affine step, we deduce that $s + v^a = -sx^{-1}u^a$. Combining with Lemmas 22.2 and 22.4, we get $s_B + v_B^a = O(\mu\|u_B^a\|)$, which is of order μ^2 if $T = \emptyset$, and $\mu^{3/2}$ if not; hence (22.15)–(22.16).

Now let us establish (22.17). Call (x^I, s^I) the point twice as far as the affine direction:

$$x^I := x + 2u^a; \quad s^I := s + 2v^a.$$

Let (x^*, s^*) be the point of $S(LCP)$ closest to (x, s). From Lemmas 22.3, 22.4 and 22.5,

$$\|(x^*, s^*) - (x^I, s^I)\| \leq \|(x^*, s^*) - (x, s)\| + \|(u^a, v^a)\| = O(\sqrt{\mu}).$$

Supposing $s^* = 0$, obtain

$$Q(x^I - x^* - \mu\tilde{x}) + R(s^I - \mu\tilde{s}) = 0,$$

hence $(x^I - x^* - \mu\tilde{x})^\top (s^I - \mu\tilde{s}) \geq 0$, or equivalently

$$(x^I - x^*)^\top s^I \geq \mu\tilde{x}^\top s^I + \mu\tilde{s}^\top (x^I - x^* - \mu\tilde{x}) = O(\mu^{3/2}),$$

and thus, using $x_T^* = 0$:

$$(x_T^I)^\top (s_T^I) \geq (x_B^* - x_B^I)^\top s_B^I + O(\mu^{3/2}) = O(\mu^{3/2}).$$

We end the proof as in the feasible case: we have

$$d^{-1}x^I + ds^I = d^{-1}x + ds + 2(d^{-1}u^a + dv^a) = 2\sqrt{xs} - 2\sqrt{xs} = 0.$$

As a result, for all $i \in I$ we have

$$0 = (d_i^{-1}x_i^I + d_i s_i^I)^2 \geq 2x_i^I s_i^I, \qquad (**)$$

and with $(*)$, $0 \geq x_i^I s_i^I \geq (x_T^I)^\top (s_T^I) \geq O(\mu^{3/2})$. Combining with $(**)$, we get

$$(d_i^{-1}x_i^I)^2 + (d_i s_i^I)^2 \leq -2x_i^I s_i^I = O(\mu^{3/2}).$$

Since $d_T \approx 1$, we deduce that $(x_i^I)^2 + (s_i^I)^2 = O(\mu^{3/2})$ and the conclusion follows.

$\square$

22.2.5 Modified Field

As in the feasible case, the affine displacement yields, if $T = \emptyset$, a new point very close to $S(LCP)$; while if $T \neq \emptyset$, the estimate of the distance remains of order $\sqrt{\mu}$. Accordingly, we will build a modified field theory analogous to that of the feasible case. Let $\hat{T}(x, s, \mu)$ be defined as follows:

$$\hat{T}(x, s, \mu) := \left\{ i = 1, \ldots, n; \ u_i^a/x_i \in \left[-\tfrac{3}{4}, -\tfrac{1}{4}\right] \text{ and } v_i^a/s_i \in \left[-\tfrac{3}{4}, -\tfrac{1}{4}\right] \right\}.$$
(22.18)

We have $\hat{T}(x, s, \mu) = T$ if $(x, s, \mu) \in \mathcal{N}_\epsilon^g$ and μ is small enough.

The *modified field* (u^M, v^M) is defined as solving

$$\begin{cases} su^M + xv^M = -x_M s_M, \\ Qu^M + Rv^M = 0, \end{cases}$$
(22.19)

where x_M and s_M are the restrictions to M of x and s, identified with their extension by 0 to the whole of $\mathbb{R}^n$; said otherwise

$$(x_M)_i = (s_M)_i = 0, \ i \notin M \ ; \ (x_M)_i = x_i, \ (s_M)_i = s_i, \ i \in M.$$

Theorem 22.7. *Assume* $M = T$. *Then:*

(i) *The modified field is of the following order:*

$$u_B^M = O(\sqrt{\mu}), \quad v_B^M = O(\mu^{3/2}), \quad u_T^M = O(\sqrt{\mu}), \quad v_T^M = O(\sqrt{\mu}).$$

(ii) *We have that*

$$x_T + u_T^M = \tfrac{1}{2}x_T + O(\mu^{3/4}); \quad s_T + v_T^M = \tfrac{1}{2}s_T + O(\mu^{3/4}).$$

Proof. The proof is identical to that of the feasible case. Indeed, the latter uses orders of the variables that are identical and the monotonicity relation $Qu^M + Rv^M = 0$, which still holds.

$\square$

Combining Theorems 22.7 and 22.9, we deduce that the point

$$(x^{\#\#}, s^{\#\#}) := (x, s) + (u^a, v^a) + (u^M, v^M)$$

is at a distance from $S(LCP)$ of order $O(\mu^{3/4})$.

22.3 Non-Feasible Predictor-Corrector Algorithm

During the affine step, the new point is chosen on the path

$$(x(\theta), s(\theta)) = (x, s) + \theta(u^\theta, v^\theta) \tag{22.20}$$

with

$$(u^\theta, v^\theta) = (u^a, v^a) + \theta f(\theta)(u^M, v^M),$$

where $f(\theta)$ is defined by (21.9): $f(\theta) = (1 + \sqrt{1 - \theta})^{-2}$. As in the feasible case, the algorithm alternates centering steps ($\theta = 1$) with affine steps, in which θ is the largest step staying inside the neighborhood of fixed size $\alpha \in]0, 1/2]$. In other words:

Algorithm 22.8. IPC (Infeasible Predictor-Corrector Algorithm)
Data: $\mu_\infty > 0$, $\alpha \in (0, 1/2]$; $w_0 := (x^0, s^0, \mu_0) \in \mathcal{V}_\alpha^g$. $k \leftarrow 0$.
REPEAT
- $w \leftarrow w^k$.
- Centralization: compute w^c; $w \leftarrow w^c$.
- Affine step: choose M. Compute (u^a, v^a) and (u^M, v^M) solving (22.3) and (22.19). Define $x(\theta), s(\theta)$ by (22.20); $\mu(\theta) := (1 - \theta)\mu$. Compute θ^a, the largest value in $]0, 1[$ such that $(x(\theta), s(\theta), \mu(\theta)) \in \mathcal{V}_\alpha^g$, $\forall\, \theta \in [0, \theta^a]$. $x^{k+1} \leftarrow x(\theta^a)$, $s^{k+1} \leftarrow s(\theta^a)$, $\mu_{k+1} \leftarrow (1 - \theta^a)\mu_k$; $k \leftarrow k + 1$.
UNTIL $\mu_k < \mu_\infty$.

Recall that the point (x^0, s^0) *dominates* a solution of (LCP) if there exists $(x^*, s^*) \in S(LCP)$ such that $x^0 \geq x^*$ and $s^0 \geq s^*$. This assumption is used in the following theorem, while it was not required in the theory of feasible algorithms.

Theorem 22.9. (i) *If $\mu_\infty > 0$, set $\bar{L} := \log(\mu_0/\mu_\infty)$. If (x^0, s^0) dominates a solution of (LCP), then Algorithm* **IPC** *stops after at most $O(n\bar{L})$ iterations (more precisely $24\alpha^{-1/2}n\bar{L}$ iterations).*
(ii) *If $\mu_\infty = 0$ and M is the estimate $\hat{T}$ stated in (22.18), then the algorithm*

identifies the set T after finitely many iterations (i.e. $M = T$), and moreover $\mu_{k+1} = O(\mu_k^{5/4})$. If, in addition, $T = \emptyset$, convergence of the sequence $\{\mu_k\}$ is quadratic.

(iii) If $\mu_\infty = 0$ and $T = \emptyset$, and if $M = \emptyset$ at each iteration, then convergence of μ_k to 0 is quadratic.

This theorem will be proved below.

22.3.1 Complexity Analysis

The centering step being the same as in the feasible case, we have that

Lemma 22.10. *If $\delta(w) \le 1/2$, then $w^c \in \mathcal{V}_\alpha^g$ and $\delta(w^c) \le \dfrac{1}{2}\delta(w)$.*

To study global convergence, it is necessary to estimate the size of the Newton displacement, and in particular of the product $u^a v^a$. The relation $(u^a)^\top v^a \ge 0$ needs no longer hold, it is therefore necessary to start the analysis from the beginning. A first result in this direction is as follows.

Lemma 22.11. *Let $(\check{u}, \check{v})$ solve*

$$\check{u} + \check{v} = f; \quad Q\check{u} + R\check{v} = g,$$

and $\tilde{x}, \tilde{s}$ be such that $Q\tilde{x} + R\tilde{s} = g$. Then

$$\begin{cases} \|\check{u}\| \le \|f\| + \|\tilde{x}\| + \|\tilde{s}\|, \\ \|\check{v}\| \le \|f\| + \|\tilde{x}\| + \|\tilde{s}\|. \end{cases}$$

Proof. (i) Let us first check the result when $g = 0$; then $\check{u}^\top \check{v} \ge 0$ so

$$\|f\|^2 = \|\check{u} + \check{v}\|^2 \ge \|\check{u}\|^2 + \|\check{v}\|^2,$$

hence $\|\check{u}\| \le \|f\|$ and $\|\check{v}\| \le \|f\|$, which implies the conclusion.

(ii) Now we pass to the general case. Let (u^1, v^1), (u^2, v^2) and (u^3, v^3) solve

$$\begin{cases} u^1 + v^1 = \tilde{x}, \\ Qu^1 + Rv^1 = 0, \end{cases} \quad \begin{cases} u^2 + v^2 = \tilde{s}, \\ Qu^2 + Rv^2 = 0, \end{cases} \quad \begin{cases} u^3 + v^3 = f, \\ Qu^3 + Rv^3 = 0. \end{cases}$$

Set

$$\begin{cases} \hat{u} := \check{u} - (\tilde{x} - u^1 - u^2 + u^3) ; \\ \hat{v} := \check{v} - (\tilde{s} - v^1 - v^2 + v^3) ; \end{cases}$$

Then

$$\begin{cases} \hat{u} + \hat{v} = 0, \\ Q\hat{u} + R\hat{v} = 0, \end{cases}$$

hence $\hat{u} = 0$, so that

$$\breve{u} = \tilde{x} - u^1 - u^2 + u^3 = v^1 - u^2 + u^3.$$

Combining with (i) applied to an upper bound of the norm of v^1, u^2 and u^3, get

$$\|\breve{u}\| \le \|v^1\| + \|u^2\| + \|u^3\| \le \|f\| + \|\tilde{x}\| + \|\tilde{s}\|,$$

and similarly for $\breve{v}$.

$\square$

In the lemma below, we use the absolute value of a vector: this is the vector formed with absolute values of components.

Lemma 22.12. *Let* $(\tilde{x}, \tilde{s})$ *satisfy* $Q\tilde{x} + R\tilde{s} = g$, *and* $(x, s, \mu) \in \mathcal{N}_\epsilon^g$. *Then*

(i)
$$\frac{\|u^\theta v^\theta\|}{\mu} \le \frac{1}{\varepsilon} \left(2\sqrt{n} + x^\top |\tilde{s}| + s^\top |\tilde{x}| \right)^2.$$

If, in addition, $(x^0, s^0, \mu_0) \in \mathcal{N}_\epsilon^g$ *dominates a solution of* (LCP), *and* $\mu \le \mu_0$, *then*

(ii)
$$\frac{\|u^\theta v^\theta\|}{\mu} \le 16\frac{n^2}{\varepsilon^3} \quad \text{and} \quad \theta^a \ge \frac{\sqrt{\alpha}}{3n}.$$

Proof. Set

$$\begin{cases} \bar{u}^\theta := d^{-1}u^\theta = d^{-1}(u^a + \theta f(\theta)u^M), \\ \bar{v}^\theta := dv^\theta \;\;= d(v^a + \theta f(\theta)v^M). \end{cases}$$

Then

$$\begin{cases} \bar{u}^\theta + \bar{v}^\theta = -\sqrt{xs} - \theta f(\theta)\sqrt{x_M s_M}, \\ QD\bar{u}^\theta + RD^{-1}\bar{v}^\theta = \mu g, \end{cases}$$

and $QD(\mu d^{-1}\tilde{x}) + RD^{-1}(\mu d\tilde{s}) = \mu g$. Lemma 22.2 implies

$$\|\bar{u}^\theta\| \le \|\sqrt{xs} + \theta f(\theta)\sqrt{x_M s_M}\| + \mu\|d^{-1}\tilde{x}\| + \mu\|d\tilde{s}\|.$$

Let us estimate the terms in the right-hand side. We have

$$\|\sqrt{xs} + \theta f(\theta)\sqrt{x_M s_M}\| \le 2\|\sqrt{xs}\| = 2\sqrt{x^\top s} \le 2\sqrt{n\mu/\varepsilon}.$$

$$\mu\|d^{-1}\tilde{x}\| = \mu \left\|\frac{s\tilde{x}}{\sqrt{xs}}\right\| \le \sqrt{\frac{\mu}{\varepsilon}}\|s\tilde{x}\| \le \sqrt{\frac{\mu}{\varepsilon}}\|s\tilde{x}\|_1 = \sqrt{\frac{\mu}{\varepsilon}}s^\top|\tilde{x}|,$$

and likewise $\mu\|d\tilde{s}\| \le \sqrt{\frac{\mu}{\varepsilon}}x^\top|\tilde{s}|$, hence

$$\|\bar{u}^\theta\| \le \sqrt{\frac{\mu}{\varepsilon}} \left(2\sqrt{n} + x^\top|\tilde{s}| + s^\top|\tilde{x}| \right);$$

Writing a similar inequality for $\|v^\theta\|$, it gives

$$\frac{\|u^\theta v^\theta\|}{\mu} = \frac{\|\bar{u}^\theta \bar{v}^\theta\|}{\mu} \le \frac{\|\bar{u}^\theta\|\|\bar{v}^\theta\|}{\mu} \le \frac{1}{\varepsilon} \left(2\sqrt{n} + x^\top|\tilde{s}| + s^\top|\tilde{x}| \right)^2,$$

hence (i). Now, from $Qx^0 + Rs^0 = h - \mu_0 g = Qx^* - \mu_0 g$, we conclude that

$$Q\left(\frac{x^* - x^0}{\mu_0}\right) + R\left(-\frac{s^0}{\mu_0}\right) = g.$$

We can therefore take $(\tilde{x}, \tilde{s}) = ((x^* - x^0), -s)/\mu_0$.

Then, if (x^0, s^0) dominates $(x^*, s^* = 0)$, we get with Lemma 22.1

$$x^\top |\tilde{s}| + s^\top |\tilde{x}| \le (x^\top s^0 + s^\top x^0)/\mu_0 \le 2\frac{n}{\varepsilon},$$

and with (i)

$$\frac{\|u^\theta v^\theta\|}{\mu} = \frac{1}{\varepsilon}(2\sqrt{n} + \frac{2n}{\varepsilon})^2 \le \frac{16n^2}{\varepsilon^3}.$$

Formula (21.11) is still valid; hence, using $\mathcal{V}_\alpha \subset \mathcal{N}_\epsilon$ for $\epsilon = 1 - \alpha$:

$$\left\|\frac{x(\theta)s(\theta)}{(1-\theta)\mu} - 1\right\| \le \frac{\alpha}{2} + \frac{\theta^2}{1-\theta}\left(\frac{\|x^M s^M\|}{\mu} + \frac{\|u^\theta v^\theta\|}{\mu}\right),$$

$$\le \frac{\alpha}{2} + \frac{\theta^2}{1-\theta}\left(\frac{\sqrt{n}}{1-\alpha} + \frac{16n^2}{(1-\alpha)^3}\right),$$

and therefore if $\theta^a \ge 1/2$

$$(\theta^a)^2 \ge \frac{\alpha}{4}\left(\frac{\sqrt{n}}{1-\alpha} + \frac{16n^2}{(1-\alpha)^3}\right)^{-1} \ge \frac{\alpha}{4} \times \frac{1}{(12n)^2}$$

so that $\theta^a \ge \dfrac{\sqrt{\alpha}}{24n}$.

$\square$

From the above lemma, and more precisely the lower estimate on θ^a, we deduce that the algorithm converges in $O(n\bar{L})$ iterations.

22.3.2 Asymptotic Analysis

The analysis is identical to the one of the feasible case (see 21.4.4).

22.4 Comments

The results presented here are due to Mizuno, Jarre and Stoer [261, 260]. Let us also mention the approaches by Bonnans and Potra [48], Potra [287], Wright [364, 365], and Zhang [375].

The complexity theory presented in Chap. 25 gives a means to estimate the order of the solutions of (LCP). In practice, it is hard to know whether the initial point dominates a solution.

A large-neighborhood non-feasible algorithm of complexity $O(n^{3/2}L)$ is presented in Bonnans, Pola, and Rebaï [47]. This reference gives bounds on the accuracy with which the directions are computed to preserve the same complexity, and if there exists a strictly complementary solution, to obtain a given asymptotic convergence rate.

23 Self-Duality

23.1 Overview

A linear problem is said to be *self-dual* if it coincides with its dual. For example, the problem

$$\operatorname*{Min}_{x \in \mathbb{R}^n} c^\top x; \quad Ax + c \geq 0,\ x \geq 0,$$

where A is $n \times n$ skew-symmetric, is self-dual. The interest of the family of self-dual problems is that one can *embed* any linear problem into a self-dual problem for which one knows a point on the central path. Moreover, computing a strictly complementary solution (of the self-dual problem) allows one either to show that the original problem has no solution, or to compute simply a solution.

Furthermore, solving the self-dual problem by a path-following method is performed by solving linear systems of the same size as those obtained by a direct resolution of the original problem. Roughly speaking, self-duality thus allows a reduction to the case where a point of the central path is known. In particular, complexity is $O(\sqrt{n}\bar{L})$. The numerical performances obtained by this process are very competitive. We present first the theory for problems with linear inequalities, and then problems in standard form.

In the case of linear complementarity problems, there also exists a theory of embedding into a problem for which an interior point is known. This theory, slightly more complex than the one of linear programming, allows also an $O(\sqrt{n}\bar{L})$ complexity estimate.

23.2 Linear Problems with Inequality Constraints

23.2.1 A Family of Self-Dual Linear Problems

Consider the following problem, where A is a $p \times n$ matrix:

$$\operatorname*{Min}_{x \in \mathbb{R}^n} c^\top x; \quad Ax \geq b; \quad x \geq 0.$$

The associated Lagrangian is

$$c^\top x + y^\top (b - Ax) - s^\top x,$$

where $y \in \mathbb{R}_+^p$ and $s \in \mathbb{R}_+^n$ are the multipliers associated with the constraints. The dual is therefore

$$\operatorname*{Max}_{y,s} b^\top y; \quad c - A^\top y = s; \quad y \geq 0, \ s \geq 0,$$

or, eliminating the variable s,

$$\operatorname*{Max}_{y} b^\top y; \quad -A^\top y + c \geq 0; \quad y \geq 0.$$

Since maximizing $b^\top y$ is equivalent to minimizing $-b^\top y$, the primal and dual formulations coincide if

$$b = -c \text{ and } A = -A^\top,$$

in other words cost and right-hand side are opposite and the constraint matrix is skew-symmetric. In this case we will say that the linear problem is *self-dual*. For future reference, the self-dual problem will be written

$$\operatorname*{Min}_{x} \hat{c}^\top \hat{x}; \quad \hat{A}\hat{x} + \hat{c} \geq 0; \quad \hat{x} \geq 0, \qquad (AD)$$

with $\hat{A}$ skew-symmetric.

The following lemma shows a very important property: a sel-dual linear problem is equivalent to a monotone complementarity problem in standard form, which has the same dimensions as the starting problem.

Lemma 23.1. *The self-dual problem* (AD) *satisfies the following properties:*
(i) *Either* $v(AD) = 0$, *or* $v(AD) = +\infty$,
(ii) $\hat{x}$ *solves* (AD) *if and only if, for some* $\hat{s} \in \mathbb{R}^n$, $(\hat{x}, \hat{s})$ *solves the linear monotone complementarity problem*

$$\hat{x}\hat{s} = 0; \quad \hat{x} \geq 0; \quad \hat{s} \geq 0; \quad \hat{s} = \hat{A}\hat{x} + \hat{c}. \qquad (23.1)$$

Proof. (i) If $v(AD) < \infty$, then (AD) is feasible; therefore its dual is feasible as well. From Corollary 19.13, their values are finite and equal. By self-duality, these values are opposite, hence $v(AD) = 0$.

(ii) If $\hat{x} \geq 0$ and $\hat{s} \geq 0$, $\hat{x}\hat{s} = 0$ is equivalent to $\hat{x}^\top \hat{s} = 0$. Eliminating $\hat{s} = \hat{A}\hat{x} + \hat{c}$, the linear complementarity problem (23.1) is thus equivalent to

$$\hat{x}^\top (\hat{A}\hat{x} + \hat{c}) = 0; \quad \hat{x} \geq 0; \quad \hat{A}\hat{x} + \hat{c} \geq 0.$$

Since A is skew-symmetric, this is in turn

$$\hat{x}^\top \hat{c} = 0, \ \hat{x} \geq 0, \ \hat{A}\hat{x} + \hat{c} \geq 0.$$

We have seen that if (AD) is feasible, $v(AD) = 0$. Hence $x \in S(AD)$ if and only if $c^\top x = 0$ and $x \in F(AD)$, wich amounts to the above relation.

To finish, let us check that (23.1) is monotone: if $v = \hat{A}u$, then $v^\top u = u^\top \hat{A}u = 0$. $\qquad \square$

Remark 23.2. If $\hat{c} \geq 0$, then $\bar{x} := 0$ is feasible and $\hat{c}^\top \bar{x} = 0$, hence $v(AD) = 0$ and $S(AD)$ contains 0. However, computing a strictly complementary solution is not trivial and we will see later that this is the type of solution that is interesting. □

23.2.2 Embedding in a Self-Dual Problem

Take again a general linear problem, written in the following format:

$$\underset{x}{\text{Min}}\, c^\top x; \quad Ax \geq b,\ x \geq 0 \tag{LP}$$

where A is $p \times n$, as well as the dual problem which is

$$\underset{y}{\text{Max}}\, b^\top y; \quad A^\top y \leq c; \quad y \geq 0.$$

We know that (x, y) is a primal-dual solution if and only if x is feasible for the primal, y for the dual, and $b^\top y \geq c^\top x$. In other words, (x, y) solves the system

$$\begin{cases} Ax & - b\tau & \geq 0, \\ - A^\top y & + c\tau & \geq 0, \\ b^\top y & - c^\top x & \geq 0, \\ y \in \mathbb{R}^p_+, & x \in \mathbb{R}^n_+, & \tau \in \mathbb{R}_+. \end{cases}$$

when $\tau = 1$. If this homogeneous system has a solution (x, y, τ) with $x \geq 0$, $y \geq 0$ and $\tau > 0$, then $(x/\tau, y/\tau)$ is a primal-dual solution of (LP).

Let us formulate a problem close to the previous one, but for which we have a feasible point. This new problem has now an auxiliary variable θ, introduced as follows. Let (x^0, y^0, s^0, u^0) satisfy:

$$x^0 \in \mathbb{R}^n_{++},\ s^0 \in \mathbb{R}^n_{++},\ y^0 \in \mathbb{R}^p_{++},\ u^0 \in \mathbb{R}^p_{++},$$

with which we associate the parameters $(\bar{c}, \bar{b}, \alpha, \beta)$ defined as follows:

$$\begin{cases} \bar{b} := b - Ax^0 + u^0, \quad \bar{c} := c - A^\top y^0 - s^0, \quad \alpha := c^\top x^0 - b^\top y^0 + 1, \\ \beta := \alpha + \bar{b}^\top y^0 - \bar{c}^\top x^0 + 1 = (y^0)^\top u^0 + (x^0)^\top s^0 + 2 > 0. \end{cases}$$

The interpretation of these variables is the following:
x^0, y^0 estimates the primal-dual solution,
u^0, s^0 estimates the gap $(Ax^0 - b, A^\top y^0 - c)$.
Note that
if x^0 is feasible for the primal: $s^0 = Ax^0 - b \Rightarrow \bar{b} = 0$,
if y^0 is feasible for the dual: $u^0 = c - A^\top y \Rightarrow \bar{c} = 0$.

Consider the linear problem

$$\text{Min } \beta\theta; \quad \begin{cases} & Ax & + \bar{b}\theta & -b\tau \geq 0, \\ & -A^\top y & -\bar{c}\theta & +c\tau \geq 0, \\ & -\bar{b}^\top y & +\bar{c}^\top x & -\alpha\tau \geq -\beta, \\ & b^\top y & -c^\top x & +\alpha\theta & \geq 0, \\ y \in \mathbb{R}^p_+, & x \in \mathbb{R}^n_+, & \theta \in \mathbb{R}_+, & \tau \in \mathbb{R}_+ . \end{cases}$$

$$(\overline{AD})$$

Its constraint matrix is skew-symmetric and its positive cost is opposite to the right-hand side term. It is therefore self-dual and has 0 value. Besides, the point

$$(y^0, x^0, \theta^0 := 1, \tau^0 := 1)$$

is feasible for $(\overline{AD})$. Incidentally, note that we obtain the format (AD), by setting

$$\hat{A} := \begin{pmatrix} 0 & A & \bar{b} & -b \\ -A^\top & 0 & -\bar{c} & c \\ -\bar{b}^\top & \bar{c}^\top & 0 & -\alpha \\ b^\top & -c^\top & \alpha & 0 \end{pmatrix}, \quad \hat{x} := \begin{pmatrix} y \\ x \\ \theta \\ \tau \end{pmatrix}, \quad \hat{c} := \begin{pmatrix} 0 \\ 0 \\ \beta \\ 0 \end{pmatrix}.$$

The dual variable associated with this starting point is

$$\hat{s}^0 = \hat{A}\hat{x}^0 + \hat{c} = \begin{pmatrix} u^0 \\ s^0 \\ 1 \\ 1 \end{pmatrix}.$$

Taking (x^0, y^0, u^0, s^0) with all components equal to 1, the point $(\hat{x}, \hat{s})$ is therefore on the central path of problem (23.1), associated with $\mu^0 = 1$. The problem can therefore be solved by a feasible path-following algorithm, applied to problem (23.1), of the type described in Chap. 21, which converges in $O(\sqrt{n}L)$ iterations. There remains to know how to use the solution thus obtained.

Lemma 23.3. Let $(\bar{y}, \bar{x}, \bar{\theta}, \bar{\tau})$ solve $(\overline{AD})$. Then $\bar{\theta} = 0$. Besides:
(i) If $\bar{\tau} > 0$, then $(\bar{x}/\bar{\tau}, \bar{y}/\bar{\tau})$ is a primal-dual solution of (LP).
(ii) If $\bar{\tau} = 0$ and if $(\bar{y}, \bar{x}, \bar{\theta}, \bar{\tau})$ is strictly complementary, then $b^\top \bar{y} > 0$ or $c^\top \bar{x} < 0$ (or both). In the first case, the primal problem is infeasible. In the second, the dual is infeasible.

Proof. From $v(\overline{AD}) = 0 = \beta\bar{\theta}$ and $\beta > 0$, we deduce that $\bar{\theta} = 0$. If $\bar{\tau} > 0$, set $(y^*, x^*) := (\bar{y}, \bar{x})/\bar{\tau}$. Reviewing one by one the constraints of $(\overline{AD})$, we obtain

$$\begin{aligned} Ax^* - b &\geq 0 \text{ primal feasibility,} \\ -A^\top y^* + c &\geq 0 \text{ dual feasibility,} \\ b^\top y^* - c^\top x &\geq 0 \text{ dual larger than primal cost.} \end{aligned}$$

This implies that (x^*, y^*) is a primal-dual solution of the original problem.

On the other hand, if $\bar{\tau} = 0$, it holds that

$$A\bar{x} \geq 0; \quad -A^\top \bar{y} \geq 0; \quad b^\top \bar{y} \geq c^\top \bar{x},$$

and the last inequality is strict for a strictly complementary solution (for it is associated with the variable $\bar{\tau} = 0$). Then $b^\top \bar{y} > 0$ or $c^\top \bar{x} < 0$. In the first case, from $-A^\top \bar{y} \geq 0$ and $b^\top \bar{y} > 0$ we deduce that the primal is infeasible (if x is primal-feasible, then $x \geq 0$ and $Ax \geq b$, hence $0 < b^\top \bar{y} \leq (Ax)^\top \bar{y} \leq x^\top A^\top \bar{y} \leq 0$, a contradiction). In the second case, the dual is infeasible. $\square$

In summary, computing a strictly complementary solution of $(\overline{AD})$ allows us either to compute a solution of the original problem, if there is one, (indeed it is not difficult to check that this solution is itself strictly complementary for the original problem) or to conclude that the primal and/or the dual is infeasible.

23.3 Linear Problems in Standard Form

Now we present a theory, similar though slightly more complex, which applies to problems in standard form.

23.3.1 The Associated Self-Dual Homogeneous System

With the problem in standard form

$$\operatorname*{Min}_{x \in \mathbb{R}^n} c^\top x; \quad Ax = b; \quad x \geq 0 \qquad\qquad (LP)$$

associate the Lagrangian

$$c^\top x - \lambda^\top (Ax - b)$$

which gives birth to the dual problem

$$\operatorname*{Max}_{\lambda \in \mathbb{R}^p} b^\top \lambda; \quad -A^\top \lambda + c \geq 0. \qquad\qquad (LD)$$

We have seen in Chap. 19 that, if $x \in F(LP)$ and $\lambda \in F(LD)$, then $c^\top x \geq b^\top \lambda$, with equality if and only if $x \in S(LP)$ and $\lambda \in S(LD)$. Computing a primal-dual solution therefore amounts to solving the problem: to find $(x, \lambda) \in \mathbb{R}_+^n \times \mathbb{R}^p$ such that

$$\begin{cases} Ax & - b = 0, \\ -A^\top \lambda & + c \geq 0, \\ b^\top \lambda - c^\top x & \geq 0. \end{cases}$$

The above problem can be considered as a linear optimization problem with 0 cost. In contrast to the previous situations, the system mixes equality and inequality constraints. Let us homogenize it by introducing a variable $\tau \geq 0$:

$$
\text{Min } 0; \quad
\left\{
\begin{array}{rl}
Ax - b\tau = 0, & \\
-A^{\top}\lambda \quad + c\tau \geq 0, & \\
b^{\top}\lambda - c^{\top}x \quad \geq 0. & \\
(\lambda, x, \tau) \in \mathbb{R}^p \times \mathbb{R}^n_+ \times \mathbb{R}_+
\end{array}
\right.
\tag{LH}
$$

Note that the associated matrix is skew-symmetric.

Lemma 23.4. *Problem* (LH) *is self-dual, and* $v(LH) = 0$. *If* $(\bar{x}, \bar{\lambda}, \bar{\tau})$ *is a strictly complementary solution of* (LH), *then*
(i) *If* $\bar{\tau} > 0$, *the point* $(\bar{x}/\bar{\tau}, \bar{\lambda}/\bar{\tau})$ *is a primal-dual solution of* (LP).
(ii) *If* $\bar{\tau} = 0$, *then at least one of the two properties below is satisfied: either* $b^{\top}\bar{\lambda} > 0$ *and* (LP) *is infeasible, or* $c^{\top}\bar{x} < 0$ *and* (LD) *is infeasible.*

Proof. Associate with (LH) the Lagrangian (note the sign convention for y)

$$
-y^{\top}(Ax - b\tau) - s^{\top}(-A^{\top}\lambda + c\tau) - \nu(b^{\top}\lambda - c^{\top}x).
$$

The resulting dual problem writes

$$
\text{Min } 0; \quad
\left\{
\begin{array}{rl}
As - b\nu = 0, & \\
-A^{\top}y \quad + c\nu \geq 0, & \\
b^{\top}y - c^{\top}s \quad \geq 0. & \\
(y, s, \nu) \in \mathbb{R}^p \times \mathbb{R}^n_+ \times \mathbb{R}_+
\end{array}
\right.
\tag{LH}
$$

It appears that the dual problem coincides with the primal. In other words, the homogenized problem (LH) is self-dual. This problem is feasible (0 is a feasible point) hence $v(LH) = 0$. Point (i) is straightforward. Concerning (ii), if $\bar{\tau} = 0$, we have by complementarity $b^{\top}\bar{\lambda} > c^{\top}\bar{x}$, hence either $b^{\top}\bar{\lambda} > 0$ or $c^{\top}\bar{x} < 0$. If, for example, $b^{\top}\bar{\lambda} > 0$ and $x \in F(LP)$, then $0 < b^{\top}\bar{\lambda} = (Ax)^{\top}\bar{\lambda} = (A^{\top}\bar{\lambda})^{\top}x \leq 0$, hence $F(LP) = \emptyset$, i.e. $v(LP) = +\infty$; same argument for the second case.

$\square$

23.3.2 Embedding in a Feasible Self-Dual Problem

Let us embed (LH) in a problem containing an additional scalar variable, and for which a feasible point is available. This problem is parameterized by an arbitrary triple

$$
(x^0, \lambda^0, s^0) \in \mathbb{R}^n_{++} \times \mathbb{R}^p \times \mathbb{R}^n_{++},
$$

and the corresponding parameters

$$
\bar{b} := b - Ax^0; \quad \bar{c} := c - A^{\top}\lambda^0 - s^0; \quad \bar{z} := c^{\top}x^0 - b^{\top}\lambda^0 + 1.
$$

It is formulated as

$$\text{Min}((x^0)^\top s^0 + 1)\theta; \quad \begin{cases} Ax \; - b\tau + \bar{b}\theta = 0, \\ -A^\top \lambda \qquad\quad + c\tau - \bar{c}\theta \geq 0, \\ b^\top \lambda - c^\top x \qquad\quad + \bar{z}\theta \geq 0, \\ -\bar{b}^\top \lambda + \bar{c}^\top x - \bar{z}\tau \qquad = -(x^0)^\top s^0 - 1. \\ (\lambda, x, \tau, \theta) \in \mathbb{R}^p \times \mathbb{R}^n_+ \times \mathbb{R}_+ \times \mathbb{R}. \end{cases} \quad (LHE)$$

Lemma 23.5.
(i) *Problem* (LHE) *is self-dual and* $v(LHE) = 0$.
(ii) *The point* $(\lambda^0, x^0, \tau^0 := 1, \; \theta^0 := 1)$ *is feasible for* (LHE).
(iii) *If* $(\bar{x}, \bar{\lambda}, \bar{\tau}, \bar{\theta})$ *is a strictly complementary solution of* (LHE), *then* $(\bar{x}, \bar{\lambda}, \bar{\tau})$ *is a strictly complementary solution of* (LH) *(and then Lemma 23.4 applies).*

Proof. Self-duality is checked by computations similar to those above. Feasibility of $(x^0, \lambda^0, \tau^0, \theta^0)$ is straightforward. It implies $v(LHE) = 0$; (i) and (ii) are proved. Let now $(\bar{x}, \bar{\lambda}, \bar{\tau}, \bar{\theta})$ be a strictly complementary solution of (LHE). Since $v(LHE) = 0$, we have $\bar{\theta} = 0$, hence the point $(\bar{x}, \bar{\lambda}, \bar{\tau})$ is feasible for (LH) and the complementarity relations write

$$\bar{x}(-A^\top \bar{\lambda} + \bar{\tau}c) = 0 \quad \text{and} \quad \bar{\tau}(b^\top \bar{\lambda} - c^\top \bar{x}) = 0.$$

We recognize the complementarity relations of (LH). These relations being satisfied strictly, the point $(\bar{x}, \bar{\lambda}, \bar{\tau})$ is indeed a strictly complementary solution of (LH). $\qquad\square$

23.4 Practical Aspects

We consider the case of problems in standard form. Set

$$\hat{A} := \begin{pmatrix} 0 & A & -b & \bar{b} \\ -A^\top & 0 & c & -\bar{c} \\ b^\top & -c^\top & 0 & \bar{z} \\ -\bar{b}^\top & \bar{c}^\top & -\bar{z} & 0 \end{pmatrix};$$

$$\hat{s} := \begin{pmatrix} s^\lambda \\ s^x \\ s^\tau \\ s^\theta \end{pmatrix}; \quad \hat{x} := \begin{pmatrix} \lambda \\ x \\ \tau \\ \theta \end{pmatrix}; \quad \hat{c} := \begin{pmatrix} 0 \\ 0 \\ 0 \\ (x^0)^\top s^0 + 1 \end{pmatrix}.$$

Arguing as in Lemma 23.1, one checks that (LHE) is equivalent to

$$xs^x = 0; \quad \tau s^\tau = 0; \quad s^\lambda = 0; \quad s^\theta = 0;$$
$$\hat{s} = A\hat{x} + c; \quad x \geq 0, \; \tau \geq 0, \; s^\lambda \geq 0, \; s^\tau \geq 0.$$

This problem generalizes the linear complementarity format, because the sign- and complementarity-constraints involve only a part of the variables. It can be reduced to a standard linear complementarity problem anyway, by elimination of λ and θ. This process allows the extension to the framework of (LHE) of the concept of central path and the associated path-following algorithms. The equation of the central path is

$$\begin{cases} \begin{pmatrix} xs^x \\ \tau s^\tau \end{pmatrix} = \mu \mathbf{1}; \quad \hat{s} = \hat{A}x + \hat{c}. \\ x \geq 0, \ \tau \geq 0, \ s^\lambda \geq 0, \ s^\tau \geq 0. \end{cases}$$

Note that, if $x^0 = \mathbf{1}$ et $s^0 = \mathbf{1}$, then the initial point is

$$\hat{x} = \begin{pmatrix} 1 \\ \lambda^0 \\ 1 \\ 1 \end{pmatrix}; \quad \hat{s} = \begin{pmatrix} 0 \\ 1 \\ 1 \\ 0 \end{pmatrix}.$$

This point lies on the central path.

Besides, the computation of the path-following direction can be decomposed as follows. Let $u(\lambda), u(x), \cdots, v(s^x), \cdots$ be the directions. Partition the computation of (u, v) as

$$s^x u(x) + xv(s^x) = \cdots,$$
$$Au(x) = bu(\tau) - \bar{b}u(\theta),$$
$$-A^\top u(\lambda) - v(s^x) = -cu(\tau) + \bar{c}u(\theta).$$

and

$$s^\tau u(\tau) + \tau v(s^\tau) = \cdots,$$
$$b^\top u(\lambda) - c^\top u(x) + \bar{z}u(\theta) - v(s^\tau) = 0,$$
$$-\bar{b}^\top u(\lambda) + \bar{c}^\top u(x) - \bar{z}u(\tau) = 0,$$

where $\cdots$ represents a certain quantity, known but depending on the algorithm. With respect to the unknowns $(u(x), v(s^\lambda), u(\lambda))$, the first system has a structure identical to that obtained in a feasible path-following algorithm. This system can be solved by giving to the parameters $(u(\tau), u(\theta))$ the values $(0, 1)$ and $(1, 0)$, which gives explicitly $(u(x), v(s^\lambda), u(\lambda))$ as a function of $(u(\tau), u(\theta))$. Substituting these values in the second system, we obtain a 3-dimensional system giving the values of $(u(\tau), u(\theta))$.

Embedding in a self-dual problem thus does not increase the dimension of the linear systems to solve.

Remark 23.6. Since the computations are actually made in floating-point arithmetic, the solution obtained for $(\overline{AD})$ is by necessity corrupted; so it can be difficult to check whether it satisfies $\bar{\tau} = 0$. A rigorous conclusion requires a purification procedure. □

23.5 Extension to Linear Monotone Complementarity Problems

We present now a partial extension of the preceding results to linear monotone complementarity problems. This extension, due to Ye [371], gave birth to the first non-feasible method converging in $O(\sqrt{n}\bar{L})$ iterations. Consider the linear monotone complementarity problem (LCP), written in standard form:

$$\begin{cases} xs = 0, \\ s = Mx + q, \\ x \geq 0, \ s \geq 0, \end{cases} \tag{23.2}$$

with $M \geq 0$. Choose $x^0 = \mathbf{1}$ and $s^0 \in \mu_0 \mathbf{1}$, $\mu_0 > 0$ so that

$$s^0 - (M + M^\top)x^0 \geq \mathbf{1} \quad \text{et} \quad s^0 \geq Mx^0 + q.$$

Set

$$\theta^0 := 1, \quad \tau^0 := 1; \quad \kappa^0 := \mu_0,$$

$$\bar{r} := s^0 - Mx^0 - q\tau^0; \quad \bar{z} := q^\top x^0 + \kappa^0,$$

$$\bar{n} := (x^0)^\top s^0 + \tau^0 \kappa^0 - (x^0)^\top Mx^0 > 0.$$

Consider the linear complementarity problem

$$\begin{cases} xs = 0, \\ \tau\kappa = 0, \\ \begin{pmatrix} s \\ \kappa \\ 0 \end{pmatrix} = \begin{pmatrix} M & q & \bar{r} \\ -q^\top & 0 & \bar{z} \\ -\bar{r}^\top & -\bar{z} & \theta \end{pmatrix} \begin{pmatrix} x \\ \tau \\ 0 \end{pmatrix} + \begin{pmatrix} 0 \\ 0 \\ \bar{n} \end{pmatrix} \end{cases} \tag{ELCP}$$

$$(x, s, \tau, \kappa) \geq 0, \ \theta \in \mathbb{R}.$$

Then $(s^0, \kappa^0, x^0, \tau^0, \theta^0)$ is an interior feasible point of this quadratic problem, which is equivalent to a linear monotone complementarity problem in standard form. In fact, $\bar{r}$ et $\bar{z}$ cannot vanish simultaneously, since $\bar{n} > 0$. One can therefore eliminate θ in one of the equations, and one obtains $2n+2$ equations; moreover, if $(x', s', \tau', \kappa', \theta')$ also satisfies the linear equations, then

$$(x - x')^\top (s - s') + (\tau - \tau')(\kappa - \kappa') = (x - x')^\top M(x - x') \geq 0,$$

hence monotonicity.

Note that, if M is skew-symmetric, the matrix associated with $(ELCP)$ is skew-symmetric as well. The problem is then very similar to those previously studied.

One can also consider (ELCP) as a liner complementarity problem in standard form. The initial point $(x^0, s^0, \tau^0, \kappa^0)$ lies on the central path. It is therefore possible to solve (ELCP) by a path-following algorithm. However, it is possible that θ changes of sign along the iterations. In fact, we have at the optimum

$$0 = x^\top s + \tau\kappa = (x^\top Mx + \tau q^\top x + \theta\bar{r}^\top x) + (-\tau q^\top x + \tau\bar{z}\theta)$$
$$= x^\top Mx + (\bar{r}^\top x + \tau\bar{z})\theta = x^\top Mx + \bar{n}\theta.$$

But $x^\top Mx \geq$ and $\bar{n} > 0$, hence $\theta \leq 0$, and $\theta = 0$ if and only if $x^\top Mx = 0$.

Then the algorithmic procedure distinguishes two cases:

(i) If a negative value of θ is obtained at some iteration, the stepsize is reduced to compute a point (x, s, τ, κ) associated with $\theta = 0$. This point will be feasible for (23.2), and close to the central path. The original problem can then be solved by a feasible-point algorithm.

(ii) Otherwise, we have $\theta \downarrow 0$. Let $(x^*, s^*, \tau_*, \kappa_*)$ be a cluster-point of $(x^k, s^k, \tau^k, \kappa^k)$. If $\tau_* > 0$ then $(x^*/\tau_*, s^*/\tau_*)$ solves (23.2). If not, and if the solution is strictly complementary, we have $\kappa_* > 0$, hence

$$s^* = Mx^* \geq 0 \text{ and } q^\top x^* < 0.$$

But

$$0 = (s^*)^\top x^* = (x^*)^\top Mx^* = \frac{1}{2}(x^*)^\top (M + M^\top)x^*.$$

The matrix $M + M^\top$ is symmetric positive semi-definite. The above relation implies that x^* is in its kernel, hence $M^\top x^* = -Mx^* \leq 0$. Let now $x \in F(LCP)$. Then

$$0 \leq (x^*)^\top (Mx + q) = x^\top M^\top x^* + q^\top x^* < 0,$$

a contradiction. We conclude that (LCP) is infeasible.

Let us summarize the above results. As seen in Lemma 20.6, a linear complementarity problem can be written in standard form (in $O(n^3)$ operations). The latter can be solved by a feasible algorithm in $O(\sqrt{n}\bar{L})$ iterations, through an embedding in a self-dual problem.

23.6 Comments

The present theory for inequality-constrained problems is taken from Jansen, Roos and Terlaky [199]. The theory for problems in standard form is due to Ye, Todd and Mizuno [372]. The extension to the framework of monotone complementarity was made by Ye [371].

Self-duality is an elegant and efficient way of fixing the question of choosing a starting point. In fact, non-feasible algorithms of the type of Chap. 22 have a complexity of at best $O(n\bar{L})$ iterations. Besides, the tests reported in Xu, Hung, and Ye [369] clearly assess the efficiency of the self-dual method, at least for linear programming. As for quadratic programming, it seems that much remains to be done.

24 One-Step Methods

24.1 Overview

In the predictor-corrector algorithms studied so far, the moves aimed at improving centralization are clearly separated from those allowing a reduction of the parameter μ. It is tempting to consider algorithmic families in which the same move reduces μ while controlling centralization.

Section 24.2 presents two feasible algorithms, based on the following idea. In the case of small neighborhood $\mathcal{V}_\alpha$, with $\alpha \leq 1/2$, we know that the centering step yields a very well-centered point w^c. This suggests a search of the new point on the segment $[w^c, w^a]$. To obtain a maximal reduction of μ, the point of $[w^c, w^a] \cap \mathcal{V}_\alpha$ closest to w^a is chosen. This point lies on the edge of $\mathcal{V}_\alpha$. Such a *largest-step algorithm* therefore computes at each iteration the value of one single scalar parameter, as in the predictor-corrector algorithm. The complexity remains unchanged: $O(\sqrt{n}L)$. If a strictly complementary solution exists, and if $\alpha \leq 1/4$, the sequence of points converges, which implies that the optimality measure converges superlinearly.

Then we present a variant, called *largest-step algorithm with safeguard*, which works in the neighborhood $\mathcal{V}_\alpha$, for $\alpha \leq 1/2$. The modification of the algorithm lies in the possibility of performing centering moves. The algorithm still has complexity $O(\sqrt{n}L)$. If there exists a strictly complementary solution, there are only finitely many centering moves (so, after some iteration, the operations become identical to those of the largest-step algorithm) and the optimality measure converges superlinearly. The advantage over the previous algorithm is that these properties hold in the neighborhood $\mathcal{V}_{1/2}$. Convergence of the sequence of points still holds, but as a consequence of specific estimates.

Section 24.3 is devoted to results of general interest, concerning *convergence of the sequence of points* computed by path-following algorithms. After discussing the problem of centering the large variables, it is shown that, in the neighborhood $\mathcal{V}_{1/4}$, the first-order estimate of the centering step decreases the logarithmic potential associated with the large variables. It follows that, under strict complementarity assumption, the sequence of points converges for any small-neighborhood algorithm ($\alpha \leq 1/4$) for which $\{\mu_k\}$ converges linearly. Finally, we give estimates of the *relative distances* in the space of large variables.

The proofs of the main results are based on previous statements, and are given in §24.4.

24.2 The Largest-Step Sethod

24.2.1 Largest-Step Algorithm

We recall the notation $w = (x, s, \mu)$, w^a and w^c being the points obtained after an affine or centering move with a step $\theta = 1$. The idea is to take as new point the element of the segment $[w^c, w^a]$ which is closest to w^a (so as to reduce μ) while staying in a small neighborhood of the central path.

Algorithm 24.1. LS (Largest Step)

> Data: $\mu_\infty > 0$, $\alpha \in (0, 1/2]$; $(x^0, s^0, \mu_0) \in \mathcal{V}_\alpha$; $k \leftarrow 0$.
>
> REPEAT
> - $w \leftarrow w^k$. Compute w^c and w^a; $w(\gamma) := \gamma w^c + (1 - \gamma)w^a$;
>   Compute the smallest value γ_k in $]0, 1[$ such that
> $w(\gamma) \in \mathcal{V}_\alpha$, $\forall\, \gamma \in [\gamma_k, 1]$.
> $w^{k+1} \leftarrow w(\gamma_k)$, $k \leftarrow k + 1$.
>
> UNTIL $\mu_k < \mu_\infty$.

Theorem 24.2. (i) *If $\mu_\infty > 0$, set $\bar{L} := \log(\mu_0/\mu_\infty)$. Then Algorithm* **LS** *stops after at most $O(\sqrt{n}\bar{L})$ iterations (more precisely $5\alpha^{-1}\sqrt{n}$ iterations).* (ii) *Suppose that (LCP) has a strictly complementary solution. If $\alpha \leq 1/4$ and $\mu_\infty = 0$, then the sequence $\{\mu_k\}$ converges superlinearly to 0.*

The proof will be given later in this chapter.

24.2.2 Largest-Step Algorithm with Safeguard

The change in the algorithm lies in the possibility of performing centering steps. This *safeguard* step is based on testing the proximity of the point obtained by taking $\gamma = 0.1$. If this proximity is small enough (more precisely smaller than 0.42) safeguarding does not occur. It does not occur either if the proximity is larger than 1. Indeed, if there exists a strictly complementary solution, it can be proved that, after finitely many iterations, this condition is never satisfied. The algorithm is then as follows:

Algorithm 24.3. SLS (Safeguarded Largest Step)

> Data: $\mu_\infty > 0$, $\alpha \in (0, 1/2]$; $(x^0, s^0, \mu_0) \in \mathcal{V}_\alpha$; $\epsilon > 0$, $k \leftarrow 0$.
>
> REPEAT

- $w \leftarrow w^k$; Compute w^c and w^a.
  Safeguard: If $\delta(0.1w^c + 0.9w^a) \in [0.42, 1]$ do: $w \leftarrow w^c$; compute w^c and w^a.
  $w(\gamma) := \gamma w^c + (1 - \gamma)w^a$;
  Compute the smallest value γ_k in $]0, 1[$ such that
  $w(\gamma) \in \mathcal{V}_\alpha, \, \forall \, \gamma \in [\gamma_k, 1]$.
  $w^{k+1} \leftarrow w(\gamma_k)$; $k \leftarrow k + 1$.
  UNTIL $\mu_k < \mu_\infty$.

Theorem 24.4. (i) *If $\mu_\infty > 0$, set $\bar{L} := \log(\mu_0/\mu_\infty)$. Then Algorithm* **SLS** *stops after at most $O(\sqrt{n}\bar{L})$ iterations (more precisely $5\alpha^{-1}\sqrt{n}$ iterations).*
(ii) *Suppose that (LCP) has a strictly complementary solution. If $\mu_\infty = 0$, the safeguard occurs finitely many times and $\{\mu_k\}$ converges superlinearly to 0.*

The proof will be given later in this chapter.

24.3 Centralization in the Space of Large Variables

We state a few lemmas concerning the centralization of large variables; they will be used to prove Theorems 24.2 and 24.4.

24.3.1 One-Sided Distance

Our analysis supposes that $S(LCP)$ is nonempty. The partition (B, N, T) was defined in Chap. 20. In the sequel we assume (LCP) to be in canonical form, i.e. $N = \emptyset$. This can be done without loss of generality, since this algorithm is invariant with respect to the corresponding transformation. We define the *proximity measure of the large variables* as follows:

$$\delta_B(x) := \|P_{Q_B X_B} \mathbf{1}\|.$$

Let us motivate this quantity by relating it to the problem of centering the large variables, stated as

$$\underset{x_B}{\text{Min}} \, \pi(x_B); \quad Q_B x_B = h - Q_T x_T - Rs.$$

The solution of this problem is called the analytic center of the corresponding feasible set. (The analytic center of the set of solutions of (LCP) was discussed in Chap. 20, and is the solution of the above problem when $x_T = 0$ and $s = 0$.) The Newton direction d_{Newton} associated with this problem, computed at the point x, is the solution of

$$\underset{d}{\text{Min}} \, -d^\top x_B^{-1} + \frac{1}{2} d^\top X_B^{-2} d; \quad Q_B d = 0.$$

The unique solution of this strictly convex problem is characterized by

$$X_B^{-2}d_{Newton} - x_B^{-1} \in \mathcal{R}(Q_B^\top); \quad Q_B d_{Newton} = 0,$$

or also

$$x_B^{-1}d_{Newton} \in 1 + \mathcal{R}((Q_B X_B)^\top); \quad Q_B X_B(x_B^{-1}d_{Newton}) = 0,$$

i.e. $x_B^{-1}d_{Newton} = P_{Q_B X_B}1$. In other words, $\delta_B(x) := \|x_B^{-1}d_{Newton}\|$ is a weighted measure of the norm of the Newton direction d_{Newton}.

The next lemma links the proximity measure δ with the proximity measure of the large variables:

Lemma 24.5. *Take* $(x, s, \mu) \in \mathcal{V}_\alpha$. *Then* $\delta_B(x) \le \delta(x, s, \mu) = \left\| \dfrac{xs}{\mu} - 1 \right\|$.

Proof. By definition of an orthogonal projection, we have that

$$\delta_B(x) = \|P_{Q_B X_B}1\| = \min\{\|1 - z\|\} \quad z \in \mathcal{R}(XQ^\top).$$

Take $(x^*, s^*) \in S(LCP)$. From $Q(x - x^*) + Rs = 0$, we deduce from Lemma 20.11 that $\mathcal{R}(X_B Q_B^\top) \ni s_B x_B/\mu$, hence the conclusion. $\square$

Set

$$\bar{d} := \sqrt{\mu x_B/s_B}; \quad \tilde{Q} := Q_B \bar{D}; \quad \hat{Q} := Q_B X_B;$$
$$u_P^c := \bar{d} P_{\tilde{Q}} \phi_B^{-1}; \quad d_{Newton} := x_B P_{\hat{Q}}1.$$

In view of Lemma 20.9, we have that $d \approx 1$ in $\mathcal{N}_\epsilon$. On the other hand (Lemma 20.12)

$$u = \gamma u_P^c + O(\sqrt{\mu}) \text{ and } v = O(\sqrt{\mu});$$
$$\text{if } T = \emptyset, \text{ then } u = \gamma u_P^c + O(\mu) \quad \text{and } v = O(\mu).$$

Accordingly, we will call u_P^c the *linear estimate* of the centering step. The result below shows that, in the neighborhood $\mathcal{V}_{1/4}$, u_P^c decreases the logarithmic potential associated with the large variables.

Proposition 24.6. *If* $(x, s) \in \mathcal{V}_\alpha$, *with* $\alpha \le 1/4$, *then*

$$\pi(x_B + u_P^c) \le \pi(x_B) - \tfrac{6}{100}\|d_{Newton}\|^2.$$

Let us first prove some lemmas. The first one is due to Gonzaga and Tapia [176].

Lemma 24.7. *Let* $g \in \mathbb{R}^n$ *satisfy* $\|g - 1\|_\infty \le \alpha$, *with* $\alpha \in (0, 1)$. *Set*

$$G = \text{diag}(g), \quad \hat{h} = P_A q, \quad h = g P_{AG}(gq).$$

Then

$$\|h - \hat{h}\| \le \alpha(1 + \alpha)\frac{2 - \alpha}{1 - \alpha}\|\hat{h}\|.$$

Proof. The relation $\hat{h} = P_A q$ implies $q \in \hat{h} + \mathcal{R}(A^T)$, hence $gq \in g\hat{h} + \mathcal{R}((AG)^T)$, and therefore

$$g^{-1}h = P_{AG}(gq) = P_{AG}(g\hat{h}).$$

Since $g^{-1}\hat{h} \in \mathcal{N}(AG)$, we have $g^{-1}(h - \hat{h}) = P_{AG}[(g - g^{-1})\hat{h}]$. Whence

$$\|g^{-1}(h - \hat{h})\| = \|P_{AG}(g - g^{-1})\hat{h}\| \le \|(g - g^{-1})\hat{h}\| \le \|g - g^{-1}\|_\infty \|\hat{h}\|.$$

As a result

$$\|h - \hat{h}\| \le \|g\|_\infty \|g^{-1}(h - \hat{h})\| \le \|g\|_\infty \|g^{-1} - g\|_\infty \|\hat{h}\|.$$

Finally

$$\|g\|_\infty \|g^{-1} - g\|_\infty \le (1 + \alpha)\left(\frac{1}{1 - \alpha} - (1 - \alpha)\right) = \alpha(1 + \alpha)\frac{2 - \alpha}{1 - \alpha},$$

and the conclusion follows.

$\square$

Let us now give a useful bound on the variation of π in a neighborhood of $\mathbf{1}$.

Lemma 24.8. *Take $h \in \mathbb{R}^n$, $\|h\| < 1$. Then*

$$\pi(\mathbf{1} + h) \le -\mathbf{1}^\top h + \frac{\|h\|^2}{2} + \frac{1}{3}\frac{\|h\|^3}{1 - \|h\|}.$$

In particular, if $\|h\| \le 1/2$, then

$$\pi(\mathbf{1} + h) \le -\mathbf{1}^\top h + \|h\|^2.$$

Proof. Since $|h_i| < 1$, we have

$$-\log(1 + h_i) = -h_i + \frac{(h_i)^2}{2} - \frac{(h_i)^3}{3} + \frac{(h_i)^4}{4} + \cdots,$$

$$\le -h_i + \frac{(h_i)^2}{2} + \frac{|h_i|^3}{3}\left(1 + \frac{3}{4}|h_i| + \frac{3}{5}|h_i|^2 + \cdots\right),$$

$$\le -h_i + \frac{(h_i)^2}{2} + \frac{|h_i|^3}{3}(1 + |h_i| + |h_i|^2 + \cdots),$$

$$= -h_i + \frac{(h_i)^2}{2} + \frac{1}{3}\frac{|h_i|^3}{1 - |h_i|},$$

hence, summing up the components,

$$\pi(\mathbf{1} + h) = -\mathbf{1}^\top h + \frac{\|h\|^2}{2} + \frac{1}{3}\sum_{i=1}^n \frac{|h_i|^3}{1 - |h_i|}.$$

But $|h_i| \le \|h\|$, therefore

$$\sum_{i=1}^{n} \frac{|h_i|^3}{1 - |h_i|} \le \sum_{i=1}^{n} \frac{\|h\|}{1 - \|h\|}|h_i|^2 = \frac{\|h_i\|^3}{1 - \|h\|},$$

hence the first result. If $\|h\| \le 1/2$, then

$$\frac{\|h\|^2}{2} + \frac{1}{3}\frac{\|h\|^3}{1 - \|h\|} \le \frac{\|h\|^2}{2} + \frac{1}{3}\frac{\frac{1}{2} \times \|h\|^2}{1 - \frac{1}{2}} = \frac{5}{6}\|h\|^2,$$

hence the conclusion.

$\square$

Proof of Proposition 24.6 First apply Lemma 24.7 with

$$g := \bar{d}x_B^{-1} = \phi_B^{-1}, \quad q := 1, \quad A := \hat{Q} = Q_B X_B.$$

We then have $AG = Q_B \bar{D} = \tilde{Q}$, and therefore

$$\hat{h} = P_{\hat{A}}q = P_{\hat{Q}}1 = x_B^{-1}d_{Newton}, \quad h = gP_{AG}(gq) = \bar{d}x_B^{-1}P_{\tilde{Q}}\phi_B^{-1} = x_B^{-1}u_P^c.$$

Since $(x, s, \mu) \in V_\alpha$, we have $\|\phi^2 - 1\|_\infty \le \|\phi^2 - 1\| \le \frac{1}{4}$, and therefore

$$\frac{3}{4}1 \le \phi^2 \le \frac{5}{4} \quad \Rightarrow \quad \frac{2}{\sqrt{5}}1 \le \phi^{-1} \le \frac{2}{\sqrt{3}}1.$$

As a result,

$$\|\phi_B^{-1} - 1\|_\infty \le \max(1 - \frac{2}{\sqrt{5}}, \frac{2}{\sqrt{3}} - 1) = \frac{2}{\sqrt{3}} - 1 < \frac{155}{1000}.$$

Applying Lemma 24.7, we obtain $\|h - \hat{h}\| \le \frac{4}{10}\|\hat{h}\|$. Lemma 24.5 implies $\|\hat{h}\| = \delta_B(x) \le 1/4$, and thus $\|h\| \le 0.35$. Using

$$\pi(x + l) - \pi(x) = \pi(1 + x^{-1}l) = \pi(1 + h)$$

and $\|h\| < 1$, we have with Lemma 24.8

$$\pi(x + l) - \pi(x) \le -1^\top h + \frac{\|h\|^2}{2} + \frac{1}{3}\frac{\|h\|^3}{1 - \|h\|}.$$

Now $1^\top \hat{h} = 1^\top P_{\hat{Q}}1 = \|P_{\hat{Q}}1\|^2 = \|\hat{h}\|^2$, hence, setting $z := h - \hat{h}$:

$$\pi(x + l) - \pi(x) \le -\|\hat{h}\|^2 - 1^\top z + \frac{\|\hat{h}\|^2 + \|z\|^2 + 2\hat{h}^\top z}{2} + \frac{1}{3}\frac{\|h\|^3}{1 - \|h\|},$$

$$= -\frac{\|\hat{h}\|^2}{2} + (\hat{h} - 1)^\top z + \frac{\|z\|^2}{2} + \frac{1}{3}\frac{\|h\|^3}{1 - \|h\|}.$$

Let us check that $(\hat{h} - \mathbf{1})^\top z = 0$. Since $\hat{h} = P_{\hat{Q}}\mathbf{1}$, we have $\hat{h} - \mathbf{1} \in \mathcal{R}(\hat{Q}^\top)$. To conclude, we show that $z \in \mathcal{R}(\hat{Q}^\top)^\perp = \mathcal{N}(\hat{Q})$. Indeed

$$\hat{Q}z = \hat{Q}(h - \hat{h}) = \hat{Q}\phi_B^{-1}P_{\tilde{Q}}(\phi_B^{-1}) - \hat{Q}P_{\hat{Q}}\mathbf{1} = \tilde{Q}P_{\tilde{Q}}(\phi_B^{-1}) = 0.$$

Using $\|z\| = \|h - \hat{h}\| \leq \frac{4}{10}\|\hat{h}\|$ we deduce

$$\pi(x + l) - \pi(x) \leq -\frac{\|\hat{h}\|^2}{2} + \frac{\|z\|^2}{2} + \frac{1}{3}\frac{\|h\|^3}{1 - \|h\|},$$

$$\leq \left(-\frac{1}{2} + \frac{0.16}{2} + \frac{1}{3}\frac{\|h\|}{1 - \|h\|}\frac{\|h\|^2}{\|\hat{h}\|^2}\right)\|\hat{h}\|^2.$$

Combining with $\|\hat{h}\| \leq 1/4$ and $\|h\| \leq 0.35$, we obtain the conclusion. $\qquad\square$

The next result gives characterizations of convergence to the analytical center of $S(LCP)$. Set $\bar{d}^k := \sqrt{\mu x_B^k / s_B^k}$ and

$$\tilde{Q}^k := Q_B\bar{D}^k; \quad \hat{Q}^k := Q_BX_B^k; \quad u_P^{c,k} := \bar{d}^k P_{\tilde{Q}^k}(\phi_B^k)^{-1}; \quad d_{Newton}^k := x_B^k P_{\hat{Q}^k}\mathbf{1}. \tag{24.1}$$

Lemma 24.9. *Take $\{(x^k, s^k, \mu_k)\} \in \mathcal{N}_\epsilon$ such that $\mu_k \to 0$. Then the following properties are equivalent:* (i) $x_B^k \to x^*$, (ii) $\pi(x_B^k) \to \pi(x_B^*)$, (iii) $d_{Newton}^k \to 0$, (iv) $u_P^{c,k} \to 0$, (v) $(u^c)^k \to 0$.

Proof. It is easy to see that (i)$\Rightarrow$(ii) (continuity of $\pi(x_B)$ at x_B^*), that (ii)$\Rightarrow$(iii) (the Newton direction in the large-variable centering problem tends to 0 at the optimum x_B^*, because the Hessian of π is invertible there), and that (iv)$\Leftrightarrow$(v) (Lemma 20.12). It therefore suffices to show (iii)$\Rightarrow$(iv)$\Rightarrow$(i).

Suppose (iii) holds. Extracting a subsequence if necessary, we can assume $x^k \to \bar{x}_B$ and $\bar{x}_B > 0$ (Lemma 20.10). Hence $d_{Newton}^k \to 0$ if and only if $P_{\hat{Q}^k}\mathbf{1} \to 0$, i.e. $\mathbf{1} \in \mathcal{R}((\hat{Q}^k)^\top) + o(1)$, and therefore $(\phi_B^k)^{-1} = \bar{d}^k(x_B^k)^{-1} \in \mathcal{R}((\hat{Q}^k)^\top) + o(1)$, hence $P_{\tilde{Q}^k}(\phi_B^k)^{-1} = o(1)$. Since $\bar{d}^k \approx 1$, we deduce that $u_P^{c,k} \to 0$, which is (iv).

Suppose (iv) holds. Again we can assume $x^k \to \bar{x}_B > 0$. Using similar arguments as above, we check that $(x_B^k)^{-1} \in \mathcal{R}(Q^\top) + o(1)$. Passing to the limit, we get $(\bar{x}_B)^{-1} \in \mathcal{R}(Q^\top)$, hence $\bar{x} = x^*$, which proves that $x^k \to x^*$, hence (i). $\qquad\square$

24.3.2 Convergence with Strict Complementarity

Consider a sequence $\{(x^k, s^k)\}$ fitting the general framework of primal-dual algorithms (Chap. 21):

$$(x^{k+1}, s^{k+1}) = (x^k, s^k) + \theta_k(u^k, v^k),$$

with $\theta_k \in [0,1]$, $\gamma_k \in [0,1]$, and (u^k, v^k) solving the Newton equations associated with the central-path equation:

$$\begin{cases} s^k u^k + x^k v^k = \gamma_k \mu_k \mathbf{1} - x^k s^k, \\ Q u^k + R v^k = 0, \end{cases}$$

Suppose $\{\mu_k\}$ summable: $\sum_k \mu_k < +\infty$ (this property has been established for predictor-corrector algorithms; we will check it for largest-step algorithms). The next theorem shows that, in the neighborhood $\mathcal{V}_{1/4}$, convergence of the sequence is guaranteed. This result will allow us too check that the largest-step algorithm in $\mathcal{V}_{1/4}$, to be described later, converges superlinearly.

Theorem 24.10. *Suppose strict complementarity holds. If $\{\mu_k\}$ is summable, and if $\{(x^k, s^k)\} \subset \mathcal{V}_\alpha$, with $\alpha \le 1/4$, then $\{(x^k, s^k)\}$ converges to a strictly complementary solution (x^*, s^*). If $\sum_{k=0}^{\infty} \theta_k \gamma_k = +\infty$, this solution is the analytic center of $S(LCP)$.*

Proof. By Lemma 20.12), $u^k = \theta_k \gamma_k u_P^{c,k} + O(\mu_k)$. Since $\{u_P^{c,k}\}$ is bounded in $\mathcal{V}_\alpha$ and μ_k is summable, $\{x^k\}$ converges when $\sum_{k=1}^{\infty} \theta_k \gamma_k < +\infty$. There remains to prove that, when $\sum_{k=1}^{\infty} \theta_k \gamma_k = +\infty$, the sequence $\{x^k\}$ converges to the analytic center of the optimal face. From Lemma 24.9, it suffices to show that $\pi(x^k)$ converges to $\pi(x^*)$. Using Proposition 24.6, the fact that $\theta_k \gamma_k \in [0,1]$, and convexity of π, we have

$$\pi(x^k + \theta_k \gamma_k u_P^{c,k}) \le \pi(x^k) - \tfrac{6}{100} \theta_k \gamma_k \|d_{Newton}^k\|^2.$$

The function π is Lipschitz-continuous in $\mathcal{V}_\alpha$, hence

$$\pi(x^{k+1}) \le \pi(x^k) - \tfrac{6}{100} \theta_k \gamma_k \|d_{Newton}^k\|^2 + O(\mu_k). \qquad (*)$$

A first consequence of this inequality is that

$$\pi(x^k) \le \pi(x^{k^0}) + O\left(\sum_{j=k^0}^{k} \mu_j \right), \quad \forall k^0 \in \mathbb{N}, \ k > k^0,$$

and therefore, since μ_k is summable,

$$\limsup \pi(x^k) \le \pi(x^{k^0}) + O\left(\sum_{j=k^0}^{\infty} \mu_j \right) \quad \Rightarrow \quad \limsup \pi(x^k) \le \liminf \pi(x^k),$$

so that $\pi(x^k)$ converges. On the other hand, since π is bounded from below on $\mathcal{N}_\epsilon$, we deduce from $(*)$, using summability of μ_k, that

$$\sum_{k=1}^{\infty} \theta_k \gamma_k \|d_{Newton}^k\|^2 < \infty.$$

Since $\sum_{k=1}^{\infty} \theta_k \gamma_k = \infty$, a subsequence $\{x^k\}_{k \in K}$ satisfies $\lim_{k \in K} \|d_{Newton}^k\|^2 = 0$. From Lemma 24.9, $\{x^k\}_{k \in K} \to x^*$. Since the whole sequence $\pi(x^k)$ converges, this implies that the sequence $\{x^k\}$ converges to the analytic center of the optimal face, which was to be proved. □

24.3.3 Convergence without Strict Complementarity

Proposition 24.6 can also be applied to the analysis of predictor-corrector algorithms in the neighborhood $\mathcal{V}_{1/4}$. Convergence of the sequence to the analytic center of the optimal face follows, without the strict complementarity assumption.

Proposition 24.11. *The predictor-corrector algorithm* **PC** *(presented in 21.3) in the neighborhood $\mathcal{V}_{1/4}$ produces a sequence converging to the analytic center of $S(LCP)$.*

Proof. Define $u_P^{c,k}$ and d_{Newton}^k as in (24.1). From Lemma 20.12, we have

$$u_B^c = u_P^{c,k} + O(\sqrt{\mu_k}); \quad u^a = O(\sqrt{\mu_k}).$$

Using the fact that $\pi(x_B)$ is Lipschitz-continuous in $\mathcal{V}_\alpha$, and combining with Proposition 24.6, we get

$$\pi(x_B^k + u_B^c) \le \pi(x_B^k) - \tfrac{6}{100}\|d_{Newton}^k\|^2 + O(\sqrt{\mu_k}).$$

The affine move having size $O(\sqrt{\mu})$, we have

$$\pi(x_B^{k+1}) \le \pi(x_B^k) - \tfrac{6}{100}\theta_k\gamma_k\|d_{Newton}^k\|^2 + O(\sqrt{\mu_k}).$$

Take $\varepsilon > 0$. From Lemma 24.9, there exists $\nu > 0$ such that we have (for k large enough)

$$\|d_{Newton}^k\|^2 \ge \nu \quad \text{if} \quad \pi(x^k) \ge \pi(x_B^*) + \varepsilon.$$

For k large enough, we will have

$$\pi(x_B^{k+1}) \le \pi(x_B^k) - \tfrac{6}{100}\theta_k\gamma_k\|d_{Newton}^k\|^2 + \min(\varepsilon, \tfrac{1}{100}\nu).$$

Depending whether $\pi(x_B^k) \le \pi(x_B^*) + \varepsilon$ or not, we deduce that

$$\pi(x_B^{k+1}) \le \max\left(\pi(x_B^*) + 2\varepsilon, \pi(x_B^k) - \tfrac{5}{100}\nu\right).$$

As long as $\pi(x_B^k) > \pi(x_B^*) + 2\varepsilon$, therefore, the values of $\pi(x_B^k)$ decrease down to a value smaller than $\pi(x_B^*) + 2\varepsilon$; afterwards, $\pi(x_B^k)$ never increases above this value. Since ε was arbitrary, we deduce that $\pi(x_B^k) \to \pi(x_B^*)$. Lemma 24.9 implies the conclusion. □

24.3.4 Relative Distance in the Space of Large Variables

The technical points below will be useful for the largest-step algorithm with safeguard. Recall that the point on the central path associated with $\mu > 0$ is denoted by $w^\mu = (x^\mu, s^\mu, \mu)$. The following concept will be useful. Take $(x, s) \in F(LCP)$. The *relative distance* in the x-space is

$$\text{dist}(x, \mu) = \left\| \frac{x - x^\mu}{x^\mu} \right\|. \tag{24.2}$$

The relative distance in the s-space is defined likewise. Since $x^\mu s^\mu = \mu \mathbf{1}$, an equivalent expression is

$$\text{dist}(x, \mu) = \left\| \frac{x s^\mu}{\mu} - \mathbf{1} \right\|. \tag{24.3}$$

The lemma below (see Gonzaga and Bonnans [175]) links relative distances and proximity measure.

Lemma 24.12. *Let* $w = (x, s, \mu) \in F(LCP)$ *be such that* $\delta(w) = \left\| \dfrac{xs}{\mu} - \mathbf{1} \right\|$ *is smaller than* $\sqrt{2}/2$. *Then*

$$\text{dist}(x, \mu) + \text{dist}(s, \mu) \leq \sqrt{2}\left(1 - \sqrt{1 - \sqrt{2}\delta(w)}\right).$$

Proof. Set

$$dx := \frac{x - x^\mu}{x^\mu}; \quad ds := \frac{s - s^\mu}{s^\mu}; \quad \delta = \delta(w).$$

We can write

$$x = x^\mu(\mathbf{1} + dx), \quad s = s^\mu(\mathbf{1} + ds).$$

The relative distance in the x-space is, using $dx^\top ds \geq 0$:

$$\text{dist}(x, \mu) = \|dx\| \leq \|dx + ds\|.$$

The proximity measure to the central path satisfies

$$\delta = \left\| \frac{xs}{\mu} - \mathbf{1} \right\| = \left\| \frac{x}{x^\mu} \frac{s}{s^\mu} - \mathbf{1} \right\| = \|dx + ds + dx\,ds\|.$$

In view of Mizuno's Lemma 20.24, we have $\|(dx)(ds)\| \leq \|dx + ds\|^2/\sqrt{8}$, and therefore

$$\delta \geq \|dx + ds\| - \|(dx)(ds)\| \geq \|dx + ds\| - \|dx + ds\|^2/\sqrt{8}.$$

Set $z = \|dx + ds\|$. Then

$$z^2 - \sqrt{8}z + \sqrt{8}\delta \geq 0.$$

The solutions of this equation are $z = \sqrt{2}\left(1 \pm \sqrt{1 - \sqrt{2}\delta}\right)$. Since $\delta \leq \sqrt{2}/2$, these solutions are real. The inequality is satisfied for

$$z = \sqrt{2}\left(1 - \sqrt{1 - \sqrt{2}\delta}\right) \leq \sqrt{2} \quad \text{or} \quad z = \sqrt{2}\left(1 + \sqrt{1 - \sqrt{2}\delta}\right) > \sqrt{2}.$$
(24.4)

Since $\text{dist}(x, \mu) + \text{dist}(s, \mu) = \|dx\| + \|ds\| \leq z$, it suffices to check that, when $\delta(w) \leq \sqrt{2}/2$, we have $\|dx + ds\| \leq \sqrt{2}$.

We argue by contradiction. Let $\tilde{w} \in F(LCP)$ be such that $\delta(\tilde{w}) \leq \sqrt{2}/2$ and $\|\tilde{dx} + \tilde{ds}\| > \sqrt{2}$.

The mapping $x, s \in F^0 \mapsto \delta^2(x, s, \tilde{\mu})$ is smooth and attains its minimum at a unique point $(x_{\tilde{\mu}}, s_{\tilde{\mu}})$. Construct a smooth curve $\alpha \mapsto (x(\alpha), s(\alpha))$ in F^0, linking $(\tilde{x}, \tilde{s})$ and $(x_{\tilde{\mu}}, s_{\tilde{\mu}})$, along which $\delta(x(\alpha), s(\alpha), \tilde{\mu})$ decreases strictly. It suffices to integrate the differential equation

$$\frac{d}{dt}(dx, ds) = (u(x, s), \; v(x, s))$$

where $(u(x, s), \; v(x, s))$ is the centering direction at (x, s). Then

$$\frac{d}{dt}\delta(x, s, \tilde{\mu}) = \frac{2}{\tilde{\mu}}\left(\frac{xs}{\tilde{\mu}} - 1\right)^{\top}(su(x, s) + xv(x, s)) = -2\delta(x, s, \tilde{\mu}),$$

so that $\delta(x, s, \tilde{\mu})$ decreases exponentially along the trajectory. Since $\|\tilde{dx} + \tilde{ds}\| > \sqrt{2}$ and $\|dx_{\tilde{\mu}} + ds_{\tilde{\mu}}\| = 0$, there is a point $(\hat{x}, \hat{s})$ on the curve such that $\delta(\hat{x}, \hat{s}, \tilde{\mu}) < \sqrt{2}/2$ and $z = \|\hat{dx} + \hat{ds}\| = \sqrt{2}$. This contradicts (24.4). $\qquad\square$

24.4 Convergence Analysis

24.4.1 Global Convergence of the Largest-Step Algorithm

Recall that, if $w = (x, s, \mu)$ is the current point of the largest-step algorithm, then $w^{\sharp} = (x^{\sharp}, s^{\sharp}, \mu_{\sharp})$ and $\delta^{\sharp}$ make up the point obtained after one iteration of the algorithm and the associated proximity. From 21.2.2, we have $\delta^{\sharp} = \frac{1}{\gamma\mu}\|uv\|$. Set

$$d := \sqrt{\frac{x}{s}}, \; \bar{u} := d^{-1}u, \; \bar{v} := dv.$$

Then

$$\bar{u} + \bar{v} = d^{-1}u + dv = \frac{1}{\sqrt{xs}}(su + xv) = \frac{1}{\sqrt{xs}}(\gamma\mu\mathbf{1} - xs) = \frac{\gamma\mu}{\sqrt{xs}} - \sqrt{xs}.$$

Since $\bar{u}^{\top}\bar{v} = u^{\top}v \geq 0$, by Mizuno's Lemma 20.24,

$$\delta^\sharp = \frac{1}{\gamma\mu}\|uv\| = \frac{1}{\gamma\mu}\|d^{-1}udv\| = \frac{1}{\gamma\mu}\|\bar{u}\bar{v}\| \leq \frac{1}{\gamma\mu\sqrt{8}}\|\bar{u}+\bar{v}\|^2,$$

$$= \frac{1}{\gamma\mu\sqrt{8}}\left\|\frac{\gamma\mu}{\sqrt{xs}} - \sqrt{xs}\right\|^2 = \frac{1}{\gamma\sqrt{8}}\left\|(\gamma-1)\sqrt{\frac{\mu}{xs}} + \sqrt{\frac{\mu}{xs}} - \sqrt{\frac{xs}{\mu}}\right\|^2,$$

$$\leq \frac{1}{\gamma\sqrt{8}}\left((\gamma-1)\left\|\sqrt{\frac{\mu}{xs}}\right\| + \left\|\sqrt{\frac{\mu}{xs}} - \sqrt{\frac{xs}{\mu}}\right\|\right)^2.$$

Using

$$\left\|\sqrt{\frac{\mu}{xs}}\right\| \leq \sqrt{n}\left\|\sqrt{\frac{\mu}{xs}}\right\|_\infty \leq \sqrt{\frac{n}{1-\alpha}}$$

and

$$\left\|\sqrt{\frac{\mu}{xs}} - \sqrt{\frac{xs}{\mu}}\right\| \leq \left\|\sqrt{\frac{\mu}{xs}}\right\|_\infty\left\|\frac{xs}{\mu} - 1\right\| \leq \frac{\alpha}{\sqrt{1-\alpha}},$$

we get for $\alpha \leq 1/2$

$$\delta^\sharp \leq \frac{1}{\gamma(1-\alpha)\sqrt{8}}[(1-\gamma)\sqrt{n}+\alpha]^2 \leq \frac{1}{\gamma\sqrt{2}}[(1-\gamma)\sqrt{n}+\alpha]^2.$$

We analyze this expression for $\gamma \in \left[\frac{1.2}{\sqrt{2}}, 1\right]$. Then $\frac{1}{\gamma\sqrt{2}} \leq \frac{1}{1.2}$, hence $\delta^\sharp \leq \alpha$ as soon as

$$[(1-\gamma)\sqrt{n}+\alpha]^2 \leq 1.2\alpha,$$

i.e.

$$\gamma \geq 1 - \frac{\sqrt{1.2\alpha}-\alpha}{\sqrt{n}}.$$

The step $\bar{\gamma}$ realizing the equality $\delta^\sharp = \alpha$ will therefore be at most (using $\sqrt{1.2\alpha} \geq 1.2\alpha$)

$$\bar{\gamma} \leq \max\left(\frac{1.2}{\sqrt{2}}, 1 - \frac{\sqrt{1.2\alpha}-\alpha}{\sqrt{n}}\right) \leq \max\left(\frac{1.2}{\sqrt{2}}, 1 - \frac{\alpha}{5\sqrt{n}}\right) = 1 - \frac{\alpha}{5\sqrt{n}}.$$

We thus obtain global convergence in $5\alpha^{-1}\sqrt{n}\bar{L}$ iterations.

24.4.2 Local Convergence of the Largest-Step Algorithm

From 21.2.2, since $\theta = 1$, we have $\gamma^\sharp = \frac{1}{\gamma\mu}\|uv\|$. Also, using $\delta^c = \frac{1}{\mu}\|u^c v^c\|$ and $\gamma \in [0,1]$:

$$\delta^\sharp = \frac{1}{\gamma\mu}\|(\gamma u^c + (1-\gamma)u^a)(\gamma v^c + (1-\gamma)v^a)\|,$$

$$= \frac{\gamma}{\mu}\|u^c v^c\| + \frac{1-\gamma}{\mu}\|u^c v^a + u^a v^c\| + \frac{(1-\gamma)^2}{\gamma\mu}\|u^a v^a\|,$$

$$\leq \gamma\delta^c + \frac{1}{\mu}\|u^c v^a + u^a v^c\| + \frac{1}{\gamma\mu}\|u^a v^a\|.$$

Under strict complementarity, we have proved (Lemmas 20.10 and 20.12) that $u^a = O(\mu)$, $v^c = O(\mu)$ and $v^a = O(\mu)$. On the other hand, we know that $\delta^c \leq \alpha/2$ (Lemma 21.5). Hence

$$\alpha = \delta^\sharp \leq \frac{\alpha}{2} + O(\|u^c\|) + O\left(\frac{\mu}{\gamma}\right).$$

Since Theorem 24.10 applies, we know that $\{x^k\}$ converges and that, if $\sum_k \gamma_k = +\infty$, then x^k converges to the analytic center of $S(LCP)$.

Then $u^c \to 0$ from Lemma 24.9. We deduce from the above relation that

$$O\left(\frac{\mu}{\gamma}\right) \geq \frac{\alpha}{2} + o(1),$$

hence $\gamma = O(\mu)$, which was to be proved.

24.4.3 Convergence of the Largest-Step Algorithm with Safeguard

The global convergence analysis follows immediately from that of the largest-step algorithm. By contrast, the asymptotic analysis requires accurate estimates of the centering of variables. Recall that the point on the central path associated with $\mu > 0$ is denoted by $w^\mu = (x^\mu, s^\mu, \mu)$.

We have seen that, at least in the small neighborhoods, a centering step considerably reduces the proximity measure. It can therefore be expected that s^c and s^μ are close enough to ensure closeness of the quantities $\mathrm{dist}(x, \mu) = \|xs^\mu/\mu - 1\|$ and $\|xs^c/\mu - 1\|$. The following lemma establishes a relation between these quantities.

Lemma 24.13. *Take $w \in \mathcal{V}_\alpha$ and let w^c be the point obtained after a centering step. Then*

$$\left\|\frac{xs^c}{\mu} - 1\right\| \leq \mathrm{dist}(s^c, \mu) + (1 + \mathrm{dist}(s^c, \mu))\,\mathrm{dist}(x, \mu).$$

Proof. Set $s^c = s^\mu(1 + ds^c)$. Then

$$\left\|\frac{xs^c}{\mu} - 1\right\| = \|dx + ds^c + dx\,ds^c\| \leq \|ds^c\| + (1 + \|ds^c\|)\|dx\|.$$

Since $\mathrm{dist}(x, \mu) = \|dx\|$ and $\mathrm{dist}(s^c, \mu) = \|ds^c\|$, the conclusion follows.

□

The next result is a direct application of the previous lemmas. It involves numerical values which will be useful in the analysis of the algorithms.

Lemma 24.14. (i) *If $\delta(w) \leq 0.5$, we have*

$$\mathrm{dist}(x, \mu) \leq 0.65, \quad \delta(w_c) \leq 0.177, \quad \mathrm{dist}(s^c, \mu) \leq 0.1895, \quad \|xs^c/\mu - 1\| \leq 0.97.$$

(ii) *If $\delta(w) \leq 0.177$ (which is the case, when $\alpha \leq 0.5$, after centering a point of $\mathcal{V}_\alpha$), then*

$$\text{dist}(x, \mu) \leq 0.1898, \ \delta(w_c) \leq 0.014, \ \text{dist}(s^c, \mu) \leq 0.014, \ \|xs^c/\mu - 1\| \leq 0.206.$$

(iii) *If $\delta(w) \leq 0.5$ and if $\text{dist}(x, \mu) \leq 0.2$ then $\|xs^c/\mu - 1\| \leq 0.43$.*

The preceding results do not assume $B = \{1, \cdots, n\}$ and are valid without strict complementarity. These two properties, however, are necessary for what follows.

Because x corresponds to large variables, and because the strict complementarity assumption holds, $S(LCP)$ has an analytic center (x^*, s^*), for which $x^* \in \mathbb{R}^n_{++}$ and $s^* = 0$. We can define

$$\text{dist}(x, 0) := \left\| \frac{x - x^*}{x^*} \right\|. \tag{24.5}$$

Since $\frac{d}{dt}x^\mu$ is the affine displacement computed at the point (x^μ, s^μ, μ), and the latter is of order μ by Lemma 20.12, $x^\mu = x^* + O(\mu)$. We deduce

$$\text{dist}(x, 0) = \left\| \frac{x - x^\mu + O(\mu)}{x^\mu + O(\mu)} \right\| = \text{dist}(x, \mu) + O(\mu). \tag{24.6}$$

Lemma 24.15. *Take $w \in \mathcal{V}_\alpha$ and $\gamma \in (0, 1]$. Then*

$$\frac{x^\sharp s^\sharp}{\gamma \mu} - 1 = \gamma \left(\frac{x^c s^c}{\mu} - 1 \right) + (1 - \gamma) \left(\frac{xs^c}{\mu} - 1 \right) + \frac{O(\mu)}{\gamma} \tag{24.7}$$

and

$$\delta(w) \leq 0.1 \, \delta(w^c) + 0.9 \left\| \frac{xs^c}{\mu} - 1 \right\| + O(\mu).$$

Moreover, the output $\tilde{w}$ of a Newton move with $\gamma = 0.1$ satisfies
(i) *If $\delta(w) \leq 0.5$ then $\delta(\tilde{w}) \leq 0.89 + O(\mu)$.*
(ii) *If $\delta(w) \leq 0.177$ then $\delta(\tilde{w}) \leq 0.19 + O(\mu)$.*
(iii) *If $\delta(w) \leq 0.5$ and $\text{dist}(x, 0) \leq 0.2$ then $\delta(\tilde{w}) \leq 0.403 + O(\mu)$.*

Remark 24.16. The next lemmas use several bounds on μ, like $\bar{\mu}_1, \tilde{\mu}_j$, etc. These quantities are constants depending on problem's data: (Q, R, h), but not on the initial point w^0. □

Lemma 24.17. *There exists $\bar{\mu}_1 > 0$ such that, if $\mu \leq \bar{\mu}_1$, then*
(a) *during the safeguarding step, we have $\delta(0.1w^c + 0.9w^a) < 1$;*
(b) *after safeguard (be it active or not),*

$$\delta(0.1w^c + 0.9w^a) < 0.42 \ \text{and} \ \|xs^c/\mu - 1\| \leq 0.49 + O(\mu),$$

(c) *when the iteration is complete, $x^\sharp - x = O(\mu)$ and $\gamma = O(\mu)$.*

Proof. Set $\tilde{w} = 0.1w^c + 0.9w^a$. Item (a) follows from Lemma 24.15(i) when $\mu \leq \tilde{\mu}_1$.

If the safeguard is not activated, we have $\delta(\tilde{w}) < 0.42$ by construction; otherwise, we have after a centering step $\delta(w) \leq 0.177$ (Lemma 24.14(i)) and then $\delta(\tilde{w}) \leq 0.19 + O(\mu)$ (Lemma 24.15(ii)). In either case, $\delta(\tilde{w}) < 0.42$ for $\mu \leq \tilde{\mu}_2 \leq \tilde{\mu}_1$. Combining with (24.7), we get when $\mu \leq \tilde{\mu}_2$,

$$0.42 \geq -0.1\,\delta(w^c) + 0.9\left\|\frac{xs^c}{\mu} - \mathbf{1}\right\| + O(\mu).$$

But $\delta(w^c) \leq 0.177$ by Lemma 24.14(i), and therefore $\|xs^c/\mu - \mathbf{1}\| \leq 0.487 + O(\mu)$, hence (b).

There remains to prove (c). From (b), we have that

$$\begin{aligned}
0.5 = \delta(w) &\leq \gamma\delta(w^c) + (1 - \gamma)\left\|\frac{xs^c}{\mu} - e\right\| + \frac{O(\mu)}{\gamma}, \\
&\leq 0.177\gamma + 0.49(1 - \gamma) + O(\mu)/\gamma, \\
&\leq 0.49 + O(\mu)/\gamma.
\end{aligned}$$

This implies $\gamma = O(\mu)$, hence the result.

$\square$

Lemma 24.18. *There exists $\bar{\mu} > 0$ such that the safeguard is activated at most once when $\mu_k \leq \bar{\mu}$.*

Proof. It suffices to check that, if the safeguard is activated once, the hypothesis of point (iii) of Lemma 24.15 applies. In that case, the safeguard is never activated again when μ_k is small enough.

By Lemma 24.14, $x^{k+1} = x^k + O(\mu_k)$. Therefore,

$$\begin{aligned}
\mathrm{dist}(x^{k+1}, 0) = \left\|\frac{x^k + O(\mu_k)}{x^*} - \mathbf{1}\right\| &= \left\|\frac{x^k}{x^*} - \mathbf{1}\right\| + O(\mu_k) \\
&= \mathrm{dist}(x^k, 0) + O(\mu_k).
\end{aligned}$$

As a result

$$\mathrm{dist}(x^{k+1}, 0) \leq \mathrm{dist}(x^k, 0) + O(\mu_k). \tag{24.8}$$

On the other hand, let k be an iteration at which the safeguard is activated, and let $\bar{k} > k$ be the first iteration at which it is activated again. From Lemma 24.14(ii), we have after centering

$$\mathrm{dist}(x^k, 0) \leq 0.19 + O(\mu_k).$$

Combining with Lemma 24.17(c), we obtain for $\bar{k} > k$

$$\mathrm{dist}(x^{\bar{k}}, 0) \leq 0.19 + O(\mu_k) + K\sum_{j=k}^{\bar{k}-1}\mu_j.$$

Since $\{\mu_k\}$ converges linearly to 0, we have that $\sum_{j=k}^{\infty} \mu_j = O(\mu_k)$, so that

$$\operatorname{dist}(x^{\bar{k}}, 0) \leq 0.19 + O(\mu_k).$$

If μ_k is small enough, namely $\mu_k \leq \bar{\mu} \leq \tilde{\mu}_1$, then $\operatorname{dist}(x^{\bar{k}}, 0) < 0.2$. Thus, the safeguard cannot be activated at iteration $\bar{k}$, and the conclusion follows.

□

24.5 Comments

The largest-step method is due to McShane [251]. The convergence analysis of the large variables in the neighborhood $\mathcal{V}_{1/4}$, which produces superlinear convergence, is due to Bonnans and Gonzaga [44]. Gonzaga [174] shows that the safeguarding mechanism allows a convergence analysis in $\mathcal{V}_{1/2}$. Finally, Gonzaga and Bonnans [175] have established estimates of the relative distance, and have studied a modified method, in which each iteration is made up of several displacements, computed with the same Jacobian. A super-quadratically convergent algorithm is then obtained.

Extensions to the large neighborhood have not been studied. The study of Algorithm 21.9 has established that, if $w \in \mathcal{N}_\epsilon$, then $w + (\epsilon^3/4n)(w^c - w)$ lies in the interior of $\mathcal{N}_\epsilon$. An algorithm can therefore be stated, in which the new point is searched on the segment $[w + (\epsilon^3/4n)(w^c - w), w^a]$. This algorithm might well have an $O(n\bar{L})$ complexity; but its superlinear convergence seems more difficult to establish.

25 Complexity of Linear Optimization Problems with Integer Data

25.1 Overview

The aim of complexity theory is to evaluate the minimal number of operations required to compute a solution of a given problem (and to determine the corresponding algorithm). This theory is far from being closed, since the answer is not known even for solving a linear system $Ax = b$, with A a matrix $n \times n$ invertible: classical factorization methods have an $O(n^3)$ complexity but certain fast algorithms have an $O(n^\alpha)$ complexity, with $\alpha < 2.5$; the minimal value of α (proved to be larger than 2) is still unknown (see Coppersmith and Winograd [87]).

Furthermore, the available complexity estimates are only upper bounds of the number of operations to be performed in the worst case; they do not necessarily reflect the actual behavior of algorithms. Despite these weaknesses, complexity theory has been a main instigator for algorithmic research; so it is useful to know its general concepts, in order to read the literature. We will present the basis of complexity theory for the linear problem in standard form

$$\underset{x \in \mathbb{R}^n}{\text{Min}} \, c^\top x; \quad Ax = b, \, x \geq 0 \qquad (LP)$$

(with $b \in \mathbb{R}^p$), when the components of (A, b, c) are integers. Rational components of (A, b, c) can be reduced to this case via an appropriate scaling. This theory gives an upper bound of $\|x\|$ and of $c^\top x$ on the feasible set $F(LP)$, as well as a lower bound of $(c^\top x - v(LP))$ when x is a non-optimal extreme point of the feasible set. It also gives a so-called purification process which, starting from a point $x \in F(LP)$ "almost" optimal, allows the computation of a solution of (LP) in $O(n^3)$ operations. Combining these results with the estimates of speed of convergence for specific algorithms, one deduces the number of iterations (barring purification, which is cheap) required by these algorithms to yield an exact solution of a linear problem.

25.2 Main Results

25.2.1 General Hypotheses

In this chapter, we will assume that (i) A has rank $p \leq n$ (eliminating dependent equalities, any problem can be reduced to this case), (ii) the feasible set $F(P)$ is nonempty and bounded, (iii) the components of (A, b, c) are integers.

25.2.2 Statement of the Results

We define the following amounts:

- ℓ is the total number of bits required to encode (A, b, c); it is the amount of memory necessary to store (A, b, c),
- $L := \ell + n + 1$ is the *length* of (LP).

Let E be a convex subset of $\mathbb{R}^n$. We say that x is an extreme point of E if $x \in E$ and if x is an endpoint of every segment contained in E and containing x; in other words

$$x \in E \text{ ; if } y, z \in E, \ \alpha \in]0, 1[\text{ and } x = \alpha y + (1 - \alpha)z, \quad \text{then } x = y = z.$$

We set

$$B(x) := \{i = 1, \cdots, n; \quad x_i > 0\}.$$

It is easy to check that any extreme point of $F(P)$ is a basic point (it cannot have more than p nonzero components).

Theorem 25.1. (i) *Any point* $x \in F(LP)$ *satisfies* $|c^\top x| \leq 2^L$, *as well as* $\sum_{i=1}^n x_i \leq 2^L$.
(ii) *If, in addition, x is an extreme point, then*

(a)
$$x_i = 0 \quad or \quad x_i > 2^{-L},$$

(b)
$$c^\top x = v(LP) \quad or \quad c^\top x > v(LP) + 2^{-2L}.$$

Theorem 25.2. *Let $x^0 \in F(LP)$ be such that $c^\top x^0 < v(LP) + 2^{-2L}$. Then there exists an algorithm, said of purification (specified later in the proof) which, starting from x^0, computes in $O(n^3)$ operations a solution of (LP).*

The theorems will be proved in the remainder of the chapter.

25.2.3 Application

Consider an algorithm in which the cost converges R-linearly to the optimal cost: a sequence x^k of feasible points is constructed such that, for a certain $\alpha \in]0, 1[$:

$$c^\top x^k - v(LP) \le \alpha^k [c^\top x^0 - v(LP)]. \tag{25.1}$$

In view of Theorem 25.1, $c^\top x^0 - v(LP) \le 2^{L+1}$. The relation $c^\top x^k - v(LP) \le 2^{-2L}$ will therefore hold as soon as

$$\alpha^k 2^{L+1} \le 2^{-2L} \Leftrightarrow k \ge \frac{3L+1}{|\log_2 \alpha|}.$$

This is the proof of

Lemma 25.3. *Let a feasible point of (LP) be known. Neglecting purification, a feasible interior-point algorithm satisfying (25.1) finds a solution in at most $(3L+1)|\log_2 \alpha|^{-1}$ (rounded up) iterations.*

If the number α has the form $1 - \varepsilon(n)$, with $0 < \varepsilon(n) \le 1$, then $|\log_2 \alpha|^{-1} = |\log_2(1 - \varepsilon(n))|^{-1} \le 1/\varepsilon(n)$, and the algorithm converges in $(3L+1)/\varepsilon(n)$ iterations.

Another result applies to the family of primal algorithms studied in Chap. 21.

Lemma 25.4. *Let a sequence $\{(x^k, s^k, \mu_k)\} \subset \mathcal{V}_\alpha$ satisfy $\mu_k \le (1-\varepsilon(n))\mu_{k-1}$. Suppose $\mu_0 = O(2^L)$. Then (neglecting purification), the algorithm converges in $O(L/\varepsilon(n))$ iterations.*

Proof. Suppose first μ_0 is arbitrary. We have that

$$c^\top x^k - v(LP) \le (x^k)^\top s^k \le \frac{n}{\varepsilon}\mu_k \le (1 - \varepsilon(n))^k \mu_0,$$

and a 2^{-2L}-optimal point is obtained as soon as $(1 - \varepsilon(n))^k \mu_0 \le 2^{-2L}$, which holds if $k = (2L + log_2(\mu_0))/\varepsilon(n)$. If $\mu_0 = O(2^L)$, we obtain $k = O(L/\varepsilon(n))$. ☐

Now we present the tools necessary to prove the above results.

25.3 Solving a System of Linear Equations

With a vector g or a matrix G with integer components, is associated the memory *length* necessary to store g or G in binary; it is denoted by $L(g)$ or $L(G)$.

Lemma 25.5. *Take $\alpha \in \mathbb{R}_{++}$. Then $\alpha \le 2^{L(\alpha)}$. In other words, $L(\alpha) \ge \log_2 \alpha$.*

Proof. Let L be a positive integer. The largest number $\alpha \in \mathbb{R}_{++}$ requiring not more than L binary positions is

$$2^{L-1} + 2^{L-2} + \cdots + 2 + 1 = 2^L - 1 \leq 2^L,$$

hence the result.

$\square$

Lemma 25.6. *Let G be an $n \times n$ invertible matrix with integer components. Then $\det G \leq 2^{L(G)}$.*

Proof. It is known that $\det G$ is the volume of a portion of $\mathbb{R}^n$: the parallelotope constructed on the vectors that are the columns of G, denoted by $G._j$. This volume is maximal when the column vectors are orthogonal, and therefore

$$\det G \leq \prod_{j=1}^{n} \|G._j\|.$$

Since $L(G) = \sum_{j=1}^{n} L(G._j)$, it suffices to prove that $\|g\| \leq 2^{L(g)}, \quad \forall g \in \mathbb{R}^n$.

Set $L_i := L(g_i)$, $i = 1, \cdots, n$. According to Lemma 25.4, we have $g_i \leq 2^{L_i}$. Since $L \geq 1$, and hence, $2^{L_i} \geq 2$, and

$$\|g\|^2 = \sum_{i=1}^{n} g_i^2 \leq \sum_{i=1}^{n} 2^{2L_i} \leq \prod_{i=1}^{n} 2^{2L_i} = 2^{2L(g)},$$

the result follows.

$\square$

Proposition 25.7. *Let G be an $n \times n$ invertible matrix and $g \in \mathbb{R}^n$, both with integer components. Then $x := G^{-1}g$ satisfies*
(i) $\max_i |x_i| \leq 2^{L(G)+L(g)}$,
(ii) $x_i \leq 0$ *or* $x_i \geq 2^{-L(G)}, \quad \forall i = 1, \cdots, n$.

Proof. Recall Cramer's formula: $x = G^{-1}g$ satisfies

$$x_i = \frac{\det(G_1, \cdots, G_{i-1}, g, G_{i+1}, \cdots, G_n)}{\det G}.$$

Proof of (i): Since G is invertible and has integer elements, we have that $|\det G| \geq 1$. The conclusion follows via Cramer's formula combined with Lemma 25.6.

Proof of (ii): Since G and g have integer components, we have from Cramer's formula:

$$x_i > 0 \Rightarrow |\det(G_1, \cdots, G_{i-1}, g, G_{i+1}, \cdots, G_n)| \geq 1.$$

If $x_i > 0$, combining with Lemma 25.6 we obtain $x_i \geq \frac{1}{|\det G|} \geq 2^{-L(G)}$.

$\square$

25.4 Proofs of the Main Results

25.4.1 Proof of Theorem 25.1

Because $F(P)$ is nonempty closed and bounded, $\sum_i x_i$ attains its maximum on a nonempty set S_1. Let $x \in S_1$ have a minimal number of nonzero components. One easily checks that $\{A_{.i}\}_{i \in B(x)}$ has rank $|B(x)|$. Proposition 25.7 can be applied to the linear system

$$\sum_{i \in E} A_{.i} x_i = b$$

(or rather to $|B(x)|$ independent rows extracted from this system). We then deduce $\|x\|_\infty \leq 2^{L(A)+L(b)}$, hence

$$\sum_{i=1}^n x_i = \|x\|_1 \leq n\|x\|_\infty \leq 2^{L(A)+L(b)+\log_2 n}.$$

On the other hand,

$$|c^\top x| \leq \|c\|\|x\| \leq 2^{L(c)} \times \sqrt{n}\|x\|_\infty \leq 2^{L(A)+L(b)+L(c)+\frac{3}{2}\log_2 n},$$

and (i) is proved.

To prove (ii), consider two extreme points x^1 and x^2. With each of them is associated a linear system with invertible matrix of rank at most p, obtained by extracting from the relation $Ax = b$ the zero components of x and then the independent rows. With Cramer's formula we deduce that x^1 and x^2 have the form $x^k = y^k / \det G^k$, $k = 1, 2$ where $y^k \in \mathbb{R}^n$ has integer components, G^k is an invertible matrix of rank at most p, and $L(G^k) \leq L$, $k = 1, 2$. Hence

$$x^1 - x^2 = \frac{y^1 \det G^2 - y^2 \det G^1}{(\det G^1)(\det G^2)} \; ; \; c^\top x^1 - c^\top x^2 = \frac{c^\top y^1 \det G^2 - c^\top y^2 \det G^1}{(\det G^1)(\det G^2)}.$$

The numerator has integer components. Thus, for $i \in \{1, \cdots, n\}$, if $c^\top x^1 \neq c^\top x^2$, then $|c^\top x^1 - c^\top x^2| \geq |(\det G^1)^{-1}(\det G^2)^{-1}| \geq 2^{-2L}$.

25.4.2 Proof of Theorem 25.2

The process in this theorem resembles the search for a basic point in Chap. 19, Proposition 19.2. We start from an interior point x^0, and we construct a sequence $\{x^k\}$ such that $\{x^k\}$ has k zero components and $c^\top x^{k+1} < c^\top x^k$, until $k = n - p$: then $\bar{x} := x^{n-p}$ is basic and $c^\top \bar{x} \leq c^\top x^0$. By Theorem 25.1(ii)b, this point $\bar{x}$ solves (LP).

To construct x^1, a direction d in the kernel of A is constructed by LU-factorization of an invertible submatrix A_B, with $|B| = p$; cost: $O(n^3)$ operations, which gives a basis. One single nonbasic component is changed;

changing d to $-d$ if necessary, one may assume $c^\top d \leq 0$. Going as far as possible in the direction d while staying feasible, one hits a constraint (at least), say j_1, which gives the point x^1. At step k of the algorithm, the simplex pivoting procedure then yields a new basic variable. Since only one column of the basis matrix changes, updating the LU-factorization can be done in $O(p^2)$ iterations (through techniques which even preserve sparsity to a large extent, see Reid [304]). The process can then be repeated.

If the basis matrix A_B is invertible, this point is extreme in $F(P)$. Otherwise, the process can be repeated to cancel components, until an extreme point is obtained.

25.5 Comments

Purification is discussed in Kortanek and Jishan [218]. The above results assume computations to be done in exact arithmetic. But the rational numbers involved in the algorithms can have a very long representation. This must be taken into account to evaluate the time necessary to solve a linear problem. A finer approach (and closer to actual computations) is to take into account the error propagation due to finite arithmetic. It can be shown that, if each number is coded with a memory of length (L), convergence speed keeps the same order of magnitude.

On the other hand, the simplex algorithm solves in finitely many operations a problem with real data. Complexity estimates have therefore been sought after, which do not involve the length of the problem, but only its dimension n, p. Some results exist, in which a concept of problem *conditioning* comes into play. The question of estimating complexity as a function of dimension only is still widely open.

26 Karmarkar's Algorithm

26.1 Overview

The algorithm of Karmarkar [204] is important from a historical point of view. In fact, it raised an impetus of research which ended in the path-following algorithms presented here. It is therefore interesting to have an idea of it.

This algorithm applies to linear problems in projective form

$$\operatorname*{Min}_{x \in \mathbb{R}^n} c^\top x; \quad Ax = 0, \ a^\top x = 1, \ x \geq 0 \qquad (PLP)$$

with A $p \times n$ matrix of rank p and $a \in \mathbb{R}^n$. The constraint $Ax = 0$ will be referred to as the homogeneous one. An essential concept of the algorithm is the potential function, homogeneous of degree 0 in x, called the *Karmarkar potential*:

$$\varphi_v(x) := n \log(c^\top x - va^\top x) - \sum_{i=1}^n \log x_i = \log(c^\top x - va^\top x)^n + \pi(x),$$

where $v \in \mathbb{R}$ is a lower estimate of $v(PLP)$ (i.e. $v \leq v(PLP)$). The algorithm minimizes φ_v, while adjusting the value of the parameter v at each iteration. Convergence of the cost to $v(PLP)$ follows. The algorithm has the remarkable property of reducing φ by *at least* $1/4$ at each iteration. The displacement is computed as follows:

1. Scaling of variable moving x to the point $\mathbf{1} := (1, \ldots, 1)^\top \in \mathbb{R}^n$.
2. Computation of the direction d, opposite to the orthogonal projection of the gradient of $x \mapsto \varphi_v(x)$ onto the kernel of the homogeneous constraint.
3. Line-search along the direction d. A point $\hat{x}$ is obtained such that $\varphi_v(\hat{x}) < \varphi_v(x)$ and $A\hat{x} = 0$ (however, $a^\top \hat{x} \neq 0$ in general).
4. Shift back: $x \leftarrow \hat{x}/a^\top \hat{x}$.

26.2 Linear Problem in Projective Form

26.2.1 Projective Form and Karmarkar Potential

We call projective form of a linear problem the (PLP)-format. It is easy to pass from the standard to the projective form. In fact, the standard form is

$$\underset{y \in \mathbb{R}^{n+1}}{\text{Min}} (c_1, \dots, c_n, 0)^\top y; \quad (A, -b)y = 0; \quad y_{n+1} = 1; \quad y \geq 0,$$

which is indeed a particular projective form.

We call *potential of Karmarkar-type* the function parameterized by $q > 0$ and $v \in \mathbb{R}$:

$$\varphi_v(x) = q \log(c^\top x - v a^\top x) - \sum_{i=1}^n \log x_i.$$

For $q = n$ we find back the Karmarkar potential. It is positively homogeneous of degree 0 in x:

$$\forall \alpha > 0, \quad \varphi_v(\alpha x) = n \log[\alpha(c^\top x - v a^\top x)] - \sum_{i=1}^n \log \alpha x_i = \varphi_v(x).$$

26.2.2 Minimizing the Potential and Solving (PLP)

We will see that minimizing φ allows the resolution of (PLP). The next lemma uses the length L (defined in Chap. 25) of a problem with integer data.

Lemma 26.1. *If $F(PLP)$ is bounded, then $\pi(x)$ is bounded from below on $F(PLP)$. In particular, if (A, a) has integer components, then*

$$\pi(x) \geq -nL \log 2.$$

Proof. If $F(PLP)$ is bounded, then $-\log x_i$ is bounded from below, and $\pi(x)$ as well. Since the function $-\log$ is convex, we have

$$-\log\left(\sum_{i=1}^n \alpha_i \beta_i\right) \leq -\sum_{i=1}^n \alpha_i \log \beta_i \quad \text{whenever } \beta \in \mathbb{R}_{++}^n, \quad \alpha \in \mathbb{R}_+^n, \quad \sum_{i=1}^n \alpha_i = 1.$$

Then

$$\frac{1}{n}\left(\pi(x) + n \log \sum_{i=1}^n x_i\right) = \frac{1}{n}\sum_{i=1}^n\left[-\log x_i + \log \sum_{j=1}^n x_j\right]$$

$$= -\sum_{i=1}^n \frac{1}{n}\log \frac{x_i}{\sum_{i=1}^n x_i} \geq -\log \frac{1}{n} = \log n \geq 0,$$

and therefore, with Theorem 25.1: $\pi(x) \geq -n \log \sum_{i=1}^n x_i \geq -nL \log 2$. $\qquad \square$

Now construct an estimate of $c^\top x - v(PLP)$ when $\varphi_v(x)$ is known.

Lemma 26.2. *Suppose $F(PLP)$ is bounded and $v \leq v(PLP)$. Then*

$$c^\top x - v(PLP) \leq c^\top x - v \leq e^{\frac{1}{n}[\varphi_v(x) - \inf\{\pi(x); x \in F(PLP)\}]}.$$

In particular, if (A, a) has integer components, then

$$c^\top x - v(PLP) \leq e^{\frac{1}{n}[\varphi_v(x) + nL \log 2]}.$$

Proof. If $x \in F(LPL)$, then $a^\top x = 1$, hence

$$n \log(c^\top x - v) = \varphi_v(x) + \sum_{i=1}^n \log x_i = \varphi_v(x) - \pi(x).$$

Thus,

$$n \log(c^\top x_0 - v) \leq \varphi_v(x) - \inf_{x \in F(PLP)} \pi(x),$$

hence the result (with a use of the previous lemma, when (A, a) has integer components). $\square$

From Lemma 26.2, to construct a minimizing sequence, it suffices to construct $\{x^k\} \in F(PLP)$ and $\{v^k\} \leq v(PLP)$ such that $\varphi_{v^k}(x^k) \to -\infty$.

To compute a 2^{-2L}-optimal solution when (A, a) has integer components (and an exact solution can then be computed via purification, see Chap. 25), it suffices to obtain

$$e^{\frac{1}{n}[\varphi_v(x) + nL \log 2]} \leq 2^{-2L}$$

i.e. $\varphi_v(x) \leq -3nL \log 2$. Karmarkar's algorithm enables the construction of such a point.

26.3 Statement of Karmarkar's Algorithm

We are now in a position to state the algorithm. We first limit ourselves to the case where the optimal cost is known. Updating a lower bound of the cost is discussed in §26.4.3. The constant L in the algorithm below represents a stopping criterion, which can be taken to the length of the problem when the latter has integer data and must be solved exactly.

Algorithm 26.3. Karmarkar's algorithm
Data: $x^0 \in F(PLP)$, $L > 0$; $k := 0$
   REPEAT
- Direction of the projected gradient: compute

$$d^k := -X^k P_{AX^k} X^k \nabla \varphi_v(x^k).$$

- Line-search. Compute $\rho^k > 0$ such that $x^{k+1} := x^k + \rho^k d^k$ satisfies

$$\varphi_v(\hat{x}^{k+1}) \leq \varphi_v(\hat{x}^k) - 1/4.$$

- Shift back: $x^{k+1} := \hat{x}^{k+1} / a^\top \hat{x}^{k+1}$.
   UNTIL $\varphi_v(x^k) > -3nL \log 2$.

26.4 Analysis of the Algorithm

26.4.1 Complexity Analysis

In the form the algorithm is stated, estimating the complexity is easy: since φ_v is positively homogeneous, we have from the line-search that

$$\varphi_v(x^{k+1}) = \varphi_v(\hat{x}^{k+1}) \leq \varphi_v(x^k) - 1/4,$$

hence $\varphi_v(x^k) \leq \varphi_v(x^0) - k/4$. For a problem with integer data, in view of Theorem 25.1:

$$\varphi_v(x^0) \leq n \log(2^L + 2^L) - \sum_{i=1}^{n} \log 2^{-L}$$
$$= n(2L \log 2 + \log 2) = n(2L + 1) \log 2 \leq 3nL \log 2.$$

To obtain a value $-3nL \log 2$ of the potential function, at most $24nL \log 2$ iterations are therefore required. Complexity is indeed $O(nL)$ iterations.

26.4.2 Analysis of the Potential Decrease

The difficulty is to check that the potential does decrease by a fixed amount in the direction d^k. The lemma below analyses this decrease.

Lemma 26.4. (i) If $\rho \|(x^k)^{-1} d^k\| \leq 1/2$, we have

$$\varphi_v(1 + \rho(x^k)^{-1} d^k) \leq \varphi_v(1) + (\rho^2 - \rho)\|(x^k)^{-1} d^k\|^2.$$

(ii) If $\rho \|(x^k)^{-1} d^k\| = 1/2$, then $\varphi_v(x^k + \rho d^k) \leq \varphi_v(x^k) - \dfrac{1}{4}$.

Proof. (i) Set $h^k := (x^k)^{-1} d^k = -P_{AX^k} X^k \varphi_v'(x^k)$. We have that

$$\pi(x^k + \rho d^k) - \pi(x^k) = \pi(1 + \rho h^k) - \pi(1). \tag{26.1}$$

Using the lower bound of Lemma 24.8, we get

$$\Delta \leq -\rho \mathbf{1}^\top h^k + \rho^2 \|h^k\|^2.$$

The first part of the cost is concave; bounding it from above by its linearization and combining with the above relation, we obtain

$$\varphi_v(x^k + \rho d^k) - \varphi_v(x^k) \leq \rho \nabla \varphi_v(x^k)^\top d^k + \rho^2 \|h^k\|^2.$$

By definition of h^k, we have $x^k \nabla \varphi_v(x^k) + h^k \in \mathcal{N}(AX^k)^\perp$, and therefore

$$\nabla \varphi_v(x^k) d^k = (x^k \nabla \varphi_v(x^k))^\top h^k = -\|h^k\|^2,$$

hence the first relation.

(ii) If $\rho\|h^k\| = \frac{1}{2}$, we deduce from (i)

$$\varphi(x^k + \rho d^k) - \varphi(x^k) \leq (\rho\|h\|^k)^2 - \rho\|h\|^k \times \|h\|^k = \frac{1}{4} - \frac{\|h\|^k}{2}.$$

If $\|h^k\| \geq 1$, the last relation follows. It therefore remains to show that, if $v = v(PLP)$, then $\|h^k\| \geq 1$. The gradient of φ_v at x^k is

$$\nabla\varphi_v(x^k) = \frac{n}{c^\top x^k - v}(c - va) - (x^k)^{-1},$$

hence

$$h^k = -P_{AX^k}x^k\nabla\varphi_v(x^k) = \frac{-n}{c^\top x^k - v}P_{AX^k}x^k(c - va) + P_{AX^k}\mathbf{1}.$$

Since $0 = Ax^k = AX^k\mathbf{1}$, we have that $P_{AX^k}\mathbf{1} = \mathbf{1}$ and

$$h^k = \mathbf{1} - \frac{n}{c^\top x^k - v}P_{AX^k}x^k(c - va). \tag{26.2}$$

Now $c^\top x^k - v > 0$; to prove $\|h^k\| \geq 1$, it suffices to show that at least one of the components of $P_{AX^k}x^k(c - va)$ is nonpositive. But if $\bar{x} \in S(PLP)$, we have

$$0 = c^\top\bar{x} - v = (c - va)^\top\bar{x} = [x^k(c - va)]^\top(x^k)^{-1}\bar{x},$$

i.e. $x^k(c - va) \perp (x^k)^{-1}\bar{x}$. Since $0 = A\bar{x} = AX^k(x^k)^{-1}\bar{x}$, and hence, $(x^k)^{-1}\bar{x} \in \mathcal{N}(AX^k)$, and since the projection is a symmetric operator,

$$[P_{AX^k}x^k(c - va)]^\top[(x^k)^{-1}\bar{x}] = [x^k(c - va)]^\top P_{AX^k}(x^k)^{-1}\bar{x},$$
$$= [x^k(c - va)]^\top(x^k)^{-1}\bar{x} = 0.$$

As $\bar{x} \geq 0$ and $a^\top\bar{x} = 1$, we have that $\bar{x} \neq 0$; $P_{AX^k}[x^k(c - va)] > 0$ is therefore impossible.

□

26.4.3 Estimating the Optimal Cost

Since the optimal cost is in general unknown, we are led to construct an increasing sequence $\{v_k\}$ such that $\lim v_k \leq v(PLP)$. We choose v_k such that the vector

$$h^k := \mathbf{1} - \frac{n}{c^\top x^k - v_k}P_{AX^k}x^k(c - v_ka)$$

has norm at least 1 (we will see how in a moment). Taking again the proof of Lemma 26.4, we then obtain

$$\varphi_{v_k}(x^{k+1}) \leq \varphi_{v_k}(x^k) - \frac{1}{4}.$$

Since φ_v is a decreasing function of v, we get

$$\varphi_{v_{k+1}}(x^{k+1}) \le \varphi_{v_k}(x^{k+1}) \le \varphi_{v_k}(x^k) - \tfrac{1}{4}$$

and therefore by induction, setting $\bar{v} := v(PLP)$:

$$\varphi_{\bar{v}}(x^k) \le \varphi_{v_k}(x^k) \le \varphi_{v_0}(x^0) - \tfrac{1}{4}k.$$

For a problem with integer data, we can take

$$v_0 = -2^L \Rightarrow \varphi_{v_0}(x^0) \le 3nL\log 2,$$

so that we obtain the same complexity estimate as when the optimal cost is supposed to be known: $24nL\log 2 = O(nL)$.

To compute v_k such that $\|h^k\| \ge 1$, it can be observed that, from the proof of Lemma 26.4, we have $\|h^k\| \ge 1$ if one of the components of $P_{AX^k}x^k(c - v^k)$ is nonpositive, which is true if $v = v(PLP)$. It therefore suffices to compute

$$\hat{c} := P_{AX^k}x^k c; \quad \hat{a} := P_{AX^k}x^k a$$

and to take

$$v_{k+1} = \begin{cases} v_k, \text{ if } \min\limits_i(\hat{c} - v_k\hat{a})_i \le 0 \\ \sup\{v^\sharp \ge v_k\}; \quad \min\limits_i(\hat{c} - v\hat{a})_i \ge 0, \ \forall v \in [v^k, v^\sharp], \text{ otherwise.} \end{cases}$$

In the second case, v_k can be computed with the formula

$$v_{k+1} = \min\{\hat{c}_i/\hat{a}_i; \quad \hat{a}_i > 0\}.$$

26.4.4 Practical Aspects

The two nontrivial steps of the algorithm are the computation of d and the line-search. The latter is able to accelerate the calculations. Let us analyze in detail the computation of d. Set

$$\delta := x^{-1}d; \quad g := X\nabla\varphi_v(x^k).$$

Then $\delta = P_{AX}g$, which is equivalent to the existence of $\lambda \in \mathbb{R}^p$ such that

$$\begin{cases} \delta + (AX)^\top\lambda = g, \\ AX\delta \quad\quad = 0. \end{cases}$$

Multiply the first equation by AX to eliminate δ; we get

$$AX^2A^\top\lambda = g.$$

Once this linear system is solved, we obtain $\delta = g - XA^\top\lambda$. The linear systems to be solved are therefore of the same type as in predictor-corrector and largest-step algorithms.

26.5 Comments

This presentation of the Karmarkar algorithm is inspired from Gonzaga [173]. It is important to stress that the same type of complexity can be obtained by computing a descent direction for the Karmarkar potential in an "affine" framework (Gonzaga [171]), i.e. using the standard form instead of the projective form. The key element of the method, allowing a polynomial complexity, is therefore the potential function, and not the projective form as it was long believed. Adler, Resende, and Veiga [1] discuss the numerical properties of the algorithm.

References

1. I. Adler, M.G.C. Resende, and G. Veiga. An implementation of Karmarkar's algorithm for linear programming. *Math. Programming*, 44:297–335, 1989.
2. M. Al-Baali. Descent property and global convergence of the Fletcher-Reeves methods with inexact line search. *IMA Journal of Numerical Analysis*, 5:121–124, 1985.
3. W. Alt. The Lagrange-Newton method for infinite-dimensional optimization problems. *Numerical Functional Analysis and Optimization*, 11:201–224, 1990.
4. W. Alt. Sequential quadratic programming in Banach spaces. In W. Oettli and D. Pallaschke, editors, *Advances in Optimization*, number 382 in Lecture Notes in Economics and Mathematical Systems, pages 281–301. Springer-Verlag, 1992.
5. W. Alt. Semi-local convergence of the Lagrange-Newton method with application to optimal control. In R. Durier and Ch. Michelot, editors, *Recent Developments in Optimization*, number 429 in Lecture Notes in Economics and Mathematical Systems, pages 1–16. Springer-Verlag, 1995.
6. W. Alt and K. Malanowski. The Lagrange-Newton method for nonlinear optimal control problems. *Computational Optimization and Applications*, 2:77–100, 1991.
7. M. Anitescu. Degenerate nonlinear programming with a quadratic growth condition. *SIAM Journal on Optimization*, 10:1116–1135, 2000.
8. M. Anitescu. On the rate of convergence of sequential quadratic programming with nondifferentiable exact penalty function in the presence of constraint degeneracy. *Mathematical Programming*, 92:359–386, 2002.
9. K.M. Anstreicher and M.H. Wright. A note on the augmented Hessian when the reduced Hessian is semidefinite. *SIAM Journal on Optimization*, 11:243–253, 2000.
10. P. Armand and J.Ch. Gilbert. A piecewise line-search technique for maintaining the positive definiteness of the updated matrices in the SQP method. *Computational Optimization and Applications*, 16:121–158, 2000.
11. P. Armand, J.Ch. Gilbert, and S. Jan-Jégou. A BFGS-IP algorithm for solving strongly convex optimization problems with feasibility enforced by an exact penalty approach. *Mathematical Programming*, 92:393–424, 2002.
12. L. Armijo. Minimization of functions having Lipschitz continuous first partial derivatives. *Pacific Journal of Mathematics*, 16:1–3, 1966.
13. K. Arrow, L. Hurwicz, and H. Uzawa. *Studies in Nonlinear Programming*. Stanford University Press, Stanford, CA, 1958.
14. K.J. Arrow and R.M. Solow. Gradient methods for constrained maxima with weakened assumptions. In K.J. Arrow, L. Hurwicz, and H. Uzawa, editors,

Studies in Linear and Nonlinear Programming. Stanford University Press, Standford, Calif., 1958.

15. A. Auslender. Numerical methods for nondifferentiable convex optimization. *Mathematical Programming Study*, 30:102–126, 1987.

16. A. Auslender and M. Teboulle. Lagrangian duality and related multiplier methods for variational inequality problems. *SIAM Journal on Optimization*, 10:1097–1115, 2000.

17. A. Auslender, M. Teboulle, and S. Ben-Tiba. Interior proximal and multiplier methods based on second order homogeneous functionals. *Mathematics of Operations Research*, 24:645–668, 1999.

18. F. Babonneau, C. Beltran, A. Haurie, C. Tadonki, and J.-P. Vial. Proximal-ACCPM: a versatile oracle based optimization method. To appear in *Computational Management Science*, 2006.

19. L. Bacaud, C. Lemaréchal, A. Renaud, and C. Sagastizábal. Bundle methods in stochastic optimal power management: a disaggregated approach using preconditioners. *Computational Optimization and Applications*, 20(3):227–244, 2001.

20. R.E. Bank, B.D. Welfert, and H. Yserentant. A class of iterative methods for solving saddle point problems. *Numerische Mathematik*, 56:645–666, 1990.

21. A. Belloni, A. Diniz, M.E. Maceira, and C. Sagastizábal. Bundle relaxation and primal recovery in Unit Commitment problems. The Brazilian case. *Annals of Operations Research*, 120:21–44, 2003.

22. A. Belloni and C. Sagastizábal. Dynamic bundle methods: Application to combinatorial optimization. Technical report, Optimization on line, 2004. http://www.optimization-online.org/DB_HTML/2004/08/925.html.

23. A. Ben-Tal and A. Nemirovski. *Lectures on Modern Convex Optimization – Analysis, Algorithms, and Engineering Applications*. MPS/SIAM Series on Optimization 2. SIAM, 2001.

24. M.W. Berry, M.T. Health, I. Kaneko, M. Lawo, R.J. Plemmons, and R.C. Ward. An algorithm to compute a sparse basis of the null space. *Numerische Mathematik*, 47:483–504, 1985.

25. D.P. Bertsekas. Multiplier methods: a survey. *Automatica*, 12:133–145, 1976.

26. D.P. Bertsekas. *Constrained Optimization and Lagrange Multiplier Methods*. Academic Press, 1982.

27. D.P. Bertsekas. *Nonlinear Programming*. Athena Scientific, 1995. Second edition, 1999.

28. J.T. Betts. *Practical Methods for Optimal Control Using Nonlinear Programming*. SIAM, 2001.

29. J.T. Betts and P.D. Frank. A sparse nonlinear optimization algorithm. *Journal of Optimization Theory and Applications*, 3:519–541, 1994.

30. L.T. Biegler, J. Nocedal, and C. Schmid. A reduced Hessian method for large-scale constrained optimization. *SIAM Journal on Optimization*, 5:314–347, 1995.

31. P.T. Boggs, A.J. Kearsley, and J.W. Tolle. A global convergence analysis of an algorithm for large-scale nonlinear optimization problems. *SIAM Journal on Optimization*, 9:833–862, 1999.

32. P.T. Boggs, A.J. Kearsley, and J.W. Tolle. A practical algorithm for general large scale nonlinear optimization problems. *SIAM Journal on Optimization*, 9:755–778, 1999.

33. P.T. Boggs and J.W. Tolle. A family of descent functions for constrained optimization. *SIAM Journal on Numerical Analysis*, 21:1146–1161, 1984.

34. P.T. Boggs and J.W. Tolle. A strategy for global convergence in a sequential quadratic programming algorithm. *SIAM Journal on Numerical Analysis*, 26:600–623, 1989.

35. P.T. Boggs and J.W. Tolle. Sequential quadratic programming. In *Acta Numerica 1995*, pages 1–51. Cambridge University Press, 1995.

36. P.T. Boggs, J.W. Tolle, and P. Wang. On the local convergence of quasi-Newton methods for constrained optimization. *SIAM Journal on Control and Optimization*, 20:161–171, 1982.

37. J.F. Bonnans. Asymptotic admissibility of the unit stepsize in exact penalty methods. *SIAM Journal on Control and Optimization*, 27:631–641, 1989.

38. J.F. Bonnans. Local study of Newton type algorithms for constrained problems. In S. Dolecki, editor, *Optimization*, number 1405 in Lecture Notes in Mathematics, pages 13–24. Springer-Verlag, 1989.

39. J.F. Bonnans. Théorie de la pénalisation exacte. *Modélisation Mathématique et Analyse Numérique*, 24:197–210, 1990.

40. J.F. Bonnans. Local analysis of Newton-type methods for variational inequalities and nonlinear programming. *Applied Mathematics and Optimization*, 29:161–186, 1994.

41. J.F. Bonnans. Exact penalization with a small nonsmooth term. *Revista de Matemáticas Aplicadas*, 17:37–45, 1996.

42. J.F. Bonnans, J.Ch. Gilbert, C. Lemaréchal, and C. Sagastizábal. A family of variable metric proximal methods. *Mathematical Programming*, 68:15–47, 1995.

43. J.F. Bonnans, J.Ch. Gilbert, C. Lemaréchal, and C. Sagastizábal. *Optimisation Numérique – Aspects théoriques et pratiques*. Number 27 in Mathématiques et Applications. Springer Verlag, Berlin, 1997.

44. J.F. Bonnans and C.C. Gonzaga. Convergence of interior point algorithms for the monotone linear complementarity problem. *Mathematics of Operations Research*, 21:1–25, 1996.

45. J.F. Bonnans and G. Launay. Sequential quadratic programming with penalization of the displacement. *SIAM Journal on Optimization*, 5:792–812, 1995.

46. J.F. Bonnans, E.R. Panier, A.L. Tits, and J.L. Zhou. Avoiding the Maratos effect by means of a nonmonotone line search II: Inequality constrained problems – Feasible iterates. *SIAM Journal on Numerical Analysis*, 29:1187–1202, 1992.

47. J.F. Bonnans, C. Pola, and R. Rebaï. Perturbed path following interior point algorithms. *Optimization Methods and Software*, 11-12:183–210, 1999.

48. J.F. Bonnans and F.A. Potra. Infeasible path following algorithms for linear complementarity problems. *Mathematics of Operations Research*, 22:378–407, 1997.

49. J.F. Bonnans and A. Shapiro. Optimization problems with perturbations – A guided tour. *SIAM Review*, 40:202–227, 1998.

50. J.F. Bonnans and A. Shapiro. *Perturbation Analysis of Optimization Problems*. Springer Verlag, New York, 2000.

51. W. Boothby. *An Introduction to Differentiable Manifolds and Differential Geometry*. Academic Press, New York, 1975.

52. J. Borwein and A.S. Lewis. *Convex Analysis and Nonlinear Optimization.* Springer Verlag, New York, 2000.

53. J.H. Bramble and J.E. Pasciak. A preconditioning technique for indefinite systems resulting from mixed approximations of elliptic problems. *Mathematics of Computation*, 50:1–7, 1988.

54. J.H. Bramble, J.E. Pasciak, and A.T. Vassilev. Analysis of the Uzawa algorithm for saddle point problems. *SIAM Journal on Numerical Analysis*, 34:1072–1092, 1997.

55. U. Brännlund. *On relaxation methods for nonsmooth convex optimization.* PhD thesis, Royal Institute of Technology - Stockholm, 1993.

56. J.R. Bunch and L. Kaufman. Some stable methods for calculating inertia and solving symmetric linear systems. *Mathematics of Computation*, 31:163–179, 1977.

57. R.S. Burachik, A.N. Iusem, and B.F. Svaiter. Enlargement of monotone operators with applications to variational inequalities. *Set-Valued Anal.*, 5(2):159–180, 1997.

58. R.S. Burachik, C. Sagastizábal, and S. Scheinberg de Makler. An inexact method of partial inverses and a parallel bundle method. *Optimization Methods and Software*, 21(3):385–400, 2006.

59. R.S. Burachik, C. Sagastizábal, and B. F. Svaiter. Bundle methods for maximal monotone operators. In R. Tichatschke and M. Théra, editors, *Ill-posed Variational Problems and Regularization Techniques*, number 477 in Lecture Notes in Economics and Mathematical Systems, pages 49–64. Springer-Verlag Berlin Heidelberg, 1999.

60. J.V. Burke. An exact penalization viewpoint of constrained optimization. *SIAM Journal on Control and Optimization*, 29:968–998, 1991.

61. J.V. Burke and S.-P. Han. A robust sequential quadratic programming method. *Mathematical Programming*, 43:277–303, 1989.

62. J.D. Buys. *Dual algorithms for constrained optimization.* PhD thesis, Rijksuniversiteit te Leiden, Leiden, The Netherlands, 1972.

63. R.H. Byrd. An example of irregular convergence in some constrained optimization methods that use the projected Hessian. *Mathematical Programming*, 32:232–237, 1985.

64. R.H. Byrd. On the convergence of constrained optimization methods with accurate Hessian information on a subspace. *SIAM Journal on Numerical Analysis*, 27:141–153, 1990.

65. R.H. Byrd, J.Ch. Gilbert, and J. Nocedal. A trust region method based on interior point techniques for nonlinear programming. *Mathematical Programming*, 89:149–185, 2000.

66. R.H. Byrd and J. Nocedal. A tool for the analysis of quasi-Newton methods with application to unconstrained minimization. *SIAM Journal on Numerical Analysis*, 26:727–739, 1989.

67. R.H. Byrd and J. Nocedal. An analysis of reduced Hessian methods for constrained optimization. *Mathematical Programming*, 49:285–323, 1991.

68. R.H. Byrd and R.B. Schnabel. Continuity of the null space basis and constrained optimization. *Mathematical Programming*, 35:32–41, 1986.

69. R.H. Byrd, R.A. Tapia, and Y. Zhang. An SQP augmented Lagrangian BFGS algorithm for constrained optimization. *SIAM Journal on Optimization*, 2:210–241, 1992.

70. A. Cauchy. Méthode générale pour la résolution des systèmes d'équations simultanées. *C. R. Acad. Sci. Paris*, 25:535–538, 1847.

71. J. Céa. *Optimisation: Théorie et Algorithmes*. Dunod, Paris, 1971.

72. Y. Chabrillac and J.-P. Crouzeix. Definiteness and semidefiniteness of quadratic forms revisited. *Linear Algebra and its Applications*, 63:283–292, 1984.

73. R.M. Chamberlain, C. Lemaréchal, H.C. Pedersen, and M.J.D. Powell. The watchdog technique for forcing convergence in algorithms for constrained optimization. *Mathematical Programming Study*, 16:1–17, 1982.

74. C. Charalambous. A lower bound for the controlling parameters of the exact penalty functions. *Mathematical Programming*, 15:278–290, 1978.

75. L. Chauvier, A. Fuduli, and J.Ch. Gilbert. A truncated SQP algorithm for solving nonconvex equality constrained optimization problems. In G. Di Pillo and A. Murli, editors, *High Performance Algorithms and Software for Nonlinear Optimization*, pages 146–173. Kluwer Academic Publishers B.V., 2003.

76. G. Chen and M. Teboulle. A proximal-based decomposition method for convex minimization problems. *Math. Programming*, 64(1, Ser. A):81–101, 1994.

77. X. Chen and M. Fukushima. Proximal quasi-Newton methods for nondifferentiable convex optimization. *Math. Program.*, 85(2, Ser. A):313–334, 1999.

78. E. Cheney and A. Goldstein. Newton's method for convex programming and Tchebycheff approximations. *Numerische Mathematik*, 1:253–268, 1959.

79. P.G. Ciarlet. *Introduction à l'Analyse Numérique Matricielle et à l'Optimisation* (second edition). Masson, Paris, 1988.

80. F.H. Clarke. *Optimization and Nonsmooth Analysis*. John Wiley & Sons, New York; reprinted by SIAM, 1983.

81. T.F. Coleman and A.R. Conn. Nonlinear programming via an exact penalty function: asymptotic analysis. *Mathematical Programming*, 24:123–136, 1982.

82. T.F. Coleman and A.R. Conn. Nonlinear programming via an exact penalty function: global analysis. *Mathematical Programming*, 24:137–161, 1982.

83. T.F. Coleman and D.C. Sorensen. A note on the computation of an orthonormal basis for the null space of a matrix. *Mathematical Programming*, 29:234–242, 1984.

84. L. Conlon. *Differentiable Manifolds – A first Course*. Birkhauser, Boston, 1993.

85. A.R. Conn, N.I.M. Gould, and Ph.L. Toint. *LANCELOT: A Fortran Package for Large-Scale Nonlinear Optimization (Release A)*. Number 17 in Computational Mathematics. Springer Verlag, Berlin, 1992.

86. A.R. Conn, N.I.M. Gould, and Ph.L. Toint. *Trust-Region Methods*. MPS/SIAM Series on Optimization. MPS/SIAM, Philadelphia, 2000.

87. D. Coppersmith and S. Winograd. On the asymptotic complexity of matrix multiplications. *SIAM J. Computation*, 11:472–492, 1982.

88. G. Corliss and A. Griewank, editors. *Automatic Differentiation of Algorithms: Theory, Implementation, and Application*. Proceedings in Applied Mathematics 53. SIAM, Philadelphia, 1991.

89. R. Correa and C. Lemaréchal. Convergence of some algorithms for convex minimization. *Mathematical Programming*, 62:261–275, 1993.

90. R.W. Cottle. Manifestations of the Schur complement. *Linear Algebra and its Applications*, 8:189–211, 1974.

91. R.W. Cottle, J.S. Pang, and R.E. Stone. *The linear complementarity problem*. Academic Press, New York, 1992.

92. M. Cui. A sufficient condition for the convergence of the inexact Uzawa algorithm for saddle point problems. *Journal of Computational and Applied Mathematics*, 139:189–196, 2002.

93. J.-C. Culioli and G. Cohen. Decomposition/coordination algorithms in stochastic optimization. *SIAM J. Control Optim.*, 28(6):1372–1403, 1990.

94. G.B. Dantzig. *Linear programming and extensions*. Princeton University Press, Princeton, N.J., 1963.

95. G.B. Dantzig and P. Wolfe. The decomposition algorithm for linear programming. *Econometrica*, 29(4):767–778, 1961.

96. W.C. Davidon. Variable metric methods for minimization. AEC Research and Development Report ANL-5990, Argonne National Laboratory, Argonne, Illinois, 1959.

97. W.C. Davidon. Variable metric method for optimization. *SIAM Journal on Optimization*, 1:1–17, 1991.

98. F. Delbos and J.Ch. Gilbert. Global linear convergence of an augmented Lagrangian algorithm for solving convex quadratic optimization problems. *Journal of Convex Analysis*, 12:45–69, 2005.

99. F. Delbos, J.Ch. Gilbert, R. Glowinski, and D. Sinoquet. Constrained optimization in seismic reflection tomography: a Gauss-Newton augmented Lagrangian approach. *Geophysical Journal International*, 164:670–684, 2006.

100. F. Delprat-Jannaud and P. Lailly. What information on the Earth model do reflection travel times hold? *Journal of Geophysical Research*, 97:19827–19844, 1992.

101. F. Delprat-Jannaud and P. Lailly. Ill-posed and well-posed formulations of the reflection travel time tomography problem. *Journal of Geophysical Research*, 98:6589–6605, 1993.

102. R.S. Dembo and T. Steihaug. Truncated-Newton algorithms for large-scale unconstrained optimization. *Mathematical Programming*, 26:190–212, 1983.

103. D. den Hertog. *Interior-point approach to linear, quadratic and convex programming*. Kluwer Academic Publishers, Boston, 1994.

104. J.E. Dennis and J.J. Moré. A characterization of superlinear convergence and its application to quasi-Newton methods. *Mathematics of Computation*, 28:549–560, 1974.

105. J.E. Dennis and J.J. Moré. Quasi-Newton methods, motivation and theory. *SIAM Review*, 19:46–89, 1977.

106. J.E. Dennis and R.B. Schnabel. A new derivation of symmetric positive definite secant updates. In *Nonlinear Programming 4*, pages 167–199. Academic Press, 1981.

107. J.E. Dennis and R.B. Schnabel. *Numerical Methods for Unconstrained Optimization and Nonlinear Equations*. Prentice-Hall, Englewood Cliffs, 1983.

108. P. Deuflhard. *Newton Methods for Nonlinear Problems – Affine Invariance and Adaptative Algorithms*. Number 35 in Computational Mathematics. Springer, Berlin, 2004.

109. G. Di Pillo and L. Grippo. A new class of augmented Lagrangians in nonlinear programming. *SIAM Journal on Control and Optimization*, 17:618–628, 1979.

110. G. Di Pillo and S. Lucidi. On exact augmented Lagrangian functions in nonlinear programming. In G. Di Pillo and F. Giannessi, editors, *Nonlinear Optimization and Applications*, pages 85–100. Plenum Press, New York, 1996.

111. G. Di Pillo and S. Lucidi. An augmented Lagrangian function with improved exactness properties. *SIAM Journal on Optimization*, 12:376–406, 2001.

112. M.P. do Carmo. *Riemannian Geometry.* Birkhauser, Boston, 1993.

113. E.D. Dolan and J.J. Moré. Benchmarking optimization software with performance profiles. *Mathematical Programming*, 91:201–213, 2002.

114. I.S. Duff and J.K. Reid. MA27 – A set of Fortran subroutines for solving sparse symmetric sets of linear equations. Technical Report AERE R10533, HMSO, London, 1982.

115. I.S. Duff and J.K. Reid. The multifrontal solution of indefinite sparse symmetric linear systems. *ACM Transactions on Mathematical Software*, 9:301–325, 1983.

116. I.S. Duff and J.K. Reid. Exploiting zeros on the diagonal in the direct solution of indefinite sparse symmetric linear systems. *ACM Transactions on Mathematical Software*, 22:227–257, 1996.

117. I. Ekeland and R. Temam. *Analyse convexe et problèmes variationnels.* Dunod-Gauthier Villars, Paris, 1974.

118. H.C. Elman and G.H. Golub. Inexact and preconditioned Uzawa algorithms for saddle point problems. *SIAM Journal on Numerical Analysis*, 31:1645–1661, 1994.

119. I.I. Eremin. The penalty method in convex programming. *Soviet Mathematics Doklady*, 8:459–462, 1966.

120. F. Facchinei. Exact penalty functions and Lagrange multipliers. *Optimization*, 22:579–606, 1991.

121. F. Facchinei and S. Lucidi. Convergence to second-order stationary points in inequality constrained optimization. *Mathematics of Operations Research*, 23:746–766, 1998.

122. A.V. Fiacco and G.P. McCormick. *Nonlinear Programming: sequential unconstrained minimization technique.* J. Wiley, New York, 1968.

123. P. Finsler. Über das vorkommen definiter und semidefiniter formen und scharen quadratischer formen. *Commentarii Mathematici Helvetica*, 9:188–192, 1937.

124. R. Fletcher. A class of methods for nonlinear programming with termination and convergence properties. In J. Abadie, editor, *Integer and Nonlinear Programming*. North-Holland, Amsterdam, 1970.

125. R. Fletcher. A FORTRAN subroutine for quadratic programming. Report R 6370, Atomic Energy Research Establishment, Harwell, England, 1970.

126. R. Fletcher. A model algorithm for composite nondifferentiable optimization problems. *Mathematical Programming Study*, 17:67–76, 1982.

127. R. Fletcher. Second order corrections for non-differentiable optimization. In D. Griffiths, editor, *Numerical Analysis*, pages 85–114. Springer-Verlag, 1982.

128. R. Fletcher. *Practical Methods of Optimization* (second edition). John Wiley & Sons, Chichester, 1987.

129. R. Fletcher and T. Johnson. On the stability of null-space methods for KKT systems. *SIAM Journal on Matrix Analysis and Applications*, 18:938–958, 1997.

130. R. Fletcher and S. Leyffer. Nonlinear programming without a penalty function. *Mathematical Programming*, 91:239–269, 2002.

131. R. Fletcher and S. Leyffer. A bundle filter method for nonsmooth nonlinear optimization. Technical report, University of Dundee Numerical Analysis Report NA 195, December, 1999. http://www-unix.mcs.anl.gov/~leyffer/papers/nsfilter.pdf.

132. R. Fletcher and M.J.D. Powell. A rapidly convergent descent method for minimization. *The Computer Journal*, 6:163–168, 1963.

133. A. Forsgren, P.E. Gill, and W. Murray. Computing modified Newton directions using a partial Cholesky factorization. *SIAM Journal on Scientific Computing*, 16:139–150, 1995.

134. M. Fortin and R. Glowinski. *Méthodes de Lagrangien Augmenté – Applications à la Résolution Numérique de Problèmes aux Limites*. Number 9 in Méthodes Mathématiques de l'Informatique. Dunod, Paris, 1982.

135. A. Frangioni. Generalized bundle methods. *SIAM Journal on Optimization*, 13(1):117–156, 2003.

136. M. Fukushima. A descent algorithm for nonsmooth convex optimization. *Mathematical Programming*, 30:163–175, 1984.

137. D. Gabay. Minimizing a differentiable function over a differential manifold. *Journal of Optimization Theory and Applications*, 37:177–219, 1982.

138. D. Gabay. Reduced quasi-Newton methods with feasibility improvement for nonlinearly constrained optimization. *Mathematical Programming Study*, 16:18–44, 1982.

139. Gauss. *Theoria motus corporum coelestium*. F. Perthes and I.H. Besser, Hamburg, 1809.

140. J. Gauvin. A necessary and sufficient regularity condition to have bounded multipliers in nonconvex programming. *Mathematical Programming*, 12:136–138, 1977.

141. J. Gauvin. *Théorie de la programmation mathématique non convexe*. Les Publications CRM, Montréal, 1992.

142. J. Gauvin. *Lecons de Programmation Mathématique*. Éditions de l'École Polytechnique de Montréal, Montréal, 1995.

143. D.M. Gay, M.L. Overton, and M.H. Wright. A primal-dual interior method for nonconvex nonlinear programming. In Y.-X. Yuan, editor, *Advances in Nonlinear Programming*. Kluwer Academic Publishers, 1998.

144. J.Ch. Gilbert. Mise à jour de la métrique dans les méthodes de quasi-Newton réduites en optimisation avec contraintes d'égalité. *Modélisation Mathématique et Analyse Numérique*, 22:251–288, 1988.

145. J.Ch. Gilbert. On the local and global convergence of a reduced quasi-Newton method. *Optimization*, 20:421–450, 1989.

146. J.Ch. Gilbert. Maintaining the positive definiteness of the matrices in reduced secant methods for equality constrained optimization. *Mathematical Programming*, 50:1–28, 1991.

147. J.Ch. Gilbert. Superlinear convergence of a reduced BFGS method with piecewise line-search and update criterion. Rapport de Recherche 2140, INRIA, BP 105, 78153 Le Chesnay, France, 1993.

148. J.Ch. Gilbert. On the realization of the Wolfe conditions in reduced quasi-Newton methods for equality constrained optimization. *SIAM Journal on Optimization*, 7:780–813, 1997.

149. J.Ch. Gilbert. Piecewise line-search techniques for constrained minimization by quasi-Newton algorithms. In Y.-X. Yuan, editor, *Advances in Nonlinear Programming*, chapter 4, pages 73–103. Kluwer Academic Publishers, 1998.

150. J.Ch. Gilbert. *Éléments d'Optimisation Différentiable – Théorie et Algorithmes*. 2006. http://www-rocq.inria.fr/~gilbert/ensta/optim.html.

151. J.Ch. Gilbert, G. Le Vey, and J. Masse. La différentiation automatique de fonctions représentées par des programmes. Rapport de Recherche n° 1557, Inria, BP 105, F-78153 Le Chesnay, France, 1991.

152. J.Ch. Gilbert and C. Lemaréchal. Some numerical experiments with variable-storage quasi-Newton algorithms. *Mathematical Programming*, 45:407–435, 1989.

153. J.Ch. Gilbert and J. Nocedal. Global convergence properties of conjugate gradient methods for optimization. *SIAM Journal on Optimization*, 2:21–42, 1992.

154. P.E. Gill and W. Murray. Newton-type methods for unconstrained and linearly constrained optimization. *Mathematical Programming*, 7:311–350, 1974.

155. P.E. Gill, W. Murray, and M.A. Saunders. SNOPT: an SQP algorithm for large-scale constrained optimization. Numerical Analysis Report 96-2, Department of Mathematics, University of California, San Diego, La Jolla, CA, 1996.

156. P.E. Gill, W. Murray, and M.A. Saunders. SNOPT: an SQP algorithm for large-scale constrained optimization. *SIAM Journal on Optimization*, 12:979–1006, 2002.

157. P.E. Gill, W. Murray, M.A. Saunders, G.W. Stewart, and M.H. Wright. Properties of a representation of a basis for the null space. *Mathematical Programming*, 33:172–186, 1985.

158. P.E. Gill, W. Murray, M.A. Saunders, and M.H. Wright. User's guide for NPSOL (version 4.0): a Fortran package for nonlinear programming. Technical Report SOL-86-2, Department of Operations Research, Stanford University, Stanford, CA 94305, 1986.

159. P.E. Gill, W. Murray, M.A. Saunders, and M.H. Wright. Constrained nonlinear programming. In G.L. Nemhauser, A.H.G. Rinnooy Kan, and M.J. Todd, editors, *Handbooks in Operations Research and Management Science*, volume 1: Optimization, chapter 3, pages 171–210. Elsevier Science Publishers B.V., North-Holland, 1989.

160. P.E. Gill, W. Murray, and M.H. Wright. *Practical Optimization*. Academic Press, New York, 1981.

161. S.T. Glad. Properties of updating methods for the multipliers in augmented Lagrangians. *Journal of Optimization Theory and Applications*, 28:135–156, 1979.

162. R. Glowinski and Q.-H. Tran. Constrained optimization in reflexion tomography: the augmented Lagrangian method. *East-West J. Numer. Math.*, 1(3):213–234, 1993.

163. J.-L. Goffin. On convergence rates of subgradient optimization methods. *Mathematical Programming*, 13:329–347, 1977.

164. J.L. Goffin and K.C. Kiwiel. Convergence of a simple sugradient level method. *Math. Program.*, 85:207–211, 1999.

165. D. Goldfarb and A. Idnani. A numerically stable dual method for solving strictly convex quadratic programs. *Mathematical Programming*, 27:1–33, 1983.

166. D. Goldfarb and M.J. Todd. Linear programming. In G.L. Nemhauser et al., editor, *Handbook on Operations Research and Management Science*, volume 1, Optimization, pages 73–170. North-Holland, 1989.

167. A.J. Goldman and A.W. Tucker. Polyhedral convex cones. In H.W. Kuhn and A.W. Tucker, editors, *Linear inequalities and related systems*, pages 19–40, Princeton, 1956. Princeton University Press.

168. A.A. Goldstein and J.F. Price. An effective algorithm for minimization. *Numerische Mathematik*, 10:184–189, 1967.

169. E.G. Gol'shteĭn and N.V. Tretyakov. *Modified Lagrangians and Monotone Maps in Optimization*. Discrete Mathematics and Optimization. John Wiley & Sons, New York, 1996.

170. G.H. Golub and C.F. Van Loan. *Matrix Computations* (second edition). The Johns Hopkins University Press, Baltimore, Maryland, 1989.

171. C.C. Gonzaga. Polynomial affine algorithms for linear programming. *Math. Programming*, 49:7–21, 1990.

172. C.C. Gonzaga. Path following methods for linear programming. *SIAM Review*, 34:167–227, 1992.

173. C.C. Gonzaga. A simple presentation of Karmarkar's algorithm. In *Workshop on interior point methods*, Budapest, 1993.

174. C.C. Gonzaga. The largest step path following algorithm for monotone linear complementarity problems. *Mathematical Programming*, 76:309–332, 1997.

175. C.C. Gonzaga and J.F. Bonnans. Fast convergence of the simplified largest step path following algorithm. *Mathematical Programming series B*, 76:95–115, 1997.

176. C.C. Gonzaga and R.A. Tapia. On the convergence of the Mizuno–Todd–Ye algorithm to the analytic center of the solution set. *SIAM J. Optimization*, 7:47–65, 1997.

177. J. Goodman. Newton's method for constrained optimization. *Mathematical Programming*, 33:162–171, 1985.

178. N. Gould, D. Orban, and Ph.L. Toint. Numerical methods for large-scale nonlinear optimization. In *Acta Numerica 2005*, pages 299–361. Cambridge University Press, 2005.

179. N.I.M. Gould. On practical conditions for the existence and uniqueness of solutions to the general equality quadratic programming problem. *Mathematical Programming*, 32:90–99, 1985.

180. G. Gramlich, R. Hettich, and E.W. Sachs. Local convergence of SQP methods in semi-infinite programming. *SIAM Journal on Optimization*, 5:641–658, 1995.

181. A. Griewank. *Evaluating Derivatives – Principles and Techniques of Algorithmic Differentiation*. SIAM Publication, 2000.

182. C.B. Gurwitz. Local convergence of a two-piece update of a projected Hessian matrix. *SIAM Journal on Optimization*, 4:461–485, 1994.

183. W.W. Hager. Stabilized sequential quadratic programming. *Computational Optimization and Applications*, 12:253–273, 1999.

184. S.-P. Han. Superlinearly convergent variable metric algorithms for general nonlinear programming problems. *Mathematical Programming*, 11:263–282, 1976.

185. S.-P. Han. A globally convergent method for nonlinear programming. *Journal of Optimization Theory and Applications*, 22:297–309, 1977.

186. S.-P. Han and O.L. Mangasarian. Exact penalty functions in nonlinear programming. *Mathematical Programming*, 17:251–269, 1979.

187. P.C. Hansen. *Rank-Deficient and Discrete Ill-Posed Problems: Numerical Aspects of Linear Inversion*. SIAM, Philadelphia, 1998.

188. M. Held and R. Karp. The traveling salesman problem and minimum spanning trees: Part II. *Mathematical Programming*, 1(1):6–25, 1971.

189. J. Herskovits. A view on nonlinear optimization. In J. Herskovits, editor, *Advances in Structural Optimization*, pages 71–116. Kluwer Academic Publishers, 1995.

190. J. Herskovits. Feasible direction interior-point technique for nonlinear optimization. *Journal of Optimization Theory and Applications*, 99:121–146, 1998.

191. M.R. Hestenes. Multiplier and gradient methods. *Journal of Optimization Theory and Applications*, 4:303–320, 1969.

192. M.R. Hestenes. *Conjugate Direction Methods in Optimization*. Number 12 in Applications of Mathematics. Springer-Verlag, 1980.

193. M.R. Hestenes and E. Stiefel. Methods of conjugate gradients for solving linear systems. *Journal of Research of the National Bureau of Standards*, 49:409–436, 1952.

194. N.J. Higham. *Accuracy and Stability of Numerical Algorithms* (second edition). SIAM Publication, Philadelphia, 2002.

195. J.-B. Hiriart-Urruty and C. Lemaréchal. *Convex Analysis and Minimization Algorithms*. Number 305-306 in Grundlehren der mathematischen Wissenschaften. Springer-Verlag, Berlin, 1993.

196. J.-B. Hiriart-Urruty and C. Lemaréchal. *Fundamentals of Convex Analysis*. Springer-Verlag, Berlin, 2001. Abridged version of *Convex analysis and minimization algorithms. I* and *II* [Springer, Berlin, 1993].

197. W. Hoyer. Variants of the reduced Newton method for nonlinear equality constrained optimization problems. *Optimization*, 17:757–774, 1986.

198. A.D. Ioffe. Necessary and sufficient conditions for a local minimum. 1: a reduction theorem and first order conditions. *SIAM Journal on Control and Optimization*, 17:245–250, 1979.

199. B. Jansen, C. Roos, and T. Terlaky. The theory of linear programming: skew symmetric self-dual problems and the central path. *Optimization*, 29:225–233, 1994.

200. A.H.G. Rinnooy Kan and G.T. Timmer. Global optimization. In G.L. Nemhauser, A.H.G. Rinnooy Kan, and M.J. Todd, editors, *Handbooks in Operations Research and Management Science*, volume 1: Optimization, chapter 9, pages 631–662. Elsevier Science Publishers B.V., North-Holland, 1989.

201. S. Kaniel and A. Dax. A modified Newton's method for unconstrained minimization. *SIAM Journal on Numerical Analysis*, 16:324–331, 1979.

202. L.V. Kantorovich and G.P. Akilov. *Functional Analysis* (second edition). Pergamon Press, London, 1982.

203. E. Karas, A. Ribeiro, C. Sagastizábal, and M.V. Solodov. A bundle-filter method for nonsmooth convex constrained optimization. *Math. Program. Ser. B*, 2006. Accepted for publication.

204. N. Karmarkar. A new polynomial time algorithm for linear programming. *Combinatorica*, 4:373–395, 1984.

205. J. E. Kelley. The cutting plane method for solving convex programs. *J. Soc. Indust. Appl. Math.*, 8:703–712, 1960.

206. L. Khachian. A polynomial algorithm in linear programming. *Soviet Mathematics Doklady*, 20:191–194, 1979.

207. L.G. Khachiyan. A polynomial algorithm in linear programming. *Doklady Adad. Nauk SSSR*, 244:1093–1096, 1979. Trad. anglaise : Soviet Math. Doklady 20(1979), 191-194.

476   References

208. K.V. Kim, Yu.E. Nesterov, and B.V. Cherkasskiĭ. An estimate of the effort in computing the gradient. *Soviet Math. Dokl.*, 29:384–387, 1984.
209. K.C. Kiwiel. An exact penalty function algorithm for nonsmooth convex constrained minimization problems. *IMA J. Numer. Anal.*, 5(1):111–119, 1985.
210. K.C. Kiwiel. *Methods of Descent for Nondifferentiable Optimization*. Lecture Notes in Mathematics 1133. Springer Verlag, Berlin, 1985.
211. K.C. Kiwiel. A constraint linearization method for nondifferentiable convex minimization. *Numerische Mathematik*, 51:395–414, 1987.
212. K.C. Kiwiel. A subgradient selection method for minimizing convex functions subject to linear constraints. *Computing*, 39(4):293–305, 1987.
213. K.C. Kiwiel. Proximity control in bundle methods for convex nondifferentiable minimization. *Mathematical Programming*, 46(1):105–122, 1990.
214. K.C. Kiwiel. Exact penalty functions in proximal bundle methods for constrained convex nondifferentiable minimization. *Math. Programming*, 52(2, Ser. B):285–302, 1991.
215. K.C. Kiwiel, T. Larsson, and P.O. Lindberg. The efficiency of ballstep subgradient level methods for convex optimization. *Mathematics of Operations Research*, 24(1):237–254, 1999.
216. V. Klee and G.L. Minty. How good is the simplex algorithm ? In O. Shisha, editor, *Inequalities III*, pages 159–175. Academic Press, New York, 1972.
217. M. Kojima, N. Megiddo, T. Noma, and A. Yoshise. *A unified approach to interior point algorithms for linear complementarity problems*. Number 538 in Lecture Notes in Computer Science. Springer Verlag, Berlin, 1991.
218. K. Kortanek and Z. Jishan. New purification algorithms for linear programming. *Naval Research Logistics*, 35:571–583, 1988.
219. F.-S. Kupfer. An infinite-dimensional convergence theory for reduced SQP methods in Hilbert space. *SIAM Journal on Optimization*, 6:126–163, 1996.
220. J. Kyparisis. On uniqueness of Kuhn-Tucker multipliers in nonlinear programming. *Mathematical Programming*, 32:242–246, 1985.
221. L. Lasdon. *Optimization Theory for Large Systems*. Macmillan Series in Operations Research, 1970.
222. C.T. Lawrence and A.L. Tits. Nonlinear equality constraints in feasible sequential quadratic programming. *Optimization Methods and Software*, 6:265–282, 1996.
223. C.T. Lawrence and A.L. Tits. Feasible sequential quadratic programming for finely discretized problems from SIP. In R. Reemtsen and J.-J. Rückmann, editors, *Semi-infinite Programming*, pages 159–193. Kluwer Academic Publishers B.V., 1998.
224. C.T. Lawrence and A.L. Tits. A computationally efficient feasible sequential quadratic programming algorithm. *SIAM Journal on Optimization*, 11:1092–1118, 2001.
225. F. Leibfritz and E.W. Sachs. Inexact SQP interior point methods and large scale optimal control problems. *SIAM Journal on Optimization*, 38:272–293, 1999.
226. C. Lemaréchal. An algorithm for minimizing convex functions. In J.L. Rosenfeld, editor, *Information Processing '74*, pages 552–556. North Holland, 1974.
227. C. Lemaréchal. An extension of Davidon methods to nondifferentiable problems. *Mathematical Programming Study*, 3:95–109, 1975.

228. C. Lemaréchal. A view of line-searches. In A. Auslender, W. Oettli, and J. Stoer, editors, *Optimization and Optimal Control*, number 30 in Lecture Notes in Control and Information Science, pages 59–78. Springer-Verlag, Heidelberg, 1981.

229. C. Lemaréchal and R. Mifflin. Global and superlinear convergence of an algorithm for one-dimensional minimzation of convex functions. *Mathematical Programming*, 24:241–256, 1982.

230. C. Lemaréchal, A.S. Nemirovskii, and Yu.E. Nesterov. New variants of bundle methods. *Mathematical Programming*, 69:111–148, 1995.

231. C. Lemaréchal, F. Oustry, and C. Sagastizábal. The $\mathcal{U}$-Lagrangian of a convex function. *Transactions of the AMS*, 352(2):711–729, 2000.

232. C. Lemaréchal, F. Pellegrino, A. Renaud, and C. Sagastizábal. Bundle methods applied to the unit-commitment problem. In J. Doležal and J. Fidler, editors, *System Modelling and Optimization*, pages 395–402. Chapman and Hall, 1996.

233. C. Lemaréchal and C. Sagastizábal. An approach to variable metric bundle methods. In J. Henry and J-P. Yvon, editors, *Systems Modelling and Optimization*, number 197 in Lecture Notes in Control and Information Sciences, pages 144–162. Springer Verlag, 1994.

234. C. Lemaréchal and C. Sagastizábal. More than first-order developments of convex functions: primal-dual relations. *Journal of Convex Analysis*, 3(2):1–14, 1996.

235. C. Lemaréchal and C. Sagastizábal. Practical aspects of the Moreau-Yosida regularization: theoretical preliminaries. *SIAM Journal on Optimization*, 7(2):367–385, 1997.

236. C. Lemaréchal and C. Sagastizábal. Variable metric bundle methods: from conceptual to implementable forms. *Mathematical Programming*, 76(3):393–410, 1997.

237. K. Levenberg. A method for the solution of certain nonlinear problems in least squares. *Quart. Appl. Math.*, 2:164–168, 1944.

238. D.C. Liu and J. Nocedal. On the limited memory BFGS method for large scale optimization. *Mathematical Programming*, 45:503–520, 1989.

239. D.G. Luenberger. *Introduction to Linear and Nonlinear Programming* (second edition). Addison-Wesley, Reading, USA, 1984.

240. L. Lukšan and J. Vlček. A bundle-Newton method for nonsmooth unconstrained minimization. *Math. Programming*, 83(3, Ser. A):373–391, 1998.

241. L. Lukšan and J. Vlček. Globally convergent variable metric method for convex nonsmooth unconstrained minimization. *J. Optim. Theory Appl.*, 102(3):593–613, 1999.

242. Z.-Q. Luo, J.-S. Pang, and D. Ralph. *Mathematical Programs with Equilibrium Constraints*. Cambridge University Press, 1996.

243. Y. Maday, D. Meiron, A.T. Patera, and E.M. Ronquist. Analysis of iterative methods for the steady and unsteady Stokes problem: application to spectral element discretizations. *SIAM Journal on Scientific Computing*, 14:310–337, 1993.

244. J.H. Maddocks. Restricted quadratic forms, inertia theorems, and the Schur complement. *Linear Algebra and its Applications*, 108:1–36, 1988.

245. Ph. Mahey, S. Oualibouch, and Pham Dinh Tao. Proximal decomposition on the graph of a maximal monotone operator. *SIAM J. Optim.*, 5(2):454–466, 1995.

246. O.L. Mangasarian and S. Fromovitz. The Fritz John necessary optimality conditions in the presence of equality and inequality constraints. *Journal of Mathematical Analysis and Applications*, 17:37–47, 1967.

247. N. Maratos. *Exact penalty function algorithms for finite dimensional and control optimization problems*. PhD thesis, Imperial College, London, 1978.

248. D.W. Marquardt. An algorithm for least-squares estimation of nonlinear parameters. *J. Soc. Indust. Appl. Math.*, 11:431–441, 1963.

249. MATHWORKS. The Matlab distributed computing engine. http://www.mathworks.com/.

250. D.Q. Mayne and E. Polak. A superlinearly convergent algorithm for constrained optimization problems. *Mathematical Programming Study*, 16:45–61, 1982.

251. K.A. McShane. A superlinearly convergent $O(\sqrt{n}L)$ iteration interior-point algorithms for linear programming and the monotone linear complementarity problem. *SIAM J. Optimization*, 4:247–261, 1994.

252. R. Mifflin. An algorithm for constrained optimization with semi-smooth functions. *Mathematics of Operations Research*, 2(2):191–207, 1977.

253. R. Mifflin. A modification and an extension of Lemaréchal's algorithm for nonsmooth minimization. *Mathematical Programming Study*, 17:77–90, 1982.

254. R. Mifflin. A quasi-second-order proximal bundle algorithm. *Mathematical Programming*, 73(1):51–72, 1996.

255. R. Mifflin and C. Sagastizábal. Proximal points are on the fast track. *Journal of Convex Analysis*, 9(2):563–579, 2002.

256. R. Mifflin and C. Sagastizábal. Primal-Dual Gradient Structured Functions: second-order results; links to epi-derivatives and partly smooth functions. *SIAM Journal on Optimization*, 13(4):1174–1194, 2003.

257. R. Mifflin and C. Sagastizábal. A $\mathcal{V}U$-proximal point algorithm for minimization. *Math. Program.*, 104(2–3):583–608, 2005.

258. R. Mifflin, D.F. Sun, and L.Q. Qi. Quasi-Newton bundle-type methods for nondifferentiable convex optimization. *SIAM Journal on Optimization*, 8(2):583–603, 1998.

259. S. Mizuno. A new polynomial time method for a linear complementarity problem. *Mathematical Programming*, 56:31–43, 1992.

260. S. Mizuno. A superlinearly convergent infeasible-interior-point algorithm for geometrical LCPs without a strictly complementary condition. *Mathematics of Operations Research*, 21:382–400, 1996.

261. S. Mizuno, F. Jarre, and J. Stoer. A unified approach to infeasible-interior-point algorithms via geometrical linear complementarity problems. *J. Applied Mathematics and Optimization*, 33:315–341, 1994.

262. S. Mizuno, M.J. Todd, and Y. Ye. On adaptative step primal-dual interior-point algorithms for linear programming. *Mathematics of Operations Research*, 18:964–981, 1993.

263. R.D.C. Monteiro and T. Tsuchiya. Limiting behavior of the derivatives of certain trajectories associated with a monotone horizontal linear complementarity problem. *Mathematics of Operations Research*, 21:793–814, 1996.

264. J.J. Moré. Recent developments in algorithms and software for trust region methods. In A. Bachem, M. Grötschel, and B. Korte, editors, *Mathematical Programming, the State of the Art*, pages 258–287. Springer-Verlag, Berlin, 1983.

265. J.J. Moré and D.J. Thuente. Line search algorithms with guaranteed sufficient decrease. *ACM Transactions on Mathematical Software*, 20:286–307, 1994.

266. J.J. Moré and G. Toraldo. Algorithms for bound constrained quadratic programming problems. *Numerische Mathematik*, 55:377–400, 1989.

267. J.J. Moré and S.J. Wright, editors. *Optimization Software Guide*, volume 14 of *Frontiers in Applied Mathematics*. SIAM Publications, 1993.

268. J.J. Moreau. Proximité et dualité dans un espace hilbertien. *Bulletin de la Société Mathématique de France*, 93:273–299, 1965.

269. W. Murray and M. L. Overton. Steplength algorithms for minimizing a class of nondifferentiable functions. *Computing*, 23(4):309–331, 1979.

270. W. Murray and F.J. Prieto. A sequential quadratic programming algorithm using an incomplete solution of the subproblem. *SIAM Journal on Optimization*, 5:590–640, 1995.

271. W. Murray and M.H. Wright. Projected Lagrangian methods based on the trajectories of penalty and barrier functions. Technical Report SOL-78-23, Department of Operations Research, Stanford University, Stanford, CA 94305, 1978.

272. A.S. Nemirovskii and D. Yudin. *Problem Complexity and Method Efficiency in Optimization*. Wiley-Interscience Series in Discrete Mathematics, 1983. (Original Russian: Nauka, 1979).

273. Y.E. Nesterov and A.S. Nemirovskii. *Interior-Point Polynomial Algorithms in Convex Programming*. Number 13 in SIAM Studies in Applied Mathematics. SIAM, Philadelphia, 1994.

274. J. Von Neumann and O. Morgenstein. *Theory of games and economic behavior*. Princeton University Press, Princeton, 1944.

275. J. Nocedal. Updating quasi-Newton matrices with limited storage. *Mathematics of Computation*, 35:773–782, 1980.

276. J. Nocedal and M.L. Overton. Projected Hessian updating algorithms for nonlinearly constrained optimization. *SIAM Journal on Numerical Analysis*, 22:821–850, 1985.

277. J. Nocedal and S.J. Wright. *Numerical Optimization*. Springer Series in Operations Research. Springer, New York, 1999.

278. J.M. Ortega and W.C. Rheinboldt. *Iterative Solution of Nonlinear Equations in Several Variables*. Academic Press, New York, 1970.

279. A. Ouorou. Epsilon-proximal decomposition method. *Math. Program.*, 99(1, Ser. A):89–108, 2004.

280. U.M. Garcia Palomares and O.L. Mangasarian. Superlinear convergent quasi-Newton algorithms for nonlinearly constrained optimization problems. *Mathematical Programming*, 11:1–13, 1976.

281. E.R. Panier and A.L. Tits. Avoiding the Maratos effect by means of a non-monotone line search I: General constrained problems. *SIAM Journal on Numerical Analysis*, 28:1183–1195, 1991.

282. E.R. Panier and A.L. Tits. On combining feasibility, descent and superlinear convergence in inequality constrained optimization. *Mathematical Programming*, 59(2):261–276, 1993.

283. E.R. Panier, A.L. Tits, and J. Herskovits. A QP-free, globally convergent, locally superlinearly convergent algorithm for inequality constrained optimization. *SIAM Journal on Control and Optimization*, 26:788–811, 1988.

284. T. Pietrzykowski. An exact potential method for constrained maxima. *SIAM Journal on Numerical Analysis*, 6:299–304, 1969.

285. E. Polak. *Optimization – Algorithms and Consistent Approximations.* Number 124 in Applied Mathematical Sciences. Springer, 1997.

286. B.T. Polyak. *Introduction to Optimization.* Optimization Software, New York, 1987.

287. F.A. Potra. An $O(nL)$ infeasible-interior-point algorithm for LCP with quadratic convergence. *Annals of Operations Research,* 62:81–102, 1996.

288. M.J.D. Powell. A method for nonlinear constraints in minimization problems. In R. Fletcher, editor, *Optimization,* pages 283–298. Academic Press, London, New York, 1969.

289. M.J.D. Powell. On the convergence of the variable metric algorithm. *Journal of the Institute of Mathematics and its Applications,* 7:21–36, 1971.

290. M.J.D. Powell. Some global convergence properties of a variable metric algorithm for minimization without exact line searches. In R.W. Cottle and C.E. Lemke, editors, *Nonlinear Programming,* number 9 in SIAM-AMS Proceedings. American Mathematical Society, Providence, RI, 1976.

291. M.J.D. Powell. Algorithms for nonlinear constraints that use Lagrangian functions. *Mathematical Programming,* 14:224–248, 1978.

292. M.J.D. Powell. The convergence of variable metric methods for nonlinearly constrained optimization calculations. In O.L. Mangasarian, R.R. Meyer, and S.M. Robinson, editors, *Nonlinear Programming 3,* pages 27–63, 1978.

293. M.J.D. Powell. A fast algorithm for nonlinearly constrained optimization calculations. In G.A. Watson, editor, *Numerical Analysis Dundee 1977,* number 630 in Lecture Notes in Mathematics, pages 144–157. Springer-Verlag, Berlin, 1978.

294. M.J.D. Powell. Nonconvex minimization calculations and the conjugate gradient method. In *Lecture Notes in Mathematics 1066,* pages 122–141. Springer-Verlag, Berlin, 1984.

295. M.J.D. Powell. Convergence properties of algorithms for nonlinear optimization. *SIAM Review,* 28:487–500, 1985.

296. M.J.D. Powell. On the quadratic programming algorithm of Goldfarb and Idnani. *Mathematical Programming Study,* 25:46–61, 1985.

297. M.J.D. Powell. The performance of two subroutines for constrained optimization on some difficult test problems. In P.T. Boggs, R.H. Byrd, and R.B. Schnabel, editors, *Numerical Optimization 1984,* pages 160–177. SIAM Publication, Philadelphia, 1985.

298. M.J.D. Powell. A view of nonlinear optimization. In J.K. Lenstra, A.H.G. Rinnooy Kan, and A. Schrijver, editors, *History of Mathematical Programming, A Collection of Personal Reminiscences,* pages 119–125. CWI North-Holland, Amsterdam, 1991.

299. M.J.D. Powell and Y. Yuan. A recursive quadratic programming algorithm that uses differentiable exact penalty functions. *Mathematical Programming,* 35:265–278, 1986.

300. B.N. Pshenichnyj. Algorithm for a general mathematical programming problem. *Kibernetika,* 5:120–125, 1970.

301. B.N. Pshenichnyj. *The Linearization Method for Constrained Optimization.* Number 22 in Computational Mathematics. Springer-Verlag, 1994.

302. B.N. Pshenichnyj and Yu.M. Danilin. *Numerical Methods for Extremal Problems.* MIR, Moscow, 1978.

303. W. Queck. The convergence factor of preconditioned algorithms of the Arrow-Hurwicz type. *SIAM Journal on Numerical Analysis,* 26:1016–1030, 1989.

304. J.K. Reid. A sparsity-exploiting variant of the Bartels-Golub decomposition for linear programming bases. *Mathematical Programming*, 24:55–69, 1982.
305. P.A. Rey and C.A. Sagastizábal. Dynamical adjustment of the prox-parameter in variable metric bundle methods. *Optimization*, 51(2):423–447, 2002.
306. S.M. Robinson. A quadratically convergent algorithm for general nonlinear programming problems. *Mathematical Programming*, 3:145–156, 1972.
307. S.M. Robinson. Perturbed Kuhn-Tucker points and rates of convergence for a class of nonlinear-programming algorithms. *Mathematical Programming*, 7:1–16, 1974.
308. S.M. Robinson. Generalized equations and their solutions, part II: applications to nonlinear programming. *Mathematical Programming Study*, 19:200–221, 1982.
309. R.T. Rockafellar. *Convex Analysis*. Number 28 in Princeton Mathematics Ser. Princeton University Press, Princeton, New Jersey, 1970.
310. R.T. Rockafellar. New applications of duality in convex programming. In *Proceedings of the 4th Conference of Probability, Brasov, Romania*, pages 73–81, 1971.
311. R.T. Rockafellar. Augmented Lagrange multiplier functions and duality in nonconvex programming. *SIAM Journal on Control*, 12:268–285, 1974.
312. R.T. Rockafellar. Augmented Lagrangians and applications of the proximal point algorithm in convex programming. *Mathematics of Operations Research*, 1:97–116, 1976.
313. R.T. Rockafellar. Monotone operators and the proximal point algorithm. *SIAM Journal on Control and Optimization*, 14:877–898, 1976.
314. R.T. Rockafellar and R.J.-B. Wets. *Variational Analysis*. Springer Verlag, Heidelberg, 1998.
315. T. Rusten and R. Winthier. A preconditioned iterative method for saddle point problems. *SIAM Journal on Matrix Analysis and Applications*, 13:887–904, 1992.
316. A. Ruszczyński. Decomposition methods in stochastic programming. In *Mathematical Programming*, volume 79, 1997.
317. C. Sagastizábal and M.V. Solodov. On the relation between bundle methods for maximal monotone inclusions and hybrid proximal point algorithms. In *Inherently parallel algorithms in feasibility and optimization and their applications (Haifa, 2000)*, volume 8 of *Stud. Comput. Math.*, pages 441–455. North-Holland, Amsterdam, 2001.
318. C. Sagastizábal and M.V. Solodov. An infeasible bundle method for nonsmooth convex constrained optimization without a penalty function or a filter. *SIAM Journal on Optimization*, 16(1):146–169, 2005.
319. R. Saigal. *Linear Programming: A Modern Integrated Analysis*. Kluwer Academic Publishers, Boston, 1995.
320. R.W.H. Sargent. The development of SQP algorithm for nonlinear programming. In L.T. Biegler, T.F. Coleman, A.R. Conn, and F.N. Santosa, editors, *Large-Scale Optimization with Applications, part II: Optimal design and Control*, pages 1–19. IMA Vol. Math. Appl. 93, 1997.
321. R.W.H. Sargent and M. Ding. A new SQP algorithm for large-scale nonlinear programming. *SIAM Journal on Optimization*, 11:716–747, 2000.
322. K. Schittkowski. The nonlinear programming method of Wilson, Han and Powell with an augmented Lagrangian type line search function, Part 1: convergence analysis. *Numerische Mathematik*, 38:83–114, 1981.

323. K. Schittkowski. NLPQL: a FORTRAN subroutine solving constrained nonlinear programming problems. *Annals of Operations Research*, 5:485–500, 1985.
324. K. Schittkowski. Solving nonlinear programming problems with very many constraints. *Optimization*, 25:179–196, 1992.
325. K. Schittkowski. *Numerical Data Fitting in Dynamical Systems*. Kluwer Academic Press, Dordrecht, 2002.
326. H. Schramm and J. Zowe. A version of the bundle idea for minimizing a nonsmooth function: conceptual idea, convergence analysis, numerical results. *SIAM Journal on Optimization*, 2(1):121–152, 1992.
327. SCILAB. A free scientific software package. http://www.scilab.org/.
328. D.F. Shanno. Conjugate gradient methods with inexact searches. *Mathematics of Operations Research*, 3:244–256, 1978.
329. D.F. Shanno and K.H. Phua. Algorithm 500, minimization of unconstrained multivariate functions. *ACM Transactions on Mathematical Software*, 2:87–94, 1976.
330. A. Shapiro and J. Sun. Some properties of the augmented Lagrangian in cone constrained optimization. *Mathematics of Operations Research*, 29:479–491, 2004.
331. N. Shor. Utilization of the operation of space dilatation in the minimization of convex functions. *Kibernetica*, 1:6–12, 1970. (English translation: *Cybernetics*, 6, 7-15).
332. N.Z. Shor. *Minimization methods for non-differentiable functions*. Springer Verlag, Berlin, 1985.
333. D. Silvester and A. Wathen. Fast iterative solution of stabilized Stokes systems part II: using general block preconditioners. *SIAM Journal on Numerical Analysis*, 31:1352–1367, 1994.
334. M. Slater. Lagrange multipliers revisited: a contribution to non-linear programming. Cowles Commission Discussion Paper, Math. 403, 1950.
335. M. V. Solodov. A class of decomposition methods for convex optimization and monotone variational inclusions via the hybrid inexact proximal point framework. *Optim. Methods Softw.*, 19(5):557–575, 2004.
336. M. V. Solodov and B. F. Svaiter. A hybrid approximate extragradient-proximal point algorithm using the enlargement of a maximal monotone operator. *Set-Valued Anal.*, 7(4):323–345, 1999.
337. M. V. Solodov and B. F. Svaiter. A hybrid projection-proximal point algorithm. *J. Convex Anal.*, 6(1):59–70, 1999.
338. M. V. Solodov and B. F. Svaiter. A unified framework for some inexact proximal point algorithms. *Numerical Functional Analysis and Optimization*, 22:1013–1035, 2001.
339. B. Speelpenning. *Compiling fast partial derivatives of functions given by algorithms*. PhD thesis, Department of Computer Science, University of Illinois at Urbana-Champaign, Urbana, IL 61801, 1980.
340. P. Spellucci. *Numerische Verfahren der nichtlinearen Optimierung*. Birkhäuser, 1993.
341. P. Spellucci. A new technique for inconsistent problems in the SQP method. *Mathematical Methods of Operations Research*, 47:355–400, 1998.
342. P. Spellucci. An SQP method for general nonlinear programs using only equality constrained subproblems. *Mathematical Programming*, 82:413–448, 1998.
343. J. E. Spingarn. Partial inverse of a monotone operator. *Appl. Math. Optim.*, 10(3):247–265, 1983.

344. M. Spivak. *A Comprehensive Introduction to Differential Geometry.* Publish or Perish, 1979.

345. T. Steihaug. The conjugate gradient method and trust regions in large scale optimization. *SIAM Journal on Numerical Analysis*, 20:626–637, 1983.

346. R.A. Tapia. Diagonalized multiplier methods and quasi-Newton methods for constrained optimization. *Journal of Optimization Theory and Applications*, 22:135–194, 1977.

347. R.A. Tapia. On secant updates for use in general constrained optimization. *Mathematics of Computation*, 51:181–202, 1988.

348. T. Terlaky, editor. *Interior Point Methods of Mathematical Programming.* Kluwer Academic Publishers, Boston, 1996.

349. T. Terlaky, J. P. Vial, and K. Roos. *Theory and algorithms for linear optimization: an interior point approach.* Wiley intersciences, New York, 1997.

350. M.J. Todd. On convergence properties of algorithms for unconstrained minimization. *IMA Journal of Numerical Analysis*, 9(3):435–441, 1989.

351. K. Tone. Revisions of constraint approximations in the successive QP method for nonlinear programming problems. *Mathematical Programming*, 26:144–152, 1983.

352. P. Tseng. Alternating projection-proximal methods for convex programming and variational inequalities. *SIAM J. Optim.*, 7(4):951–965, 1997.

353. R.J. Vanderbei. *Linear Programming: Foundations and extensions.* Kluwer Academic Publishers, Boston, 1997.

354. S.A. Vavasis. *Nonlinear Optimization – Complexity Issues.* Oxford University Press, New York, 1991.

355. R. Verfürth. A combined conjugate gradient-multigrid algorithm for the numerical solution of the Stokes problem. *IMA Journal of Numerical Analysis*, 4:441–455, 1984.

356. K. Veselić. Finite catenary and the method of Lagrange. *SIAM Review*, 37:224–229, 1995.

357. M. Wagner and M.J. Todd. Least-change quasi-Newton updates for equality-constrained optimization. *Mathematical Programming*, 87:317–350, 2000.

358. A. Wathen and D. Silvester. Fast iterative solution of stabilized Stokes systems part I: using simple diagonal preconditioners. *SIAM Journal on Numerical Analysis*, 30:630–649, 1993.

359. R.B. Wilson. *A simplicial algorithm for concave programming.* PhD thesis, Graduate School of Business Administration, Harvard University, Cambridge, MA, USA, 1963.

360. P. Wolfe. A duality theorem for nonlinear programming. *Quarterly Applied Mathematics*, 19:239–244, 1961.

361. P. Wolfe. Convergence conditions for ascent methods. *SIAM Review*, 11:226–235, 1969.

362. P. Wolfe. A method of conjugate subgradients for minimizing nondifferentiable functions. *Mathematical Programming Study*, 3:145–173, 1975.

363. H. Wolkowicz, R. Saigal, and L. Vandenberghe, editors. *Handbook of Semidefinite Programming – Theory, Algorithms, and Applications.* Kluwer Academic Publishers, 2000.

364. S.J. Wright. A path-following infeasible-interior-point algorithm for linear complementarity problems. *Optimization Methods and Software*, 2:79–106, 1993.

365. S.J. Wright. An infeasible interior point algorithm for linear complementarity problems. *Mathematical Programming*, 67:29–52, 1994.

366. S.J. Wright. *Primal-dual interior-point methods*. SIAM, Philadelphia, 1996.

367. S.J. Wright. Superlinear convergence of a stabilized SQP method to a degenerate solution. *Computational Optimization and Applications*, 11:253–275, 1998.

368. Y. Xie and R.H. Byrd. Practical update criteria for reduced Hessian SQP: global analysis. *SIAM Journal on Optimization*, 9:578–604, 1999.

369. X. Xu, P.F. Hung, and Y. Ye. A simplification of the homogeneous and self-dual linear programming algorithm and its implementation. *Annals of Operations Research*, 62:151–172, 1996.

370. Y. Ye. *Interior point algorithms*. Wiley-Interscience Series in Discrete Mathematics and Optimization. John Wiley & Sons Inc., New York, 1997.

371. Y. Ye. On homogeneous and self-dual algorithm for LCP. *Mathematical Programming*, 76:211–222, 1997.

372. Y. Ye, M.J. Todd, and S. Mizuno. An $O(\sqrt{n}L)$-iteration homogeneous and self-dual linear programming algorithm. *Mathematics of Operations Research*, 19:53–67, 1994.

373. Y. Yuan. An only 2-step Q-superlinearly convergence example for some algorithms that use reduced Hessian informations. *Mathematical Programming*, 32:224–231, 1985.

374. W.I. Zangwill. Non-linear programming via penalty functions. *Management Science*, 13:344–358, 1967.

375. Y. Zhang. On the convergence of a class of infeasible interior-point methods for the horizontal linear complementarity problem. *SIAM J. Optimization*, 4:208–227, 1994.

376. G. Zoutendijk. Nonlinear programming, computational methods. In J. Abadie, editor, *Integer and Nonlinear Programming*, pages 37–86. North-Holland, Amsterdam, 1970.

Index

Universitext